Wavelets and Wavelet Transform Systems and Their Applications

Cajetan M. Akujuobi

Wavelets and Wavelet Transform Systems and Their Applications

A Digital Signal Processing Approach

 Springer

Cajetan M. Akujuobi
Prairie View A&M University
Prairie View, TX, USA

Solution manual can be downloaded at https://link.springer.com/book/9783030875275

ISBN 978-3-030-87530-5   ISBN 978-3-030-87528-2   (eBook)
https://doi.org/10.1007/978-3-030-87528-2

This Springer imprint is published by the registered company Springer Nature Switzerland AG
The registered company address is: Gewerbestrasse 11, 6330 Cham, Switzerland

Dedicated to my family:
My Wife Caroline and my Children Obinna
and Chijioke. I also dedicate this book to all
my students, particularly all of my many
master's and doctoral students from all over
the world.

Preface

Overview

This book is all about wavelets, wavelet transforms, and how they can be applied to solve problems in different fields of study. The question asked often is, what are wavelets? The answer is that wavelets are waveforms of limited duration that have average values of zero. In comparison to sinusoids, wavelets do have a beginning and an end, while sinusoids theoretically extend from minus to plus infinity. Sinusoids are smooth and predictable and are good at describing constant frequency which otherwise can be called stationary signals. In the case of wavelets, they are irregular, of limited duration, and often non-symmetrical. They are better at describing anomalies, pulses, and other events that start and stop within the signal.

This book on wavelets and wavelet transform systems and their applications has grown out of my teaching "Wavelets and Their Applications" graduate course and my research activities in the fields of digital signal processing and communication systems for many decades. The notes on which this book is based on have been used for a one-semester graduate course entitled "Wavelets and Their Applications" that I have taught for several decades at Prairie View A & M University. The book chapters have increased to 21 because of additional new materials considered and therefore can be used for a two-semester course as well. The materials have been updated continuously because of active research in the application of wavelets and wavelet transforms to several areas of science and engineering and lots of research with my graduate students in the areas of wavelet applications.

We live in the Information Age where information is analyzed, synthesized, and stored at a much faster rate using different techniques such as different wavelets and wavelet transforms. For many decades, wavelets and wavelet transforms have received much attention in the literature of many communities in the areas of science and engineering. There are different types of wavelets. These wavelets are used as analyzing tools by pure mathematicians (in harmonic analysis, for the study of

Calderon-Zygmund operators), by statisticians (in nonparametric estimation), and by electrical engineers (in signal analysis).

In physics, wavelets are used because of their applications to time-frequency or phase-space analysis and their renormalization concepts. In computer vision research, wavelets are used for "scale-space" methods. In stochastic processes, they are used in application of self-similar processes. Because of wavelets and wavelet transforms' connections with multirate filtering, quadrature mirror filters, and sub-band coding, they have found home in the digital signal processing community. The image processing community uses wavelets because of their applications in pyramidal image representation and compression. In harmonic analysis, wavelets are used because of the special properties of wavelet bases, while the speech processing community uses wavelets because of their efficient signal representation, event extraction, and the mimicking of the human auditory system.

Most importantly, wavelet analysis tools can be used as an adaptable exchange to Fourier transform analysis and representation. While there may be many books written over the past decades in the area of wavelets, it is hard to find a wavelet book that is not heavily into rigorous mathematical equations, and in most cases, little or no real applications. In addition, most of the books are not written as textbooks for classroom teaching and to make students understand what wavelets are and how to apply them to solving societal problems.

In this book, it is very simplified to real applications in solving societal problems. It is not buried into rigorous mathematical formulas. The book is very suitable as a textbook for upper-level undergraduate study and graduate studies. The practicing engineers in industry will find the book very useful. Not only can the book be used for training of future digital signal processing engineers, it can also be used in research, developing efficient and faster computational algorithms for different multi-disciplinary applications. Engineering and scientific professionals can use this book in their research and work-related activities.

In actuality, wavelets provide a common link between mathematicians and engineers. Topics such as decomposition and reconstruction algorithms, subdivision algorithms, fast numerical computations, frames, time-frequency localizations, and continuous- and discrete-wavelet transforms are covered for their use of wavelets and wavelet transforms. In addition, topics such as fractals and fractal transforms, mixed signal systems, sub-band coding, image compression, real-time filtering, radar applications, transient analysis, medical imaging, segmentation, blockchain systems, information security, and vibration in aeroelastic systems are some of the areas covered in the book. Applications of many of these wavelets and wavelet transform analyses are developed across disciplines in the book. This book, entitled *Wavelets and Wavelet Transform Analysis and Applications: A Signal Processing Approach* is a unique book because of its in-depth treatment of the applications of wavelets and wavelet transforms in many areas across many disciplines. The book does this in a very simplified and understandable manner without the mathematical rigor that scares many people away from the field. It uses lots of diagrams to illustrate points being discussed. In addition, the concepts introduced in the book are reinforced with review questions and problems. MATLAB codes and algorithms are introduced in the book to give readers hands-on experience. *As a disclaimer, because some of the codes were written in different times based on different versions of MATLAB, it is anticipated that the codes may need revisions before they can be adapted to the recent versions of MATLAB. The readers are encouraged to do whatever revisions they deem necessary for their work.*

As evident from the outline, the book is divided into seven parts. It begins with the fundamental concepts of wavelets, wavelet transform, and Fourier transform in Chapter 1. The areas of similarities and dissimilarities are also discussed. Part 1 discusses wavelets, wavelet transforms, generations of wavelets, and similarities between wavelets and fractals. Part 2 deals with wavelets and wavelet transform applications to mixed signal systems. Part 3 addresses wavelets and wavelet transform application to compression. Part 4 explores wavelets and wavelet transforms to medical applications. Part 5 covers wavelet and wavelet transform application to segmentation. Part 6 discusses wavelet and wavelet transform application to cybersecurity systems. Part 7 deals with wavelet and wavelet transform application to detection, identification, discrimination, and estimation.

In Chap. 2, we describe what wavelets are and why we do look at wavelets. We describe 19 different types of wavelets with illustrations of how they can be represented. The generations of wavelets are discussed in Chap. 3. These generations of wavelets are also compared. The different dimensions of wavelet transforms are discussed in Chap. 4. These include the one and two dimensions of wavelet transforms. The similarities between wavelets and fractals are discussed in Chap. 5. The test point selection using wavelet transformation for digital-to-analog converters (DACs) is discussed in Chap. 6 while in Chap. 7, we discuss the wavelet-based dynamic testing of analog-to-digital converters (ADCs). In Chap. 8, we discuss the wavelet-based static testing of ADCs. The mixed signal systems testing automation using discrete wavelet transform–based techniques are discussed in Chap. 9.

In Chap. 10, we discuss wavelet-based compression using nonorthogonal and orthogonally compensated W-Matrices. Wavelet-based application to image and data compression is discussed in Chap. 11 while the application of wavelets to video compression is discussed in Chap. 12. We explore wavelet application to electrocardiogram (ECG) medical signal in Chap. 13. The application of wavelets to image segmentation is covered in Chap. 14 while the hybrid wavelet and fractal-based segmentation is covered in Chap. 15. We cover wavelet-based application to information security in Chap. 16 and the application of wavelets to biometrics in Chap. 17. The emerging application of wavelets to blockchain technology systems is covered in Chap. 18. We discuss wavelet-based signal detection, identification, discrimination, and estimation in Chap. 19. In Chap. 20, we cover wavelet-based identification, discrimination, detection, and parameter estimation of radar signals. The application of wavelets to vibration detection in an aeroelastic system is covered in Chap. 21.

The prerequisite for taking the course is senior or graduate level standing accompanied by background in digital signal processing, communication systems, and mathematical sciences, with a basic knowledge of linear algebra and vector space. The book can be used in giving short and long seminars on wavelets and wavelet transform applications. In addition, it may serve as a reference for engineers and researchers in communication systems, digital signal processing, and other scientific disciplines. There is a solution manual prepared for this book which will be very helpful to professors who have adopted this book for their classes.

Acknowledgments

In the many decades that I have taught "Wavelets and Their Applications" and done research-using wavelets and wavelet transforms, I have taught many students and have supervised many undergraduate, master's, and doctoral students. I thank each and everyone of these students for their contributions. My special thanks go to the following former students and current students of mine: Lan Hu, Cary Smith, Emad Awada, Shumon Alam, Ehijele Unuigbe, Brandee Rogers, Jaymars T. Davis, Basil A. Kandah, Nana K. Ampah, Jie Shen, Dexin Zhang, Ben Franklin, Olusegun Odejide, Michael C. Ndinechi, Collins Achcampong, Dan Sims, Jr., Midge Hill, James Spain, Ashley Kelsey, Khalid Ferdous, Augustine Ajuzie, Omonowo (David) Momoh, Kelechi Eze, Pankaj Chhetri, Emmanuel Okereke, Faith Nwokoma, Shemar Hunter, Odinaka Ekwonnah, Qamiyon Marshall, Bernice Hoedzoade, and many others. I thank every student who has taken my ELEG 6333 Wavelet and Their Applications Course at Prairie View A&M University. They have all made me a better teacher and a better researcher doing work using wavelets. The encouragement and support of all my students have made it possible for me to write this book. The numerous research works many of them did with me have helped me in writing this book, and I sincerely thank each of them.

I thank all of my sponsors over these many decades including, Texas Instruments, Sprint Corporation, LL3-Communications, Litton Advanced Space Systems, Northrop Grumman, National Science Foundation, Los Alamos National Laboratory, Argonne National Laboratory, U.S. Department of Education, U.S. Department of Defense, NASA, Lockheed Martin, and many others for supporting my work over these many years. My students and I benefited a lot from the support. This book would not have been possible without this support. I specially thank some of the members of the Mathematics and Computer Science Division (MCS) at Argonne National Laboratory. These are Man Kam Kwong, Sohail Zafar, Biqun Lin, and Williams Reynolds. Special thanks goes to Man Kam Kwong for his total support, collaboration, and guidance in the W-Matrices project. The Argonne Division of Educational Programs administered the program with funding provided by the U.-S. Department of Energy.

I thank Prairie View A&M University (PVAMU) and The Texas A&M University Board of Regents that approved the Center of Excellence for Communication Systems Technology Research (CECSTR) as one of the Board of Regents Approved Centers on the Campus of PVAMU and within the entire Texas A&M System. I thank all the student researchers and faculty colleagues at CECSTR and the College of Engineering, especially my colleagues in the Department of Electrical and Computer Engineering for all their support. My special thanks go to our former president, Dr. George Wright; former provost, Dr. Thomas-Smith; and former vice president for research and a great friend Dr. Willie Trotty. Special thanks to my former dean of the College of Engineering, Dean Bryant, for all his support. Without the help and support of these giants, there would not be CECSTR, and therefore there would not have been the kind of research work that may have resulted in some of the information in this book.

Finally, I thank my wife Caroline Chioma Akujuobi and my two sons Obinna and Chijioke Akujuobi for their patience, encouragement, and support throughout the time I was preparing the manuscript for this book.

Contents

Part II   Wavelet and Wavelet Transform Applications to Mixed Signal Systems

About the Author

Cajetan M. Akujuobi received his O.N.D. from Institute of Management and Technology Enugu, Nigeria in 1974; his B.S. degree from Southern University, Baton Rouge, Louisiana, in 1980; and his M.S. degree from Tuskegee University, Alabama, in 1982, all in electrical and electronics engineering. He received his MBA degree from Hampton University, Hampton, Virginia, in 1987. In 1995, he received his Ph.D. degree from George Mason University, Fairfax, Virginia, in electrical engineering with specialization in signal/image/video processing & communication systems.

He is a full professor in the Department of Electrical & Computer Engineering and the former Vice President for Research, Innovation and Sponsored Programs at Prairie View A&M University (PVAMU). He served as dean in two different universities – the dean for graduate studies at PVAMU and founding STEM dean at Alabama State University (ASU). He is the founder and the executive director of the Center of Excellence for Communication Systems Technology Research (CECSTR), a Texas A&M Board of Regents approved center where he has been able to attract research-funding exceeding over $25 Million. With his grants, he created the $1 Million TI Endowed Professorship and the CECSTR $600,000.00 Endowed Undergraduate and Graduate Assistantships.

He is the founder and the principal investigator for the SECURE Cybersecurity Center of Excellence at PVAMU where he has received over $7 million research award. Under his leadership as the Vice President for Research, Innovation and Sponsored Programs at Prairie View A&M University (PVAMU), he was instrumental in bringing to PVAMU five new Chancellors' Research Initiative (CRI) research centers worth over $35 million. He also grew the research expenditure of the university by over 10% annually during his tenure as the Vice President for Research, Innovation and Sponsored Programs. At ASU, he founded two new research centers, and was the founding executive director of the Center of Excellence for Communication Systems & Image/Signal/Video Processing (CECSIP) and the founding executive director of the STEM Center of Excellence for Modeling & Simulation Research (SCEMSR). He has worked in such corporations as Texas

Instruments, Advanced Hardware Architecture, Schlumberger, Data Race Corporation, Spectrum Engineering, Intelsat, and Bell Laboratories.

Prof. Akujuobi developed and taught the Wavelets and Their Applications course at PVAMU for over 20 years. His research interests are in wavelets and wavelet transform analysis and applications, cybersecurity, smart and connected cities, and DSP solutions. In addition, his research interests include communication systems, compressive sensing, signal/image/video processing, broadband communication systems, and mixed signal systems. He was a participant and collaborative member of the ANSI TIEI.4 Working Group that had the technical responsibility of developing the T1.413, Issue 2 ADSL standard. He has received several professional and community related honors in teaching, research, and service and has published extensively including writing books and book chapters. Two of the books he published with Dr. M. N. O. Sadiku, are *Introduction to Broadband Communication Systems* and *Solutions Manual for Introduction to Broadband Communication Systems*, both published by Chapman & Hall/CRC and Sci-Tech Publication, Boca Raton, Florida.

Prof. Akujuobi is the current chair of the IEEE Houston Section Life Members Group. He is also a Life Senior Member of the Institute of Electrical and Electronic Engineers (IEEE), senior member of the Instrument Society of America (ISA), and member of the American Society for Engineering Education (ASEE), Sigma XI, the Scientific Research Society, and the Texas Society for Biomedical Research (TSBR) Board of Directors. He is a licensed Professional Engineer in the State of Texas, USA. He is one of the founding corporate members of the IEEE Standards Association (IEEE-SA) and Industry Advisory Committee (IAC). He is listed in Who's Who in Science and Engineering, Who's Who in the World, Who's Who in America, Who's Who in American Education, and Who's Who in Industry & Finance.

Abbreviations

DFT	Discrete Fourier Transform
FFT	Fast Fourier Transform
STFT	Short Time Fourier Transform
CWT	Continuous Wavelet Transform
DWT	Discrete Wavelet Transform
IDWT	Inverse Discrete Wavelet Transform
FPGA	Field Programmable Gate Array
CPU	Computer Programming Unit
ECG	Electrocardiogram
SHA	Spline Harmonic Analysis
DHA	Discrete Harmonic Analysis
MRA	Multiresolution Analysis
FGWTs	First Generation Wavelets
SGWT	Second Generation Wavelet Transform
TGWs	Third Generation Wavelets
TGWTs	Third Generation Wavelet Transforms
DFB	Directional Filter Banks
NGWTs	Next Generation Wavelet Transforms
PSNRs	Peak Signal to Noise Ratios
LP	Lowpass
HP	Highpass
IFS	Iterated Function System
fBm	Fractional Brownian motion
QMF	Quadrature Mirror Filter
ISDN	Integrated Service Digital Network
VLSI	Very Large Scale Integration
ADCs	Analog to Digital Converters
DACs	Digital to Analog Converters
CUT	Circuit Under Test

KCL	Kirchhoff's Current Law
KVL	Kirchhoff's Voltage Law
QRF	QR Factorization
INL	Integral Nonlinearity
DNL	Differential Nonlinearity
RMS	Root-Mean-Square
ENOB	Effective Number of Bits
SNR	Signal-to-Noise Ratio
LSB	Least Significant Bit
GE	Gain Error
NSR	Noise-to-Signal Ratio
SINAD	Signal-to-Noise-and Distortion
GUI	Graphical User Interface
DUT	Device Under Test
CR	Compression Rate
MSE	Mean Square Error
BPP	Bits Per Pixel
JPEG	Joint Pictures Experts Group
DCT	Discrete Cosine transform
HVS	Human Visual System
VQ	Vector Quantization
EZWT	Embedded Zero Wavelet Trees
SPIHT	Set Partitioning in Hierarchical Trees
WDR	Wavelet Difference Reduction
ASWDR	Adaptively Scanned Wavelet Difference Reduction
EBCOT	Embedded Block Coding with Optimized Truncation
PSNR	Peak Signal to Noise Ratio
LIS	List of Insignificant Set
HDTV	High Definition Television
MEMC	Motion Estimation and Motion Compensation
MPEG	Motion Pictures Expert Group
SVC	Scalable Video Coding
WSVC	Wavelet-Based SVC
CIF	Common Interchange Format or Common Intermediate Format
FCIF	Full Common Intermediate Format
QCIF	Quarter CIF
MAP	Motion Aligned Prediction
MAU	Motion Aligned Update
GoFs	Group of Frames
ITU-T	International Telecommunication Union Telecommunication
CBR	Constant Bit Rate
GOB	Group of Blocks
MC	Motion Compensation
VCEG	Video Coding Experts Group

SD	Standard Definition
HD	High Definition
QoS	Quality of Service
MRI	Magnetic Resonance Imaging
EMG	Electromyogram
MAXERR	Maximum Squared Error
CIA	Confidentiality, Integrity and Availability
IS	Information Security
RAID	Redundant Array of Independent Disks
DoS	Denial of Service
IoT	Internet of Things
EDR	Endpoint Detection and Response
NTA	Network Traffic Analysis
CTI	Cyber Threat Intelligence
TTP	Tactics, Techniques and Procedures
IoC	Indicator of Compromise
INSD	Information Network Security Database
NUBASI	Non-Uniform Block Adaptive Segmentation on Information
SIW	Symlet Information Wavelet
AES	Advanced Encryption Standard
DES	Data Encryption Standard
NIST	National Institute of Standards and Technology
NSA	National Security Agency
RSS	Randomized Secret Sharing
RNUBASI	Reverse Non-Uniform Block Adaptive Segmentation on Image
DLT	Distributed Ledger Technology
PSK	Phase Shift Keying
FSK	Frequency Shift Keying
QAM	Quadrature Amplitude Modulation
CNR	Carrier-to-Noise Ratio
ROC	Receiver Operating Characteristic
AWGN	Additive White Gaussian Noise
BPSK	Binary Phase Shift Keying
QPSK	Quadrature Phase Shift Keying
dB	Decibel
AMRTDS	Advanced Microwave Receiver Technology Development System
LANL	Los Alamos National Laboratory
MSTRS	Miniaturized Satellite Test Reporting System
SGLS	Space Ground-Link Subsystem
QMF	Quadrature Mirror Filters
PLT	Per-Level Thresholding
GT	Global Thresholding

Symbols

$\psi_{a, b}$	Wavelet transform function
$\psi(x)$	"Mother" wavelet
a	Wavelet scaling factor
b	Center location of the wavelet function
$g^{\omega, t}$	Fourier transform function
ω	Frequency value (Fourier transform)
t	Time period
MHz	Megahertz
Kb	Kilobytes
X	Received signal
ca0 – ca3	Approximation coefficients levels 0 – 3
cd1 – cd3	Details coefficients levels 1 – 3
γ	Threshold values
H_0	Priori – Hypothesis of no signal present
H_1	Priori – Hypothesis of signal present
S	Known transmitted signal
N	Zero-mean white Gaussian noise
σ^2	Signal variance
$x_1 \ldots x_{10}$	Received signal divided into ten segments
C	Wavelet coefficients
$c_1 \ldots c_{10}$	Wavelet coefficients of each signal segments
$p(H_0)$	Probability of null signal present
$p(H_1)$	Probability of signal present
P_D	Probability of detection
P_{FA}	Probability of false alarm
P_M	Probability of missed detection
P_{NULL}	Probability of a null signal presents
P_{ERR}	Probability of error
D_0	Decision based on threshold value that transmitted signal is NULL
D_1	Decision based on threshold value that transmitted signal is present
X(h)	**h-th bin** of signal X
Φ	Phase of a signal channel
Φ_{err}	Phase channel error
w	Appropriate window
w^*	Complex conjugate of w
τ	A shift or translation parameter
ψ	Wavelet function
φ	Scaling function
$h_0(n)$	Coefficients of the corresponding lowpass filter
$h_1(n)$	Coefficients of a highpass filter
$g[n]$	Coefficient
$f(t)$	Signal function

$L^2(R)$	Vector space of measurable, square-integrable one-dimensional function $f(x)$
$L^2(R^2)$	Vector space of measurable, square-integrable function $f(x, y)$
1D	One dimension
2D	Two dimension
3D	Three dimension
Q	Orthogonal matrix
R	Upper triangular matrix
n_{max}	Maximum number of bits
P_e	Probability of error
N_{cn}	Number of points misclassified in class n and
N_{total}	Total number of reference points
thr	Threshold

Chapter 1
Fundamental Concepts

People who say it cannot be done should not interrupt those who are doing it.

                       – George Bernard Shaw (1856–1950)

1.1 Introduction

Wavelets originated from the field of mathematics. However, it was not until the past several decades that the interest in the signal processing community grew in using wavelets and in discovering the many applications of wavelets and wavelet transforms (Mallat, 1998; Miaou & Lin 2002; Oliveira & Bretas, 2011). In this chapter, the basic concepts of the wavelet idea are explored. How the concept of wavelets and its transforms differ from the traditional methods of Fourier series and its transforms are discussed in this chapter. Since there are many similarities and dissimilarities between wavelet transforms and Fourier transforms, these are also topics covered in the chapter.

1.2 Fourier Transform and Analysis

One of the traditional methods in signal processing is Fourier analysis. The Fourier transform of the function $x(t)$ is written as shown in Eq. (1.1).

$$\hat{x}(\omega) = \int_{-\infty}^{\infty} x(t)e^{-i\omega t}\,dt \tag{1.1}$$

where $x(t)$ is the signal to be transformed. The inverse Fourier transform is computed as follows:

© Springer Nature Switzerland AG 2022
C. M. Akujuobi, *Wavelets and Wavelet Transform Systems and Their Applications*,
https://doi.org/10.1007/978-3-030-87528-2_1

$$x(t) = \frac{1}{2\pi} \int\limits_{-\infty}^{\infty} \hat{x}(\omega)e^{i\omega t}d\omega \tag{1.2}$$

The Fourier series is

$$C_j = \int\limits_{-\infty}^{\infty} x(t)e^{-i\omega_j t}dt \tag{1.3}$$

where

$$x(t) = \sum_{j=-\infty}^{\infty} C_j e^{i\omega_j t} \tag{1.4}$$

We can also have discrete Fourier transform (DFT) pairs for which the fast Fourier transform (FFT) is a fast computational algorithm. However, Fourier methods are only appropriate for stationary settings. For non-stationary or transient settings, the short-time Fourier transform (STFT) became a standard tool for signal analysis.

1.3   Short-Time Fourier Transform

The short-time Fourier transform (STFT) of $x(t)$ can be defined as follows:

$$\hat{x}(\tau, \omega) = \int\limits_{-\infty}^{\infty} x(t)w * (t - \tau)e^{-i\omega t}dt \tag{1.5}$$

where w represents appropriate window, w^* is the complex conjugate of w, and τ is a shift or translation parameter. Equation (1.5) is a windowed Fourier transform. The STFT process is repeated with translated versions of the window function throughout the duration of the signal. However, STFT has the following shortcomings (see Fig. 1.1):

(i) The resolution of the frequency structure at wavelengths longer than the window width, i.e., for low frequencies, is very poor.

(ii) The long window averages energy over the window width, which results in the STFT being very poor at localizing high frequencies.

(iii) The Heisenberg inequality (time-bandwidth product bounded by $((4\pi)^{-1})$, i.e., $(\Delta t \cdot \Delta f^2 (4\pi)^{-1}))$, limits the STFT time and frequency resolutions.

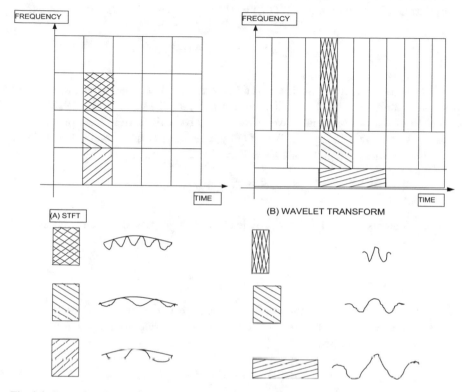

Fig. 1.1 Bases functions and time-frequency resolution of the STFT and wavelet transform (WT). The tiles show the essential concentration in the time-frequency plane of a given basis function

Therefore, there is a need to find a solution to these shortcomings. The solution can be found using the idea of wavelet transform described in Sect. 1.4.

1.4   The Wavelet Transform Idea

The advent of wavelet transform has revolutionized the way signals are represented and analyzed, quite unlike the traditional methods. The wavelet transform represents signals locally in time and frequency. Unlike the wavelet transform, the Fourier transform represents signals locally in frequency and globally in time. Properly chosen, the wavelet transform also represents a compactly supported basis in both frequency and time.

The basic wavelet transform idea is to use a transient waveform as in the STFT but increase time resolution by keeping a relative bandwidth constant as frequency increases, i.e., making bandwidth proportional to frequency. Practically, this means choosing a prototype mother wavelet, $\psi(t)$, and generating a family of (affine) wavelets $\psi_{a,\tau}(t)$ as shown in Eq. (1.6) by dilations (α) and translations (τ) of $\psi(t)$.

$$\psi_{a,\tau}(t) = \frac{1}{\sqrt{a}} \psi\left(\frac{t-\tau}{a}\right) \qquad (1.6)$$

Wavelet bases with compact support can be derived from filter banks, which were originally motivated by the multiresolution analysis of Mallat (1989, 1998). Construction of the orthonormal wavelet bases may be derived from a set of recursive equations as shown in Eqs. (1.7) and (1.8). The wavelet ψ is obtained as a linear combination of scaled and shifted versions of scaling function φ.

$$\psi(t) = \sum_{nez} h_1(n)\varphi(2t - n) \qquad (1.7)$$

The coefficients $h_1(n)$ are the coefficients of a high-pass filter, and the scaling function satisfies

$$\varphi(t) = \sum_{nez} h_1(n)\varphi(2t - n) \qquad (1.8)$$

The coefficients $h_0(n)$ are the coefficients of the corresponding low-pass filter. Equations (1.7) and (1.8) are called *the two-scale difference equations*. These coefficients are related by

$$h_1(n) = (-1)^n h_0(1 - n) \qquad (1.9)$$

There are two types of wavelet transforms—the continuous wavelet transform (CWT) and the discrete wavelet transform (DWT). We describe these two types of wavelet transforms in Sect.1.5.

1.5   Types of Wavelet Transforms

Wavelets are used in different applications by means of two types of transforms—the continuous wavelet transform and the discrete wavelet transform. These two concepts are covered in this section.

1.5.1   The Continuous Wavelet Transform (CWT)

The continuous wavelet transform (CWT) is defined by

$$\rho(a, \tau) = \int_{-\infty}^{\infty} x(t)\psi^*_{a,\tau}(t)dt, \tag{1.10}$$

while the inverse wavelet transform is given by

$$x(t) = \frac{1}{c} \int_{-\infty}^{\infty} \int_{-\infty}^{\infty} \rho(a, \tau)\psi_{a,\tau}(t)\frac{dad\tau}{a^2} \tag{1.11}$$

where

$$\psi_{a,\tau}(t) = \frac{1}{\sqrt{a}}\psi\left(\frac{t-\tau}{a}\right) \tag{1.12}$$

In Eqs. (1.10) and (1.11), $x(t)$ can be expanded in a wavelet series in the same way we deal with Fourier series. For narrow width or high-frequency wavelets, $a < 1$, while for wider width or low-frequency wavelets, $a > 1$ in Eq. (1.12). The continuous wavelet transform has its shortcomings, the problems of redundancy and impracticality for some applications. The shortcoming problems of the continuous wavelet transform are solved by using the discrete wavelet transform.

1.5.2 The Discrete Wavelet Transform (DWT)

The problems of redundancy and impracticality mentioned in Sect. 1.5.1 can be resolved by discretizing the time-scale parameters a and τ, i.e., by letting

$$a = a^{\frac{j}{2}}_0 \tag{1.13}$$

and

$$\tau = kTa^{\frac{j}{2}}_0 \tag{1.14}$$

which transforms Eqs. (1.12, 1.13, 1.14 and 1.15). The wavelet becomes

$$\psi_{jk}(t) = a^{\frac{-j}{2}}_0\psi\left(a^{-j}_0 t - kT\right) \tag{1.15}$$

with wavelet coefficients given by

$$C_{jk} = \int\limits_{-\infty}^{\infty} x(t)\psi_{jk}^{*}(t)\mathrm{d}t \tag{1.16}$$

If a_0, T, and $\psi(t)$ have the appropriate properties, we might expect as in the case of Fourier series that

$$x(t) = \sum_{j}^{\infty} \sum_{k}^{\infty} C_{jk}\psi_{jk}(t) \tag{1.17}$$

If a_0 is arbitrarily close to 1, this latter double sum is a Riemann sum approximation to the double integral in Eq. (1.11), and hence, it is reasonable to believe that such conditions are possible. To make wavelets computationally feasible, $a_0 = 2$.

1.6 Wavelet Frames and Bases

Wavelets have frames and basis. If you let $x(t)$ be an element of a function space, say $L^2(R)$, C_{jk} the inner product of x and ψ_{jk}, i.e., $C_{jk} = \ <x, \psi_{jk}>$. A frame can be defined as a sequence $\{x_n\}$ in $L^2(R)$ with the property that for A, B positive numbers

$$A\|x\|^2 \leq \sum_{n}^{\infty} |<x, x_n>|^2 \leq B\|x\|^2 \tag{1.18}$$

where A and B are called frame bounds. If $A = B$, the frame is said to be tight. If no value of $\{x_n\}$ is dropped from the frame, the frame is said to be exact.

If $A = B = 1$, and the frame is tight and exact, then the frame is an orthonormal basis. For the wavelet case, we are taking $\{x_n\}$ to be $\{\psi_{jk}\}$ so that orthonormal means

$$\int\limits_{0}^{\infty} \psi_{jk}(t)\psi_{lm}(t)\mathrm{d}t = 1 \ \ \text{for}(\ j = 1, k = m) \tag{1.19}$$
$$= 0, \ \ \text{otherwise}$$

Thus, we can construct an orthonormal basis (an exact frame with constant 1), resulting to an approximation of any square-integrable function by a series, which resolves both time and scale (frequency). An important contribution of Daubechies and Mallat is the development of a theoretical foundation of orthogonal wavelet basis. Wavelets can be made orthogonal to their own dilation and translation with discrete dilation and translation parameters. Only a discrete wavelet transform gives nonredundant signal representation. The orthogonal discrete wavelet transform decomposition and reconstruction may be done by recursion algorithms. These algorithms do for the discrete wavelet transform what the fast Fourier transform (FFT) does for the discrete Fourier transform, without using explicit wavelet transform inner products. The wavelet transform need not be orthogonal. With continuous dilation and translation variables a and τ, the wavelet transform may be highly redundant. This is a desirable feature for some application solution.

1.7 Constructing Orthonormal Wavelet Bases with Compact Support

In the construction of an orthonormal wavelet bases with compact support, the major task is finding the mother wavelet functions, ψ. We wish the wavelets to have the following properties:

(i) Compact support so that dilation and translation make sense.
(ii) Be dyadic, i.e., $a_0 = 2$ with $T = 1$.

There is no oversampling because of these choices, which make orthonormal bases possible. With the exception of the Haar basis, the mother wavelet, ψ, cannot be an even function and have compact support. A function f is compactly supported if $f(t) = 0$ if $t < a$ or $t > b$, where $-\infty < a < b < \infty$. Additional constraint on construction of wavelets of compact support is the requirement that wavelets be continuous functions, perhaps possessing continuous derivatives also. This can be achieved if the filter meets certain regularity constraints and leads to good approximation properties of wavelet bases.

1.8 Similarities Between Wavelets and Fourier Transforms

Wavelet and Fourier transforms have areas of similarities. The discrete wavelet transform (DWT) and the fast Fourier transform (FFT) are both linear operations. They tend to generate data structures that contain $log_2 n$ segments of various lengths. Usually, they fill and transform them into different data vectors of lengths 2^R. Similar also are the mathematical properties of the matrices involved in the transforms.

The inverse transform matrix for both the DWT and the FFT is the transpose of the original. Because of that, both transforms can be viewed as a rotation in function space to a different domain. In the case of the FFT, this new domain contains basis functions that are sines and cosines. In the case of the wavelet transform, this new domain contains more complicated functions called wavelets, mother wavelets, or analyzing wavelets.

The DWT and the FFT have other similarities such as their basis functions that are localized in frequency. It makes the mathematical tools such as power spectra (how much power is contained in a frequency interval) and scale grams useful at picking out frequencies and calculating power distributions.

1.9 Dissimilarities Between Wavelets and Fourier Transforms

The discrete wavelet transform (DWT) and the fast Fourier transform (FFT) have a very interesting dissimilarity in the sense that individual wavelet functions are *localized in space* while the Fourier sine and cosine functions are not. The idea of the localization feature of wavelets in space together with the localization in frequency makes many functions and operators using wavelets *sparse* when transformed into the wavelet domain. Because of this wavelet sparseness property, it has allowed wavelets to be used in many applications such as removing noise from time series, detecting features in images, and data compression.

One possibility of appreciating the time-frequency resolution differences between wavelet transform and Fourier transform is the basis function coverage of the time-frequency plane. In Fig. 1.1, we can see the idea of the windowed Fourier transform (WFT), where the window is simply a square wave. The square wave window truncates the sine or cosine function to fit a window of a particular width. Because a single window is used for all frequencies in the WFT, the resolution of the analysis is the same at all locations in the time-frequency plane. The advantage of the wavelet transform is that the windows vary depending on the width of the signal to be analyzed.

Summary
1. Wavelets originated from the field of mathematics.
2. One of the traditional methods in signal processing is Fourier analysis.
3. The STFT process is repeated with translated versions of the window function throughout the duration of the signal.
4. The wavelet transform represents signals locally in time and frequency.
5. Unlike the wavelet transform, the Fourier transform represents signals locally in frequency and globally in time.
6. The continuous wavelet transform has its shortcomings, the problems of redundancy and impracticality for some applications.
7. An important contribution of Daubechies and Mallat is the development of a theoretical foundation of orthogonal wavelet basis.
8. In the construction of an orthonormal wavelet bases with compact support, there is no oversampling because of these choices, which make orthonormal bases possible.
9. Wavelet and Fourier transforms have areas of similarities.
10. The discrete wavelet transform (DWT) and the fast Fourier transform (FFT) have a very interesting dissimilarity in the sense that individual wavelet functions are *localized in space* while the Fourier sine and cosine functions are not.

Review Questions

1.1 Wavelets did not originate from the field of mathematics.

(a) True
(b) False

1.2 Fourier methods are only appropriate for stationary settings.

(a) True
(b) False

1.3 What is the standard tool for signal analysis in non-stationary or transient settings?

(a) STFT
(b) Wavelet transform
(c) Fourier transform
(d) STFT and wavelet transform
(e) None of the above

1.4 Wavelet bases with compact support can be derived from filter banks which are originally motivated by the multiresolution analysis.

(a) True
(b) False

1.5 The wavelet transform represents signals locally in time and frequency.

(a) True
(b) False

1.6 To make wavelets computationally feasible, $a_0 = 2$.

(a) True
(b) False

1.7 The shortcomings of the continuous wavelet transform are solved by.

(a) Using the Fourier transform
(b) Using Fourier transform and discrete wavelet transform
(c) Using discrete wavelet transform
(d) None of above

1.8 Only a discrete wavelet transform gives nonredundant signal representation.

(a) True
(b) False

1.9 The orthogonal discrete wavelet transform decomposition and reconstruction may not be done by recursion algorithms.

(a) True
(b) False

1.10 With the exception of the Haar basis, the mother wavelet, ψ, cannot be an even function and have compact support.

(a) True
(b) False

Answers: 1.1b, 1.2a, 1.3d, 1.4a, 1.5a, 1.6a, 1.7c, 1.8a, 1.9b, 1.10a.

Problems
1. Are Fourier methods appropriate for non-stationary settings?
2. (a) Define the STFT process of a signal $x(t)$.
   (b) What are the shortcomings of the STFT in the processing of information?
3. Briefly describe your understanding of the wavelet transform idea.
4. (a) Describe the wavelet two-scale difference equations. (b) How are the low-pass and high-pass wavelet coefficients related?
5. How are the problems of redundancy and impracticality in wavelet transforms solved?
6. Define what a wavelet frame is and under what condition can it be tight, exact, and orthonormal?
7. What are the similarities between wavelet and Fourier transforms?
8. What are the dissimilarities between wavelet and Fourier transforms?

References

Mallat, S. (1989). A theory for multiresolution signal decomposition: the wavelet representation. *IEEE Transactions on Pattern Analysis and Machine Intelligence, 11*, 674–693.

Mallat, S. (1998). *A wavelet tour of signal processing*. Academic Press. ISBN 0-12-466606-X.

Miaou, S. G., & Lin, C. L. (2002). A quality-on-demand algorithm for wavelet-based compression of electrocardiogram signals. *IEEE Transactions on Biomedical Engineering, 49*(3), 233–239.

Oliveira, M. O., & Bretas, A. S. (2011). Application of discrete wavelet transform for differential protection of power transformers. In H. Olklcomen (Ed.), *Discrete wavelet transforms-biomedical applications* (pp. 349–356). In Tech.

Part I
Wavelets, Wavelet Transforms and Generations of Wavelets

Chapter 2
Wavelets

"It always seems impossible until it's done."

— Nelson Mandela.

2.1 Introduction

Wavelets are functions generated from one basic function ψ called the mother wavelet by dilations and translations of ψ.

These dilations and translations can be represented as $\psi^{a,\ \tau}(t) - |a|^{-1/2}\psi\{(t - \tau)/a\}$. In this case, t is a one-dimensional variable, and τ and a are shift and scale parameters, respectively. For low frequency wavelets, $a > 1$, while for high-frequency wavelets, $a < 1$. The mother wavelet ψ satisfies the condition $\int\psi(t)\,dt = 0$ (Swelden 1995; Chui 1992; Mallat 1989). Wavelets have proven in recent years to be an important theory of mathematics that is now leading to advances in signal analysis (Mallat 1989; Swelden 1998), scientific calculation, medical imaging, and image compression and many other significant areas of functionalities that are mentioned in this chapter and many other chapters in this book.

In scientific calculation, for example, wavelets may help decipher phenomena such as turbulence leading to improved aircraft design. In medical imaging, wavelets are used by scientists to produce magnetic resonance pictures of the body faster and more accurately. For data compression, wavelets identify the key features of any image, allowing engineers to reconstruct an image with a tiny fraction of the information in the original.

In this chapter, we cover what wavelets are and why we use wavelets. The different types of wavelets and their functionalities are also covered.

2.2 What Are Wavelets and Why Do We Look at Wavelets?

Even though we have defined wavelets in Sect. 2.1, the answer to the question about what wavelets are can be based on the properties of wavelets. They are as follows:

© Springer Nature Switzerland AG 2022
C. M. Akujuobi, *Wavelets and Wavelet Transform Systems and Their Applications*,
https://doi.org/10.1007/978-3-030-87528-2_2

- Wavelets are set of basis functions—limited duration, average value of zero, usually irregular and asymmetric.
- They are used in a way similar to Fourier transform.
- Wavelets are local in time and scale domains, which is in contrast to Fourier transform.
- They are capable of dealing with transitory signals.
- The algorithms are flexible and fast—highly suitable for embedded applications.

There are several reasons why we should look at wavelets in developing algorithms for the analysis of signals, especially, if they are transient signals. Some of the reasons are:

- In communications, we have continuous, stationary, smooth signals (uplink communications) combined with asymmetric, transient signals (chirp radar). For the proper analysis of those transient signals, we need wavelet analysis.
- Fourier analysis, which is what the traditional signal processing algorithms are based on, does not do well with transient signals.
- Wavelet analysis does well with transient signals.
- Wavelet algorithms allow for embedding in an instrument such as field-programmable gate array (FPGA) or serial instruction CPU.
- Wavelet transform analysis is capable of identifying the local attributes of a signal in the time and frequency domain.
- It is appropriate for broadened time interims, where high accuracy is needed at low frequencies and vice versa.

2.3   Types of Wavelets

The following examples of wavelets are discussed in this section: Haar, Daubechies, Morlet, Meyer, Mexican hat, Berlage, and a class of biorthogonal wavelets. The other types of wavelets discussed in this chapter include Shannon, Symlet, Coiflet, Gabor, and Spline. The construction methods for these wavelets are also discussed.

2.3.1   The Haar Wavelet

The name Haar wavelet originated from Alfréd Haar's work in the early twentieth century. Alfred Haar was a Hungarian mathematician who introduced the Haar wavelet in 1909. Historically, it is the first wavelet discovered and what is now known as the beginning of the wavelet idea. He suggested in 1909 an alternative system to Fourier system, hence the Haar measure, Haar wavelet, and Haar transform all named in his honor. His doctoral research was supervised by Hilbert and he graduated in June 1909. His type of wavelet is the first system of compactly supported wavelets, a complete orthonormal system for $L^2(R)$, which Alfred Haar introduced in an appendix to his dissertation in 1910. These functions consist simply of a short positive pulse followed by a short negative pulse. The Haar wavelet shown in Fig. 2.1 is represented mathematically as shown in Eq. (2.1).

$$
\begin{aligned}
\psi(t) &= 1, \ \ 0 \le t < \frac{1}{2}, \\
&= -1, \frac{1}{2} \le t < 1 \\
&= 0, \ \ \text{otherwise}
\end{aligned}
\tag{2.1}
$$

where $\psi(t)$ is the Haar wavelet orthonormal basis.

Haar wavelet is an orthogonal type of wavelet. It is the simplest type of wavelet that is applied to many areas of signal, image, and video processing. Some other areas of application include image compression, detection of microseismic signal arrivals, acoustic signal compressions, and solving integral and differential equations. The Haar wavelet is best suited for the analysis of rectangular shaped signals and images. The coefficients of the Haar wavelet are listed in Appendix A. The properties of Haar wavelet are summarized as follows:

- The function is linear combination of $\varphi(x)$, $\varphi(2x)$, $\varphi(2^2x)$, $\varphi(2^kx)$, ... and their shifting.

version, where k is an integer.

- The function can be a linear combination of $\psi(x)$, $\psi(2x)$, $\psi(2^2x)$, $\psi(2^kx)$, ... and.

their shifting version, where k is an integer.

- The set of function $2^{j/2}\varphi(2^jx - k)$, $k \in Z$) is orthogonal.

2.3.2   The Daubechies Wavelet

In 1986, Baroness Ingrid Daubechies, a Belgian physicist and mathematician while working as a guest researcher at the Courant Institute of Mathematical Sciences, discovered by using quadrature mirror filter technology compactly supported continuous wavelets that would require only a finite amount of processing. It was not until 1987 that what is known today as the family of Daubechies wavelets was made known to the research community. She is best known for her work in image wavelets and image processing. The Daubechies family of wavelets are orthogonal, discrete, and compactly supported orthonormal wavelets. Her wavelets have remarkably rich structures and they made discrete wavelet analysis very practicable.

The wavelets have the ability to be implemented using simple digital filtering ideas. The work of Daubechies was carried out in the framework of the theory of wavelets introduced by Morlet and Grossmann and developed by Myers. The Morlet and Meyer wavelets are discussed in Sects. 2.3.3 and 2.3.4 of this chapter,

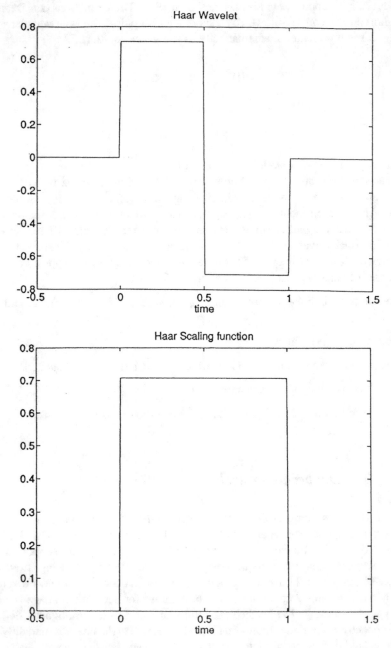

Fig. 2.1 The Haar wavelet and scaling function

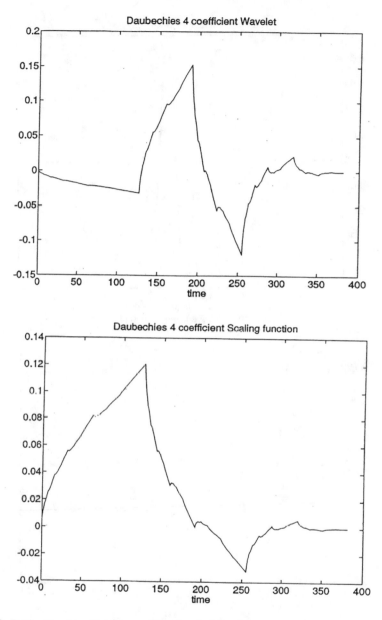

Fig. 2.2 Daubechies-4 coefficient wavelet and scaling function

respectively. Prior to her work and the pioneering effort of Haar, the only known
wavelet systems had unbounded support. The Daubechies family of wavelets have
extremal phases and the highest number of vanishing moments for a given support
width. The associated scaling filters are minimum-phase filters.

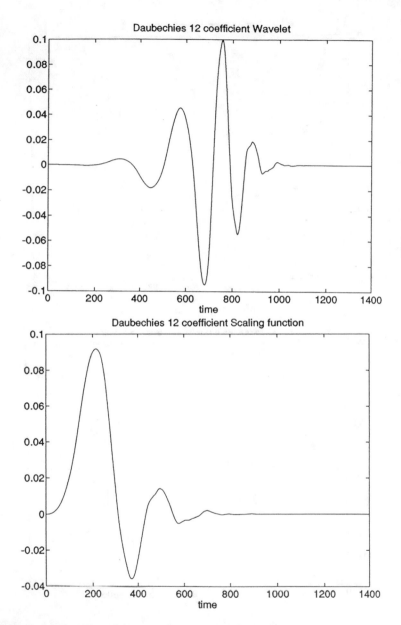

Fig. 2.3 Daubechies-12 coefficient wavelet and scaling function

The Daubechies-4, Daubechies-12, and Daubechies-18 wavelets as an example are shown in Figs. 2.2, 2.3, and 2.4, respectively. They are a family of wavelet systems whose scaling functions and primary wavelets are continuous. Figures 2.2, 2.3, and 2.4 also illustrate some of the different filter lengths of the Daubechies family of wavelets and the scaling functions. Examples of the Daubechies wavelet

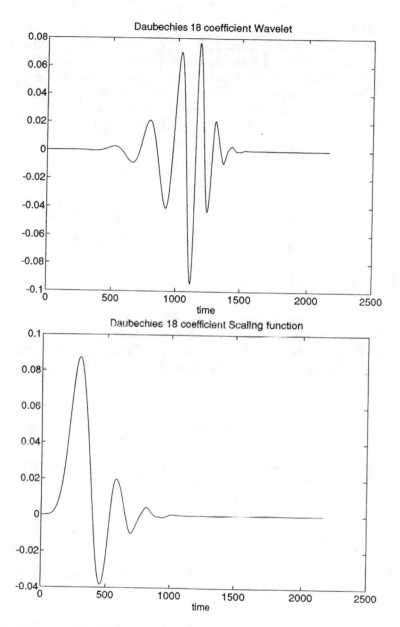

Fig. 2.4 Daubechies-18 coefficient wavelet and scaling function

coefficients are as follows: $h_1(0) = (1 + \sqrt{3})/(4\sqrt{2})$, $h_1(1) = (3 + \sqrt{3})/(4\sqrt{2})$, $h_1(2) = (3 - \sqrt{3})/(4\sqrt{2})$, $h_1(3) = (1 - \sqrt{3})/(4\sqrt{2})$. Other Daubechies wavelet filter length coefficients, the wavelet diagrams, and their scaling functions are listed in Appendix A.

Some of the application areas for the Daubechies family of wavelets are medical image compression, fractal problems, problem of self-similarity properties of signals, image enhancement, and restoration. It can also be used in filtering, denoising, discontinuities, image compression and processing, seismic and geophysics, medicine and biomedical images, physiological, communications, fractals and facial feature extraction, ultrasonic waves, and audio and speech coding. The mathematical representation of the Daubechies wavelet is as shown in Eq. (2.2).

$$\psi(t) = \sqrt{2} \sum_{k=0}^{L-1} g_k \emptyset(2x - k). \tag{2.2}$$

where $\psi(t)$ is the wavelet, g_k is the wavelet coefficient, and $\emptyset$ is the scaling function.

2.3.3   The Morlet Wavelet

Jean Morlet was a French geophysicist. In 1975, he pioneered work in the field of wavelet analysis. The Morlet wavelet is a combination of Gaussian window with a sinusoidal function. It has a good time and frequency localization feature. It is one of the most widely used wavelets. The Morlet wavelet is represented mathematically as shown in Eq. (2.3).

$$\psi(t) = \exp\left(2i\pi f_0 t\right) \exp\left(\frac{-t^2}{2}\right) \tag{2.3}$$

The real part of Eq. (2.3) is an even cos-Gaussian function. When the spectrum of the Gaussian functions is shifted to f_0 and $-f_0$, the result is the Fourier spectrum of the Morlet wavelet as shown in Eq. (2.4).

$$\psi(f) = 2\pi\left\{ \exp\left[-2\pi^2(f - f_0)^2\right] + \exp\left[-2\pi^2(f + f_0)^2\right]\right\} \tag{2.4}$$

Equation (2.4) is both real and positive valued. Mathematically, you can represent the Morlet wavelet also using Eq. (2.5).

$$\psi(t) = \left(1 + e^{-\sigma^2} - 2e^{-\frac{3}{4}\sigma^2}\right)^{-\frac{1}{2}} \pi^{-\frac{1}{4}} e^{-\frac{1}{2}t^2} \left(e^{i\sigma t} - e^{-\frac{1}{2}\sigma^2}\right) \tag{2.5}$$

where t is time and σ is the standard deviation.

The Morlet wavelet shown in Fig. 2.5 is used for sound pattern analysis. Morlet and others have used it for seismic wave analysis in oil exploration. It can be used also to discriminate abnormal heartbeat behavior in the electrocardiogram (ECG). Since the variation of the abnormal heartbeat is a non-stationary signal, this signal is

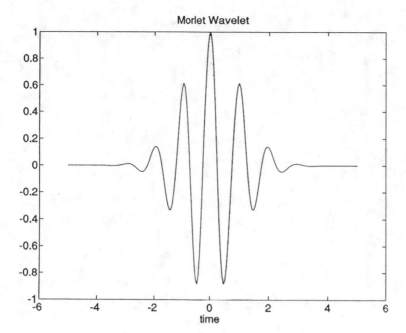

Fig. 2.5 The Morlet wavelet

suitable for wavelet based analysis. The Morlet wavelet is a continuous type of wavelet.

2.3.4   *The Meyer Wavelet*

Yves Meyer is a French mathematician and among the founders of wavelet theory. He proposed the Meyer wavelets; hence, the wavelet is named in his honor. The Meyer wavelet shown in Fig. 2.6 is defined as shown in Eq. (2.6). This real function has rapid decay and is symmetric.

$$\psi(t) = 2 \int\limits_{0}^{\infty} \sin\left[\omega(f)\right] \cos\left[2\pi\left(t - \frac{1}{2}\right)f\right] df \tag{2.6}$$

The wavelets in Eq. (2.6) with discrete dilations and translations construct an orthonormal basis. The Fourier transform of Meyer wavelet is defined as shown in Eq. (2.7).

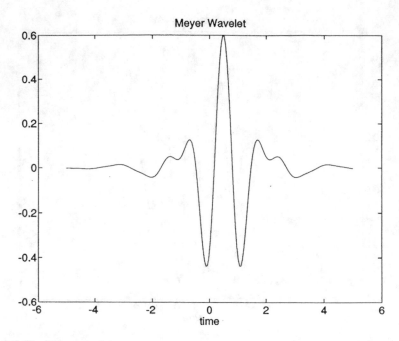

Fig. 2.6 The Meyer wavelet

$$\psi(f) = \exp(-i\pi f) \sin[\omega(f)] \tag{2.7}$$

Equation (2.7) requires an even and symmetric function $\omega(f)$. The Fourier transform of the Meyer wavelet is a compactly supported C^∞ function with rapid polynomial decay. It is used in image processing, biomedical signal compression, and image recognition. The Meyer wavelet can also be used in differential equations analysis, image edge detection, signal processing and filtering, image compression and processing, seismic and geophysics, medicine, biomedical, physiological, communications, fractals and facial feature extraction, ultrasonic waves, and audio and speech coding.

2.3.5 The Mexican Hat Wavelet

Originally, the Mexican hat wavelet was known as the Ricker wavelet, named after Norman H. Ricker (1896–1980), an American geophysicist. The wavelet is the negative normalized second derivative of a Gaussian function. The wavelet, however, is popularly known as the Mexican hat wavelet, especially in America, because its shape represents a sombrero. The Mexican hat wavelet shown in Fig. 2.7 is defined as shown in Eq. (2.8). The function is even and real valued. The Fourier

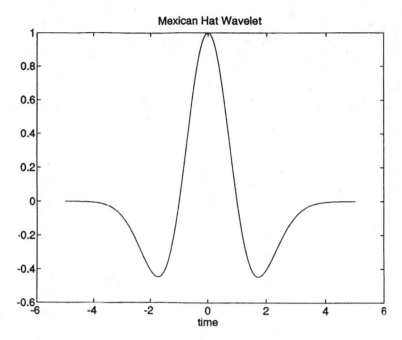

Fig. 2.7 The Mexican hat wavelet

transform of the Mexican hat wavelet is shown in Eq. (2.9). The transform is also even and real valued. It is represented mathematically using Eq. (2.10).

$$\psi(t) = \left(1 - |t|^2\right) \exp\left(-|t|^2/2\right) \tag{2.8}$$

$$\psi(f) = 4\pi^2 f^2 \exp\left(-2\pi\, f^2\right) \tag{2.9}$$

$$\psi(t) = C \times e^{\left(-t^2/2\right)\times\left(1-x^2\right)}, where\, C = 2\!\!\left/\!\left(\left(\sqrt{3}\right)\times\pi\{1/4\}\right)\right. \tag{2.10}$$

It was Gabor who first introduced the Mexican hat wavelet. The Laplacian operator or the Mexican hat wavelet is widely applied for the zero-crossing multiresolution edge detection. This is because the Mexican hat wavelet primary filters need not be orientation-dependent. It is used mainly in 2D image processing. It is also known as the Marr wavelet in honor of David Marr for his work in early vision activities. Mexican hat is proposed for nonlinear function approximation, which is used for the realization of the algebraic nonlinear components. Ricker used the wavelet as a zero-phase embedded wavelet in modeling and synthetic seismogram. It is frequently employed to model seismic data, computational electrodynamics, and denoising of strong noise signals with a low signal-to-noise ratio.

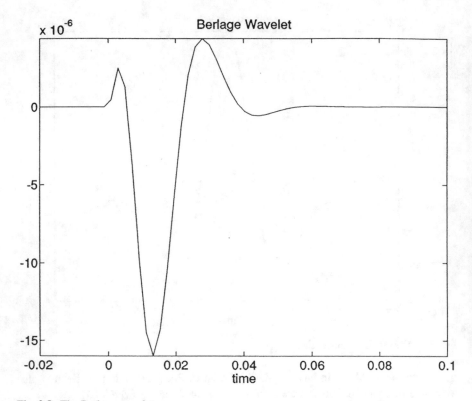

Fig. 2.8 The Berlage wavelet

2.3.6   The Berlage Wavelet

A certain class of wavelets known as the Berlage wavelets have been featured very
prominently in seismic modeling studies. The Berlage wavelet shown in Fig. 2.8 is
defined mathematically as shown in Eq. (2.11).

$$\psi(t) = AH(t)t^n e^{-\alpha t} \cos\left(2\pi f_0 t + \varphi_0\right) \tag{2.11}$$

where:

A is amplitude (constant).
$H(t)$ is the Heaviside unit step function, i.e.,
$H(t) = 0$ for $t \le 0$ and $H(t) = 1$ for $t > 0$.
α is exponential decay factor,
n is time exponent (nonnegative real constants),
f_0 is the initial frequency (fundamental frequency),
φ_0 is the initial phase angle.

This family of wavelets is useful in theoretical or computational studies where causality of the incident seismic signal is looked as a condition that is very important.

2.3.7 The Biorthogonal Wavelets

Biorthogonality provides additional degree of freedom so that both perfect reconstruction and linear phase filters are realized simultaneously. This permits the representation of any arbitrary signal x as a superposition of wavelets as shown in Eqs. (2.12), (2.13), and (2.14) respectively.

$$\psi_{jk}{}^d(t) = 2^{-j/2}\psi^d\left(2^{-j}t - k\right) \tag{2.12}$$

where:

$\psi^d{}_{jk}$ is the dual basis of the wavelet basis ψ_{jk}
j is the scaling factor,
k is the shift parameter.

Therefore,

$$x(t) = \sum_{jk} c_{jk}\,\psi^d{}_{jk}(t) \tag{2.13}$$

where

$$c_{jk} =< \psi_{jk}, x >$$

However, the inner product $<\psi_{jk}, x>$ can be defined as shown in Eq. (2.13).

$$< \psi_{jk}, x >= \int \psi_{jk}(t)x(t)dt \tag{2.14}$$

The biorthogonal wavelet bases idea was initially introduced by Cohen et al. The biorthogonal nature of the filter bank allows different filter lengths in the analysis section. This causes slightly unequal split of the signal spectrum and may cause some degradation of quantized and encoded signals. However, biorthogonal wavelets allow exactly linear phase and, therefore, have application in image coding and have been used extensively. Figure 2.9 shows the biorthogonal 18 coefficient analysis wavelet and analysis scaling function. Figure 2.10 shows the biorthogonal 18 coefficient synthesis wavelet and synthesis scaling function. While there are biorthogonal sets that are both even and odd in the analysis and synthesis coefficients, respectively, in the Vetterli and Herley biorthogonal sets, both the analysis and synthesis filter coefficients are even—18. The Vetterli and Herley biorthogonal coefficients are listed in Appendix A.

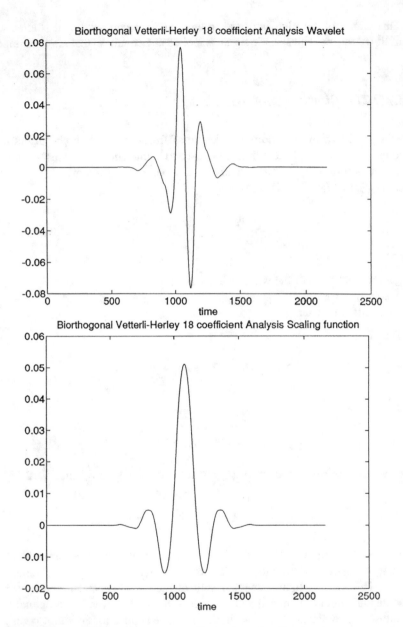

Fig. 2.9 The biorthogonal 18 coefficient analysis wavelet and scaling function

2.3.8   *The Shannon Wavelet*

Shannon wavelet is also known as sinc wavelet. Its functions are derived from the Nyquist-Shannon sampling theorem in honor of Harry Nyquist and Claude Shannon.

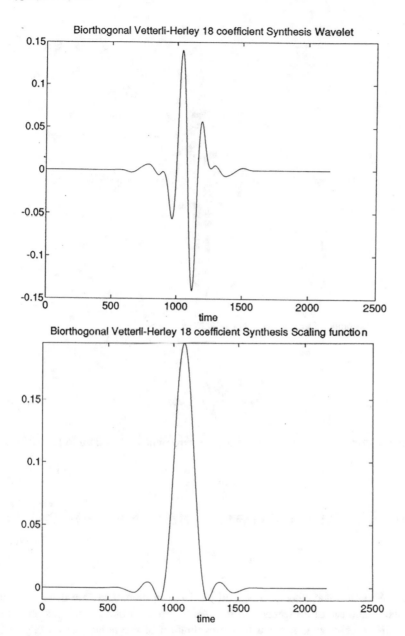

Fig. 2.10 The biorthogonal 18 coefficient synthesis wavelet and scaling function

Claude Elwood Shannon (April 30, 1916–February 24, 2001) was an American mathematician, electrical engineer, and cryptographer known as "the father of information theory." It is a continuous type of wavelet and is represented mathematically as shown in Eq. (2.15).

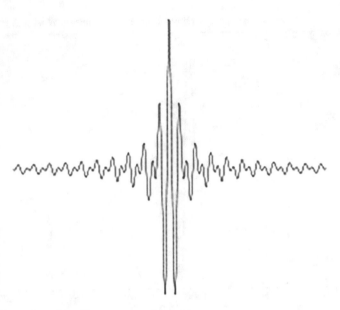

Fig. 2.11 Illustration of the Shannon type of wavelet

$$\psi(t) = \frac{\sin\left(\frac{\pi}{2}t\right)}{\frac{\pi}{2}t} \cos\left(\frac{3\pi}{2}\right)t \qquad (2.15)$$

The Shannon wavelet father function is represented as shown in Eq. (2.16).

$$\phi(x) = \text{sinc}(x) = \frac{\sin(\pi x)}{\pi x} \qquad (2.16)$$

And the wavelet function is represented as shown in Eq. (2.17).

$$\psi = \frac{\sin(2\pi x) - \sin(\pi x)}{\pi x} \qquad (2.17)$$

The Shannon wavelets are not well localized in time. They are discontinuous in frequency and therefore spread out in time. Figure 2.11 shows an illustration of the Shannon wavelet. It is applied to the evaluation of the radar cross section of the conducting and resistive surfaces. It can also be used in signal analysis, signal reconstruction, and solving boundary value problems of fractional differential equations, which are used in modeling physical and engineering processes. It is generally used in an approach where the analysis of functions ranging in multi-frequency bands is required.

2.3.9   The Symlet Wavelet

The Symlet wavelets are nearly asymmetrical wavelets proposed by Daubechies as modifications to the Daubechies family of wavelets, thus making the properties of the two wavelet functions similar. They originated from Daubechies wavelets by way of modifications with increased symmetry. Figure 2.12 shows an illustration of a Symlet wavelet. It is known as Daubechies' least asymmetry wavelet. It is used for denoising and load forecasting, removal of Gaussian additive noise from speech signal, and denoising the temperature data. The Symlet wavelet is also used in signal processing, filtering, image compression and processing, seismic and geophysics, medicine and biomedical, physiological, communications, fractals and facial feature extraction, ultrasonic waves, and audio and speech coding.

2.3.10   The Coiflet Wavelet

The Coiflet wavelets are orthogonal wavelets for which ψ have several vanishing moments and $\emptyset$ several vanishing moments after the zeroth one. This wavelet is named in honor of Ronald Coifman who requested Ingrid Daubechies to design such a wavelet to have scaling function with vanishing moments. Ronald Raphael Coifman is the Phillips Professor of Mathematics at Yale University. Coifman earned a doctorate from the University of Geneva in 1965, supervised by Jovan Karamata.

The Coiflet wavelets are discrete wavelets. The wavelets are near symmetric; the wavelet function has $N/3$ vanishing moment and scaling function $N/3 - 1$ and has been used in many applications using Calderon-Zygmund operators. The differences

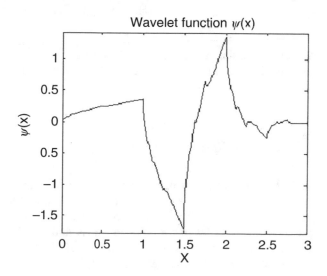

Fig. 2.12 Illustration of the Symlet wavelet (Sym2). (Source: https://www.researchgate.net/)

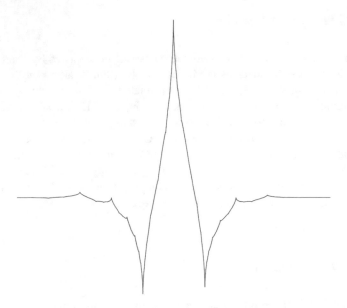

Fig. 2.13 Illustration of the Coiflet wavelet with two vanishing moments. (Source: https://upload.
wikimedia.org/wikipedia/commons/thumb/d/da/Wavelet_Coif1.svg/1024px-Wavelet_Coif1.svg.
png

with the Daubechies wavelets are that the Daubechies wavelets have $N/2-1$
vanishing moments, while the Coiflet scaling functions have $N/3-1$ zero moments
and their wavelet functions have $N/3$. Fig. 2.13 shows the Coiflet wavelet with two
vanishing moments. The Coiflet wavelet is represented mathematically as shown in
Eq. (2.18).

$$\psi(t) = \sum_{n} 2g[n]\varnothing(2t - n) \tag{2.18}$$

where $g[n]$ is the coefficient and $\varnothing$ is the scaling function. The Coiflet wavelet is
used in the compression of image data, digital signal processing, and computer
vision.

2.3.11   The Spline Wavelet

In the Spline family of wavelets, you have the B-Spline wavelet. It is a special class
of Spline wavelet that is constructed using a Spline function and having compact
supports. A Spline, in mathematical terms, refers to a numeric function that is
piecewise defined by polynomial functions and which acquires a high degree of
smoothness at the places where the polynomial pieces meet (called knots). Eq. (2.19)
shows the mathematical representation of the.

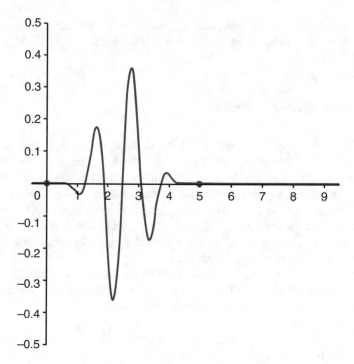

Fig. 2.14 Illustration of the Spline wavelet. (Source: https://upload.wikimedia.org/wikipedia/commons/thumb/6/66/CardinalBSplineWaveletOfOrder3.png/250px-CardinalBSplineWavelet OfOrder3.png)

Spline wavelet. The wavelets were introduced by C.K. Chui and J.Z. Wang and are based on a certain Spline interpolation formula. The Spline interpolation is a form of interpolation where the interpolant is a special type of piecewise polynomial called a Spline. They exist in different types and order. An example is the cardinal and other compactly supported types. Figure 2.14 shows an illustration of the Spline wavelet.

$$\text{B-Spline} - N_1(x) = \begin{cases} 1 & 0 \leq x < 1 \\ 0 & \text{otherwise} \end{cases} \tag{2.19}$$

$$N_m(x) = \int_0^1 N_{m-1}(x-t)dt, \ \text{for } m > 1$$

The significant property of the Spline wavelets, in addition to the symmetry, is that their dual wavelets belong to the same spaces as the original ones. The Spline wavelets are used in Spline harmonic analysis (SHA). SHA in some sense bridges the gap between the continuous and discrete harmonic analysis (DHA). It is a rather universal technique applicable to a great variety of numerical problems, not necessarily to wavelet analysis. Application of the SHA techniques to wavelet analysis is found to be remarkably fruitful. This approach led to the construction of a rich diversity of wavelet bases as well as wavelet packet ones.

2.3.12 The Gabor Wavelet

Dennis Gabor was born as Günszberg Dénes in Budapest, Hungary. He lived from June 5, 1900, to February 9, 1979. The Gabor wavelet is named in his honor. The Gabor wavelets are wavelets he invented using complex functions constructed to serve as a basis for Fourier transforms in information theory applications. His wavelets are very similar to the Morlet wavelets. The Gabor wavelet has an important property in the sense that it minimizes the product of its standard deviations in the time and frequency domain, which means that the uncertainty in information carried by the Gabor wavelet is minimized.

Figure 2.15 shows an illustration of the Gabor wavelet with $a = 2$, $x_0 = 0$, and $k_0 = 1$ and x from -5 to 5. The 1D Gabor wavelet equation is as shown in Eq. (2.20). It is a Gaussian modulated by a complex exponential.

$$f(x) = e^{-(x-x0)2/a^2} e^{-ik0(x-x0)} \tag{2.20}$$

where x is from -5 to 5, $a = 2$, $x_0 = 0$, and $k_0 = 1$.

The Gabor wavelets are nonorthogonal; therefore, the efficient decomposition into the basis is difficult. The application of the Gabor wavelets ranges from image processing to analyzing neurons in the human visual system.

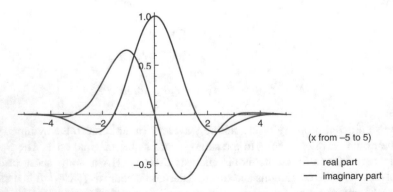

Fig. 2.15 Illustration of the Gabor wavelet. (Source: https://upload.wikimedia.org/wikipedia/commons/b/b5/Gabor_Wavelet%2C_a%3D2%2C_k%3D1.gif

2.3.13 The Lemarie-Battle Wavelet

Lemarie in 1988 and Battle in 1987 independently derived what is known as the Lemarie-Battle wavelets in their honor. The wavelets are constructed from a class of cardinal B-Splines. They are orthonormal wavelets and have compact support. Figure 2.16 shows the representation of the Lemarie-Battle wavelet. They are normally expressed in frequency domains. Therefore, the Fourier transform of a function of t, say $F(t)$, is denoted by $F(w)$. The Lemarie and Battle wavelets were derived by orthonormalizing B-Splines. The scaling function $\Phi_m(t)$ corresponding to the Lemarie-Battle wavelet $\psi_m(t)$ is given by

$$\phi_m(t) = \sum_{k \in z} \alpha_{m,k} B_m(t - k) \tag{2.21}$$

where $B_m(t)$ is the mth-order central B-Spline and the coefficients $\alpha_{m,k}$ satisfy that

$$\sum_{k \in Z} \alpha_{m,k} e^{-jk\omega} = 1 / \sqrt{\left(\sum_{k \in Z} B_{2m}(k) e^{-jk\omega} \right)} \tag{2.22}$$

In addition to other applications, the Lemarie-Battle wavelets are used in harmonic analysis. The Lemarie-Battle scaling coefficients are listed in Appendix A.

2.3.14 The Mallat Wavelet Multiresolution Analysis

Stéphane Georges Mallat was born on October 24, 1962. He is a French applied mathematician. He developed the adaptation of Meyer wavelets to discrete format resulting to the multiresolution analysis (MRA) construction for compactly supported wavelets. Stéphane Georges Mallat made fundamental contributions to the development of wavelet theory in the late 1980s and early 1990s. His MRA wavelet construction discussed fully in Chap. 4 made the implementation of wavelets practical for engineering applications by demonstrating the equivalence of wavelet bases and conjugate mirror filters used in discrete, multirate filter banks in signal processing. In addition, he developed (with Sifen Zhong) the wavelet transform modulus maxima method for image characterization, a method that uses the local maxima of the wavelet coefficients at various scales to reconstruct images. Mallat's MRA additionally is applied in the fields of mathematics, signal processing, music synthesis, and image segmentation. Mallat wavelet coefficients are listed in Appendix A. The Mallat 23 Coefficient Symmetric Wavelet is as shown in Fig. 2.17.

$$s^J$$

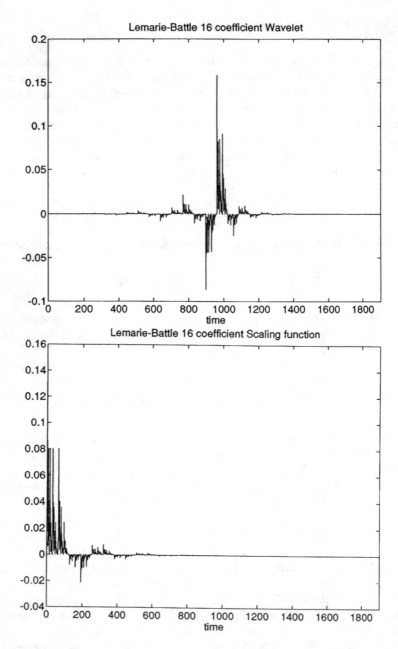

Fig. 2.16 Lemarie-Battle wavelet

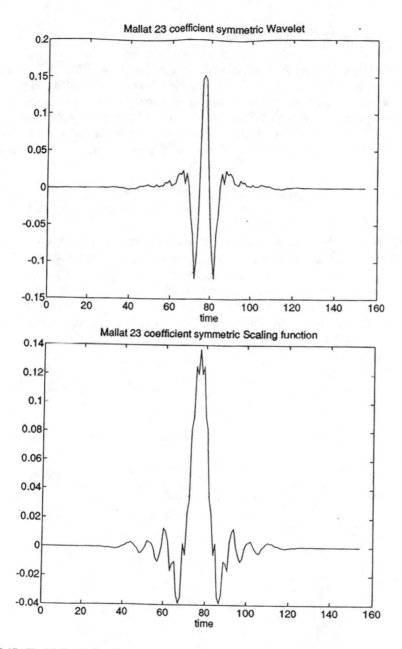

Fig. 2.17 The Mallat 23 Coefficient Symmetric Wavelet

2.3.15   The Poisson Wavelet

In 1995–1996, Karlene A. Kosanovich, Allan R. Moser, and Michael J. Piovoso introduced the Poisson wavelet. This wavelet is associated with a Poisson probability distribution and Poisson kernel. The signal is represented as a discrete sum of Poisson transform coefficients. This wavelet is used to represent the signal in the scale translation domain of the wavelet transform and in the order of the Poisson transform. The Poisson wavelet for n positive integer is as shown in Eq. (2.23).

$$\psi_n(t) = \begin{cases} \left(\dfrac{t-n}{n!}\right) t^{n-1} e^{-t}, & \text{for } t \geq 0 \\ 0 & \text{for } t < 0 \end{cases} \tag{2.23}$$

The plot for the Poisson wavelets corresponding to $n = 1, 2, 3, 4$ is shown in Fig. 2.18. The Poisson wavelet is less localized in time but more localized in frequency. These Poisson wavelets are much less oscillatory than many other commonly used wavelets and decay exponentially as $t \rightarrow \infty$. The nature of the Poisson wavelet is different from the other wavelets. It is only defined to be nonzero in the positive real axis. The Poisson wavelets are neither compact nor orthogonal. For these wavelets, each value of n provides the different non-compact wavelets.

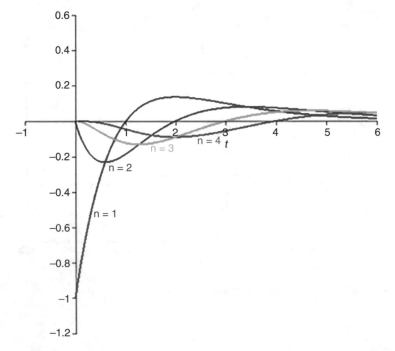

Fig. 2.18 Poisson wavelets corresponding to n = 1, 2, 3, 4. (Source: "Wavelet @ en.wikipedia. org." [Online]. Available: https://en.wikipedia.org/wiki/Wavelet)

The wavelets increase in the value of *n* and are more localized in frequency and less localized in time. The product of time and frequency localized is improved. This is the continuous and non-stationary nature of the wavelet. The use case of the Poisson wavelet is during multiresolution analysis and system identification. This transform is used to identify the real model systems.

2.3.16 Mathieu Wavelets

The Mathieu wavelets came into existence from the Mathieu equation, which is a linear second-order differential equation with periodic coefficients. Mathieu functions are used in a wide variety of physical phenomena, such as diffraction, amplitude distortion, inverted pendulum, the stability of a floating body, radio-frequency quadrupole, and vibration in a medium with a modulated density. Mathieu wavelets are elliptic cylinder discrete wavelets which provide a multiresolution analysis. The magnitude of smoothing is related to the Mathieu function. The Mathieu equation is a linear second-order differential equation with periodic coefficients. This wavelet supports the orthogonality. Mathieu wavelet function is expressed as shown in Eq. (2.24).

$$\psi(\omega) = e^{-j\omega/2} H^* \left(\frac{\omega}{2} - \pi\right) \phi\left(\frac{\omega}{2}\right) \tag{2.24}$$

where H(ω) is a periodic function. The waveform for the Mathieu wavelets is shown in Fig. 2.19. This new family of wavelets is an interesting tool for analyzing optical

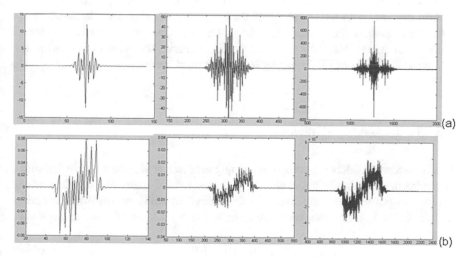

Fig. 2.19 FIR-based approximation of Mathieu wavelets. (Source: "Wavelet @ en.wikipedia.org." [Online]. Available: https://en.wikipedia.org/wiki/Wavelet)

fibers due to its "elliptical" symmetry. In addition, they are beneficial when examining the molecular dynamics of charged particles in electromagnetic traps such as the Paul trap or the mirror trap for neutral particles.

2.3.17 Strömberg Wavelet

In 1983, Jan-Olov Strömberg shared with the world his type of wavelets. It is known as the Jan-Olov Strömberg wavelet. This wavelet is the first smooth orthonormal continuous wavelet. The Strömberg wavelet function is represented as shown in Eq. (2.25).

$$\psi(x) == \frac{\psi^{(1)}(x)}{\psi^{(1)}{}_2} \tag{2.25}$$

where:

$$\psi^{(1)}(x) = \Delta(x) - e(x)G^{-1}w$$

$e(x)$ is the vector constructed with the elements of the basis of the Gram matrix G, w is the scalar product performed in $L^2(R)$ of each element of the basic with

$\Delta(x)$

Figure 2.20 shows the plot for Strömberg continuous wavelet, which is stationary and orthonormal in nature. This wavelet oscillates about the t-axis and exponential decay over the increase in time. It is used in the field of signal processing. It is particularly useful in the recovery of the exact original signal while avoiding the effects of noise and unnecessary distortions. The wavelets are also used in harmonics analysis for image and data compression. In addition, the wavelet is used in the automatic detection of local bearing in rotating machines.

2.3.18 Legendre Wavelet

The Legendre wavelets are compactly supported discrete wavelets derived from the Legendre polynomials. This is also known as spherical harmonic wavelets. Finite impulse response filters are used for the analytical study of harmonic spherical wavelets. The Legendre wavelets are as shown in Eq. (2.26).

$$\psi_{n,m}(t) = \psi(k, \widehat{n}, m, t) \tag{2.26}$$

where:

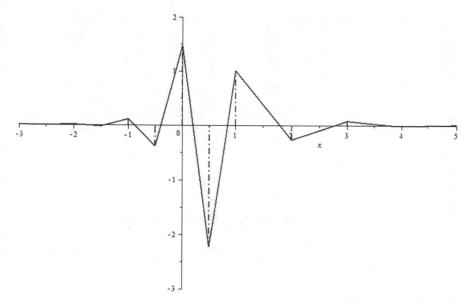

Fig. 2.20 Strömberg wavelets of order 0. (Source: "Wavelet @ en.wikipedia.org." [Online]. Available: https://en.wikipedia.org/wiki/Wavelet)

$\widehat{n} = 2n\text{-}1$, n = 1, 2, 3, ……, 2^{k-1}
k is any positive integer,
m is the order for Legendre polynomials,
t is normalized time.

In defining Eq. (2.26) for the interval [0,1], we get Eq. (2.27).

$$\psi_{n,m}(t) = \begin{cases} \sqrt{m + \dfrac{1}{2}}\, 2^{k/2} P_m\left(2^k t - \widehat{n}\right), & \text{for } \dfrac{\widehat{n}-1}{2^k} \leq t < \dfrac{\widehat{n}+1}{2^k} \\ 0, & \text{otherwise} \end{cases} \qquad (2.27)$$

where:

m = 0, 1, 2, …, M-1,
n = 1, 2, 3, .., 2^{k-1}.

The coefficient $\sqrt{m + \frac{1}{2}}$ is for orthogonality. The waveform diagram for the Legendre wavelet is given as shown in Fig. 2.21. This wavelet supports the orthogonality for the predefined boundary, and also, they are compactly supported within that boundary. Since the Legendre wavelet is in polynomial form, it is highly used in solving the initial value problems. In addition, it is widely used in many mathematics and computer-based applications.

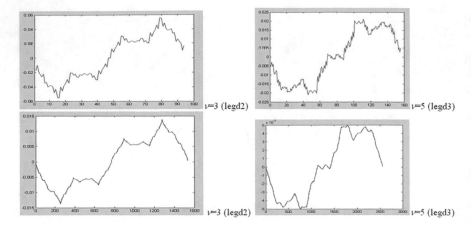

Fig. 2.21 The shape of Legendre wavelets of degree v = 3 and 5. (Source: "Wavelet @ en. wikipedia.org." [Online]. Available: https://en.wikipedia.org/wiki/Wavelet)

2.3.19   Beta Wavelet

The Beta wavelet is a continuous type of wavelets with compact support and it is the soft variety of the Haar wavelets. It is related to Beta distribution using the blur derivatives. Since the Beta wavelet has one cycle, it is called unicycle wavelets. The Beta wavelets are smooth, but the completeness and orthogonality properties that are found in the Haar wavelets are not present on the Beta wavelets. Beta wavelets have the properties of being regular, smooth, and unicycle, having analytical formulation. The wavelet relies on the central limit theorem. It has compact support, which is defined by the boundary between two points [a, b] and zeroes outside this boundary. The nature of the Beta wavelet is oscillatory. The Beta wavelet function is represented as shown in Eq. (2.28) where P(t| α, β) is unimodal. The Beta wavelet only has one cycle, either positive or a negative half cycle, where α and β are parameters that depend on the boundary [a, b].

$$\psi_{beta}(t|\alpha,\beta) = (-1)\frac{dP(t|\alpha,\beta)}{dt} \tag{2.28}$$

The waveform of the Beta wavelets is as shown in Fig. 2.22. The Beta wavelets are used in numerous real applications. The lossy image compression is one of the use cases of Beta wavelets. Beta wavelet and its derivatives are used in biomedical signal compression, image recognition, seismic, and image recognition. It is used in the image and signal processing fields.

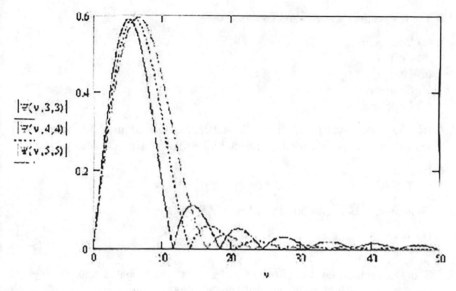

Fig. 2.22 Beta wavelets waveform. (Source: "Wavelet @ en.wikipedia.org." [Online]. Available: https://en.wikipedia.org/wiki/Wavelet)

Summary

1. In this chapter, we covered what wavelets are and why we use wavelets. The different types of wavelets and their functionalities are also covered.
2. Wavelets are set of basis functions—limited duration; average value of zero; usually irregular and asymmetric.
3. Wavelets are used in various fields of application.
4. Some of major wavelet types are Haar, Daubechies, Morlet, Meyer, Mexican hat, Berlage, and a class of biorthogonal wavelets.
5. The other types of wavelets are Shannon, Symlet, Coiflet, and Spline.
6. Stéphane Georges Mallat's MRA wavelet construction made the implementation of wavelets practical for engineering applications by demonstrating the equivalence of wavelet bases and conjugate mirror filters used in discrete, multirate filter banks in signal processing.
7. The Poisson wavelets are neither compact nor orthogonal.
8. The Beta wavelet is a continuous type of wavelets with compact support and it is the soft variety of the Haar wavelets.

Review Questions

2.1 Wavelets are functions generated from one basic function ψ called the mother wavelet by dilations and translations of ψ.

   (a) True
   (b) False

2.2 The first known wavelet and simplest was?

   (a) Daubechies
   (b) Coiflet
   (c) Haar
   (d) Shannon
   (e) Legendre

2.3 Biorthogonal wavelets provide additional degree of freedom so that both perfect reconstruction and linear phase filters can be realized simultaneously.

   (a) True
   (b) False

2.4 Haar wavelet is not an orthogonal type of wavelet.

   (a) True
   (b) False

2.5 The Daubechies family of wavelets have extremal phases and the highest number of vanishing moments for a given support width.

   (a) True
   (b) False

2.6 The Mexican hat wavelet was originally known as the _____wavelet, named after an American geophysicist who lived from 1898 to 1980.

   (a) James
   (b) Flicker
   (c) Rucker
   (d) Ricker

2.7 Which family of wavelets is useful in theoretical or computational studies where causality of the incident seismic signal is looked as a condition that is very important.

   (a) Haar
   (b) Daubechies
   (c) Symlet
   (d) Berlage
   (e) Poisson

2.8 Which wavelet is also known as sinc wavelet?

   (a) Haar
   (b) Daubechies
   (c) Shannon
   (d) Spline
   (e) Beta

2.9 A Spline, in mathematical terms, refers to a numeric function that is piecewise defined by polynomial functions and which acquires a high degree of smoothness at the places where the polynomial pieces meet (called knots).

(a) True
(b) False

2.10 The Fourier transform of the Meyer wavelet is not a compactly supported C^∞ function with rapid polynomial decay.

(a) True
(b) False

Answers: 2.1a, 2.2c, 2.3a, 2.4b, 2.5a, 2.6d, 2.7d, 2.8c, 2.9a, 2.10b.

Problems
2.1. What are wavelets?
2.2. Why do we look at wavelets?
2.3. Describe the Haar wavelet.
2.4. What areas of application can the Haar wavelet be applied?
2.5. Describe the Daubechies wavelet.
2.6. What the areas of application where the Daubechies wavelets can be used?
2.7. Describe the Morlet wavelet.
2.8. What are the application areas of the Morlet wavelet?
2.9. Describe the Meyer wavelet.
2.10. What are the application areas of the Meyer wavelet?
2.11. Describe the Mexican hat wavelet.
2.12. How can the Mexican hat wavelet be used?
2.13. Describe the Berlage wavelet.
2.14. How can the Berlage be applied?
2.15. Describe the Biorthogonal wavelet.
2.16. How can biorthogonal wavelets be used?
2.17. Describe the Shannon wavelet.
2.18. How can the Shannon wavelets be used?
2.19. Describe the Symlet wavelet.
2.20. How can the Symlet wavelets be used?
2.21. Describe the Coiflet wavelets.
2.22. How can the Coiflet wavelets be used?
2.23. Describe the Spline wavelets.
2.24. How can the Spline wavelets be used?
2.25. Describe briefly, by using both Haar and Daubechies-2 wavelets, what is the rough definition of wavelets.
2.26. What is the orthogonality of Haar and Daubechies-2 wavelets?
2.27. What are the two-scale equations of scaling/refinable functions and wavelets?
2.28. What are the vanishing moments of them?
2.29. Describe the Lemarie-Battle wavelets.
2.30. How can the Lemarie-Battle wavelets be used?

2.31. Classify the Poisson, Beta, Strömberg, Legendre, and Mathieu wavelets as orthogonal, biorthogonal, CWT, DWT, stationary, and non-stationary.

2.32. Describe how the Poisson, Beta, Strömberg, Legendre, and Mathieu wavelets can be used in signal processing applications.

References

Chui, C. K. (1992). *An introduction to wavelets*. Academic Press.

Mallat, S. (1989). A theory for multiresolution signal decomposition: The wavelet representation. *IEEE Transactions on Pattern Analysis and Machine Intelligence, 11*, 674–693.

Swelden, W. (1995). The lifting scheme: a new philosophy in biorthogonal wavelet constructions. In *Wavelet applications in signal and image processing III* (pp. 68–79).

Swelden, W. (1998). The lifting scheme: A construction of second generation wavelets. *SIAM Journal on Mathematical Analysis, 29*(2), 511–546.

Chapter 3
Generations of Wavelets

"Teach the young people how to think, not what to think."
Sidney Sugarman.

3.1 Introduction

Ever since the early 1900s when the idea of wavelets first showed up with the development of the Haar wavelets, much was not done in applying wavelets to solving signal processing problems until about 80 years later with the development of Daubechies family of wavelets and Mallat wavelet transformation ideas in the later part of the 1980s. It was at this time that several applications began to manifest itself as a result of the intensive work done by researchers in the mathematical, scientific, signal processing, and many other diversified fields such as geophysics and the medical communities. The name wavelets emerged from the literature of geophysics, through France. The word *onde* led to *ondelette*. The translation of *wave* led to *wavelet*.

As the properties of these generations of wavelets became more acceptable to the application analysis of different types of information, discovering how to apply more of these generations of wavelets became more necessary than ever. Therefore, over several decades, different generations of wavelets have been discovered based on the application areas. This chapter not only covers these different generations of wavelets, what necessitated them, and how they can be applied to solving societal problems but classified the different wavelets generationally as well. This chapter also covers a brief history of the wavelets and the shortcomings of the different generations of wavelets.

© Springer Nature Switzerland AG 2022
C. M. Akujuobi, *Wavelets and Wavelet Transform Systems and Their Applications*,
https://doi.org/10.1007/978-3-030-87528-2_3

3.2   Brief History of Wavelets

Wavelets are classified into different generations based on the type of application and the times in history they were developed. A brief account of the historical births of these wavelets is discussed in this section.

- In 1873—Karl Weierstrass.

  - Family of scaled overlapping copies of a basis function.

- In 1910—Alfred Haar.

  - Orthonormal system of compact functions (Haar basis).

- In 1930—Physicist Paul Levy.

  - Investigated Brownian motion (random signal) and concluded Haar basis is better than Fourier transform.

- In the 1930s—Littlewood–Paley and Stein.

  - Calculated the energy of the function.

- In 1946—Dennis Gabor.

  - Nonorthogonal wavelet basis with unlimited support.

- In 1960—Guido Weiss and Ronald Coifman.

  - Studied simplest element of function space called atom.

- In the 1980s—A. Grossman, J. Morlet, and I. Daubechies.

  - Signal Analysis with wavelets.

- In 1980—Grossman (physicist) and Morlet (engineer).

  - Broadly defined wavelet in terms of quantum mechanics.

- In 1985—Stephen Mallat.

  - Defined wavelet for his digital signal processing work for his Ph.D.

- In 1988—Ingrid Daubechies.

  - Used Mallat's work and constructed set of wavelets.

- In 1989—Stéphane Mallat and Yves Meyer.

  - Discovered the idea of multiresolution analysis (MRA).

- Since 1990 to present.

  - There has been many rediscoveries of wavelets in different fields.

3.3   First-Generation Wavelets

One of the examples of the traditional basis expansion models used in digital signal processing which is very indispensable in many signal processing domains includes the Fourier transform. However, over the past several decades, it has been found that there are some limitation issues that may not allow the practicality of the traditional basis expansion models. These limitation issues include:

(L1) Need for space localization—The Fourier transform is known for its localization in frequency. However, because it has global support of the basis functions, that has necessitated the prevention of its localization in space. This issue has affected many of its applications, and in particular, the need for local behavior of signals is of great interest.

(L2) Need for algorithms for faster transformations—In recent years, the advance of data acquisition technology outpaced the available computing power significantly, making the fast Fourier transform with its $O(n \log n)$ complexity a bottleneck in many applications.

(L3) Need for more flexibility—The traditional basis expansion models provided no or almost no flexibility. It made it almost impossible usually to adapt a representation to most of the current problems. An important reason for this lack of flexibility is the orthogonal nature of the traditional basis expansion models.

(L4) Need for arbitrary domains—The traditional basis representations can only represent functions defined in the Euclidean spaces R^n. Many real-world problems have embeddings $X \subset R^n$ as domain, and it is desirable to have a representation which can be easily adapted for these spaces.

(L5) Need for weighted measures and irregularly sampled data—The traditional transforms are usually not employed on spaces with weighted measures or when the input data is irregularly sampled.

To overcome these limitations (L1–L5), it became necessary to find better tools to help in the processing of information, hence the motivation in the development of wavelets such as Haar, Meyer, and Daubechies wavelets. An important distinction between traditional basis expansion models and wavelets is that there is not a single set of basis functions that defines a wavelet. Instead, the members of a family of representations with vastly different properties are denoted as wavelets.

There are three possible properties that are common to all of them and they are as follows:

(P1) The sequence $\{f_k\}_{k=1}^{m}$ forms a basis or a frame of L_p.

(P2) The elements of $\{f_k\}_{k=1}^{m}$ are localized in both space and frequency.

(P3) Fast algorithms for the analysis, synthesis, and processing of signals in its basis representation exist.

These three properties (P1–P3) and the flexibility they leave are the key to the efficiency and versatility of wavelets like Daubechies family of wavelets, Haar wavelets, and many others. Some of the first nontrivial wavelets that have been developed are the Daubechies family of wavelets and the Shannon, Coiflet, and Meyer wavelet. These, and most other wavelets developed in the 1980s, are the first-generation wavelets (FGWTs) whose construction requires the Fourier transform and whose basis functions have to be (dyadic) scales and translations of one particular mother basis function. The first-generation wavelet transforms (FGWTs) are used in identifying pure frequencies, denoising signals, detecting discontinuities and breakdown points, detecting self-similarity, and compressing images.

The limitations L3 to L5 (more flexibility, arbitrary domains, and weighted measures and irregularly sampled data) thus still apply for first-generation wavelets. The work by Mallat in 1989 and Swelden in 1996 and 1998 overcame these restrictions and led to the development of second-generation wavelets.

Example 3.1 Name and describe a first-generation wavelet.

Solution The first-generation wavelets are those wavelets which essentially need the Fourier transform, and the basis functions are dyadically scalable with translation property of one particular mother basis function. Their transforms are known as first-generation wavelet transforms (FGWTs). These are the first nontrivial wavelets developed around the 1980s.

The Morlet wavelet, named after Jean Morlet, is one of the first-generation wavelets. It was originally formulated by Goupillaud, Grossmann, and Morlet in 1984 as a constant $\kappa\sigma$ subtracted from a plane wave and then localized by a Gaussian window as shown in Eq. (3.1).

$$\Psi_\sigma(t) = C_\sigma \pi^{-\frac{1}{4}} e^{-\frac{1}{2}t^2} \left(e^{i\sigma t} - \kappa_\sigma \right) \tag{3.1}$$

where $\kappa_\sigma = e^{-\frac{1}{2}\sigma^2}$ is defined by the admissibility criterion and the normalization constant $c\sigma$ is as shown in Eq. (3.2).

$$C_\sigma = \left(1 + e^{-\sigma^2} - 2e^{-\frac{3}{4}\sigma^2} \right)^{-\frac{1}{2}} \tag{3.2}$$

The Fourier transform of the Morlet wavelet is shown in Eq. (3.3).

$$\widehat{\Psi}_\sigma(\omega) = C_\sigma \pi^{-\frac{1}{4}} \left(e^{-\frac{1}{2}(\sigma-\omega)^2} - \kappa_\sigma e^{-\frac{1}{2}\omega^2} \right) \tag{3.3}$$

The "central frequency" $\omega\Psi$ is the position of the global maximum of $\widehat{\Psi}_\sigma(\omega)$ which, in this case, is given by the solution of Eq. (3.4).

$$(\omega_\Psi - \sigma)^2 - 1 = \left(\omega_\Psi^2 - 1 \right) e^{\sigma\omega_\Psi} \tag{3.4}$$

The parameter σ in the Morlet wavelet allows trade between time and frequency resolutions. Conventionally, the restriction $\sigma > 5$ is used to avoid problems with the Morlet wavelet at low σ (high temporal resolution). The Morlet wavelet is as shown

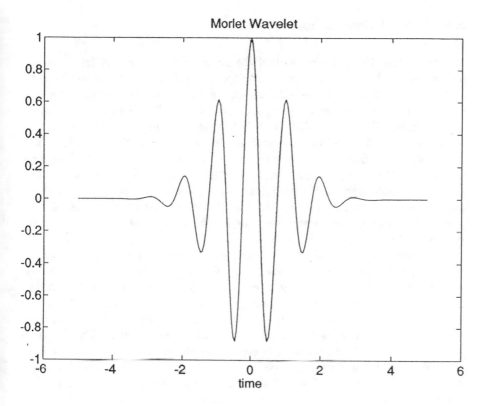

Fig. 3.1 The Morlet wavelet

in Fig. 3.1. For signals containing only slowly varying frequency and amplitude modulations like audio, it is not necessary to use small values of σ. In this case, $\kappa\sigma$ becomes very small $\sigma > 5 \Rightarrow \kappa_\sigma < 10^{-5}$ and is, therefore, often neglected. Under the restriction $\sigma > 5$, the frequency of the Morlet wavelet is conventionally taken to be $\omega_\psi \simeq \sigma$.

Both the scaling function as low-pass filter and the wavelet function as high-pass filter must be normalized by a factor $1/\sqrt{2}$. The wavelet coefficients of the Morlet wavelets are derived by reversing the order of the scaling function coefficients and then reversing the sign of every second one (C6 wavelet = {−0.022140543057, 0.102859456942, 0.544281086116, −1.205718913884, 0.477859456942, 0.102859456942}). Mathematically, this looks like $B_k = (-1)^k C_{N-1-k}$ where k is the coefficient index, B is a wavelet coefficient, and C is a scaling function coefficient. N is the wavelet index, i.e., 6 for C6.

3.4   Second-Generation Wavelets

In 1995 and in 1998, a new method for the construction of wavelets, which is independent of Fourier transform, was proposed by Swelden (1995) and Swelden (1998). It was called the lifting scheme, the second-generation wavelets, or integer wavelet transform. The second-generation wavelets originated with the concept of lifting scheme (Kingsbury 2001), to maintain the time-frequency localization and fast algorithms instead of Fourier domain to deploy in geometrical applications. This should replace translation and dilation as well as any Fourier analysis. The basic algorithm of the lifting scheme is to split even samples; then they are adjusted to serve as the coarse version of the original signal data x_k in even set $\{x_k: k \text{ even}\}$ and odd set $\{x_k: k \text{ odd}\}$, then predict odd signal using even part to detect the missing parts called details, and update even samples for adjustment to serve the coarse version of the original signal.

The block diagram of the (SGWT) and the illustration of the basic algorithm are as shown in Fig. 3.2 (a) and (b), respectively. The input signal f can be split into odd $\gamma 1$ and even $\lambda 1$ samples using shifting and downsampling. The detail coefficients $\gamma 2$ are then interpolated using the values of $\gamma 1$ and the prediction operator on the even values: $\gamma 2 = \gamma 1 - P(\lambda 1)$. The next stage (the updating operator) alters the approximation coefficients using the detailed ones: $\lambda 2 = \lambda 1 + U(\gamma 2)$. The functions prediction operator P and updating operator U as shown in Fig. 3.2 (a) effectively define the wavelet used for decomposition. This concept can be deployed as in the discrete wavelet transform (DWT) to create a filter bank with a number of levels. The variable tree used in wavelet packet decomposition can also be used.

Wavelets can be categorized into DWT and continuous wavelet transform (CWT) as discussed in Chap. 1. The basis functions of DWTs are defined over a discrete space which becomes continuous only in the limit case, whereas the basis functions of CWTs are continuous but require discretization if they are to be used on a computer. The second-generation wavelets permit the representation of functions in $L^2(R)$, the space of functions with finite energy, in a very general setting $L^2(X, \sum, \mu)$, where $X \subseteq \cdot R^n$ is a spatial domain, $\sum$ denotes a α-algebra defined over X, and μ is a (possibly weighted) measure on $\sum$. The inner product defined over X will be denoted as $\{.\}$. A multiresolution analysis $M = \{V_j \subset L^2 | j \in \zeta \subset Z\}$ consisting of a sequence of nested subspaces Vj on different levels j is employed to define the basis functions.

The second-generation wavelet transforms (SGWTs) are extensively used for lossy data compression, in geographical data analysis and computer graphics. They are efficient in coding and implementation of the compression algorithms. They are flexible for topography of the Earth as to create wavelets that live on a sphere. In this way, the topographic data of the Earth can be compressed and manipulated much like a 1D signal. Furthermore, a 3D object that is meshed by triangles can be analyzed using the multiresolution scheme if we allow ourselves to think of the triangulation as the domain of the problem and the coordinates of the vertices as function values. The lifting approach could provide a viable alternative to the fast

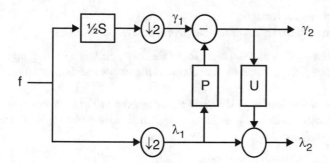

Fig. 3.2   (**a**) Block diagram of SGWT (**b**) Illustration of basic algorithm

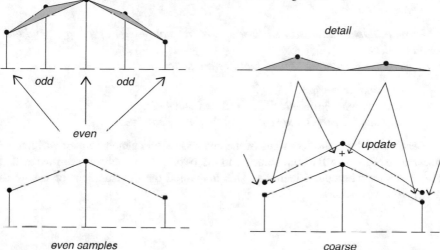

Fig. 3.2 (continued)

multipole method that is very successful for scattering problems. Some of the wavelets used in the second-generation wavelet transforms (SGWTs) are packed integer wavelet, fast lifting wavelet, **directional lifting-based wavelet,** weighted adaptive lifting-based wavelet, and the spatially adaptive integer lifting wavelet.

3.5   The Shortcomings with the First and Second Generations of the Wavelet Models

The first and second generations of wavelets are not suitable for morphological features. This therefore meant that the first and second generations of wavelets provided no accurate wavelet techniques to extract morphological features such as (Jiang and Blunt 2004):

- *Linear/curved scratches.*
- *Plateaux with direction/objective properties, as follows:*

- *Automotive—Plateau honed surfaces that are produced by two machining processes. The final surface always includes the rough machining information (deep valleys).*
- *Biomedical engineering—Surface topographies in different material heads that are normally combined with morphological features from different wear stage in service, such as:*

    - *Regular shallow scratches.*
    - *Random deeper scratches.*
    - *Arc scratches.*

- *Morphological features in surface texture are not always of a large amplitude and isolated.*
- *In many cases, morphological features are as follows:*

    - *Direction properties.*
    - *They mix with harmonic roughness and waviness.*
    - *Have similar amplitude scales and frequency bands.*

Because of these shortcomings of the first and second generations of wavelets, it became necessary to find the solution to the best way to analyze information that may have morphological features. This motivated the development of the third-generation wavelets.

3.6   Third Generation of Wavelets

The third-generation wavelets (TGWs) (Jiang and Blunt, 2004) give affine invariance, with independence of the reference frame for the measurements, and perfect reconstruction, limit redundancy, and have efficient computation. In general, the third-generation wavelet representation uses the complex wavelet transform based on Z-transform theory of linear time invariant sampled systems. The complex wavelets and complex biorthogonal wavelets are usually the wavelets of choice. The main approach to complex biorthogonal wavelet operation is to use two-channel filter banks in the real and imaginary parts. A Q-shift filter design technique is used to construct the low-pass finite infinite response (FIR) filter to satisfy a linear phase and perfect reconstruction condition as shown in Eq. (3.5).

$$H_{L2}(z) = H_L\!\left(z^2\right) + z^{-1}H_L\!\left(z^2\right)H_L(z)H_L\!\left(z^{-1}\right) + H_L\!\left(-z^{-1}\right)H_L(-z) = 2 \quad (3.5)$$

The engineering and bioengineering surfaces shown in Figs. 3.3 and 3.4 demonstrate the special properties of the third-generation wavelet model (Jiang et al., 2008). The figures are raw measured surfaces, including the form, waviness, and

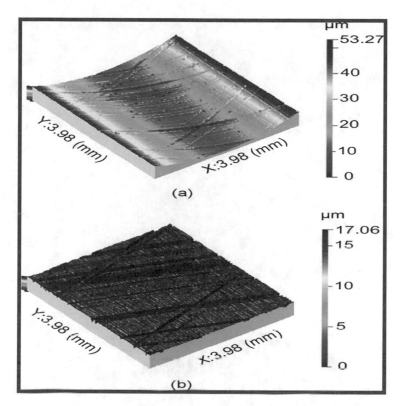

Fig. 3.3 Cylinder surface analysis using the complex wavelet. (**a**) Raw measured surface, (**b**) extracted feature surface. (Source: Jiang et al., 2008). With permission from Elsevier

roughness components that almost submerge the main features of the deep valleys. The deep valleys are reconstructed using complex wavelets in which not only the form and some harmonic components are removed very effectively, but also the shape of the scratches are retained very well. There is no affine aliasing; no new artifacts and the edge of the deep valleys are preserved perfectly.

The third-generation wavelet transforms (TGWTs) are extensively deployed in the area of computer vision to exploit the concept of visual contexts. First the candidate regions are focused, where objects of interest can be achieved, and then additional features are computed through TGWT for that region of interest. Accurate detection and recognition of smaller objects are done based on these computed features. Furthermore, the TGWTs are applied to detect the activated voxels of cortex and additionally the temporal independent component analysis to haul out the desired independent sources by Bayesian filter.

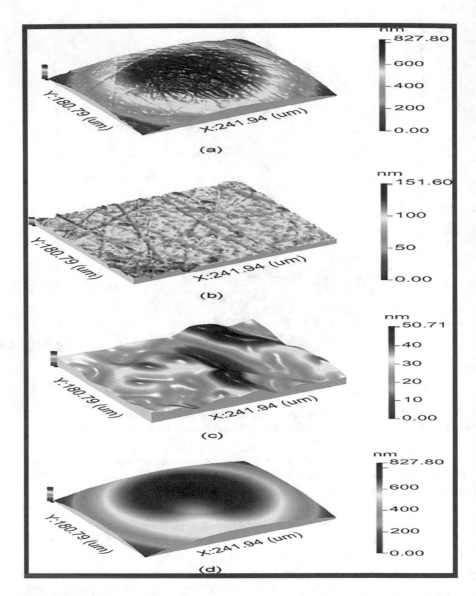

Fig. 3.4 Wavelet decomposition of a precision-lapped ceramic sphere used as a femoral bearing surface in a replacement hip joint. (**a**) Raw measured surface; (**b**) roughness surface; (**c**) wavy surface; (**d**) form surface (Source: Jiang et al., 2008). With permission from Elsevier

3.7   Next-Generation Wavelets

The next-generation wavelet transforms (NGWTs) are those next wavelets and transforms which can optimize the peak signal-to-noise ratios (PSNRs) and error-free, lossless, and advanced multi-level resolution. These wavelets will be more advanced in terms of efficiency and performance. These are application specific in different areas such as human vision characterization, frequency localization, feature extraction, seismic analysis, biomedical analysis, etc. Contourlets form a multiresolution directional tight frame designed to efficiently approximate images made of smooth regions separated by smooth boundaries. The contourlet transform has a fast implementation based on a Laplacian pyramid decomposition followed by directional filter banks (DFB) applied on each band-pass subband. Instead of using the Laplacian pyramid, a new pyramid structure for the multiscale decomposition is used in the new contourlets which is conceptually similar in the steerable pyramid. The difference from the Laplacian pyramid is that the new multiscale pyramid can employ a different set of low-pass and high-pass filters for the first level and all the other levels. This is a crucial step in reducing the frequency domain aliasing of the DFB. The new multiscale pyramid is to cancel the aliasing components of the DFB.

3.8   Comparisons of the Different Generations of Wavelets

The different generations of wavelets have been covered in Sects. 3.3, 3.4, 3.5 and 3.6. In this section, we show in tabular form (Table 3.1) the differences and applications of these wavelet generations. It is important to note that in each generation, there are specific applications tailored to those wavelets of that generation. As advancements continue to grow in terms of possibilities to new applications, those motivations helped and will continue to help in the development of the second, third, and even the next generations.

Summary
1. Different generations of wavelets have been discovered over several decades based on the application areas.
2. Wavelets are classified into different generations based on the type of application and the times in history they were developed.
3. These generations show examples of traditional basis expansion models in digital signal processing which are very indispensable in many signal processing domains that include Fourier transform and Laplace transform.
4. Some limitation issues may not allow the practicality of the traditional basis expansion models in some of these generations of wavelets.
5. An important distinction between traditional basis expansion models and wavelets is that there is not a single set of basis functions that defines a wavelet. Instead, the members of a family of representations with vastly different properties are denoted as wavelets.

Table 3.1 The comparisons of the different generations of wavelets

No	Generations of wavelets	Differences	Applications
1	First generation	These wavelets can be used for periodic or infinite signals, but they cannot be optimized in bounded domains	The wavelet transform is used in self-similarity detection, image compressions, signal denoising, and identification of pure frequencies and detection of discontinuities. Other applications are acoustic signal compressions, fingerprint image compression, image processing, enhancement, and restoration. Fractal analysis and denoising noisy data
2	Second generation	Fastest for moderately short filters, but one needs to first find the factorizations of the filter bank matrices, but these factors are very well documented for JPEG2000 wavelets	Generations fastest for moderately short filters, but one needs to first find the factorizations of the filter banks matrices. These factors are very well documented for JPEG2000 wavelets. They are used tremendously for efficient coding in compression algorithms, computer graphics, geographical data analysis, and lossy data compressions. The FBI has used the CDF wavelets in fingerprint compression scans. With these wavelets, compression ratios of about 20 to 1 could be achieved. They have been applied in multiresolution analysis, system identifications, and parameter estimations
3	Third generation	They do not oscillate and do not show aliasing and degrees of shift variance in their magnitudes. They exhibit a 2D attribute of the signal to be transformed and produce redundancies	They are applied in sparse representation, multiresolution, and useful features characterizations based on the image structures. They are also used in the medical profession because they provide intuitive bridges between time and frequency data, which could clarify interpretations of complex head trauma spectra produced with Fourier transform. They are also used in music industry for transcriptions of music since they produce precise results that are not possible earlier with the Fourier transforms. It is capable of capturing short bursts of repeating and alternating music notes
4	Next generations	They will be too specific and too constrained because these wavelet transforms are still very much at the	The applications will include human vision characterizations, frequency localizations, feature extractions, analysis of seismic

(continued)

Table 3.1 (continued)

No	Generations of wavelets	Differences	Applications
		research stage and meant for specific applications	information, and analysis of bio-medical information, and they are used in sensor networks, cellular networks, etc.

6. The Daubechies family of wavelets and the Meyer wavelet and most of the other wavelets developed in the 1980s are the first-generation wavelets whose construction requires the Fourier transform and whose basis functions have to be (dyadic) scales and translations of one particular mother basis function.
7. In 1995, Swelden proposed a new method for the construction of wavelets, which is independent of Fourier transform. It was called the lifting scheme, the second-generation wavelets, or integer wavelet transform.
8. The first and second generations of wavelets are not suitable for morphological features.
9. The third-generation wavelets give affine invariance, with independence of the reference frame for the measurements, and also perfect reconstruction, limit redundancy, and have efficient computation.
10. The next-generation wavelet transforms (NGWTs) are those next wavelets and transforms which can optimize the peak signal-to-noise ratios (PSNRs) and error-free, lossless, and advanced multi-level resolution.

Review Questions

3.1. The Haar and Daubechies wavelets are part of the first generation of wavelets.

   (a) True.
   (b) False.

3.2. In 1980 Grossman (physicist) and Morlet (Engineer) broadly defined wavelet in terms of quantum mechanics.

   (a) True.
   (b) False.

3.3. Space localization is not one of the limitations of the traditional models of signal processing.

   (a) True.
   (b) False.

3.4. The following wavelets are part of the first-generation wavelets.

   (a) Haar.
   (b) Meyer.
   (c) Daubechies.
   (d) All of the above.

3.5. The first-generation wavelet transforms are not used in identifying pure frequencies, denoising signals, detecting discontinuities and breakdown points, detecting self-similarity, and compressing images.

(a) True.
(b) False.

3.6. In 1996, a new method for the construction of wavelets, which is independent of Fourier transform, was proposed by Swelden (1996). It was called the lifting scheme, the second-generation wavelets, or integer wavelet transform.

(a) True.
(b) False.

3.7. The first and second generations of wavelets were suitable for morphological features.

(a) True.
(b) False.

3.8. The third-generation wavelet transforms (TGWTs) are extensively deployed in the area of _____to exploit the concept of visual contexts.

(a) Restoration.
(b) Segmentation.
(c) Computer vision.
(d) All of the above.

3.9. The third generation of wavelets do not oscillate and do not show aliasing and degrees of shift variance in their magnitudes.

(a) True.
(b) False.

3.10. The next-generation wavelet transforms (NGWTs) are those next wavelets and transforms which can optimize the peak signal-to-noise ratios (PSNRs) and error-free, lossless, and advanced multi-level resolution.

(a) True.
(b) False.

Answers: 3.1a, 3.2a, 3.3b, 3.4d, 3.5b, 3.6a, 3.7b, 3.8c, 3.9a, 3.10a.

Problems
3.1. What are first-generation wavelets?
3.2. Name at least five of the first-generation wavelets.
3.3. Describe the Coiflet wavelet using both mathematical and diagrammatic representations in your description.
3.4. List some of the coefficients for the Coiflet wavelet.
3.5. What is the major drawback of the first-generation wavelets?
3.6. What are the major advantages of the first-generation wavelets?

3.7. What are second-generation wavelets?

3.8. How does the basic algorithm of the second-generation wavelets operate?

3.9. Name at least five of the second-generation wavelets.

3.10. What are the major advantages of the second-generation wavelets?

3.11. What are the major disadvantages of the second-generation wavelets?

3.12. What are the shortcomings of the first and second generations of wavelets?

3.13. What are third-generation wavelets?

3.14. What are the special properties of the third-generation wavelets?

3.15. How can the third-generation wavelets be deployed?

3.16. What are the major advantages of the third-generation wavelets?

3.17. What are the major disadvantages of the third-generation wavelets?

3.18. Name at least five of the next-generation wavelets.

3.19. What are next-generation wavelets?

3.20. Name at least five of the possible next-generation wavelets.

3.21. What are the major advantages of the next-generation wavelets?

3.22. What are the major disadvantages of the next-generation wavelets?

References

Jiang, X., & Blunt, L. (2004). Third generation wavelet for the extraction of morphological features from micro and nano scalar surfaces. *Wear, 257*, 1235–1240.

Jiang, X., Scott, P., & Whitehouse, D. (2008). Wavelets and their applications for surface metrology. *Elsevier, CIRP Annals – Manufacturing Technology, 57*(1), 555–558. https://doi.org/10.1016/j.cirp.2008.03.110

Kingsbury, N. (2001). Complex wavelets for shift invariant analysis and filtering of signals. *Journal of Applied and computational Harmonic Analysis, 10*(3), 234–253.

Swelden, W. (1995). The lifting scheme: A new philosophy in biorthogonal wavelet constructions. In *Wavelet applications in signal and image processing III*, (pp. 68–79).

Swelden, W. (1998). The lifting scheme: A construction of second generation wavelets. *SIAM Journal on Mathematical Analysis, 29*(2), 511–546.

Chapter 4
Wavelet Transforms

"There are no secrets to success. It is the result of
preparation, hard work, and learning from failure."

Colin Powell

4.1 Introduction

A wavelet transform of a signal $f(t)$ is the decomposition of the signal into a set of basis functions consisting of contractions, expansions, and translations of a mother function $\psi(t)$, called the wavelet.

What happens is that a given function or continuous-time signal is divided into different scale components by the specific wavelet. In most cases, it is possible to assign a frequency range to each scale components. The representation of the specific function by specific wavelets is called a wavelet transform. The multiscale technique is used in the processing of signals using wavelet transforms. Wavelets are adjustable and adaptable (Burrus et al., 1998; Akansu & Haddad, 1992). Since there are various wavelet bases, they can be designed to fit into different applications. Wavelet transform (WT) can be either continuous or discrete. If the signal is in itself a sequence of numbers, or samples of a function of continuous variables, a wavelet expansion of the signal is referred to as discrete time wavelet transform. The algorithm for the discrete wavelet transform is much simpler than the integral function of the continuous wavelet transform. In this chapter, the multiscale analysis technique is covered. Using the idea of the multiscale analysis technique, we apply it to the wavelet transforms of 1D and 2D signals.

4.2 The Multiscale Wavelet Transform

If we let Z and R be the set of integers and real numbers, respectively. Let $L^2(R)$ be the vector space of measurable, square-integrable one-dimensional function $f(x)$. In the two-dimensional case, $L^2(R^2)$ denotes the vector space of measurable, square-integrable function $f(x, y)$. The multiscale analysis and representation technique

© Springer Nature Switzerland AG 2022
C. M. Akujuobi, *Wavelets and Wavelet Transform Systems and Their Applications*,
https://doi.org/10.1007/978-3-030-87528-2_4

Fig. 4.1 Multiscale
analysis and
representation idea

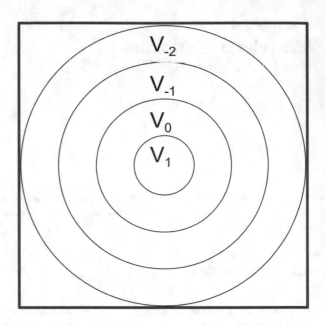

provide signals with containment, completeness, scaling, and basis/frame properties
as reflected in Eqs. 4.1–4.4, respectively (Mallat, 1989a, b). The multiscale analysis
and representation of $L^2(R)$ are increasing sequences $V_j, j \in Z$, of closed subspaces
of $L^2(R)$.

$$Containment: \quad \subset V_3 \subset V_2 \subset V_0 \subset V_{-1} \subset V_{-2} \subset V_{-3} \subset \quad (4.1)$$

with the following properties:

$$Completeness: \quad \cap_{jeZ} V_j = \{0\}, \cup_{jeZ} V_j = L^2(R^n); \quad (4.2)$$

$$Scaling: \quad f(X) \in V_j \leftrightarrow f(2x) \in V_{j+1}; \quad (4.3)$$

$$The\ Basis/Frame\ Property: \quad f(x) \in V_0 \leftrightarrow f(x\text{-}k) \in V_0 \quad (4.4)$$

for each $j \in Z$.

There is a scaling function $\varphi(t) \in V_0$ such that the functions $\varphi(x - k), k \in Z^n$ form a
Riesz basis of V_0 (see Appendix B for the definition of the Riesz basis). The
sequence V_j is called the multiscale approximation. As j increases, V_j becomes
space of lower resolution as shown in Fig. 4.1. For each $j \in Z$, let W_j be the
orthogonal complementary subspace of V_j in V_{j-1}; that is, $V_{j-1} = V_j \_ W_j$ and
$W_j \perp W_{j'}$ if $j \neq j'$. (If $j > j'$, for example, then $W_j \subset V_{j'} \perp W_{j'}$.) It follows that, for
$j < J$,

$$V_j = V_j \bigoplus_{k=0}^{J-j+1} W_{J-k} \qquad (4.5)$$

where all these subspaces are orthogonal. By virtue of Eq. (4.2), this implies that

$$L^2(R) = \bigoplus_{j \in Z} W_j \qquad (4.6)$$

is a decomposition of $L^2(R)$ into mutually orthogonal subspaces. Hence, every $f \in L^2(R)$ has a (unique) orthogonal decomposition

$$f = \sum_{j \in Z} g_j, g_j \in W_j \qquad (4.7)$$

The multiscale technique decomposes high-resolution signals and images into hierarchy of pieces or components, each more detailed than the next. Thus, the multiscale concept which is based on the wavelet theory of successive approximation, or successive refinement, of the image (see Fig. 4.2) provides the mechanism for transmitting various grades of images depending upon quality of reconstruction sought and the limitation set by channel capacity.

The successive type of approximation is particularly well adapted for computer vision applications such as signal coding, texture discrimination, image filtering, edge detection, matching algorithms, and fractal analysis. The application of the multiscale technique to signal processing is applied to all of the wavelet-based applications discussed in the various chapters of this textbook. In Sect. 4.3, the one-dimensional (1D) wavelet transform is discussed.

Example 4.1 A simplified version of the wavelet transform process.

Given the number sequence (11, 9, 5, 7), determine the wavelet transform of the sequence.

Solution
- First step: sequence (11, 9, 5, 7).

  - Pair-wise average → {(11 + 9)/2, (5 + 7)/2} = (10, 6).
  - Lost detail information → {(11–10), (5–6)} = (**1–1**).
  - [10 + 1 = **11**, 10–1 = **9**, 6 + (−1) = **5**, 6-(−1) = **7**]

- Next step: sequence (10, 6).

  - Average → {(10 + 6)/2} = (**8**).
  - Detail → (8–6) = (**2**).

- - > Wavelet transformation (**8, 2, 1, −1**).

Thus, you can see the concept of wavelet transform, which is based on the wavelet theory of successive approximation, or successive refinement.

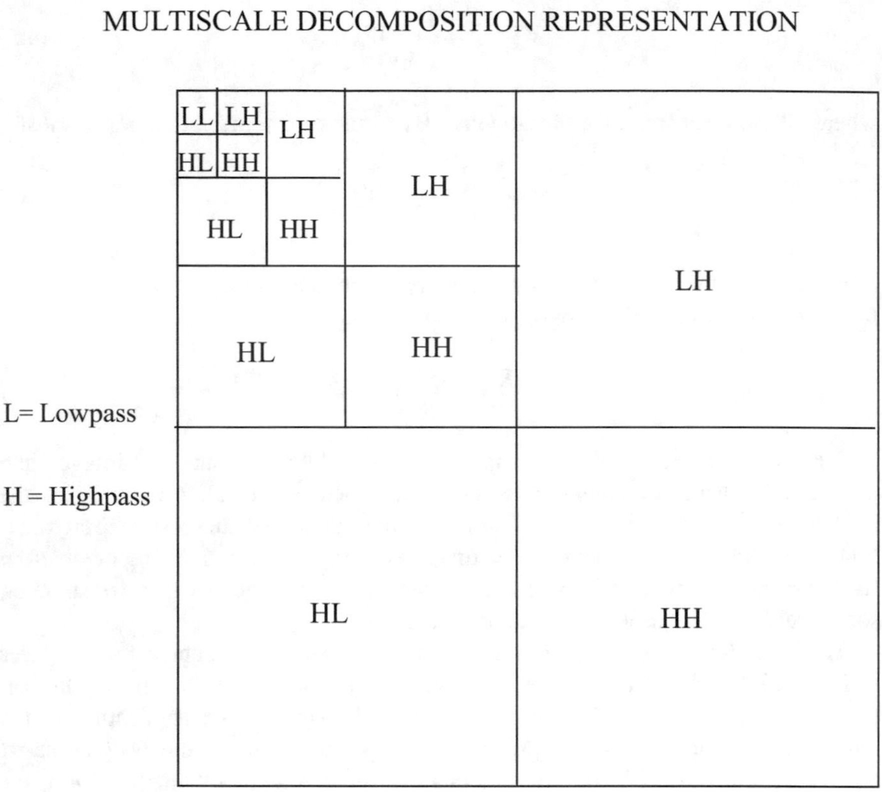

MULTISCALE DECOMPOSITION REPRESENTATION

Fig. 4.2 Multiscale decomposition representation

4.3   One-Dimensional Wavelet Transform

The wavelet transform **Wf(u,s)** of a function or signal $f(t)$, at the scale s and position u, is computed by correlating $f(t)$ with a wavelet atom as shown in Eq. (4.8).

$$Wf(u, s) = \int_{-\infty}^{\infty} f(t) \frac{1}{\sqrt{s}} \psi^* \left(\frac{t - u}{s} \right) dt \qquad (4.8)$$

In the one-dimensional (1D) wavelet transform, the signal is decomposed into its low-pass and high-pass segments, an approximation and a detail parts. It makes use of the low-pass and high-pass filter coefficients generated from the wavelet functions. We use Fig. 4.3 as the model diagram for the algorithm of the discrete wavelet transform, showing the decomposition and reconstruction of the signal $X[n]$. It illustrates how a signal is separated into its low-pass coefficients h_0 and high-pass coefficients h_1.

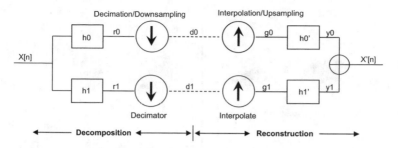

Fig. 4.3 Wavelet transform system for a 1D signal

The decomposition process involves the convolution of the signal elements with the respective wavelet low-pass and high-pass filter coefficients. Thereafter, the signal is taken through a decimation procedure in the decomposition process. In the decimation process, the signal is decimated by 2 which means taken every other sample of the elements of the convolved signal set. The involvement of the reconstruction process starts with the interpolation of the input from the decomposed low-pass and high-pass segments of the 1D wavelet transformation model as shown in Fig. 4.3. The interpolation process means adding zeros in the signal set space created by the decimation process. The result is convolved with the synthesis low-pass h'_0 and high-pass h'_1 filter coefficients in the reconstruction section of the 1D wavelet transformation model. The signal is then recombined to give the reconstructed signal $X'[n]$. As we discussed in Chap. 2, there are several types of wavelets among which are Haar, Daubechies, Symlet, Morlet, Meyer, and many others. The ones that are used as an example in this chapter are the Haar and Daubechies wavelets. Example 4.2 demonstrates what goes on in the 1D wavelet transformation using an arbitrary input and wavelet filter coefficients.

Example 4.2

Given the following simplified coefficients for $h_0[n]$ and signal input $x[n]$ for a 1D wavelet transform system model shown in Fig. 4.4:

$h_0[n]$ *coefficients:* $\{1, 1\}$
Input signal $x[n]$: $[1, 2, 2, 1]$

The high-pass filter coefficient $h_1[n]$ is related by the low-pass coefficient $h_0[n]$ by

$$h_1[n] = (-1)^n h_0[n].$$

Perform a mathematical theoretical calculation to demonstrate the 1D wavelet transformation of the given signal, and compare the original signal with the reconstructed signal. We call the separation between the decomposition section and the reconstruction section the channel using the communication systems

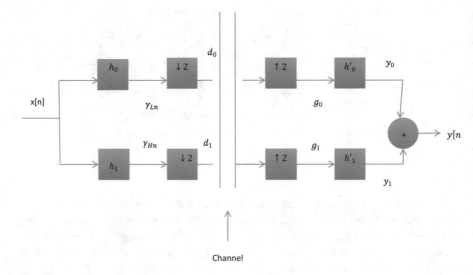

Fig. 4.4 Wavelet transform system for a 1D signal used for Example 4.1

analogy. The decomposition section becomes the transmitting section, while the reconstruction section becomes the receiving section.

Solution
Let $h_0[n]$ be low-pass filter coefficient.

Let $h_1[n]$ be high-pass filter coefficient.

• The high-pass filter is related to the low-pass as shown in Eq. (4.9).

$$h_1[n] = (-1)^n h_0[n] \tag{4.9}$$

• Given $h_0[n] = \{1, 1\}$.
• We use Equation (4.9) to determine the equivalent $h_1[n]$ coefficients.

$$h_1[0] = (-1)^0(1) = 1 \times 1 \rightarrow \mathbf{1}$$

$$h_1[1] = (-1)^1(1) = -1 \times 1 \rightarrow -\mathbf{1}$$
$$\rightarrow h_1[n] = \{1, -1\}$$

• Given $x[n] = \{1, 2, 2, 1\}$
• Convolve $h_0[n]$ with $x[n]$ that implies that $h_0[n] * x[n]$

$$\rightarrow \{1, 2, 2, 1\} * \{1, 1\}$$

$$\gamma_0 = \quad 1, 2, 2, 1$$
$$1, 1$$

$$(1 \times 0) + (1 \times 1) + (0 \times 2) + (0 \times 1) \rightarrow \gamma_0 = \mathbf{1}$$

$$\gamma_1 = 1, 2, 2, 1$$
$$1, 1$$

$$(1 \times 1) + (1 \times 2) + (0) + (0) \rightarrow \gamma_1 = (1 + 2) = \mathbf{3}$$

$$\gamma_2 = 1, 2, 2, 1$$
$$1, 1$$

$$(0) + (1 \times 2) + (1 \times 2) + (0) \rightarrow \gamma_2 = (2 + 2) = \mathbf{4}$$

$$\gamma_3 = 1, 2, 2, 1$$
$$1, 1$$

$$(0) + (0) + (1 \times 2) + (1 \times 1) \rightarrow \gamma_3 = (2 + 1) = \mathbf{3}$$

$$\gamma_4 = 1, 2, 2, 1$$
$$1, 1$$

$$(0) + (0) + (0) + (1 \times 1) + 0 \rightarrow \gamma_4 = \mathbf{1}$$

$\rightarrow$Result after low-pass filtering (convolution) is

$$\gamma_{Ln} = x[n] * h_0[n] = \{1, 3, 4, 3, 1\}$$

- What process do we use in the decimation process? Here we leave the first element value of γ_{Ln} and take the next one off and continue in that order, i.e., take every other number, which is the decimation by 2 process.

$$\rightarrow \{1, 3, 4, 3, 1\} \therefore \gamma_{Ln} \rightarrow \{3, 3\} = d_0$$

- What process do we use in the interpolation process? In the interpolation process, we insert *0 s* where we originally decimated during the decimation process.

$$\{3, 3\} \rightarrow \{0, 3, 0, 3\} = g_0$$

- Then we convolve g_0 with $h'_0[n]$, i.e., $g_0[n] * h'_0[n]$

$$\rightarrow \{0, 3, 0, 3\} * \{1, 1\} = \{0, 3, 3, 3, 3\}$$

$$\rightarrow y_0 = \{0, 3, 3, 3, 3\}$$

- We repeat same process with the high-pass side as we did in the low-pass side.

$$\rightarrow x_n[n] * h_1[n] \rightarrow \{1, 2, 2, 1\} * \{-1, 1\} = \gamma_{Hn}$$

$$\rightarrow \gamma_{Hn} = \{1, 1, 0, -1, -1\}$$

Decimate γ_1

$$\rightarrow d_1 = \{1, 1, 0, -1, -1\} = \{1, -1\} \text{ after decimation}$$

- Interpolate $\rightarrow g_1 = \{0, 1, 0, -1\}$
- Then convolve $g_1[n]$ *with* $h'_1[n]$, i.e., $g_1[n] * h'_1[n]$

$$\rightarrow \{0, 1, 0, -1\} * \{1, -1\} = y_1$$

$$\rightarrow y_1 = \{0, -1, 1, 1, -1\}$$

$\therefore$ OUTPUT $y[n] = y_0 + y_1$

$$= \{0, 3, 3, 3, 3\} + \{0, -1, 1, 1, -1\}$$

$$= \{0, 2, 4, 4, 2\}$$

- To get exact output for perfect reconstruction.

$$y[n] = 2 \times x[n-1] \text{ because of delay}$$

- We therefore normalize $y[n]$ by dividing it by 2 and removing the leading and trailing zeros. We get as the reconstructed signal, exactly the original signal that we transformed. Therefore, y[n] = {1, 2, 2, 1}, meaning perfect reconstruction.

4.3.1 Relationship Between the High-Pass and the Low-Pass Coefficients

As stated in Chap. 1, Section 1.4, the high-pass filter coefficient $h_1[n]$ is related to the low-pass coefficient $h_0[n]$. This relationship is represented in many textbooks and in the literature in so many forms as shown in Eqs. (4.9–4.13).

$$h_1(n) = (-1)^{(n-1)} h_0(n-1) \tag{4.10}$$

$$h_1(N-n-1) = (-1)^{(N-n-1)} h_0(N-n-1) \tag{4.11}$$

$$h_1(n) = (-1)^n h_0(1-n) \tag{4.12}$$

$$h_1(N - 1 - n) = (-1)^{(N-1-n)} h_0(N\text{-}1\text{-}n) \qquad (4.13)$$

Irrespective of which of the Eqs. (4.9–4.13) we use, it is necessary that we must be consistent so that we can get the desired transformation outcome in our application. As an example, we cannot use Eq. (4.9) in the transformation of a signal in the decomposition side in any application and then change halfway in the reconstruction side to using any of the other Eqs. (4.10–4.13) in the other half of the transformation application process. We cannot achieve the correct outcome of near perfect reconstruction if we are not consistent.

4.3.2   Multiple Stage Decomposition and Reconstruction Idea

In the 1D stage decomposition, which is also called the analysis process, it can be iterated, with successive approximations being decomposed in turn. That means one signal is broken down into many lower-resolution components. This is called the wavelet decomposition tree. A signal's wavelet decomposition tree yields valuable information both at the low-pass and high-pass frequencies. In Fig. 4.5, we show the 1D wavelet decomposition/analysis process with two stages as well as in the reconstruction/synthesis process. This 1D wavelet transform idea can be extended to nth stages.

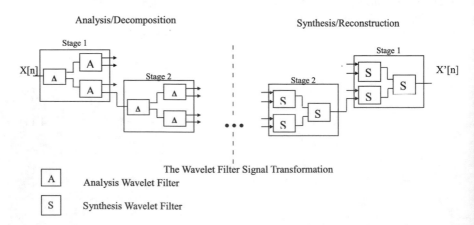

Fig. 4.5 1D multiple stages of signal decomposition and reconstruction algorithm system

4.3.3 Determination of the Number of Stages for Decomposition and Reconstruction

Since the decomposition/analysis and reconstruction/synthesis processes are iterative, in theory, it can be continued indefinitely. The size of the signal is one of the determining factors as to what stage to stop in either of the wavelet transform processes. As shown in Fig. 4.5, we have only two stages in the analysis and in the reconstruction sections. This can be iteratively extended to any nth stage. In reality, the decomposition can proceed until the individual details consist of a single sample or pixel. For example, in a 256 × 256 image, we can decompose the image up to the seventh stage. In practice, we will select a suitable number of stages based on the size and nature of the signal application requirements. In most cases, the application suitability determines how many stages the signal can be decomposed.

4.4 Two-Dimensional Wavelet Transform

The algorithmic decomposition system model for the 2D wavelet transform is as shown in Fig. 4.6. The signal X is convolved with the low-pass (LP) filter coefficient resulting to LP1 and high-pass (HP) filter coefficient resulting to HP1 horizontally. After decimation/downsampling by 2, we get LP2 and HP2 respectively. The same process is done vertically, in a column-wise decomposition process. The results after

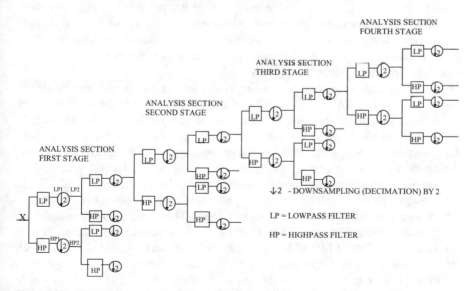

Fig. 4.6 The 2D multistage image decomposition algorithmic system model

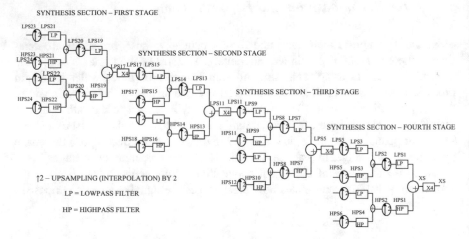

Fig. 4.7   The 2D multistage image reconstruction algorithmic system model

the convolution processes are LP3, HP3, LP4, and HP4 respectively. The result after the vertical decimation processes are LP5, HP5, LP6, and HP6 respectively. After the horizontal and vertical decomposition processes, that concludes the first stage of the decomposition process. Any one of Eqs. (4.9–4.13) could be used in determining the relationship between the low-pass and high-pass filter coefficients. For the second, third, and fourth stages of the decomposition processes, the same procedure that we used in the first stage is repeated in each case. That iteration process can continue to the nth stage. It should be noted that what is projected as input from stage 1 to stage 2 of the decomposition section is always the low-pass—low-pass side of the decomposition section. This is because the low-frequency components of the decomposition are where the significant portion of the energy of the signal resides. The high-frequency side is mostly noise.

The algorithmic reconstruction system model for the 2D wavelet transform is as shown in Fig. 4.7. This reconstruction or synthesis process is the reversal of the decomposition/analysis process. We start with LPS23, HPS23, LPS24, and HPS24 respectively as the input to the interpolation system of the reconstruction section. Instead of decimation, we have to interpolate/upsample. The interpolation process is performed by inserting zeros in the spaces created by the decimation process. The outputs of the interpolation processes are LPS21, HPS21, LPS22, and HPS22 respectively. The convolution process is performed horizontally to have outputs of LPS20 and HPS20 following the previous steps. These two summed outputs are interpolated resulting to outputs LPS19 and HPS19, which are then convolved with the filter coefficients of LP and HP vertically and summed to result to LPS17, which serves as the input to the second stage of the reconstruction/synthesis stage. This then completes the first stage of the reconstruction process. The same process is iterated to the fourth stage as shown in Fig. 4.7. The reconstructed signal is

XS. Perfect reconstruction is expected. As shown in Akujuobi et al. (1993), wavelet-based multiresolution evaluations can be done using images.

Example 4.3
Given a 256×256 contour image of the Lena image X, decompose the image using the example of the 2D multistage image decomposition algorithmic system model of Fig. 4.7. Show the outputs of LP1, LP2, and LP5. Use Eq. (4.13) to demonstrate the relationship between the low-pass and high-pass filter coefficients. Use the Haar wavelet filter coefficient of Eq. (2.1) in Chap. 2.

Solution Using the information given in Example 4.3, Fig. 4.8 shows the contour plot results of the decomposition algorithm for the first stage showing the input image X, LP1, LP2, and LP5 respectively. As we can see from Fig. 4.8, X is the 256×256 image input. LP1 is still 256×256 in size as the input. It is what we get after the first convolution process on the low-pass analysis side without the any decimation/downsampling. LP2 has the size of 128×256 because it has only been downsampled horizontally at this point. In LP5, the size of the image is 128×128 because it has now been downsampled both horizontally and vertically by 2.

Example 4.4
From the decomposed outcome of Example 4.3 solution, reconstruct the image, and show the contour image of the reconstructed Lena image to demonstrate the outputs of LPS5, LPS7, LPS8, and LPS11 of the third stage. Use Eq. (4.13) to demonstrate the relationship between the low-pass and high-pass filter coefficients. Use the Haar wavelet filter coefficient of Eq. (2.1) in Chap. 2.

Solution Using the information given in Example 4.4, Fig. 4.9 shows the contour plot results of the reconstructed Lena image for the third stage showing the outputs of LPS5, LPS7, LPS8, and LPS11 of the third stage. In LPS5 and LPS7, we see that the size of the image is 128×128 because the 64×64 image has been upsampled both horizontally and vertically by 2. In LPS8, the size of the image is 64×128 which means that the 64×64 image input has only been upsampled by 2 vertically. In LPS11, that is the 64×64 low-pass synthesis input to the third stage coming from the second stage of the synthesis process.

Summary
1. A wavelet transform of a signal $f(t)$ is the decomposition of the signal into a set of basis functions consisting of contractions, expansions, and translations of a mother function $\psi(t)$, called the wavelet.
2. The representation of the specific function by specific wavelets is called a wavelet transform.
3. The algorithm for the discrete wavelet transform is much simpler than the integral function of the continuous wavelet transform.
4. The multiscale analysis and representation technique provide signals with containment, completeness, scaling, and basis/frame properties.
5. The multiscale technique decomposes high-resolution signals and images into hierarchy of pieces or components, each more detailed than the next.

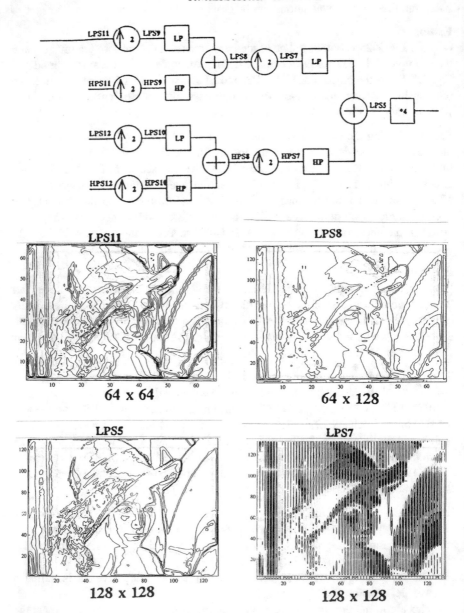

Fig. 4.8 Some contour plot results of the decomposition algorithm for the first stage

6. In the one-dimensional (1D) wavelet transform, the signal is decomposed into two, an approximation and a detail.
7. Since the analysis process is iterative, in theory, it can be continued indefinitely.

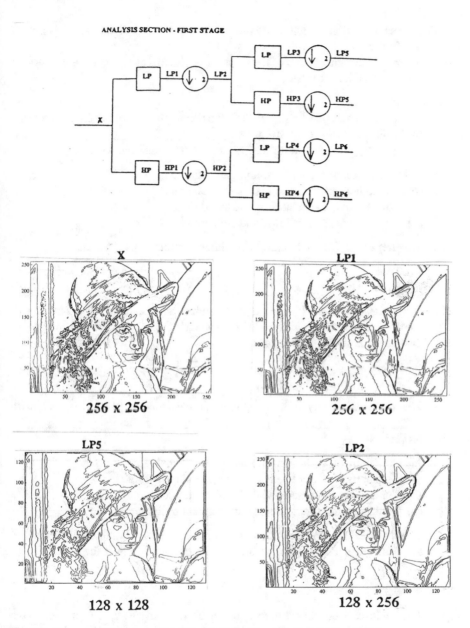

Fig. 4.9   Some contour plot results of the reconstruction algorithm for the third stage

8. The size of the signal is one of the determining factors as to what stage to stop in the decomposition process.
9. In the 1D stage decomposition which could be called the analysis process, it can be iterated, with successive approximations being decomposed in turn, so that one signal is broken down into many lower-resolution components.

10. This reconstruction or synthesis process is the reversal of the decomposition/ analysis process.
11. The size of the signal is one of the determining factors as to what stage to stop in the decomposition process.

Review Questions

4.1. A wavelet transform of a signal $f(t)$ is the decomposition of the signal into a set of basis functions consisting of _____, _____, and _____ of a mother function $\psi(t)$, called the wavelet.

(a) Contractions, expansions, and translations.
(b) Contradictions, expansions, and translations.
(c) Contractions, expectations, and translations.
(d) Contractions, expansions, and transitions.

4.2. Wavelet transform (WT) cannot be either continuous or discrete.

(a) True.
(b) False.

4.3. If the signal is in itself a sequence of numbers, or samples of a function of continuous variables, a wavelet expansion of the signal is referred to as discrete time wavelet transform.

(a) True.
(b) False.

4.4. The multiscale analysis and representation technique provide signals with _____, _____, _____, and _____ properties.

(a) Containment, complexion, scaling, and basis/frame.
(b) Containment, completeness, scaling, and basic/fume.
(c) Contact, completeness, scaled, and basic/frame.
(d) Containment, completeness, scaling, and basis/frame.

4.5. The multiscale technique decomposes high-resolution signals and images into hierarchy of pieces or components, each more detailed than the next.

(a) True.
(b) False.

4.6. The multiscale concept which is based on the wavelet theory of successive approximation, or successive refinement, of the image provides the mechanism for transmitting various grades of images depending upon quality of reconstruction sought and the limitation set by channel capacity.

(a) True.
(b) False.

4.7. The decomposition section of the wavelet transform can be called the same as:

(a) Synthesis.
(b) Decimation.
(c) Analysis.
(d) Interpolation.

4.8. The reconstruction section of the wavelet transform can be called the same as:

(a) Decimation.
(b) Synthesis.
(c) Analysis.
(d) Interpolation.

4.9 The size of the signal is not one of the determining factors as to what stage to stop in the decomposition process.

(a) True.
(b) False.

4.10. In the wavelet transform algorithm, the reconstruction/synthesis processes are the reversal of the decomposition/analysis processes.

(a) True.
(b) False.

Answers: 4.1a, 4.2b, 4.3a, 4.4d, 4.5a, 4.6a, 4.7c, 4.8b, 4.9b, 4.10a.

Problems

4.1. Given the number sequence (13, 11, 7, 9), determine the basic wavelet transform idea of the sequence.

4.2. Design a one stage wavelet analysis and synthesis transform algorithmic model capable of transforming a 1D signal to various low-pass and high-pass filter components.

4.3. Perform the theoretical calculation for the one stage wavelet analysis transform capable of transforming the 1D signal $X[n]$ to various low-pass and high-pass filter components. Use the Haar wavelet filter coefficients. Let $h_1(n) = (-1)^{(n-1)}h_0(n-1)$. Let the Haar wavelet coefficients be $(1/2, -1/2)$ and let the input signal $X[n]$ be [1, 2, 3, 4, 5].

4.4. Perform a theoretical calculation for a one stage wavelet analysis transform capable of transforming a 1D signal to various low-pass and high-pass filter components. Use the Haar wavelet filter coefficients. Let $h_1(N - n - 1) = (-1)^{(N-n-1)}h_0(N - n - 1)$. Let the Haar wavelet coefficients be $(1/2, -1/2)$ and let the input signal $X[n]$ be [1, 2, 3, 4, 5].

4.5. Compare the original signal with the reconstructed signal based on using:

(i) $h_1(n) = (-1)^{(n-1)}h_0(n-1)$
(ii) $h_1(N - n - 1) = (-1)^{(N-n-1)}h_0(N - n - 1)$, respectively, based on the theoretical calculations of problems 4.3 and 4.4.

4.6. Name at least four different ways in which the high-pass filter coefficient $h_1[n]$ can be related to the low-pass coefficient $h_0[n]$.

4.7. Perform the theoretical calculation for the one stage wavelet decomposition/ analysis and reconstruction/synthesis transforms capable of transforming the 1D signal $X[n]$ to various low-pass and high-pass filter components. Use the Daubechies-4 wavelet filter coefficients. Let $h_1(n) = (-1)^{(n-1)}h_0(n-1)$. Let the Daubechies-4 wavelet coefficients be h_0 = (0.482962913145, 0.836516303738, 0.224143868042, -0.129409522551), and let the input signal $X[n]$ be [1, 2, 3, 4, 5].

4.8. Perform the theoretical calculation for the one stage wavelet decomposition/ analysis and reconstruction/synthesis transforms capable of transforming the 1D signal $X[n]$ to various low-pass and high-pass filter components. Use the Daubechies-4 wavelet filter coefficients. Let $h_1(N-n-1) = (-1)^{(N-n-1)}h_0(N-n-1)$. Let the Daubechies-4 wavelet coefficients be h_0 = (0.482962913145, 0.836516303738, 0.224143868042, -0.129409522551), and let the input signal $X[n]$ be [1, 2, 3, 4, 5].

4.9. What conclusions can you deduce from your answers from problems 4.8 and 4.9?

4.10. Compare the original signal with the reconstructed signal based on using:

(i) $h_1(n) = (-1)^{(n-1)}h_0(n-1)$
(ii) $h_1(N-n-1) = (-1)^{(N-n-1)}h_0(N-n-1)$, respectively, based on the theoretical calculations of problems 4.7 and 4.8.

4.11. Describe briefly, by using both Haar and Daubechies-2 wavelets, what is the rough definition of wavelets.

4.12. What is the orthogonality of Haar and Daubechies-2 wavelets?

4.13. What are the two-scale equations of scaling/refinable functions and wavelets?

4.14. What are the vanishing moments of the Haar and Daubechies wavelets?

4.15. Describe briefly, also by using both Haar and Daubechies-2 wavelets, what are upsampling by 2 and downsampling by 2.

4.16. What are the Haar and Daubechies wavelets' corresponding low-pass and high-pass filters?

4.17. What are the decomposition and reconstruction algorithms?

4.18. What are Haar and Daubechies FIR quadrature mirror filter (QMF) filter banks?

References

Akansu, A. N., & Haddad, R. A. (1992). *Multiresolution signal decomposition, transforms, subbands, wavelets*. Academic Press.

Akujuobi, C. M., Parikh, V., & Baraniecki, A. Z. (1993). Performance evaluation of multiresolution image analysis using wavelet transform. In *Proceedings of modelling and simulation, IASTED International Conference* (pp. 188–192).

Burrus, C., Gopinath, R., and Guo, H. (1998). Introduction to wavelets and wavelet transforms, a primer. .

Mallat, S. (1989a). A theory for multiresolution signal decomposition: The wavelet representation. *IEEE Transactions on Pattern Analysis and Machine Intelligence, 11*(7), 674–693.

Mallat, S. (1989b). Multifrequency channel decomposition of images and wavelet models. *IEEE Transactions on Acoustics Speech and Signal Processing, 37*(12), 2091–2110.

Chapter 5
Similarities Between Wavelets and Fractals

"Champions keep playing until they get it right."

Billie Jean King.

5.1 Introduction

This chapter covers the similarities that exist between wavelets and fractals. Fractals and fractal transforms are covered along with the different types of fractals. We covered wavelets and the different types of wavelets in Chap. 2. In Chap. 4, we covered wavelet transforms. In this chapter, we cover the similarities between wavelets and fractals as they relate to their various properties and their different application areas.

Wavelets are functions generated from one basic function ψ called the mother wavelet by dilations and translations of ψ. These dilations and translations can be represented as $\psi^{a,\,\tau}(t) = |a|^{-1/2}\psi\{(t - \tau)/a\}$. In this case, t is a one-dimensional variable, and τ and a are shift and scale parameters, respectively. For low-frequency wavelets, $a > 1$, while for high-frequency wavelets, $a < 1$. The mother wavelet ψ satisfies the condition $\int \psi(t)dt = 0$. Wavelets have proven in recent years to be an important theory of mathematics that is now leading to advances in signal analysis, scientific calculation, medical imaging, image compression, and many other areas.

In scientific calculation, for example, wavelets may help decipher phenomena such as turbulence leading to improved aircraft design. In medical imaging, wavelets are used by scientists to produce magnetic resonance pictures of the body faster and more accurately. For data compression, wavelets identify the key features of any image, allowing engineers to reconstruct an image with a tiny fraction of the information in the original.

One of the advantages of the wavelet transform over Fourier transform is that the wavelet transform provides an insight that simultaneously combines features of the time domain and the frequency domain. This feature can also be found in Wigner distribution. However, the wavelet transform is of the first order, while Wigner distribution is of the second order. This means that linear superposition can be accommodated by the wavelet transform, while this is not the case in Wigner

C. M. Akujuobi, *Wavelets and Wavelet Transform Systems and Their Applications*,
https://doi.org/10.1007/978-3-030-87528-2_5

distribution. Fourier analysis is briefly discussed in Chap. 1 to help understand the wavelet theory better.

Fractals are subsets of geometric spaces, such as (C, spherical) and (R^2, Euclidean). In the case of a deterministic fractal, it may be defined as a fixed point of a contractive transformation on the space of non-empty compact subsets, with the Hausdorff metric distance. Fractals manifest a high degree of visual complexity. Many natural objects exhibit the property that as one views the object at greater resolutions, more self-similar structure is revealed. Fractal sets enjoy this property, because of the fixed-point property. It is important, therefore, to start this section with the basic properties of fractals. Let F represent a fractal. Falconer's (Falconer, 1990) basic properties of F are as follows:

(i) *F has a fine structure, i.e., detail on arbitrarily small scales.*
(ii) *F is too irregular to be described in traditional geometrical language, both locally and globally.*
(iii) *Often F has some form of self-similarity, perhaps approximate or statistical.*
(iv) *Usually, the fractal dimension of F (defined in some way) is greater than its topological dimension.*
(v) *In most cases of interest, F is defined in a simple way, perhaps, recursively.*

These basic properties of fractals are what Falconer used as his complete definition of fractals. Fractals can be looked at in so many other different ways. They can be looked at as geometric shapes that have two special properties: self-similarity and fractional dimensionality. They can be geometrical figures in which an identical motif repeats itself on an ever diminishing scale. The self-similarity idea (Akujuobi, Baraniecki, 1994) is discussed in Sect. 5.2.

5.2   The Self-Similarity Idea in Wavelets and Fractals

Figure 5.1 shows the self-similarity idea using of the Sierpinski fractal. A structure is said to be self-similar if it can be broken down into arbitrarily small pieces, each of which is a small replica of the entire structure. This idea is discussed in Sect. 5.6.1.

The mathematical framework for the treatment of the irregular, seemingly complex shapes found in nature, from the small-scale structure of disordered systems to coastlines, mountain ranges, clouds, and the distribution of stars in the night sky, is provided by fractals. In fractal geometry, some of the building blocks originated in the exactly self-similar mathematical "monsters" (such as the Koch curve and Sierpinski gasket) of the early 1900s.

There is a deep connection between wavelets and fractals. Wavelets and fractals came into existence from the same mathematical community, hence the mathematical sophistication involved in each of them. Both are used as analysis tools in many applications of science and engineering, especially electrical engineering. Both exhibit self-similarity properties at different resolutions.

In fractals as well as in wavelets, similarity transformations can be replaced by affine

Fig. 5.1 The self-similarity
idea using the Sierpinski
fractal

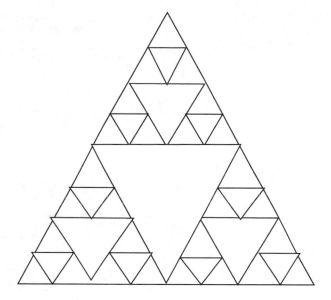

transformations in two and higher dimensions. This technique is used in wavelet and
fractal constructions.

Scaling properties exist in both wavelets and fractals. Wavelets and fractals may
also be applied to image/data compression, texture segmentation and classification,
pattern recognition, image construction and reconstruction, scientific calculation,
medical imaging, and weather satellite data measurements. The idea of multiscale
analysis and representation can be implemented using wavelets as well as fractals.
The duality principle applies to both wavelets and fractals. Both can be represented
using geometrical objects. Fractal dimension can be computed from wavelet representation. Fractal signals can be estimated from noisy measurements using wavelets.
Wavelets have compactly supported bases while there is compact support in a fractal
space. In fact, fractals offer an extremely compact method for describing objects and
formations.

One of the application areas common to both wavelets and fractals is the coding
of images. Wavelets have been used extensively in the coding of images. Fractals
have also been used extensively in the coding of images. There are already wavelet
and fractal packet libraries being used in image compression and analysis. Figures 5.2 and 5.3 show the pictorial representations of the similarity categories.
Some of these areas of similarities between wavelets and fractals are discussed in
detail in Sects. 5.5 and 5.6.

Fig. 5.2 Similarities
between wavelets and
fractals based on their
properties

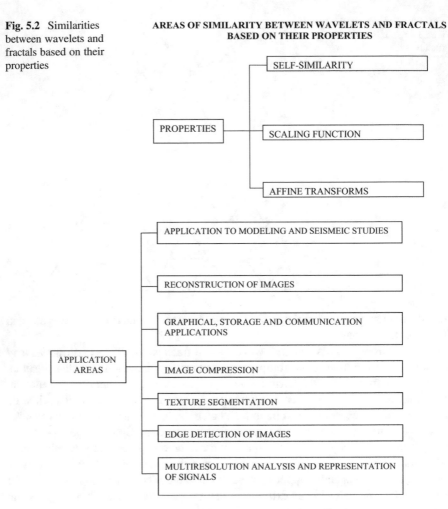

Fig. 5.2 Similarities between wavelets and fractals based on their properties

Fig. 5.3 Similarities between wavelets and fractals based on their application

5.3   Fractal Dimension Idea

Fractals have parameters associated with them that can help in their comparison. These are referred to generally as fractal dimensions (Barnsley, 1993). There are several types of fractal dimensions: Hausdorff dimension, self-similarity dimension, information dimension, topological dimension, box counting dimension, Euclidean dimension, and capacity dimension. These different types of dimensions are all related. However, fractal dimensions provide an objective means for comparing fractals.

Example 5.1

Let (Q,d) be a complete metric space. Then $\Gamma(Q)$ denotes the space whose points are the compact subsets of Q, other than the empty set.

Let $A \in \Gamma(Q)$ where (Q, d) is a metric space. For each $\varepsilon > 0$, let $N(A,\varepsilon)$ denote the smallest number of closed balls of radius $\varepsilon > 0$ needed to cover A. If Eq. (5.1)

$$D = \text{Lim}_{\varepsilon \to \infty}\left(\frac{Ln(N(A,\varepsilon))}{Ln\left(\frac{1}{\varepsilon}\right)}\right) \tag{5.1}$$

exists, then D is called the fractal dimension of A. We also use the notation $D = D(A)$, as "A has fractal dimension D."

5.4 Iterated Function System Code

An iterated function system (IFS) code is an IFS

$$\{t_n, p_n : n = 1, 2, 3 . . . . . N\} \tag{5.2}$$

where:

t_n are affine transformations,
p_n are associated probabilities of the affine transformations,

such that the average contractivity condition is obeyed (Barnsley, 1993). IFS codes can be stored as compact sets of numbers. These sets then form libraries of sets that can be called to produce corresponding fractals. This idea leads to significant savings of computer memory. This is particularly important when working with personal computers. In addition, the economic impact cannot be overlooked. IFS codes are useful in many application areas such as mapping, biological modeling, computer graphics, desktop publishing, advertising, geophysics, graphic design, and synthetic speech (Barnsley et al., 1988).

5.5 Types of Fractals

There are different types of fractals. Some of these are the random, scaling, Koch, Sierpinski, and Cantor fractals. These fractals and their constructs are fully discussed in this section.

5.5.1   Random Fractals

In the early 1900s, some of the building blocks of fractal geometry originated in the exactly self-similar mathematical "monsters" such as the Koch curve and Sierpinski gasket. While such exact deterministic constructs serve as useful tools in building an understanding and intuition about scaling properties, the fractal shapes found in nature possess a statistical rather than exact self-similarity. The fractional Brownian motion (fBm) of Mandelbrot and Wallis is one of the most useful mathematical models for the random fractals found in nature such as mountainous terrain and clouds (see Figs. 5.4 and 5.5). In addition, one can better understand anomalous diffusion and random walks on fractals using the idea of fractional Brownian motion. Fractional Brownian motion is a non-stationary, self-affine process defined by Eq. (5.3).

$$B(t) = \frac{1}{\Gamma(H + 0.5)}$$

$$\times \left[\int_{-\infty}^{0} \left(|t - s|^{H-0.5} - |s|^{H-0.5} \right) dB(s) + \int_{-\infty}^{0} \left(|t - s|^{H-0.5} db(s) \right) \right] \quad (5.3)$$

Self-affinity implies that the relation

$$B_H(bt) - B_H(0) \approx b^H \{B_H(t) - B_H(0)\} \quad (5.4)$$

which is valid in distribution for any value of b. Fractional Brownian motion can be regarded as a generalization of the well-known Brownian motion process, for which $H = 1/2$. The parameter H lies in the range $0 < H < 1$ and quantifies the roughness of the curves $B(t)$. H which is the Hurst coefficient is related to the fractal dimension by $D = 2 - H$. The self-affinity implications are as shown in Eq. (5.4).

5.5.2   Scaling Fractals

The scaling fractals are a set of fractals that have length-area-volume relationships. In using standard analysis, from the fact that the circumferential length of a circle of radius R is equal to $2\pi R$, and the area of the disc bounded by the circle is πR^2, it follows that

$$(\text{length}) = 2\pi^{1/2}(\text{area})^{1/2} \quad (5.5)$$

Among squares, the corresponding relation is

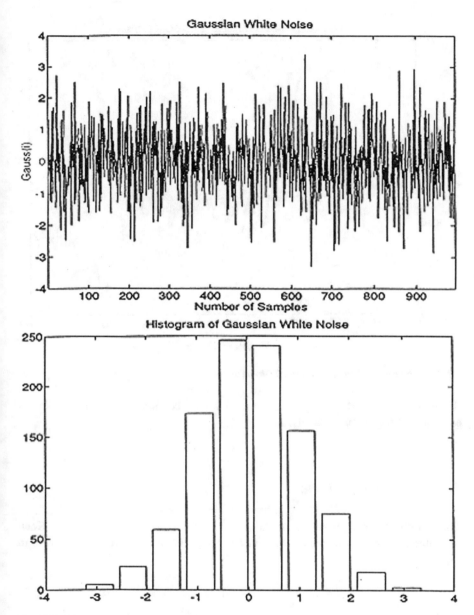

Fig. 5.4 The Brownian noise (Gaussian noise) and its histogram showing the Gaussian distribution-like shape

$$(\text{length}) = 4(\text{area})^{1/2} \tag{5.6}$$

In the fractal length-area relation, if we consider a collection of geometrical similar islands with fractal coastlines of dimension $D > 1$. The standard ratio

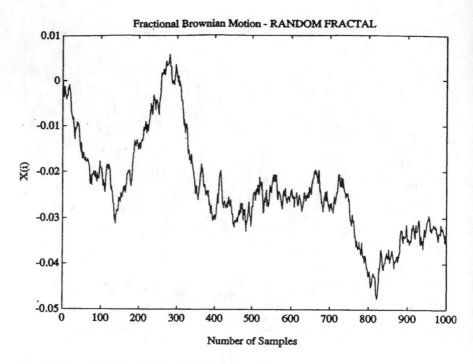

Fig. 5.5 The random fractal—fractional Brownian motion

(length)/(area) will definitely have a useful fractal counterpart. Figure 5.6 shows examples of the scaling fractals.

5.5.3 The Koch Fractal

The Koch fractal came into existence around 1904. Helge von Koch (a mathematician) constructed his fractal using the general formula shown in Eqs. (5.7) and (5.8).

$$\zeta = t_0 + t_1.4 + t_2.4^2 + \ldots + t_{p-1}.4^{p-1} \tag{5.7}$$

$$x = a(t_0) + a(t_1) + a(t_2) + \ldots + a(t_{p-1}) \tag{5.8}$$

where:

ζ is the index of the line segment in quaternary (base 4)
p is the level of approximation (order),

88

SCALING FRACTALS

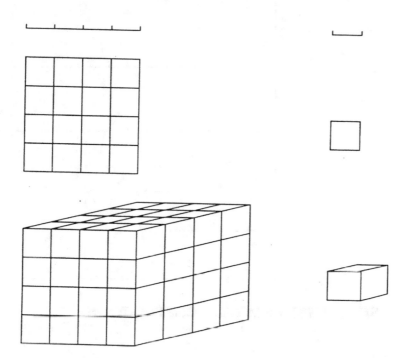

Fig. 5.6 The scaling fractals

χ is the sum of the directions corresponding to the numbers in the quaternary expansion.

The Koch fractal shown in Fig. 5.7 can be used to model natural scenes, for example, islands.

5.5.4 The Sierpinski Fractal

The idea of the Sierpinski fractal started in 1915 when a Polish mathematician named Vaclav Sierpinski (1882–1969) came up with a nice variation on the ternary tree. This came to be known as the Sierpinski Sieve or Sierpinski fractal. An example of the Sierpinski fractal is shown in Fig. 5.8.
This type of fractal is constructed with an equilateral triangle which is divided into four smaller equilateral triangles. The middle triangle is then removed, producing the triangular hole.

Fig. 5.7 The Koch fractal

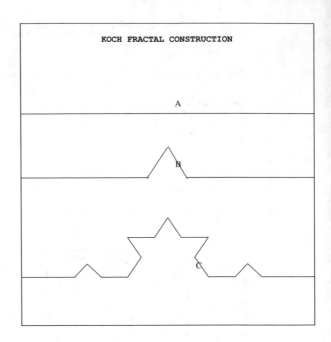

SIERPINSKI FRACTAL CONSTRUCTION

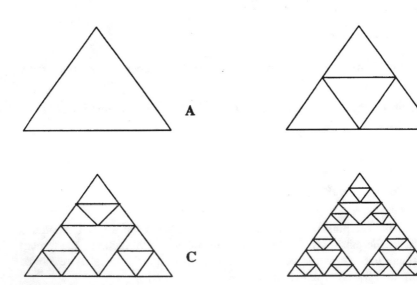

Fig. 5.8 The Sierpinski fractal

5.5.5 *The Cantor Fractal*

The Cantor fractal may be the oldest fractal. It was discovered by Cantor as far back as 1870. It can be constructed from a line segment in which the middle point is removed in the next resolution of the line segment without removing the endpoints. Let C be the Cantor fractal set. The subset of the metric space [0, 1] can be obtained using the Barnsley idea by removing successively the middle one-third open sub-intervals. This process is illustrated in Fig. 5.9. A nested sequence of the Cantor fractal is constructed of closed sets

$$L_n \subset L_{N-1} \ldots \ldots L_4 \subset L_3 \subset L_2 \subset L_1 \subset L_0$$

where:

$L_0 = [0, 1]$
$L_1 = [0, 1/3] \cup [2/3, 3/3]$
$L_2 = [0, 1/9] \cup [2/9, 3/9] \cup [6/9, 7/9] \cup [8/9, 9/9]$
$L_3 = [0, 1/27] \cup [2/27, 3/27] \cup [6/27, 7/27] \quad \cup [8/27, 9/27] \cup [18/27, 19/27] \cup [20/27, 21/27] \cup [24/27, 25, 27] \cup [26/27, 27/27] L_4 = L_3$ with the middle one-third of each interval in L_3 removed
$L_N = L_{N-1}$ with the middle one-third of each interval in L_{N-1} removed.

The Cantor fractal set C can therefore be defined as

CANTOR SET FRACTAL CONSTRUCTION

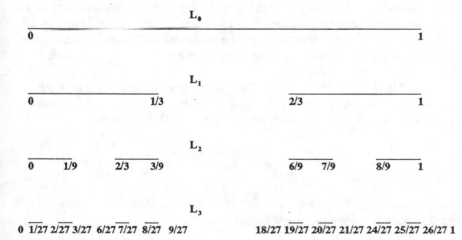

Fig. 5.9 The Cantor fractal

$$C = \cap_{N=0}^{\infty} L_N. \tag{5.9}$$

where:

C is the Cantor fractal.
n is the number of resolutions,
L is the line segment.

There is a striking resemblance between the Cantor fractal technique and the multiscale technique that is discussed in Chap. 4. In this chapter, we only addressed the similarity issues. Suffice it to say that this kind of fractal can be used to demonstrate the idea of a multiscale techniques.

5.6   Areas of Similarities Between Wavelets and Fractals Based on Their Properties

This section discusses the similarities between wavelets and fractals based on their properties. The property areas discussed include self-similarity, scaling functions, and affine transforms.

5.6.1   Self-Similarity

"Self-similarity" is a distinctive feature of most fractals and wavelets: a small portion of the figure resembles some larger part when magnified, either exactly or very closely. Mandelbrot and Wallis (1968) pointed out in a wide variety of natural and man-made phenomena this self-similarity characteristic. Self-similarity can also be called scale invariance. Because computer-generated fractal and wavelet-based images have similar patterns on many different scales, relatively little code is all that is usually needed to create them.

It was Mandelbrot (Mandelbrot, 1982) who demonstrated that certain natural textures can be modeled with Brownian fractal noise. The fractal Brownian noise F (x) is a stochastic (ergodic) random process whose local differences

$$/F(x) - F\ (x + \Delta x)/10$$
$$\|\Delta x\|^H$$

have a probability distribution function $g(x)$ which is Gaussian as illustrated in Fig. 5.4. Such a random process is self-similar, i.e.,

$$\forall r > 0, F(x), r^H F(rx)$$

are statistically identical. Hence, a realization of *F(x)* looks similar at any scale, and for any resolution, r^H is the scaling factor, and *H* is the Hurst exponent whose range is between *0* and *1*. The implementation of the multiscale idea using wavelets (Mallat, 1989) and fractals (Jones, et al. 1991) presents a very unique opportunity of appreciating the self-similarity properties of these two important signal processing tools.

5.6.2   Scaling Function

Wavelets and fractals both make use of scaling functions. In wavelets, there are two functions (Argoul et al. 1989) (UWE 1989), the scaling function and the mother wavelet that form the foundation of wavelet computations. These functions are defined recursively, as linear combinations of scaled and shifted versions of scaling function $\varphi(t)$ which is as shown in Eq. (5.10).

$$\phi(t) = \Sigma_{nez} h_1(n) \phi(2t - n) 12 \tag{5.10}$$

such that

$$\int_R \phi(t) dt = 1$$

and the wavelet $\psi(t)$ is defined as shown in Eq. (5.11).

$$\psi(t) = \Sigma_{nez}(-1)^n h_1(n+1)\phi(2t-n)14 \tag{5.11}$$

The coefficients $h_1(n)$ in Eqs. (5.10) and (5.11) are the wavelet coefficients. The scaling function also appears in the wavelet multiscale analysis, i.e., looking at the signal at different scales—different resolutions. A similar idea is used in the case of fractals. In fractals, the scaling or self-similarity property is closely connected with intuitive notion of dimension. A line segment, for example, which is one dimensional, possesses a similar scaling property. It can be divided into N identical parts each of which is scaled down by the ratio **r** = 1/N. The scaling factor **r** = 1/ √ N for a two-dimensional object (square area in a plane) and $1/^3 \sqrt{N}$ for a three-dimensional object (solid cube). For a D-dimensional self-similar object, it can be divided into N smaller objects of itself each of which is scaled down by a factor **r** where **r** = $1/^D \sqrt{N}$ or

$$N = 115 \mathbf{r}^D \tag{5.12}$$

Conversely, the fractal or similarity dimension of a self-similar object of N parts scaled by a ratio r from whole is given by

$$D = \frac{\log (N)\ 16}{\log \left(\frac{1}{r}\right)\ 17}$$ (5.13)

5.6.3   Affine Transforms

Affine transformations are defined as combinations of rotations, scalings, and translations of the coordinate axes in n-dimensional space. The concept of affine transforms is found in both fractals and wavelets. Equations (5.14) and (5.15) show examples of fractal affine transformations in two dimensions.

$$T(a, b) = \left(\frac{1}{2}a + \frac{1}{4}b + 1, \frac{1}{4}a + \frac{1}{2}b + 2\right)\ 18$$ (5.14)

Equation (5.15) is the matrix form of Eq. (5.14).

$$T\left|\begin{matrix}a\\b\end{matrix}\right| = \left|\begin{matrix}.5 & .25\\.25 & .5\end{matrix}\right|\left|\begin{matrix}a\\b\end{matrix}\right| + \left|\begin{matrix}1\\2\end{matrix}\right|$$ (5.15)

Fractal affine transforms are mostly useful in the rotations, scalings, and translations of the coordinate axes in n-dimensional space.
In the case of the wavelet, affine wavelet transforms are as shown in Eq. (5.16) for some mother wavelet ψ.

$$\Psi_{(a,\tau)}(t) = \frac{1}{\sqrt{a}}\Psi\left(\frac{T - \tau}{a}\right), 19$$ (5.16)

In Eq. (5.16), a and τ represent the dilations (scalings) and the translations of ψ, respectively. The affine wavelet transform is used to overcome poor time resolution of high frequencies.

5.7   Areas of Similarities Between Wavelets and Fractals Based on Their Application Areas

In this section, we discuss the similarities between wavelets and fractals based on their application areas (Akujuobi & Baraniecki, 1992a, b). The application areas covered in this section include modeling and seismic studies, reconstruction of images, and graphical, storage, and communication applications. Other areas include image compression, texture segmentation, edge detection of images, geometrical objects representation, and the multiscale analysis and representation of signals.

5.7.1   Application to Modeling and Seismic Studies

Wavelets and fractals can both be applied to the field of seismic studies. Symmetric wavelets, such as the Berlage and Morlet wavelets, are used for such studies. The Berlage and Morlet wavelets have been discussed in Chap. 2. It is interesting to note that a wide range of the seismic waveforms observed in both fields and the processed data might themselves be random fractals. Therefore, both wavelets and fractals offer sufficient flexibility for modeling and measuring these seismic waveforms.

Random fractals, on the other hand, using the idea of midpoint displacement, can also model soil erosion and seismic patterns in trying to understand the changes in fault zones. Random fractals such as the fractional Brownian motion (fBm) have been discussed in Sect. 5.5.1 of this chapter. The idea of midpoint displacement to model images of planets, moons, clouds, mountains, trees, continents, and planets is one of the advantages of this technique (Voss, 1985). The fractional Brownian motion (fBm) of Mandelbrot and Wallis (1968) is one of the most useful mathematical models for the random fractals found in nature such as mountainous terrain and clouds.

5.7.2   Construction and Reconstruction of Images

Wavelets are used to determine filter coefficients (Daubechies, 1998) with either the quadrature mirror filter (QMF) banks (Kundu and Chen, 1991) or the multiscale analysis banks (Mallat, 1989) and to achieve perfect reconstruction from decomposed images. The QMF bank is a multirate digital filter bank. It consists of an analysis and synthesis filter banks, each of which further comprises of a pair of low-pass as well as high-pass filters and decimators and interpolators. The filters at the analysis filter bank band limit the input signal so that the effect of aliasing will be avoided after the signal is transmitted and recovered. The decimators reduce the number of samples to be transmitted after filtering. The frequencies of the sampling are, thus, reduced by a factor of two. At the synthesis filter bank, the interpolators insert zeros between received samples before the combination and recovery of the signals. The synthesis filters further eliminate imaging effect.

Fractals, like wavelets, can also achieve similar reconstruction results (Barnsley and Sloan, 1988; Kolata, 1991). The fractal example of the reconstruction of an iterative function system (IFS)—compressed image—is in the use of the random iteration algorithm (Barnsley and Sloan, 1988). The tolerance setting used during the collage mapping stage determines the accuracy of the reconstructed image.

5.7.3 Graphical, Storage, and Communication Applications

Photographic and natural scene images are encoded and stored using fractal-based techniques at a ratio exceeding 1,000,000 to 1. They are transmitted in encoded form at video rates over Integrated Services Digital Network (ISDN) lines and other networks for integrated data and voice communications (Barnsley and Sloan, 1988). Sophisticated procedures like the combinatorial searching algorithms that allow full color images to be encoded can be used to automate the fractal-based collage mapping stage of these fractal images.

Similarly, photographic and natural scene images are also encoded and stored using wavelet-based techniques at a ratio exceeding 100 to 1. Wickerhauser and many others (Kolata, 1991) propose that images such as fingerprint images can be encoded, stored, and transmitted over telephone lines using wavelets.

5.7.4 Image Compression

Image compression using fractals is a relatively useful technique. The technique has yielded compression ratios in excess of 1,000,000 to 1 (Barnsley and Sloan, 1988). The technique, however, is computationally very demanding.

Fractals are used in image processing techniques such as color separation, edge detection, spectrum analysis, and texture-variation analysis. The image is divided into segments. These segments are looked up in a library of fractals. The library does not contain literal fractals that would require astronomical amounts of storage. Instead, the library contains relatively compact sets of numbers, called IFS codes, that will reproduce the corresponding fractals. We have discussed IFS codes in Sect. 5.4. Once all the segments are looked up in the library, and their IFS codes found, the original digitized image can then be discarded and the codes kept, thus achieving a very high image compression ratio.

Wavelets, like fractals, are recent techniques that have been applied to achieve image compression. Wavelet transform methods for image compression produce a multi-level representation in which the geometric structure of the image is preserved within each subband or level. Wavelet compression methods incorporate several useful features like reduction of aliasing distortion, inherent parallelism, elimination of blocking segmentation of the image, possibility of efficient layout onto very large-scale integration (VLSI), and localization of visual "relevant" information. Wavelet methods tend to produce random noise, which is far much less offensive to the human visual system than the aliasing noise produced by the Fourier-based spectral techniques. A compression ratio of over 100–1 can be achieved using a wavelet technique (Zettler et al., 1991). Using a 3D wavelet transform, video signals and images can be compressed to higher ratios of about 156–1 (Lewis & Knowles, 1990). The technique of using discrete wavelet transform in compression is very efficient in the sense that it does not produce aliasing noise.

5.7.5   Texture Segmentation

Fractal functions appear to provide a good description of surface textures and their images; thus, it is natural to use the fractal model for texture classification and image segmentation. The concept of local fractal dimension is used for texture segmentation. Methods to calculate the fractal dimension of discrete signals have been prescribed to characterize the local fractal behavior at each image point (Pentland, 1984). A 3D information can be well-extracted from 2D image modeled in terms of fractal surfaces. Texture analysis accommodates the problems of edge detection, image segmentation, and calculating the local dimensionality. Chang and Kuo (1992) have shown that using a tree-structured wavelet transform is a natural and effective tool for achieving texture analysis and segmentation of images. A texture correct classification of over 95% of the original texture has already been achieved.

Wavelet-based techniques have been applied successfully to the problem of texture segmentation (Teshome, 1991). This was done using the idea of a compactly supported two-dimensional wavelet transform. Such transformation decomposes the textural information into orientation sensitive multiple channels. Information in each channel is reduced to a discriminant feature by appropriate statistics. Classification is then achieved through metric measures and semirecursive Markov-based learning models. This technique when implemented has the possibility of producing a probability of misclassification error of less than 0.2% in its application.

5.7.6   Edge Detection of Images

One prominent area of application of wavelets and fractals is in the area of edge detection of images. Edge localization occurs when an edge detector determines the location of an edge in an image. For the wavelet application, these edges are then detected from the local maxima of the image wavelet transform. The edges that are important for image visualization are then selected for coding purposes. The idea of wavelet application is extended to the detection of multiscale edges. This has helped in the characterization of signals, including images from multiscale edges.

Fractals are also applied to the detection of edges of images. First, the image is divided into non-overlapping blocks, and then the fractal dimension is estimated using the method of blanket covering. Second, using wavelet transforms, edge points are then detected. The dilation parameter is controlled by the estimated fractal dimension.

5.7.7   Geometrical Objects Representation

Wavelets and fractals can be represented using geometrical objects. The space of $N = 3$ wavelet conditions that can be realized as a "pinched torus" is an example, where N is generically one-half the number of nonzero wavelet coefficients.

Fractal interpolation functions, on the other hand, are elementary functions in that they are of a geometrical character, that they can be represented concisely by "formulas," and that they can be computed rapidly. The graphs of these functions are used to approximate image components such as profiles of mountain ranges, and the tops of clouds.

5.7.8   The Multiscale Analysis and Representation of Signals

Let Z and R denote the set of integers and real numbers, respectively, and $L^2(R^n)$ be the vector space of measurable, square-integrable two-dimensional functions $f(x,y)$. The multiscale analysis and representation of $L^2(R^n)$ (Mallat, 1989) are increasing sequences $V_j, j \in Z$, of closed subspaces of $L^2(R^n)$.

$$\subset V_3 \subset V_2 \subset V_0 \subset V_{-1} \subset V_{-2} \subset V_{-3} \subset \ldots \ldots \tag{5.17}$$

with the following properties:

$$I_{jeZ} V_j = 0, Y_{jeZ} V_j = L^2(R^n) \tag{5.18}$$

$$f(x) \in V_{j+1} f(2x) \in V_{j+1} 21 \tag{5.19}$$

$$f(x) \in V_0 \leftrightarrow f(x-k) \in V_0 \, 22 \tag{5.20}$$

for each $j \in Z$.

There is a function $g \in V_0$ such that the functions $g(x - k)$, $k \in Z$ form a Riesz basis of V_0 (see Appendix B for discussion on Riesz basis). The sequences V_j and T_j are called the multiscale approximation. As j increases, V_j and T_j become spaces of lower resolution (see Fig. 5.10). The multiscale concept may be viewed as successive approximation, or successive refinement, of the signal. This type of approximation is particularly well adapted for computer vision applications such as signal coding, texture discrimination, edge detection, matching algorithms, and fractal analysis.

The multiscale idea has also been used in the statistical analysis of computer-generated fractal imagery. In this case, convolution filters designed to detect elementary image structures, such as blobs and bars, are applied over a range of scales. The outcome of that can be displayed as statistical distributions, normalized to scale

THE MULTIRESOLUTION ANALYSIS AND REPRESENTATION IDEA

(a)

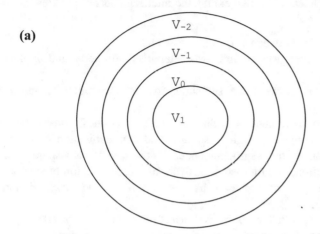

AS j INCREASES Vj BECOMES A SPACE OF LOWER RESOLUTION

THE FRACTAL MULTIRESOLUTION IDEA (BINARY TREE)

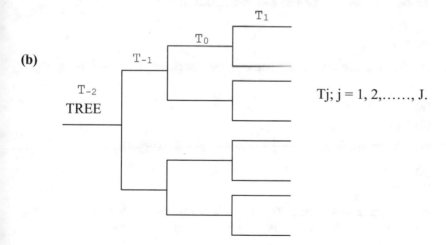

$T_j; j = 1, 2, \ldots\ldots, J.$

AS j INCREASES T_j BECOMES A SPACE OF LOWER RESOLUTION

Fig. 5.10 The multiresolution technique idea for (**a**) wavelets and (**b**) fractals

invariant form using fractal dimensions, which characterize the fractal image. Cantor fractals discussed in Sect. 5.5.5 also exhibit the multiscale technique idea.

Summary
1. There are similarities between wavelets and fractals.
2. These similarities are grouped into two categories—property and application areas.
3. In the property areas, we have self-similarity, scaling function, and affine transforms.
4. The application areas include modeling and seismic studies; construction and reconstruction of images; graphical, storage, and communication applications; image compression; texture segmentation; edge detection of images; geometrical objects representation; and multiscale analysis and representation of signals.
5. There are different types of fractals such as the random, scaling, Koch, Sierpinski, and Cantor fractals.
6. Fractal dimensions provide an objective means for comparing fractals.
7. IFS codes are useful in many application areas such as mapping, biological modeling, computer graphics, desktop publishing, advertising, geophysics, graphic design, and synthetic speech.

Review Questions

5.1. Wavelets are functions not generated from one basic function ψ called the mother wavelet by dilations and translations of ψ.

   (a) True.
   (b) False.

5.2. Fractals are subsets of geometric spaces, such as (C, spherical) and (R^2, Euclidean).

   (a) True.
   (b) False.

5.3. Fractals do not manifest a high degree of visual complexity.

   (a) True.
   (b) False.

5.4. A structure is said to be self-similar if it can be broken down into arbitrarily small pieces, each of which is a small replica of the entire structure.

   (a) True.
   (b) False.

5.5. One of the application areas common to both wavelets and fractals is the coding of images.

   (a) True.
   (b) False.

5.6. Wavelets and fractals do not have deep connections.

   (a) True.
   (b) False.

5.7. The fractional Brownian motion (fBm) of Mandelbrot and Wallis is one of the most useful mathematical models for the random fractals found in nature such as mountainous terrain and clouds.

   (a) True.
   (b) False.

5.8. The concept of affine transforms can be found in both fractals and wavelets.

   (a) True.
   (b) False.

5.9. Wavelets and fractals cannot both be applied to the field of seismic studies.

   (a) True.
   (b) False.

5.10. Wavelets and fractals can be represented using geometrical objects.

   (a) True.
   (b) False.

Answers: 5.1b, 5.2a, 5.3b, 5.4a, 5.5a, 5.6b, 5.7a, 5.8a, 5.9b, 5.10a.

Problems

5.1. What are wavelets and fractals?

5.2. What are some of the deep connections between wavelets and fractals?

5.3. (a) Name the different types of fractal dimensions. (b) Why is fractal dimension important?

5.4. (a) What is an IFS code? (b) How important are IFS codes?

5.5. Name the different types of fractals.

5.6. What is a random fractal?

5.7. What are scaling fractals?

5.8. Describe the Koch fractal.

5.9. Describe the Sierpinski fractal.

5.10. Describe the Cantor fractal.

5.11. Name three similarity areas between wavelets and fractals based on their properties.

5.12. Describe the self-similarity property of wavelets and fractals.

5.13. Describe the scaling function similarity property of wavelets and fractals.

5.14. Describe affine transforms similarity property of wavelets and fractals.

5.15. Name at least ten similarity application areas of wavelets and fractals.

5.16. Describe modeling and seismic studies similarity application of wavelets and fractals.

5.17. Describe the reconstruction of images similarity application of wavelets and fractals.

5.18. Describe the graphical, storage, and communication similarity application of wavelets and fractals.
5.19. Describe the image compression similarity application of wavelets and fractals.
5.20. Describe the texture segmentation similarity application of wavelets and fractals.
5.21. Describe the edge detection of images similarity application of wavelets and fractals.
5.22. Describe the geometrical objects representation similarity application of wavelets and fractals.
5.23. Describe the multiscale analysis and representation of signals similarity application of wavelets and fractals.

References

Akujuobi, C. M., & Baraniecki, A. Z. (1992a). Wavelets and fractals: Overview of their similarities based on application areas. In *Proceedings of the IEEE international symposium on time-frequency and time-scale analysis* (pp. 197–200).

Akujuobi, C. M., & Baraniecki, A. Z. (1992b). Wavelets and fractals: A comparative study. In *Proceedings of the IEEE workshop on statistical signal and Array processing* (pp. 42–45).

Akujuobi, C. M., & Baraniecki, A. Z. (1994). A comparative analysis of wavelets and fractals. In C. T. Leondes (Ed.), *2D and 3D digital signal processing techniques and applications* (Vol. 67, pp. 143–197). Academic Press Inc.

Argoul, F., Arneodo, A., Elezgaray, J., & Grasseau, G. (1989). Wavelet transform of fractal aggregates. *Physics Letters A, 135*(6), 327–336.

Barnsley, M. (1993). *Fractals everywhere* (2nd ed., p. 1993). Academic Press Inc.

Barnsley, M. E., & Sloan, A. D. (1988). *A better way to compress images*. Byte.

Barnsley, M. F., Devaney, R. L., Mandelbrot, B. B., Peitgen, H. O., Saupe, D., & Voss, R. F. (1988). *The science of fractal images*. Springer-Verlag.

Chang, T., & Kuo, C. C. J. (1992). A wavelet transform approach to texture analysis. *IEEE ICASSP-92, IV*, 661–664.

Daubechies, I. (1998). Orthonormal bases of compactly supported wavelets. *Comm-Pure and Applied Math, 41*, 969–996.

Falconer, K. (1990). *Fractal geometry – Mathematical foundations and applications*. Wiley.

Jones, J.G., Thomas, R.W., Earwicker, P.G., and Addison, S. (1991). Multiresolution statistical analysis of computer-generated fractal imagery. Graphical models and image processing, Academic Press, , 53, 4.

Kolata, G. (1991). *New technique stores images more efficiently*. New York Times.

Kundu, A., & Chen, J. (1991). *Structural texture recognition using QMF Bank based subband decomposition* (pp. 2693–2696). IEEE ICASSP-91.

Lewis, G., & Knowles, A. S. (1990). Video compression using 3D wavelet transform. *Electronic Letters, 20*(6).

Mallat, S. (1989). A theory for multiresolution signal decomposition: The wavelet representation. *IEEE Transactions on Pattern Analysis and Machine Intelligence, 11*(7), 674–693.

Mandelbrot, B. B. (1982). *The fractal geometry of nature*. Freeman.

Mandelbrot, B. B., & Wallis, J. W. (1968). Fractional Brownian motions, fractional noises, and applications. *SIAM Review, 10*, 422–437.

Pentland, A. P. (1984). Fractal-based description of natural scenes. *IEEE Transaction on Pattern Analysis and Machine Intelligence, 14*(6).

Teshome, H. (1991). *Multichannel wavelets decomposition for texture segmentation*. Ph.D. Thesis, Stevens Institute of Technology.

UWE (1989). *UltraWave explorer user's manual*. Aware Inc., AD890223.1.

Voss, R. I. (1985). Random fractal forgeries. In R. A. Eamshaw (Ed.), *Fundamental algorithms for computer graphics*. Springer-Verlag.

Zettler, W. R., Huffman, J., & Linden, D. C. E. (1991). Application of compactly supported wavelets to image compression. *Aware Technical Report*, AD900119.

Part II
Wavelet and Wavelet Transform
Applications to Mixed Signal Systems

Chapter 6
Test Point Selection Using Wavelet Transforms for Digital-to-Analog Converters

6.1 Introduction

In many mixed signal systems such as analog-to-digital converters (ADCs) and digital-to-analog converters (DACs) testing applications, the test point space, i.e., the range of candidate test points, is much larger than the number of test points at which it is either necessary or economically feasible to make measurements. Not all test points are equally useful, so some selection criterion must be developed to ensure that a necessary, sufficient, and robust set is chosen. In this chapter, we will cover the idea of test point selection using wavelet transforms for mixed signal systems—a DAC system is used as an example.

In addition, we cover what we call the Stenbakken and Souders algorithm. We then cover the wavelet test point selection algorithm and the implementation of the programmatic testing methods. This test point selection technique is demonstrated using an 8-bit DAC. We accomplish, using this test point algorithm and wavelet transform, the prediction of the integrated nonlinearity (INL) of the DAC by measuring only 11 test points out of many points. The measures of a restricted number of test points are sufficient to calculate the coefficients and then to predict the INL. Many other researchers have also applied this idea in the testing of mixed signal systems (Huang, 2000; Akujuobi et al., 2003).

"Life isn't about finding yourself. Life is about creating yourself."
—George Bernard Shaw

© Springer Nature Switzerland AG 2022
C. M. Akujuobi, *Wavelets and Wavelet Transform Systems and Their Applications*,
https://doi.org/10.1007/978-3-030-87528-2_6

6.2   The Stenbakken and Souders Algorithm

In 1987, Gerard N. Stenbakken and T. Michael Souders came up with the idea of the test point selection method (Stenbakken & Sounders, 1987). This method consists of performing the following tasks:

- First, we develop a linear error model as shown in Eq. (6.1).

$$y = A \cdot x \tag{6.1}$$

where:

y is the $m * 1$ vector of error response at m candidate test points
x is the $n * 1$ vector of underlying variables
A is the $m * n$ linear coefficient matrix model that relates the error response to the underlying variables

- Second, select the test points using QR factorization (QRF) as shown in Eq. (6.2). The QRF is discussed more in Appendix C.

$$\tilde{y} = \tilde{A} \cdot x \tag{6.2}$$

where:

$\tilde{y}$ is the $n * 1$ vector
$\tilde{A}$ is the $n * n$ reduced matrix

- Third, make n measurements.
- Fourth, estimate parameters x as shown in Eq. (6.3).

$$\hat{x} = \tilde{A}^{-1} \cdot \tilde{y} \tag{6.3}$$

- Fifth, predict response at all m candidate test points using Eq. (6.4).

$$\hat{y} = A \cdot \hat{x} \tag{6.4}$$

where:

$\hat{y}$ is the $m * 1$ vector of predicted response
$\hat{x}$ is the $n * 1$ vector
A is the $m * n$ matrix.

The wavelet transform discussed in Chap. 4 is known for its ability to find local features of a signal from its subband decomposition. Since the input-output relationship of a linear circuit can be represented by a matrix; it is possible to make use of the coefficients of the wavelet transform for constructing the linear matrix representation of the device under test. In the next section, we cover the test point selection method implemented using wavelet transform.

6.3   Wavelet Transform Test Point Selection Algorithm

In the development of the algorithm for the wavelet-based test point selection, it is important to understand the type of circuit model to use.

6.3.1   Circuit Model

The important key of the test point selection method is how to select a set of points to test the circuit from a given test pattern, that is, the model of circuit under test (CUT). The model describes the relationship between the circuit input and its corresponding output, i.e., the relationship between input x and output y is as shown in Eq. (6.1).

6.3.2   Three Basic Model Types

The matrix A is usually called the sensitivity matrix. It can be obtained via several ways.

(a) **Physical models:** The physical models are obtained based on differential circuit equations like Kirchhoff's current law (KCL) or Kirchhoff's voltage law (KVL).
(b) **A priori models:** A priori models are the ones that use a set of vectors to represent the test circuits in a general case.
(c) **Empirical models:** These models are basically the same as "a priori models" except that the vector sets are not obtained by some known functions like step functions. They are obtained by analyzing the data from exhaustive testing of representative units coming off the production line.

6.3.3   Circuit Model by Wavelet Transforms

If we can establish a correct model of the circuit, we can find the feature of the signal. This can be done by studying a few test outputs and describing their signatures as in empirical models, or by picking a certain set of the outputs and using them as the basis signatures. The signature of a DAC cannot be used for a filter. The other way to find the signatures is through certain types of transforms such as the wavelet transforms. Because of the advantages of wavelet transform in signal analysis, we use the discrete wavelet transform (DWT) to decompose the signal into different scales of the basic signal, where each of the decomposed signals is independent of each other.

6.3.4   Choosing Test Points by QR Factorization

The goal of the test point selection method is to reduce the test points in order to reduce the testing time and cost. The test point selection method is based on the QR factorization (QRF). The QRF decomposes the A matrix (sensitivity matrix) into a product of an orthogonal matrix Q and an upper triangular matrix R, i.e., as shown in Eq. (6.5).

$$A_{p \times m} = Q_{p \times n} R_{n \times m} \tag{6.5}$$

We can select the most important row vectors in R and discard the rest. This is done by discarding the row vectors with smaller diagonal element in R.

6.4   Implementation of the Test Method in Programming

The test method is implemented in a programmatic way. We use this method in an 8-bit digital-to-analog converter (DAC). Sections 6.4.1–6.4.9 respectively represent the different steps of the method used in the test point selection for the 8-bit DAC system. Figure 6.1 shows the flowchart for the programming algorithm.

6.4.1   Measured INL Data of 8-Bit DAC

The integral nonlinearity (INL) of a DAC is a global measure defined as the maximum deviation of the transfer characteristics from the ideal values represented as a straight line drawn from zero to full scale. For testing a DAC for INL, an integral nonlinearity-like signal is generated to extract the features of the DAC transfer characteristics. The INL expresses both the maximum deviation and all the other deviations from the ideal values at the other input nodes. We use the measured INL data, as shown in Fig. 6.2 and Table 6.1, which is obtained from actual measurement, to build the model of the tested DAC. These INL data are from only one device exclusively.

Fig. 6.1 Wavelet-based test
point selection method

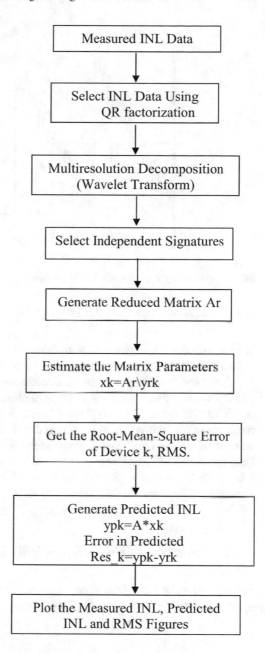

6.4.2   Selection of a Set of Maximally Independent INL

The optimal solution can be obtained by a QR factorization. We use the pivoting
technique of QR factorization to select a set of INL that is maximally independent.
The number of selected INL is determined so as to minimize the error in prediction.

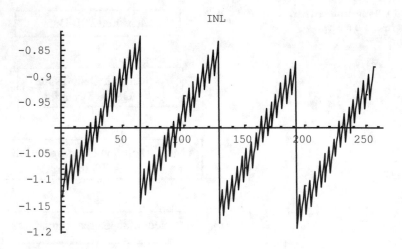

Fig. 6.2 INL of a DAC. (Source: Texas Instruments)

6.4.3   Multiresolution Decomposition

We have discussed the wavelet transform multiscale/multiresolution decomposition in Chap. 4. Wavelet filters are used in the multiresolution decomposition process. We apply the wavelet transform to decompose the selected INL to obtain a set of coefficients. In this algorithm implementation, the Haar and Daubechies wavelet coefficients are used. We obtain the different predicted errors using different wavelet coefficients, as shown in Figs. 6.3 and 6.4 respectively. We compare the Haar wavelet transform with Daubechies wavelet transform. The better accuracy coming from the predicted error comes from using the Haar wavelet transform. On each selected INL, we apply the multiresolution decomposition to decompose the sum of signatures. Thus, we obtain a set of signatures.

6.4.4   Selection of the Independent Signatures (Coefficients)

We apply the QR factorization on the matrix containing all signatures to choose the optimal number of signatures to obtain the model A.

6.4.5   Generation of a Reduced Matrix **Ar** from **Q** and **R** Matrices

The generation of a reduced matrix Ar from Q and R matrices is done by discarding all the rows in R with $|R_{jj}| < e$, where e is a threshold value predetermined by the

Table 6.1 256 INL data of an 8-bit DAC

−1.1343E+00	−9.9424E-01	−1.1447E+00	−1.0046E+00
−1.1085E+00	−9.6843E-01	−1.1189E+00	−9.7882E-01
−1.0948E+00	−9.5467E-01	−1.1052E+00	−9.6506E-01
−1.0690E+00	−9.2885E-01	−1.0793E+00	−9.3924E-01
−1.1184E+00	−9.7828E-01	−1.1288E+00	−9.8868E-01
−1.0926E+00	−9.5247E-01	−1.1030E+00	−9.6286E-01
−1.0788E+00	−9.3871E-01	−1.0892E+00	−9.4910E-01
−1.0530E+00	−9.1289E-01	−1.0634E+00	−9.2329E-01
−1.1071E+00	−9.6696E-01	−1.1175E+00	−9.7736E-01
−1.0813E+00	−9.4115E-01	−1.0916E+00	−9.5154E-01
−1.0675E+00	−9.2739E-01	−1.0779E+00	−9.3778E-01
−1.0417E+00	−9.0157E-01	−1.0521E+00	−9.1197E-01
−1.0911E+00	−9.5101E-01	−1.1015E+00	−9.6140E-01
−1.0653E+00	−9.2519E-01	−1.0757E+00	−9.3558E-01
−1.0515E+00	−9.1143E-01	−1.0619E+00	−9.2182E-01
−1.0257E+00	−8.8561E-01	−1.0361E+00	−8.9601E-01
−1.0708E+00	−9.3066E-01	−1.0812E+00	−9.4105E-01
−1.0449E+00	−9.0484E-01	−1.0553E+00	−9.1524E-01
−1.0312E+00	−8.9108E-01	−1.0416E+00	−9.0147E-01
−1.0054E+00	−8.6526E-01	−1.0158E+00	−8.7566E-01
−1.0548E+00	−9.1470E-01	−1.0652E+00	−9.2509E-01
−1.0290E+00	−8.8888E-01	−1.0394E+00	−8.9928E-01
−1.0152E+00	8.7512E-01	−1.0256E+00	−8.8552E-01
−9.8941E-01	−8.4931E-01	−9.9980E-01	−8.5970E-01
−1.0435E+00	−9.0338E-01	−1.0539E+00	−9.1377E-01
−1.0177E+00	8.7756E-01	−1.0281E+00	−8.8796E-01
−1.0039E+00	−8.6380E-01	−1.0143E+00	−8.7419E-01
−9.7809E-01	−8.3799E-01	−9.8848E-01	−8.4838E-01
−1.0275E+00	8.8742E-01	−1.0379E+00	−8.9781E-01
−1.0017E+00	−8.6160E-01	−1.0121E+00	−8.7200E-01
−9.8795E-01	−8.4784E-01	−9.9834E-01	−8.5824E-01
−9.6213E-01	−8.2203E-01	−9.7253E-01	−8.3242E-01
−1.1832E+00	−1.0431E+00	−1.1936E+00	−1.0535E+00
−1.1574E+00	−1.0173E+00	−1.1678E+00	−1.0277E+00
−1.1436E+00	−1.0035E+00	−1.1540E+00	−1.0139E+00
−1.1178E+00	−9.7770E-01	−1.1282E+00	−9.8809E-01
−1.1672E+00	−1.0271E+00	−1.1776E+00	−1.0375E+00
−1.1414E+00	−1.0013E+00	−1.1518E+00	−1.0117E+00
−1.1277E+00	−9.8756E-01	−1.1381E+00	−9.9795E-01
−1.1018E+00	−9.6174E-01	−1.1122E+00	−9.7214E-01
−1.1559E+00	−1.0158E+00	−1.1663E+00	−1.0262E+00
−1.1301E+00	−9.9000E-01	−1.1405E+00	−1.0004E+00
−1.1163E+00	−9.7623E-01	−1.1267E+00	−9.8663E-01
−1.0905E+00	−9.5042E-01	−1.1009E+00	−9.6081E-01
−1.1400E+00	−9.9985E-01	−1.1504E+00	−1.0102E+00
−1.1141E+00	−9.7404E-01	−1.1245E+00	−9.8443E-01
−1.1004E+00	−9.6028E-01	−1.1108E+00	−9.7067E-01
−1.0746E+00	−9.3446E-01	−1.0850E+00	−9.4486E-01
−1.1196E+00	−9.7950E-01	−1.1300E+00	−9.8990E-01
−1.0938E+00	−9.5369E-01	−1.1042E+00	−9.6408E-01
−1.0800E+00	−9.3993E-01	−1.0904E+00	−9.5032E-01

(continued)

Table 6.1 (continued)

−1.0542E+00	−9.1411E-01	−1.0646E+00	−9.2451E-01
−1.1037E+00	−9.6355E-01	−1.1140E+00	−9.7394E-01
−1.0778E+00	−9.3773E-01	−1.0882E+00	−9.4813E-01
−1.0641E+00	−9.2397E-01	−1.0745E+00	−9.3436E-01
−1.0383E+00	−8.9816E-01	−1.0487E+00	−9.0855E-01
−1.0923E+00	−9.5223E-01	−1.1027E+00	−9.6262E-01
−1.0665E+00	−9.2641E-01	−1.0769E+00	−9.3680E-01
−1.0528E+00	−9.1265E-01	−1.0631E+00	−9.2304E-01
−1.0269E+00	−8.8683E-01	−1.0373E+00	−8.9723E-01
−1.0764E+00	−9.3627E-01	−1.0868E+00	−9.4666E-01
−1.0506E+00	−9.1045E-01	−1.0610E+00	−9.2085E-01
−1.0368E+00	−8.9669E-01	−1.0472E+00	−9.0709E-01
−1.0110E+00	−8.7088E-01	−1.0214E+00	−8.8127E-01

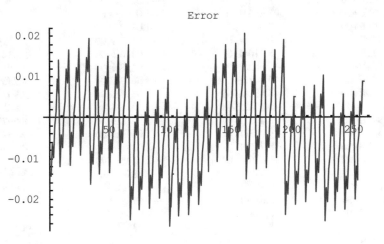

Fig. 6.3 Predicted error using Haar wavelet

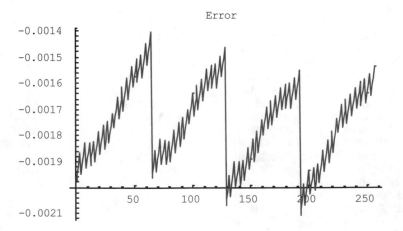

Fig. 6.4 Predicted error using Daubechies wavelet

program, and the corresponding column of Q. In this 8-bit DAC device testing, we reduce the 256 of Table 6.1 to 12 test points (using Haar wavelet) and 11 test points (using Daubechies wavelet). The samples of the selected test points are listed below for the Haar and Daubechies wavelets respectively.

- **12 test points selected using Haar wavelet:**

```
1 signature
1 test point selected :
2 signatures
2 test points selected :
3 signatures
4 test points selected :
4 signatures
8 test points selected :
5 signatures
9 test points selected :
6 signatures
12 test points selected :
7 signatures
12 test points selected :
```

- **11 test points selected using Daubechies wavelet:**

```
1 signature
1 test point selected :
2 signatures
3 test points selected :
3 signatures
4 test points selected :
4 signatures
6 test points selected :
5 signatures
8 test points selected :
6 signatures
9 test points selected :
7 signatures
11 test points selected :
```

6.4.6   Estimation of the Matrix Parameters of the Device

The complete number of selected test points and the corresponding reduced matrix conclude the pre-production operations. These operations are performed only once for each production device or type. For each device, k measurements are made at the selected test points and subtracted from the nominal or ideal performance to give the reduced measurement error vector y_{rk} for device k. The model parameters for that device x_k are estimated using Eq. (6.6).

$$x_k = A_r^{-1} y_{rk} \tag{6.6}$$

6.4.7   Generation of the Predicted INL y_{pk} for all M Candidate Test Points

The selection of the predicted INL is performed using Eq. (6.7).

$$y_{pk} = A \cdot x_k \tag{6.7}$$

The difference between the predicted INL at the measured test points and measured values is given in Eq. (6.8).

$$e_k = y_{pk} - y_{rk} \tag{6.8}$$

6.4.8   Getting the Root-Mean-Square for the Device k

The root-mean-square (RMS) for the device k is given by Eq. (6.9).

$$RMS_k = \sqrt{\frac{\sum_j (e_{kj})^2}{N}} \tag{6.9}$$

6.4.9   Plotting the Measured INL, Predicted INL, and RMS Figures

We can directly figure out the difference of the measured INL and the predicted INL and know if this method is good and which wavelet transform is better. Figures 6.5, 6.6, 6.7, 6.8, 6.9, 6.10, 6.11, and 6.12 are the measured INL, predicted INL, and *RMS* figures using Haar and Daubechies wavelets. These steps are implemented with the Mathematica software and a package of Mathematica named Wavelet Explorer. The error in the prediction, i.e., the difference between the predicted errors and the measured errors, is much less, about $-0.0021 \sim -0.0014$ LSB, which is acceptable to manufacturers.

Summary
1. Not all test points are equally useful, so some selection criterion must be developed to ensure that a necessary, sufficient, and robust set is chosen.
2. A method of test point selection is proposed and is successfully implemented using an 8-bit DAC mixed signal system.
3. The prediction of the INL is accomplished measuring only 11 test points for Haar wavelet and 12 test points for Daubechies wavelets instead of using 256 test points.
4. The measures of a restricted number of test points are sufficient to calculate the coefficients and then to predict the INL.
5. The error in the prediction, i.e., the difference between the predicted errors and the measured errors, is much less, about $-0.0021 \sim -0.0014$ LSB, which is accepted by manufacturers.

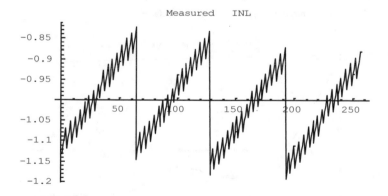

Fig. 6.5   Measured INL

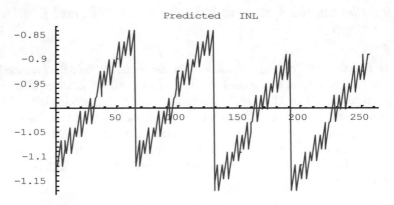

Fig. 6.6 Predicted INL using Haar wavelet

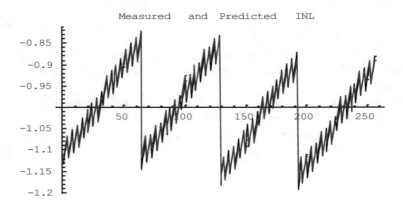

Fig. 6.7 Measured and predicted INL using Haar wavelet

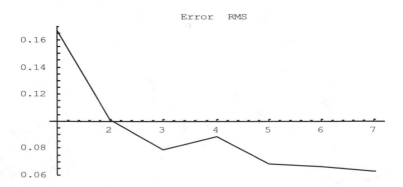

Fig. 6.8 RMS of predicted INL using Haar wavelet

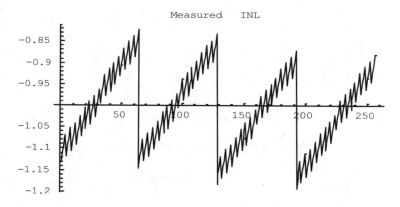

Fig. 6.9 Measured INL

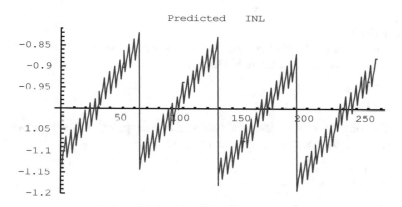

Fig. 6.10 Predicted INL using Daubechies wavelet

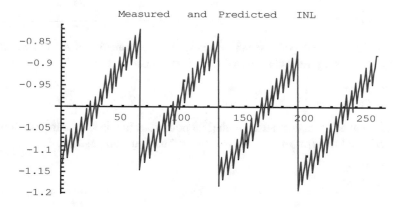

Fig. 6.11 Measured and predicted INL using Daubechies wavelet

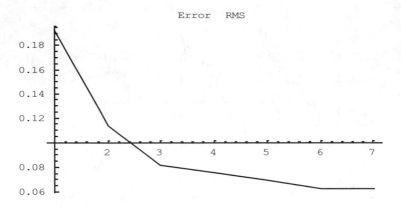

Fig. 6.12 RMS of predicted INL using Daubechies wavelet

Review Questions

6.1. The measures of a restricted number of test points are sufficient to calculate the coefficients and then to predict the INL.

(a) True.
(b) False.

6.2. The measures of a restricted number of test points are not sufficient to calculate the coefficients and then to predict the INL.

(a) True.
(b) False.

6.3. The idea of test point selection using wavelet transforms for mixed signal systems using a DAC system saves time.

(a) True.
(b) False.

6.4. The idea of test point selection using wavelet transforms for mixed signal systems using a DAC system does not save cost.

(a) True.
(b) False.

6.5. The wavelet transform used in the test point selection is known for its ability to find local features of a signal from its subband decomposition.

(a) True.
(b) False.

6.6. The error in the prediction is the difference between the predicted errors and the measured errors.

(a) True.
(b) False.

6.7. The INL does not express both the maximum deviation and all the other deviations from the deal values at the other input nodes.

(a) True.
(b) False.

6.8. The optimal solution in the selection of a set of maximally independent INL can be obtained by a QR factorization.

(a) True.
(b) False.

6.9. The error in the prediction, i.e., the difference between the predicted errors and the measured errors, is much less, about $-0.0021 \sim -0.0014$ LSB, which is accepted by manufacturers.

(a) True.
(b) False.

6.10. On each selected INL, we do not apply the multiresolution decomposition to decompose the sum of signatures.

(a) True.
(b) False.

Answers: 6.1a, 6.2b, 6.3a, 6.4b, 6.5a, 6.6a, 6.7b, 6.8a, 6.9a, 6.10b.

Problems

6.1. Why is test point selection using wavelet transforms for mixed signal systems such as DACs important?

6.2. What is the Gerard N. Stenbakken and T. Michael Souders algorithm for the test point selection method?

6.3. Why is the wavelet transform important in the test point selection method?

6.4. Why is it important to understand the type of circuit model to use in the development of the algorithm for the wavelet-based test point selection?

6.5. Describe the three basic model types that can be used in the modeling of the circuit under test (CUT).

6.6. Why do we consider modeling the circuit using wavelet transform?

6.7. What is the goal of the test point selection method?

6.8. How can you choose the test points by QR factorization?

6.9. What is the importance of the measured data in the test point selection process of INL using a DAC as an example?

6.10. How can you select a set of maximally independent INL?

6.11. What is the significance of using the multiscale/multiresolution technique?

6.12. How can you select the independent signatures?

6.13. How can you generate the reduced matrix Ar from Q and R matrices?

6.14. How can you estimate the matrix parameters of the device under test?

6.15. How can you generate the predicted INL y_{pk} for all m candidate test points?

6.16. Define the root-mean-square (RMS) for the device k.

References

Akujuobi, C.M., Sadiku, M.N.O., Lian, J., & Hu, L. et al. (2003). "Test-point selection method for mixed signal systems using discrete wavelet transform", *Proceedings GSPX-ISPC*, Dallas, March 31 – April 3, 2003.

Huang, H. (2000). "Test point-selection using wavelet transforms for mixed-signal circuit", Master's Thesis, Texas A&M University, pp. 1–60, May 2000.

Stenbakken G.N. and Sounders, T.M (1987). "Test-point selection and testability measures via QR factorization of linear models", IEEE Transactions on Instrumentation and Measurement, Vol. IM-36, No. 2, pp. 406–410, 1987.

Chapter 7
Wavelet-Based Dynamic Test of ADCS

> *"Not everything that can be counted counts and not*
> *everything that counts can be counted."*
>
> –Albert Einstein

7.1   Introduction

The fast Fourier transform is a traditional tool for most of the analog-to-digital converter (ADC) dynamic tests, such as effective number of bits (ENOB), integral nonlinearity (INL), and differential nonlinearity (DNL) (Burns & Roberts, 2004). The wavelet transform has many advantages when compared to the fast Fourier transform (FFT) (Akujuobi, 1998, 2000). In this chapter, we explore the testing of ENOB and the DNL of mixed signal systems using wavelet transforms.

In ENOB testing using the FFT, we estimate the mean square error due to quantization from frequency domain data. Using the discrete wavelet transform (DWT) method for testing the ENOB of ADCs yields instantaneous ENOB values and DNL. In the late 1990s, the works of people like Yamaguchi and Soma established how wavelet transform is used in the testing of ADCs (Yamaguchi & Soma, 1997). In this chapter, we explore some of the challenging issues in the testing of wavelet-based ADCs and how the challenges can be resolved. Since 1997 and from the early to late 2000, many works have been done using wavelets in the testing of ADCs and mixed signal systems in general (Akujuobi & Hu, 2002; Marshall & Akujuobi, 2002; Akujuobi & Awada, 2007; Akujuobi et al., 2008; Awada & Akujuobi, 2018a, b).

7.2   Measuring ENOB Using the Conventional Method

The well-known conventional method for ENOB testing is the fast Fourier transform (FFT) and for DNL testing is the sinusoidal histogram. We discuss in the next section how the ENOB can be measured using FFT.

© Springer Nature Switzerland AG 2022   123
C. M. Akujuobi, *Wavelets and Wavelet Transform Systems and Their Applications*,
https://doi.org/10.1007/978-3-030-87528-2_7

7.2.1   *Measuring the ENOB Using FFT Method*

The output samples from the device under test (DUT) which in this case is an ADC are transformed into the frequency domain by applying the FFT to the output signal. We then estimate the signal-to-noise ratio (SNR) of the ADC. It should be noted that any loss of the ENOB in this process should be because of the reflection due to the contribution of the quantization noise through SNR estimates.

7.2.2   *Computing SNR Through FFT*

Assume that, in the frequency band of interest, the analog test stimulus is of the form shown in Eq. (7.1).

$$x_{in}(t) = A_x \cos{(\omega_x t + \varphi_x)} \tag{7.1}$$

where:

$x_{in}(t)$= analog test stimulus
A_x= the amplitude of the signal test stimulus
ω_x= the test frequency

If we let the ADC output to be as shown in Eq. (7.2).

$$y_{out}(n) = s(n) + \eta(n) \tag{7.2}$$

where:

$s(n)$ is the signal
$\eta(n)$ is the noise

For optimum accuracy, a sample record $y_{out}(n)$ consisting of M samples must contain an integer number of whole cycles of the sine wave. To compute the SNR, first calculate $y_{out}(k)$, which is the M-point FFT of $y_{out}(n)$ and is given as shown in Eq. (7.3).

$$Y_{out}(k) = \sum_{n=0}^{M-1} y_{out}(n) e^{-j(2\pi/M)kn} \tag{7.3}$$

If we let the desired frequency component ω_x be the j-th element of $Y_{out}(k)$. Based on Parseval's relation for the FFT (see Appendix D) in the textbook, together with classical statistical theory, an estimate of the variance of the signal $s(n)$ (which is also the signal power $\widehat{P}_s$) is given by Eq. (7.4).

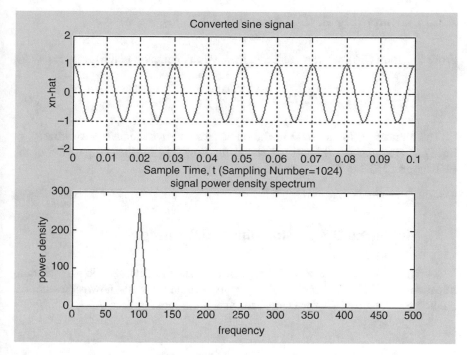

Fig. 7.1 Converted sine signal and signal power density spectrum

$$\widehat{\sigma}_s^2 = \widehat{P}_s = \frac{2}{(M-1)M}|Y(j)|^2. \tag{7.4}$$

An unbiased estimate of the noise power σ_η^2 (which we shall designate $\widehat{P}_\eta$) is given by Eq. (7.5).

$$\widehat{\sigma}_\eta^2 = \widehat{P}_\eta = \frac{2}{(M-1)M}\sum_{k=1}^{(M-1)/2}|Y(k)|^2, k \neq j. \tag{7.5}$$

Combining Eqs. (7.4) and (7.5) yields the SNR for the test frequency ω_x as shown in Eq. (7.6).

$$SNR = 10\log_{10}\left\{\frac{|Y(j)|^2}{\sum\limits_{k=1}^{(M-1)/2}|Y(k)|^2, k \neq j}\right\} \tag{7.6}$$

Figure 7.1 shows the sine signal converted by ADC and the signal power density spectrum of the sine signal.

7.2.3   Computing ENOB Through SNR

From Equation (7.6), we can get SNR; then we can compute ENOB from Eq. (7.7).

$$ENOB = \frac{SNR[db] - 1.76}{6.02}[bit] \tag{7.7}$$

The numbers 1.76 and 6.02 in the numerator and denominator, respectively, of Eq. (7.7) are constants. Figure 7.2 shows the flowchart for the computation of the FFT-based ENOB.

7.3   Measuring DNL Using Sinusoidal Histogram

It is common to use sine wave histogram tests for the determination of the non-linearities of ADCs. The sinusoidal histogram method is a well-known conventional method for testing ADCs for DNL. This idea is explored in Sect. 7.3.1.

Fig. 7.2 Computation of ENOB using FFT

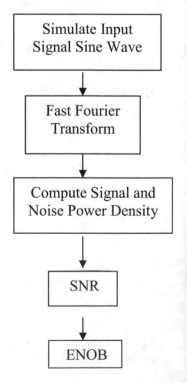

7.3.1 Differential and Integral Nonlinearity

The differential nonlinearity (DNL) and integral nonlinearity (INL) for any particular values for the gain and the offset are defined as shown in Eqs. (7.8)–(7.11), respectively.

$$DNL[k] = \frac{G \cdot W[k] - Q}{Q},$$ (7.8)

$$DNL = \max |DNL[k]|,$$ (7.9)

$$INL[k] = \frac{\varepsilon[k]}{Q},$$ (7.10)

$$INL = \max |INL[k]|.$$ (7.11)

where:

$W[k] = T[k + 1] - T[k]=$ the kth code bin width
$G=$ the gain, nominally 1
$Q = V/(2^N - 2)=$ the average code bin width
$\varepsilon[k]=$ the residual error

7.3.2 Sinusoidal Histogram Measurement

A sine wave that slightly overdrives the ADC is sampled many times. The data are collected as series of R records each of which contains M samples. Each record is taken with the same constant sampling rate. The record length and ratio of the sampling rate to the signal frequency are chosen so that the phases of the samples are uniformly distributed between 0 and 2π. If the range of the ADC is not symmetrical about 0 V, a constant, approximately equal to the mid-scale voltage of the ADC, must be added to the sine wave.

Let $h[i]=$ total number of samples received in code bin i, and let

$$ch[k] = \sum_{i=0}^{k} h[i],$$ (7.12)

The applied signal is of the form as shown in Eq. (7.13).

$$v[t] = A \sin [\omega t + \varphi] + d$$ (7.13)

The frequency and phase of the sine wave are not used in the data analysis. The values of A and d are assumed to be known, but they need not be. Error in the values for A and d will affect the values calculated for the gain and the offset of the ADC but

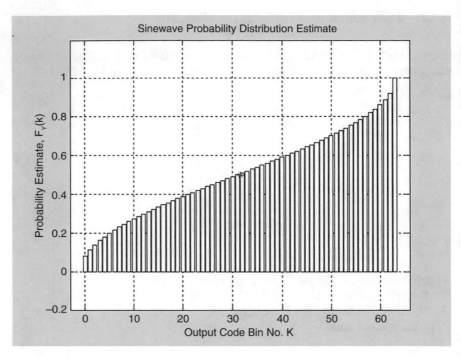

Fig. 7.3 Sinusoid histogram

will not affect values for DNL or INL at all. The transition levels $T[k]$ are estimated from the data using Eq. (7.14).

$$T[k] = d - A \cos \left[\frac{\pi ch[k-1]}{S} \right]. \tag{7.14}$$

where

$S = M \cdot R =$ total number of samples.

The code bin widths are given in Eq. (7.15).

$$W[k] = T[k+1] - T[k]. \tag{7.15}$$

Figure 7.3 is the sinusoid histogram of which a sine wave overdrives the 6-bit ADC. The amplitude of the sine wave is 1, the frequency is 100 Hz, and the sample number is 2^{13}.

If the values of A and d are unknown, approximate values can be obtained from Eq. (7.14) and approximate values for the first and last (or any two) transition levels. Values for gain and offset may then be determined by any desired method, and INL and DNL can be determined from Eqs. (7.8) through (7.11). In Sect. 7.4, we discuss how ENOB and DNL can be measured using discrete wavelet transform.

7.3.3   Measuring ENOB and DNL of ADC Using Discrete Wavelet Transform

The idea of using wavelets instead of the traditional methods of measuring ENOB and DNL is discussed in this section. The Haar and Daubechies types of wavelets are used in the measurement. Figure 7.4 shows the flowchart for the wavelet-based computation of ENOB and DNL of an ADC. Figure 7.5 shows the block diagram of the algorithm using DWT. We discuss in Sect. 7.4.1 how we can measure the instantaneous ENOB and DNL of an ADC.

Fig. 7.4 Computation of ENOB and DNL using DWT

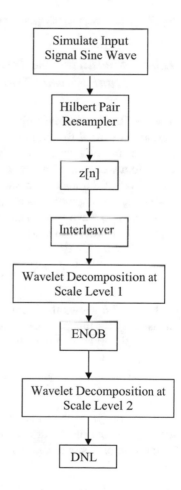

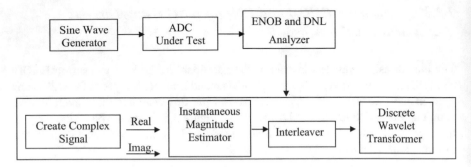

Fig. 7.5 Block diagram of algorithm using DWT

7.3.4 Measuring the Instantaneous ENOB and DNL Using Haar Wavelet Transform

The ENOB estimate given by Eq. (7.7) is a mean value, not instantaneous value. The deviations caused by different bit failures are simply summed into the total noise power in the FFT test. In the dynamic testing of an ADC, it is more desirable to estimate a worst-case ENOB scenario than a mean ENOB. This corresponds to the estimation of the maximum value of the quantization noise. Similarly, we can use instantaneous magnitude as the probing signal to get the instantaneous DNL.

If a cosine wave is input to an ADC, the output samples $\widehat{x}[n]$ are the sum of the input cosine wave and a non-idealities term $\frac{\Delta}{2}\varepsilon[n]$.

$$\widehat{x}[n] = A \cos\left(2\pi f_0 n + \varphi\right) + \frac{\Delta}{2}\varepsilon[n]. \tag{7.16}$$

Figure 7.6 shows the input signal $s[n]$, quantization signal $x[n]$, and converted signal $\widehat{x}[n]$, for this particular case.

The Hilbert transform of Eq. (7.16) is as shown in Eq. (7.17).

$$\widehat{x}[m] = H(x[n]) + e[m] = A \sin\left(2\pi f_0 n + \varphi\right) + \frac{\Delta}{2}\varepsilon[m] \tag{7.17}$$

Therefore, a Hilbert-Pair resampler can construct the complex-valued signal $\widehat{x}[n] + j\widehat{x}[m]$ out of the sampled real signal $\widehat{x}[n]$. The instantaneous magnitude $|z[n]|$ of this complex signal $\widehat{x}[n] + j\widehat{x}[m]$ is as shown in Eq. (7.18).

$$| z[n] | \equiv \sqrt{\widehat{x}[n]^2 + \widehat{x}[m]^2}$$
$$= A + \frac{\Delta}{2}\left\{\varepsilon[n] \cos\left(2\pi f_0 n + \varphi\right) + \varepsilon[m] \sin\left(2\pi f_0 n + \varphi\right)\right\}. \tag{7.18}$$

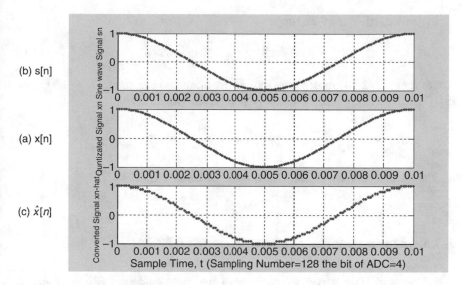

Fig. 7.6 (**a**) Sine wave signal, (**b**) quantization signal, and (**c**) converted signal

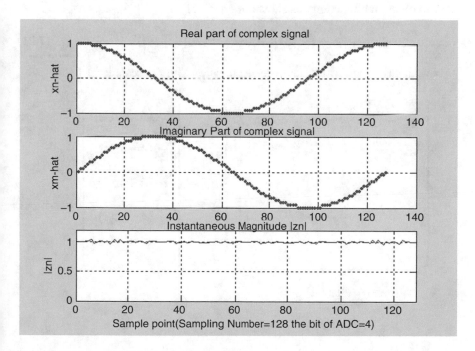

Fig. 7.7 Waveforms of $\widehat{x}[n]$, $\widehat{x}[m]$, and $|z[n]|$

Thus, the modulated signal term of $|z[n]|$ carries the information describing any size failure of the ADC under test. The waveforms of the $\widehat{x}[n]$, $\widehat{x}[m]$, and $|z[n]|$ are shown in Fig. 7.7.

In general, $|z[n]|$ is in the range

$$A - \alpha\Delta \leq |z[n]| \leq A + \alpha\Delta \tag{7.19}$$

where $\alpha = \sqrt{2}/2$ for most cases (but in our simulation example, $\alpha = 1/3$). From Eq. (7.18), the maximum deviation of $|z[n]|$ from value A is

$$|z[n] - A| = \left|\frac{\Delta}{2}\{\varepsilon[n]\cos(2\pi f_0 n + \varphi) + \varepsilon[m]\sin(2\pi f_0 n + \varphi)\}\right| \tag{7.20}$$

The maximum value of Eq. (7.20) is the worst-case ENOB.

7.3.5 Measuring the Instantaneous ENOB

We apply the Haar wavelet transform to get the worst-case ENOB. First, we interleave the instantaneous magnitude signal $|z[n]|$ with amplitude A of an ideal ADC to obtain the function f as shown in Eq. (7.21).

$$f \equiv (A, |z[1]|, A, |z[2]|, \cdots, A, |z[n]|, \cdots) \tag{7.21}$$

We can easily perform a Haar wavelet transform on f at scale level 1.

$$\begin{bmatrix} \frac{\sqrt{2}}{2} & \frac{\sqrt{2}}{2} & & & \\ \frac{\sqrt{2}}{2} & -\frac{\sqrt{2}}{2} & & & \\ & & \frac{\sqrt{2}}{2} & \frac{\sqrt{2}}{2} & \\ & & \frac{\sqrt{2}}{2} & -\frac{\sqrt{2}}{2} & \\ & & & & \ddots \end{bmatrix} * f = \begin{bmatrix} \frac{\sqrt{2}}{2}(A + |z(1)|) \\ \frac{\sqrt{2}}{2}(A - |z(1)|) \\ \frac{\sqrt{2}}{2}(A + |z(2)|) \\ \frac{\sqrt{2}}{2}(A - |z(2)|) \\ \vdots \end{bmatrix}$$

$$\Rightarrow \begin{bmatrix} \frac{\sqrt{2}}{2}(A + |z(1)|) \\ \frac{\sqrt{2}}{2}(A + |z(2)|) \\ \vdots \\ \frac{\sqrt{2}}{2}(A - |z(1)|) \\ \frac{\sqrt{2}}{2}(A - |z(2)|) \\ \vdots \end{bmatrix} \tag{7.22}$$

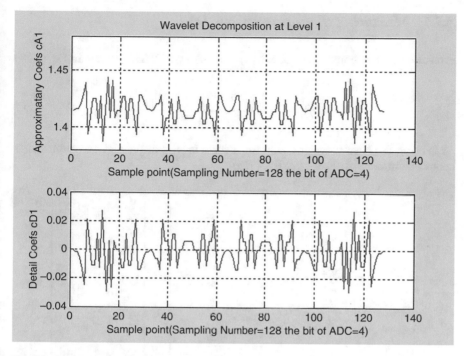

Fig. 7.8 Coefficients of Haar wavelet decomposition at scale level 1

The high-pass components at the scale level 1 of the Haar wavelet decomposition are $\frac{\sqrt{2}}{2}(A - |z[1]|)$, $\frac{\sqrt{2}}{2}(A - |z[2]|)$, $\cdots$, $\frac{\sqrt{2}}{2}(A - |z[n]|)$. The coefficients of the high-pass and low-pass of Haar wavelet decomposition at scale level 1 are shown in Fig. 7.8.

Therefore, from Eq. (7.20), we know that the largest component in the high pass at the scale level 1 is the worst-case ENOB. The largest component defines the dynamic range (DR) of the ADC under test as shown in Eq. (7.23).

$$DR \equiv -20 \log_{10}\left[\frac{\alpha}{2}\Delta\right] = -20 \log_{10}\left[\frac{\alpha}{2}\left(\frac{1}{2^{\widehat{B}-1}}\right)\right] = -20 \log_{10}\left[\frac{\alpha}{2^{\widehat{B}}}\right] \ [dB]. \quad (7.23)$$

The number of bits $\widehat{B}$ is directly calculated from the dynamic range DR as shown in Eq. (7.24).

$$\widehat{B} = \frac{DR + 20 \log_{10}(\alpha)}{20 \log_{10}(2)} \ \ [\text{bit}]. \quad (7.24)$$

7.3.6   Measuring the Instantaneous DNL

From the instantaneous magnitude, $|z[n]|$, we can get DNL as shown in Eq. (7.25).

$$DNL(n) = \frac{\max\left\{\left\|\,|z[n]| - |z[n+1]|\,\right\|\right\}}{\Delta_{ideal}} - 1 \qquad (7.25)$$

In using the Haar wavelet transform at scale level 2 of the high pass to instantaneous magnitude, $|z[n]|$, we get Eq. (7.26).

$$
\begin{bmatrix}
1 & & & & & \\
 & 1 & & & & \\
 & & \ddots & & & \\
 & & \frac{\sqrt{2}}{2} & \frac{\sqrt{2}}{2} & & \\
 & & \frac{\sqrt{2}}{2} & -\frac{\sqrt{2}}{2} & & \\
 & & & & \frac{\sqrt{2}}{2} & \frac{\sqrt{2}}{2} \\
 & & & & \frac{\sqrt{2}}{2} & -\frac{\sqrt{2}}{2} \\
 & & & & & & \ddots
\end{bmatrix}
\times
\begin{bmatrix}
\frac{\sqrt{2}}{2}(A + |z[1]|) \\
\frac{\sqrt{2}}{2}(A + |z[2]|) \\
\vdots \\
\frac{\sqrt{2}}{2}(A - |z[1]|) \\
\frac{\sqrt{2}}{2}(A - |z[2]|) \\
\frac{\sqrt{2}}{2}(A - |z[3]|) \\
\frac{\sqrt{2}}{2}(A - |z[4]|) \\
\vdots
\end{bmatrix}
$$

$$
=
\begin{bmatrix}
\frac{\sqrt{2}}{2}(A + |z[1]|) \\
\frac{\sqrt{2}}{2}(A + |z[2]|) \\
\vdots \\
\frac{2A - |z[1]| - |z[2]|}{2} \\
\frac{|z[2]| - |z[1]|}{2} \\
\frac{2A - |z[3]| - |z[4]|}{2} \\
\frac{|z[4]| - |z[3]|}{2} \\
\vdots
\end{bmatrix}
\Rightarrow
\begin{bmatrix}
\frac{\sqrt{2}}{2}(A + |z[1]|) \\
\frac{\sqrt{2}}{2}(A + |z[2]|) \\
\vdots \\
\frac{2A - |z[1]| - |z[2]|}{2} \\
\frac{2A - |z[3]| - |z[4]|}{2} \\
\vdots \\
\frac{|z[2]| - |z[1]|}{2} \\
\frac{|z[4]| - |z[3]|}{2} \\
\vdots
\end{bmatrix}
\qquad (7.26)
$$

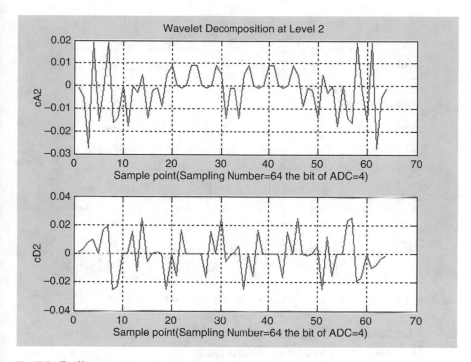

Fig. 7.9 Coefficients of Haar wavelet decomposition at scale level 2

The high-pass components at scale level 2 of Haar wavelet transform decomposition are $\frac{|z[2]| - |z[1]|}{2}$, $\frac{|z[4]| - |z[3]|}{2}$, $\cdots$. The coefficients of the high pass and low pass of Haar wavelet decomposition at scale level 2 are shown in Fig. 7.9.

Using Eq. (7.25) we can estimate DNL of an ADC under test by finding the maximum value of the high-pass filter output at scale level 2 using Eq. (7.27).

$$DNL(n) = \frac{2}{\Delta_{ideal}} \max \left\{ \left| \frac{\Delta[n]}{2} \right| \right\} - 1 \quad [\text{LSB}] \tag{7.27}$$

where $\Delta[n] = |z[n]| - |z[n + 1]|$..

Figure 7.10 shows the DNL measurement results using the Haar wavelet as compared to the conventional sinusoidal histogram method.

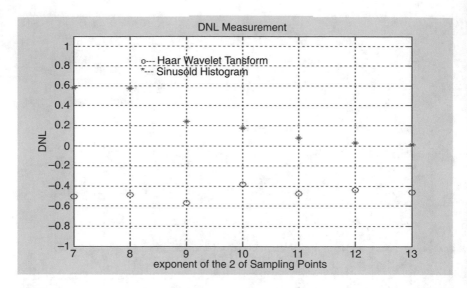

Fig. 7.10 DNL measurement using Haar wavelet and sinusoid histogram

7.4 Measuring ENOB and DNL of ADC Using Daubechies-4 Wavelet Transform

Comparing with the FFT test method, using Haar wavelet transform can shorten test times and produce superior test quality. The Haar wavelet is the simplest transform in all the different types of wavelets. In order to compare for better results, we can extend the test method presented in Sect. 7.4.1 to the one using Daubechies-4 wavelet transform. We explore the measuring of ENOB and DNL of ADCs using Daubechies-4 wavelet transform and the differences between the two kinds of wavelets in measuring ENOB and DNL of ADCs.

7.4.1 Daubechies-4 Wavelet Transform Using MATLAB

In MATLAB, wavelets are performed via filters. The illustration of the wavelet transform decomposition technique is as shown in Fig. 7.11 which we used for both the Haar and the Daubechies-4 decomposition processes where:

$$\boxed{\text{S}}$$

is convolved with the low-pass or high-pass filters.

Fig. 7.11 Illustration of the wavelet transform decomposition algorithm

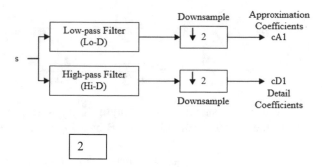

keeps the even indexed elements (we call this operation downsampling).

The length of each filter is equal to 2 m. If $n = \text{length}(S)$, the coefficients cA1 and cD1 are of length $floor\left(\frac{n-1}{2}\right) + m$.

For a finite signal, the wavelet transforms require unknown samples. This is the boundary problem associated with the wavelet transform. This boundary problem is solved by extending the given finite signal to an infinite signal. There exist a number of different ways to extend a finite signal, resulting in different solutions to the boundary problem. We use the periodic extension mode, i.e., the given finite signal is periodically extended to an infinite signal.

Example 7.1

Let the given signal be as shown in Eq. (7.28).

$$S = [S_1, S_2, S_3, \ldots, S_N] \tag{7.28}$$

By the periodic extension mode, the new infinite signal will be as shown in Eq. (7.29).

$$S' = [\ldots, S_1, S_2, S_3, \ldots, S_N, S_1, S_2, S_3, \ldots, S_N, S_1, S_2, S_3, \ldots, S_N, \ldots] \tag{7.29}$$

If we denote the low-pass filter and the high-pass filter of Daubechies-4 transform by h and g, respectively. We know that the length of low-pass filter h and high-pass filter g is 4, i.e.,

$$h = [h_1, h_2, h_3, h_4] \tag{7.30}$$

$$g = [g_1, g_2, g_3, g_4] \tag{7.31}$$

The values of the low-pass filter h and high-pass filter g are given as shown in Eqs. (7.31) and (7.32), respectively.

$$h_1 = \frac{1 + \sqrt{3}}{4\sqrt{2}}, \quad h_2 = \frac{3 + \sqrt{3}}{4\sqrt{2}}, \quad h_3 = \frac{3 - \sqrt{3}}{4\sqrt{2}}, \quad h_4 = \frac{1 - \sqrt{3}}{4\sqrt{2}}, \tag{7.32}$$

$$g_1 = \frac{\sqrt{3} - 1}{4\sqrt{2}}, \quad g_2 = \frac{3 - \sqrt{3}}{4\sqrt{2}}, \quad g_3 = \frac{3 + \sqrt{3}}{4\sqrt{2}}, \quad g_4 = \frac{\sqrt{3} + 1}{4\sqrt{2}}. \tag{7.33}$$

For a given signal x with length N (assume it is even), we perform the Daubechies-4 transform in MATLAB by using the periodic extension, i.e.,

$$[a, d] = dwt(x, 'db2', 'mode', 'ppd') \tag{7.34}$$

The low-pass and high-pass a and d, both with length $\frac{N}{2} + 1$, are given as shown in Eqs. (7.35) and (7.36), respectively.

$$a = \begin{bmatrix} h_3 & h_4 & 0 & 0 & 0 & 0 & \cdots & 0 & 0 & h_1 & h_2 \\ h_1 & h_2 & h_3 & h_4 & 0 & 0 & \cdots & 0 & 0 & 0 & 0 \\ 0 & 0 & h_1 & h_2 & h_3 & h_4 & \cdots & 0 & 0 & 0 & 0 \\ \vdots & \vdots & \vdots & \vdots & \vdots & \vdots & \cdots & \vdots & \vdots & \vdots & \vdots \\ 0 & 0 & 0 & 0 & 0 & 0 & \cdots & h_1 & h_2 & h_3 & h_4 \\ h_3 & h_4 & 0 & 0 & 0 & 0 & \cdots & 0 & 0 & h_1 & h_2 \end{bmatrix} * x \tag{7.35}$$

$$d = \begin{bmatrix} g_3 & g_4 & 0 & 0 & 0 & 0 & \cdots & 0 & 0 & g_1 & g_2 \\ g_1 & g_2 & g_3 & g_4 & 0 & 0 & \cdots & 0 & 0 & 0 & 0 \\ 0 & 0 & g_1 & g_2 & g_3 & g_4 & \cdots & 0 & 0 & 0 & 0 \\ \vdots & \vdots & \vdots & \vdots & \vdots & \vdots & \cdots & \vdots & \vdots & \vdots & \vdots \\ 0 & 0 & 0 & 0 & 0 & 0 & \cdots & g_1 & g_2 & g_3 & g_4 \\ g_3 & g_4 & 0 & 0 & 0 & 0 & \cdots & 0 & 0 & g_1 & g_2 \end{bmatrix} * x \tag{7.36}$$

7.4.2 Measuring the ADC Instantaneous ENOB and DNL Using Daubechies-4 Wavelets

In the measuring of the instantaneous ENOB and DNL, if we let f be the signal as shown in Eq. (7.37).

$$f = [|z[1]|, A, A, A, |z[2]|, A, A, A, \cdots, |z[n]|, A, A, A] \tag{7.37}$$

with length 4 N, where:

$|z[n]|$ = the instantaneous magnitude signal
A = the amplitude.

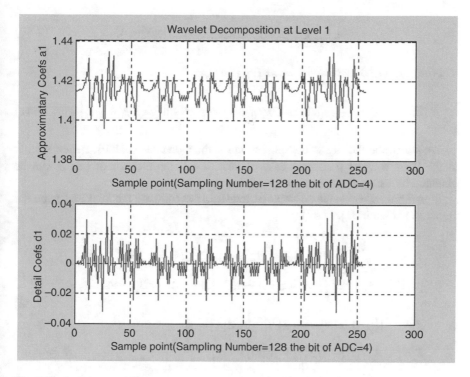

Fig. 7.12 Low-pass and high-pass coefficients of Daubechies-4 wavelet decomposition at scale level 1

We perform one scale Daubechies-4 wavelet decomposition with periodic extension mode on the signal f, i.e., using Eq. (7.35). The application is as shown in Eq. (7.38).

$$[a1, d1] = dwt(f, 'db2', 'mode', 'ppd') \tag{7.38}$$

The coefficients of the high-pass and low-pass of Daubechies-4 wavelet decomposition at scale level 1 are shown in Fig. 7.12. From Eqs. (7.33) and (7.36), the high-pass $d1$ is given as shown in Eq. (7.39).

$$d1 = \left[\frac{3+\sqrt{3}}{4\sqrt{2}}(A-|z[1]|), \frac{1-\sqrt{3}}{4\sqrt{2}}(A-|z[1]|), \frac{3+\sqrt{3}}{4\sqrt{2}}(A-|z[2]|), \frac{1-\sqrt{3}}{4\sqrt{2}}(A-|z[2]|), \cdots,\right.$$
$$\left.\frac{3+\sqrt{3}}{4\sqrt{2}}(A-|z[N]|), \frac{1-\sqrt{3}}{4\sqrt{2}}(A-|z[N]|), \frac{3+\sqrt{3}}{4\sqrt{2}}(A-|z[1]|)\right]^T,$$

$$\tag{7.39}$$

We perform downsampling on the high-pass coefficients $d1$ and denote the resulting vector by s as shown in Eq. (7.40).

$$s = \frac{\sqrt{3}-1}{4\sqrt{2}}[|z[1]|-A, |z[2]|-A, \cdots, |z[N]|-A] \tag{7.40}$$

From Eq. (7.20) shown below as Eq. (7.41),

$$|Z[N] - A| = \left|\frac{\Delta}{2}\{\varepsilon[N]\cos(2\pi f_0 + \varphi) + \varepsilon[M]\sin(2\pi f_0 N + \varphi)\}\right| \tag{7.41}$$

We know that the largest component in s is the worst-case ENOB. Hence, similar to the Haar wavelet transform case of Sect. 7.4.1, the number of bits $\widehat{B}$ can be obtained by using the largest component of s.

In using the Daubechies-4 wavelet transform to estimate the DNL, let f_1 be the signal as shown in Eq. (7.42).

$$
\begin{aligned}
f_1 = \frac{\sqrt{3}-1}{4\sqrt{2}}\Big[&\ |z[1]|-A, |z[2]|-A, |z[2]|-A, |z[2]|-A, \\
&\ |z[2]|-A, |z[3]|-A, |z[3]|-A, |z[3]|-A, \\
&\ \cdots, \\
&\ Z[N-1]-A, Z[N]-A, Z[N]-A, Z[N]-A, \\
&\ Z[N]-A, Z[1]-A, Z[1]-A, Z[1]-A \Big]
\end{aligned} \tag{7.42}
$$

We repeat the Daubechies-4 wavelet transform with periodic extension mode on the signal f_1, as shown in Eq. (7.43).

$$[a2, d2] = dwt(\ f_1, 'db2', 'mode', 'ppd') \tag{7.43}$$

Figure 7.13 shows the low-pass and high-pass of Daubechies-4 wavelet transform on signal f_1 at the level 2. By downsampling the high-pass d_2, we have

$$s_1 = \left(\frac{\sqrt{3}-1}{4\sqrt{2}}\right)^2 [|z[1]| - |z[2]|, |z[2]| - |z[3]|, \cdots, |z[N-1]| - |z[N]|, |z[N]| - |z[1]|] \tag{7.44}$$

The largest component of s_1 gives the worst DNL. See Eq. (7.27) for the formula to compute the DNL. The DNL measurement using Daubechies-4 is as shown in Fig. 7.14 as compared to the conventional sinusoidal histogram method.

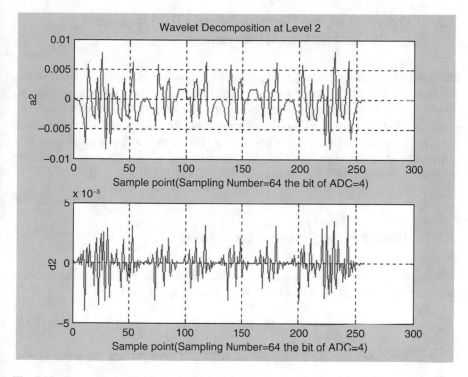

Fig. 7.13 Low-pass and high-pass coefficients of Daubechies-4 wavelet decomposition at scale level 2

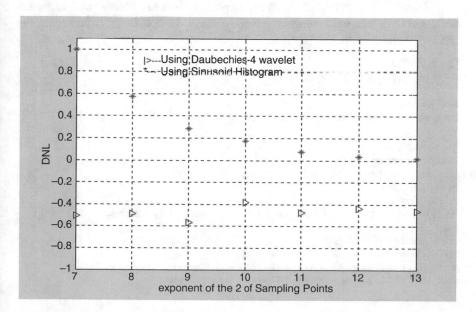

Fig. 7.14 DNL measurement using Daubechies-4 wavelet and sinusoid histogram

7.5   Extensions of the Wavelet-Based ADC Dynamic Test Algorithms and Key Observations

In this section, we explore and discuss certain implementations that we observed that might have the potential to give errors. It is left to the reader to adapt as you see fit. The key areas that are discussed are in the implementation of the Hilbert transform, the range of the $|z[n]|$, and the implementation using the Haar wavelets. The dynamic testing was extended to varying bit sizes of ADCs as well and compared with FFT measurements. The algorithms of the dynamic testing of ADCs using discrete wavelet transforms (DWTs) are illustrated and discussed in this section.

7.5.1   Hilbert Transform Implementation

It is important to examine what happens when certain ready-made software command functions are used. As an example, assuming in Eq. (7.45)

$$x[n] = A \cos (2\pi ft) \tag{7.45}$$

Its Hilbert transform imaginary part should be in theory as shown in Eq. (7.46).

$$x[m] = A \sin (2\pi ft) \tag{7.46}$$

However, using "Hilbert" function command in some commercial softwares could result to incorrect imaginary part. We use the following code to test the "Hilbert" command in MATLAB.

Example 7.2
• **MATLAB Program 7.1.**

```
% HILBERT COMMAND BASED ON READY-MADE COMMERCIAL SOFTWARES

A=1.0; %A amplitude
f=100; %frequency
ts=1024; %number of sampling points
T=1/f; % period
for i=1:ts
 tr(i)=(i-1)*T/(ts-1); % coordinates of sampling points
end
xn=A*cos(2*pi*f*tr);
xm=hilbert(xn);
plot(t(1:1024),real(xm(1:1024))), hold on, grid
plot(t(1:1024),imag(xm(1:1024)),':'), hold off
```

Example 7.3
• **MATLAB Program 7.2.**

```
%Corrected Hilbert transform code
clear
FS=2; % full scale
bit=8; % value of bits
 h=bit;
delta=FS/(2^bit-1); % sept size
A=1.0; %A amplitude
f=100; %frequency
ts=128; %number of sampling points
T=1/f; % period
for i=1:ts;
 tr(i)=(i-1)*T/(ts-1); % coordinates of sampling points
end
t=0:T/ts:T;
xn=A*cos(2*pi*f*tr);
xm=A*sin(2*pi*f*tr);
plot(xn);
hold on
plot(xm,'r.'),grid;
```

Figure 7.15 shows the plotted graph of the real and imaginary parts of the Hilbert transform as a result of the MATLAB Program 7.1. Figure 7.16 shows the wrongly computed ENOB using the MATLAB Program 7.1. Clearly, the imaginary part of Hilbert transform shown in Fig. 7.15 is not really the imaginary we want. Figure 7.16 shows the wrong result of ENOB by using the command "Hilbert" to get the imaginary part. Therefore, we create the MATLAB program of Example 7.3 for the imaginary part that can be directly used for the correct real and imaginary part of Hilbert transform as shown in Fig. 7.17 with a measured ENOB shown in Fig. 7.18 close to the ideal value.

7.5.2 The Range of $|z[n]|$

In Eq. (7.19), $|z[n]|$ is considered to be in the range as shown in Eq. (7.47).

$$A - \frac{\sqrt{2}}{2}\Delta \leq |z[n]| \leq A + \frac{\sqrt{2}}{2}\Delta. \tag{7.47}$$

However, in the algorithm, we simulated the sine input signal and quantization signal. The quantization error is very small, and the bigger the number of ADC bits, the less the quantization error. The $||z[n]| - A|$ is much less than $\left|\frac{\sqrt{2}}{2}\Delta\right|$. If we still use this range to estimate the ENOB, we could get an error, as Fig. 7.19 shows. Clearly, we can see the difference between the ENOB and ideal (actual) ADC bits. However,

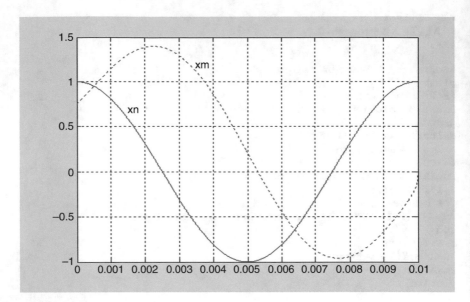

Fig. 7.15 Plot of the Hilbert transform

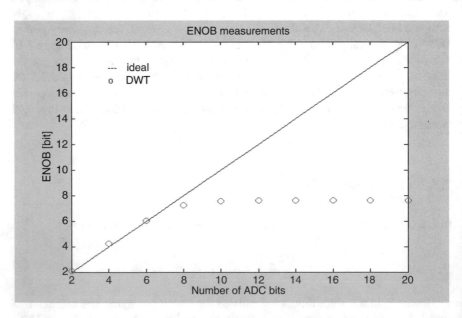

Fig. 7.16 Example of wrong ENOB by using the command "Hilbert" as shown in MATLAB Program 7.1

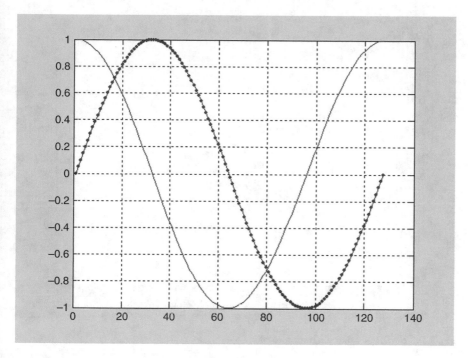

Fig. 7.17 Corrected Hilbert transform (real, blue; imaginary, red)

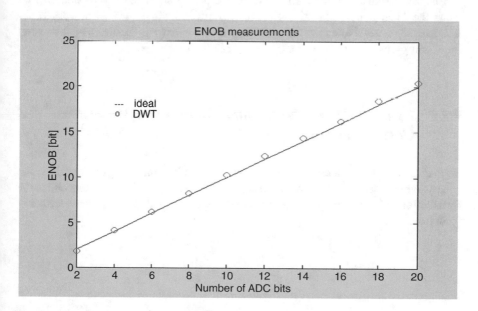

Fig. 7.18 Example of the measured ENOB by using the corrected "Hilbert" as shown in MATLAB Program 7.2

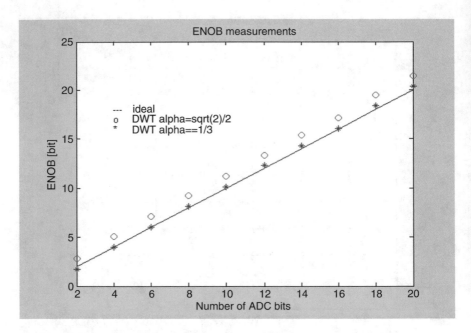

Fig. 7.19 ENOB of using different α (alpha)

when the range of the $\|z[n]\| - A\|$ is changed to $\left|\frac{1}{3}\Delta\right|$ in the algorithm, we get more satisfactory results. Figures 7.20, 7.21, 7.22 and 7.23 show the ENOB obtained by using the Haar and Daubechies-4 wavelet transform for $\alpha = 1/3$ and for $\alpha = \sqrt{2}/2$ respectively. Note that, unlike Haar wavelet transform, Daubechies-4 wavelet transform gives better results for $\alpha = \sqrt{2}/2$ than for $\alpha = 1/3$.

7.5.3   Using Different Formulations of the Haar Wavelet Coefficients in the Algorithm

The ADC testing measurements may change depending on how you formulate the Haar wavelet coefficients in the algorithm. As an example, you may use either the formulations of Equations (7.48) and (7.49) respectively as the Haar wavelet transform function.

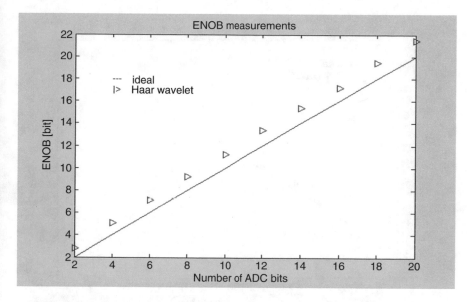

Fig. 7.20   ENOB measurements using Haar wavelet for $\alpha = 1/3$

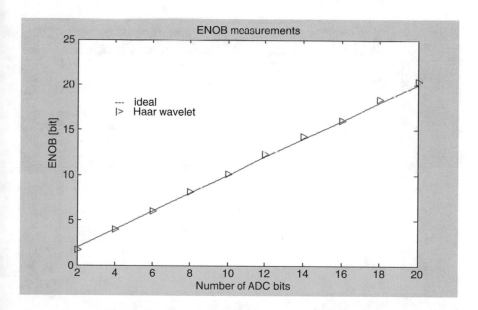

Fig. 7.21   ENOB measurements using Haar wavelet for $\alpha = \sqrt{2}/2$

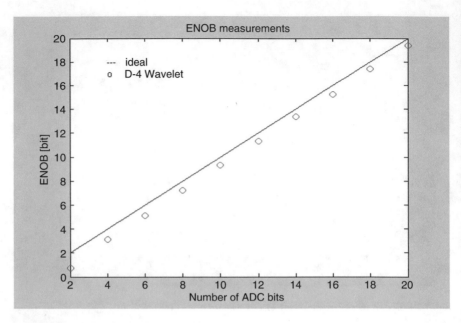

Fig. 7.22 Measurement of ENOB using Daubechies-4 wavelet for $\alpha = 1/3$

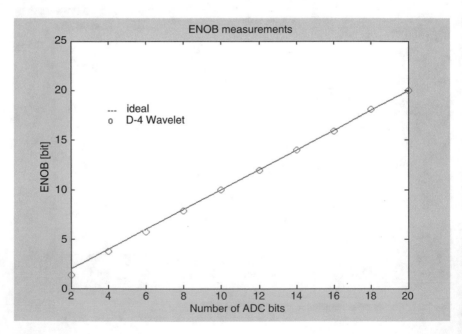

Fig. 7.23 ENOB measurement using Daubechies-4 wavelet for $\alpha = \sqrt{2}/2$

$$
\text{Formulation} = 1
\begin{bmatrix}
\dfrac{1}{2} & \dfrac{1}{2} & & & \\
\dfrac{1}{2} & -\dfrac{1}{2} & & & \\
& & \dfrac{1}{2} & \dfrac{1}{2} & \\
& & \dfrac{1}{2} & -\dfrac{1}{2} & \\
& & & & \ddots \\
& & & & & \ddots
\end{bmatrix}
\tag{7.48}
$$

$$
\text{Formulation 2}
\begin{bmatrix}
\dfrac{\sqrt{2}}{2} & \dfrac{\sqrt{2}}{2} & & & \\
\dfrac{\sqrt{2}}{2} & -\dfrac{\sqrt{2}}{2} & & & \\
& & \dfrac{\sqrt{2}}{2} & \dfrac{\sqrt{2}}{2} & \\
& & \dfrac{\sqrt{2}}{2} & -\dfrac{\sqrt{2}}{2} & \\
& & & & \ddots \\
& & & & & \ddots
\end{bmatrix}
\tag{7.49}
$$

We modified the α of the ENOB and DNL in Eqs. (7.24) and (7.27), respectively, in the calculations.

7.6   Comparative Analysis of the Measurements for ADC

In this section, we discuss simulations done using three measurement methods on the same ADCs, and then we compare their results. The algorithms are implemented in MATLAB. In both FFT and DWT methods, we used the same input signals. The input signal is a computer simulated sine wave signal with amplitude 1.0 and frequency 100 Hz.

7.6.1   ENOB Measurements

We got the ENOB values by simulating a fault-free (ideal) ADC without any extraneous noise added. The number of bits of the ADC was varied from 4 to 20. The results are shown in Table 7.1 and Fig. 7.24. The "o" indicates the ENOB estimates obtained using the Haar wavelet, and "*" indicates the estimates obtained using the standard FFT test method.

Table 7.1 ENOB measurement

Ideal bits	ENOB using DWT	Error of ENOB (DWT) and ideal	Percent of error	ENOB using FFT	Error of ENOB (FFT) and ideal	Percent of error
2	1.7237	0.2763	13.815	1.3917	0.6083	30.415
4	4.0104	−0.0104	0.26	3.8206	0.1794	4.485
6	6.0586	−0.0586	0.977	5.9523	0.0477	0.795
8	8.1541	−0.1541	1.93	8.0360	−0.036	0.45
10	10.1505	−0.1505	1.505	9.9550	0.045	0.45
12	12.2811	−0.2811	2.343	11.8627	0.1373	1.144
14	14.2664	−0.2664	1.903	14.1834	−0.1834	1.31
16	16.0770	−0.0770	0.4813	15.9684	0.0316	0.1975
18	18.3637	−0.3637	2.021	17.9962	0.00380	0.0211
20	20.3399	−0.3399	1.7	19.9011	0.0989	0.4945

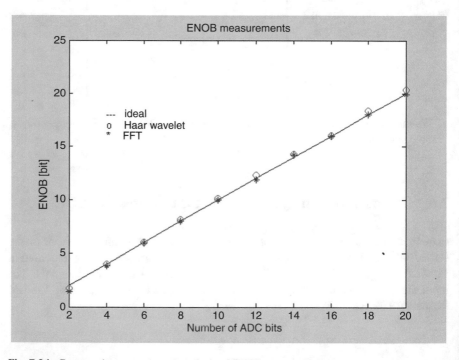

Fig. 7.24 Comparative measurement analysis of ENOB

In Table 7.2, the ENOB measurements include Daubechies-4 wavelets in the comparative analysis with Haar wavelets, FFT, and the ideal ADC. The plot of Table 7.2 is as shown in Fig. 7.25. In the ENOB testing, both results of wavelet-based and the FFT-based testing show excellent agreement with the theoretical ideal straight line. We know that the multiplication complexity of FFT is N*log (N).

Table 7.2 ENOB measurement

Ideal bits	ENOB using Haar wavelet	Error	Percent of error	ENOB using D-4 wavelet	Error	Percent of error	ENOB using FFT	Error	Percent of error
2	1.7237	0.2763	13.815	1.8086	0.1914	9.57	1.3917	0.6083	30.415
4	4.0104	−0.0104	0.26	4.0954	−0.0954	2.385	3.8206	0.1794	4.485
6	6.0586	−0.0586	0.977	6.1435	−0.1435	2.3917	5.9523	0.0477	0.795
8	8.1541	−0.1541	1.93	8.2391	−0.2391	2.9887	8.0360	−0.036	0.45
10	10.1505	−0.1505	1.505	10.2355	−0.2355	2.355	9.9550	0.045	0.45
12	12.2811	−0.2811	2.343	12.3661	−0.3661	3.0508	11.8627	0.1373	1.144
14	14.2664	−0.2664	1.903	14.3514	−0.3514	2.51	14.1834	−0.1834	1.31
16	16.0770	−0.0770	0.4813	16.1620	−0.1620	1.0125	15.9684	0.0316	0.1975
18	18.3637	−0.3637	2.021	18.4487	−0.4487	2.4928	17.9962	0.00380	0.0211
20	20.3399	−0.3399	1.7	20.4249	−0.4249	2.1245	19.9011	0.0989	0.4945

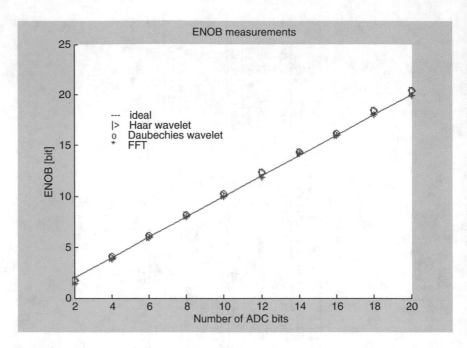

Fig. 7.25 Comparative measurement analysis of ENOB with ideal ADCs, Haar wavelet, Daubechies-4 wavelet, and FFT

However, from the wavelet filters, it is easy to see that the multiplication complexity is 2*N and 4*N for Haar and Daubechies-4 wavelet transforms, respectively. Hence the wavelet transform gives almost the same results as the FFT but uses much shorter time.

7.6.2 DNL Measurements

We can measure the DNL values using the Haar and Daubechies-4 wavelet transforms. The number of bits of the ADC is varied from 4 to 20. The results are shown in Table 7.3. Figure 7.14 shows the results of DNL using Daubechies-4 wavelet. Compare it with the result of using Haar wavelet, we see that both Daubechies-4 and Haar wavelets give the same result for DNL.

From Figs. 7.14 and 7.26, we see that the estimated DNL by using Haar and Daubechies-4 wavelets is almost unchanging. This makes it possible to reduce the number of sampling points and still get the DNL.

We can also measure the DNL values using the sinusoid histogram. In this case, the number of bits of the ADC is 6, and the sampling number is varied from 2^7 to 2^{13}. The results are shown in Table 7.4.

Table 7.3 DNL measurement using wavelet transform

Bits of ADC	2	4	6	8	10
DNL(LSB)	−0.6114	−0.5180	−0.5040	−0.4000	−0.4714
Bits of ADC	12	14	16	18	20
DNL(LSB)	−0.4821	−0.5305	−0.5305	−0.5274	−0.4972

Table 7.4 DNL measurement using sinusoid histogram

Sampling no.	2^7	2^8	2^9	2^{10}	2^{11}	2^{12}	2^{13}
DNL(LSB)	0.5787	0.5734	0.2444	0.1772	0.0796	0.0312	0.0117

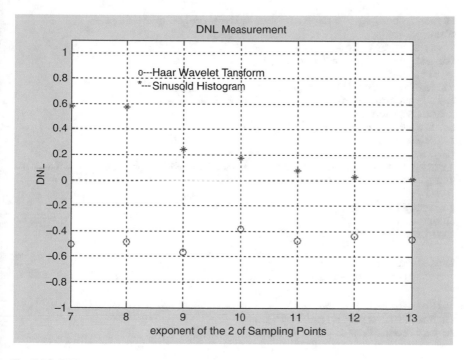

Fig. 7.26 DNL measurement

The plot of the results is shown in Fig. 7.26. "o" indicates the DNL estimates obtained using the Haar wavelet transform, and "*" indicates the estimates obtained using the sinusoid histogram. Figure 7.14 shows the results of DNL using Daubechies-4 wavelet. Compare it with the result of using Haar wavelet in Fig. 7.26; we see that both Daubechies-4 and Haar wavelets give the same result for DNL. From Figs. 7.14 and 7.26, we see that the estimated DNL by using Haar and Daubechies-4 wavelets is almost unchanging. This makes it possible to reduce the number of sampling points and still get the DNL.

7.7　The MATLAB Program for the ENOB and DNL Measurements

The algorithms are implemented in MATLAB. In both FFT and DWT methods, the same input signal was used. The input signal is sine wave signal simulated by computer. Its amplitude is 1.0 and frequency is 100 Hz. The MATLAB program is shown as Example 7.4—MATLAB Program 7.3.

Example 7.4
* **MATLAB Program 7.3.**

```
%Dynamic test ADC's ENOB and DNL using wavelet transform
clear
FS=2; % full scale
for bit=2:2:20; % value of bits
 h(bit)=bit
delta=FS/(2^bit-1); % sept size
A=1.0; %A amplitude
f=100; %frequency
ts=128; %number of sampling points
T=1/f; % period
for i=1:ts
 tr(i)=(i-1)*T/(ts-1); % coordinates of sampling points
end
t=0:T/ts:T;
sn=A*cos(2*pi*f*t);
xn=A*cos(2*pi*f*tr);
xm=A*sin(2*pi*f*tr);

%xn(8)=xn(8)+delta*0.25;
%xn(100)=xn(100)+delta*0.2;

delta2=delta/2;
xn_hat=delta2*round(xn/delta2);
xm_hat=round(xm/delta2)*delta2;

%% Do Hilbert transform

%plot the graphs of the sinusoid input signal and real and imaginary
parts of Hilbert transform
figure(1)
subplot(311) %plot input signal
plot(t,sn,'.'),grid
%axis([0 ns*1e6/fs -1.2*A 1.2*A])
ylabel('Sine wave Signal sn')
title([''])

subplot(312)
plot(tr,xn,'.'),grid
%axis([1e6*[t(ns/2) T+t(ns/2)] -1.2*A 1.2*A])
```

```
%xlabel(['Sample Time, t (Sampling Number=',num2str(ts)])
ylabel('Quantizated signal xn')
title('')

subplot(313)
plot(tr,xn_hat,'.'),grid
%axis([1e6*[t(ns/2) T+t(ns/2)] -1.2*A 1.2*A])
xlabel(['Sample Time, t (Sampling Number=',num2str(ts),' the bit of
ADC=',num2str(bit),')'])
ylabel('Converted signal xn-hat')
title('')

%get |zn| function
zn=sqrt(xn_hat.^2+xm_hat.^2);
z=A-zn;

figure(2)
hold off
subplot(311) %plot real part signal of one cycle
plot(xn_hat,'.'),grid
%axis([1e6*[t(ns/2) T+t(ns/2)] -1.2*A 1.2*A])
%xlabel(['Sample Time, t (Sampling Number=',num2str(ts),')'])
ylabel('xn-hat')
title('Real part of complex signal')
subplot(312) %plot imaginary part signal of one cycle
plot(xm_hat,'.'),grid
%axis([1e6*[t(ns/2) T+t(ns/2)] -1.2*A 1.2*A])
%xlabel(['Sample Time, t (Sampling Number=',num2str(ts),')'])
ylabel('xm-hat')
title('Imaginary Part of complex signal')
subplot(313) %plot complex signal xn+jxm
plot(zn,'-'),grid
axis([0 length(tr) 0 1.2*A])
xlabel(['Sample point (Sampling Number=',num2str(ts),' the bit of
ADC=',num2str(bit),')'])
ylabel('|zn|')
title('Instantaneous Magnitude |zn|')

%get interleaved signal f
for i=1:4:length(tr)
 f(i)=zn(floor(i/4+1));
 f(i+1)=A;
 f(i+2)=-A;
 f(i+3)=A;
end %interleaved signal f

%perform a one-step decomposition of the f using the Haar wavelet
transform
[ca1,cd1]=dwt(f,'haar'); %generate the coefficients of the level
1 approximation(cA1) and detail(cD1)
```

```
figure(3) %plot the coefficients of the level 1 approximation(cA1) and
detail(cD1)
hold off
subplot(211)
plot(ca1,'-'),grid %plot the approximation coefficients at scale level 1
%axis([1e6*[t(ns/2) T+t(ns/2)] -1.2*A 1.2*A])
xlabel(['Sample point(Sampling Number=',num2str(ts),' the bit of
ADC=',num2str(bit),')'])
ylabel('Approximatary Coefs cA1')
title('Wavelet Decomposition at Level 1')
subplot(212)
plot(cd1,'-'),grid %plot the detail coefficients at scale level 1
%axis([1e6*[t(ns/2) T+t(ns/2)] -1.2*A 1.2*A])
xlabel(['Sample point(Sampling Number=',num2str(ts),' the bit of
ADC=',num2str(bit),')'])
ylabel('Detail Coefs cD1')
title('')

%compute instantaneous ENOB
w=max(abs(cd1))/sqrt(2);
%B_hat(bit)=-log10(w)/log10(2)-0.5

alpha=1/3;
B_hat(bit)=log10(alpha/w)/log10(2)

%for n=1:length(tr)
% dBMag(n)=20*log10(abs(A-zn(n))/2);
% d(n)=log2(n);
%end

%compute instantaneous DNL
[ca2,cd2]=dwt(cd1,'haar'); %generate the coefficients of the level
2 approximation(cA2) and detail(cD2)
v=max(abs(cd2));
DNL(bit)=((2/delta)*v)-1

figure(4)
hold off
subplot(211)
plot(ca2,'-'),grid %plot the approximation coefficients at scale level 2
%axis([d(1) d(tr) 0 -120])
xlabel(['Sample point(Sampling Number=',num2str(ts/2),' the bit of
ADC=',num2str(bit),')'])
ylabel('cA2')
title('Wavelet Decomposition at Level 2')
subplot(212)
plot(cd2,'-'),grid %plot the detail coefficients at scale level 2
%axis([d(1) d(tr) 0 -120])
xlabel(['Sample point(Sampling Number=',num2str(ts/2),' the bit of
ADC=',num2str(bit),')'])
ylabel('cD2')
```

```
title('')
end

%Dynamic test ADC's ENOB and DNL using Fourier transform

A=1.0; %A amplitude
FS=2*A; % full scale
for bit=2:2:20;
 h(bit)=bit; % value of bits
delta=FS/(2^bit-1); % sept size
f=100; % frequency
ns=128;
ts=f*ns; % frequency of sampling
T=1/f; %period
%T=0.8;
for i=1:ts
 tr(i)=(i-1)*T/(ts-1); % coordinates of sampling points
end
tt=0:1/ts:10*T;
%tt=linspace(0,1,ts);
sn=A*cos(2*pi*f*tt);
xn=A*cos(2*pi*f*tt);%+randn(1,length(tt))/10;

%xn(8)=xn(8)+delta*0.25;
%xn(100)=xn(100)+delta*0.2;

xn_hat=delta*round(xn/delta); %simulated ADC output signal

%perform Fourier analysis
N=1024;
y=fft(xn_hat,N);
p=y.*conj(y)/N;
ff=ts*(0:511)/N;

figure(5)
subplot(211)
plot(tt,xn_hat,'-'),grid
%axis([1e6*[t(ns/2) T+t(ns/2)] -1.2*A 1.2*A])
xlabel(['Sample Time, t (Sampling Number=',num2str(N),')'])
ylabel('xn-hat')
title('Converted sine signal')

subplot(212)
plot(ff,p(1:512))
axis([0 500 0 300])
xlabel('frequency')
ylabel('power density')
title('signal power density spectrum')

[pmax,ind]=max(p);
p1=0;
for ii=1:(N-1)/2
```

```
p1=p1+p(ii);
end
if ind>(N-1)/2
 SNR(bit)=10*log10(pmax/p1);
else
 SNR(bit)=10*log10(pmax/(p1-pmax))
end

ENOB(bit)=(SNR(bit)-1.76)/6.02
end

figure(6)
h=[2,4,6,8,10,12,14,16,18,20];
plot(h,h,'-',h,B_hat(h),'o',h,ENOB(h),'*');
xlabel('Number of ADC bits')
ylabel('ENOB [bit]')
title('ENOB measurements');
text(4,18,'--- ideal');
text(4,17,'o DWT');
text(4,16,'* FFT');
```

7.8   Differences Between Wavelet Transform Techniques and the Conventional Techniques in a Tabular Format

Wavelet transform technique
Because of the fact that the wavelet transform is a recursive algorithm, a VLSI implementation of it, in either hardware or firmware, can provide a considerable saving through multiple applications on the same circuit block
The compactness of the wavelet transformation algorithm can enable a VLSI wavelet transform to be built directly into ADC systems, as part of a built-in self-test (BIST) capability
The wavelet transform technique measures the energy of the nonlinearity, thus enabling the analysis of causes by using the magnitude of this energy measure

Conventional transform technique
Non-recursive algorithms

Non-compact

The conventional technique calculates the maximum value of the DNL. This makes it difficult to establish cause-and-effect information such as identifying the location of circuit deviations that are causing a problem

The wavelet technique makes it possible to accurately simulate the performance of each analog cell designed into a system, regardless of the number of causes of nonlinearity. This allows the validation of an ADC design before it is manufactured

The calculation of DNL using the conventional technique does not make it possible to guide the cell design to higher quality using the DNL. This is the largest nonlinearity in the overall ADC system. This makes design optimization very difficult

The wavelet transform technique has the ability to localize non-idealities, therefore, making it possible to separate amplitude levels of non-idealities from each other

Since averaging is used in most of these conventional techniques, it makes it almost impossible to separate non-idealities localized to particular amplitude levels from each other. This is particularly true in the case of FFT testing where the input-output characteristics of an ADC are tested by performing FFT on the output signal. This allows an output code corresponding to a given amplitude to be average with other different output codes

Using Haar wavelet, for example, if the number of points to be transformed is 32 (1024), the required number arithmetic right shifts and additions are 124 (4092) and 62 (2046), respectively

Using FFT, the required number of real multiplications and additions would be 124 (9212) and 276 (16884), respectively

The Haar wavelet transform requires fewer additions for N greater than 32, and it does not require any multiplications

FFT requires more additions for N greater than 32 and requires multiplications

Discontinues can easily be isolated

The frequency range of influence of an impulse is so large that it is hard to isolate discontinues in the signal using the FFT

Using the idea of the wavelet transform technique to simultaneously compute the instantaneous ENOB and the instantaneous DNL of an ADC is a novel idea. Yamaguchi and others see this method as effective in testing high-resolution ADCs

FFT is one of the conventional techniques. It has been around for years and does not add more value to testing ADCs as wavelet transform would according to Yamaguchi et al.

The wavelet transform has the ability to describe the energy density of a signal simultaneously in time and frequency. It is computationally very efficient for the intended application also

FFT does not have the ability to describe the energy density of a signal simultaneously in time and frequency. It is global in nature

The wavelet transformation has the capability of using different types of basis functions

The Fourier transform basis functions are complex sinusoids

The wavelet transform technique is suitable for evaluating nonlinearities in an ADC. It measures the energy of the nonlinearity, thereby enabling the analysis of causes of defects by using the magnitude of this energy measure. The technique also provides a means of validating an ADC design before it is manufactured

In contrast with the FFT technique, is difficult to establish cause-and-effect information, such as identifying the location of the circuit deviations that are causing a problem. The FFT technique also does not have the capability to guide the cell design to higher quality using the DNL

Summary
1. The fast Fourier Transform is a traditional tool for most of the analog-to-digital converter (ADC) dynamic tests, such as effective number of bits (ENOB), integral nonlinearity (INL), and differential nonlinearity (DNL).
2. It is common to use sine wave histogram tests for the determination of the nonlinearities of ADCs.
3. A sine wave that slightly overdrives the ADC is sampled many times.
4. In the dynamic testing of an ADC, it is more desirable to estimate a worst-case ENOB scenario than a mean ENOB.
5. Comparing with the FFT test method, using Haar wavelet transform can shorten test times and produce superior test quality.
6. The ADC testing measurements may change depending on how you formulate the Haar wavelet coefficients in the algorithm.
7. The wavelet transform gives almost the same results as the FFT but uses much shorter time.
8. We can measure the DNL values using the Haar and Daubechies-4 wavelet transforms.
9. The estimated DNL by using Haar and Daubechies-4 wavelets is almost unchanging. This makes it possible to reduce the number of sampling points and still get the DNL.

Review Questions
7.1. The fast Fourier transform is a traditional tool for most of the analog-to-digital converter (ADC) dynamic tests, such as:

(a) Effective number of bits (ENOB)
(b) Integral nonlinearity (INL)
(c) Differential nonlinearity (DNL)
(d) All of the above

7.2. The wavelet transform has lots of advantages when compared to the fast Fourier transform (FFT) when compared to the testing of ADCs.

(a) True
(b) False

7.3. In ENOB testing using the FFT, the mean square error due to quantization is estimated from:

(a) Frequency domain data
(b) Phase domain data
(c) Time domain data
(d) None of the above

7.4. Using the discrete wavelet transform (DWT) method for testing the ENOB and DNL of ADCs does not yield instantaneous ENOB values and DNL.

(a) True.
(b) False

7.5. Any loss of the ENOB in this process of testing and ADC should be as a result of the reflection due to the contribution of the quantization noise through SNR estimates.

(a) True
(b) False

7.6. It is not common to use sine wave histogram tests for the determination of the nonlinearities of ADCs.

(a) True
(b) False

7.7. The sinusoidal histogram method is a well-known conventional method for testing ADCs for DNL.

(a) True.
(b) False.

7.8. The Haar wavelet transform is not the simplest wavelet transform in all the different types of wavelets.

(a) True
(b) False

7.9. We cannot measure the DNL values of ADCs using the Haar and Daubechies-4 wavelet transforms.

(a) True
(b) False

7.10. For a finite signal, the wavelet transforms require unknown samples.

(a) True
(b) False

Answers: 7.1d, 7.2a, 7.3a, 7.4b, 7.5a, 7.6b, 7.7a, 7.8b, 7.9b, 7.10a.

Problems

7.1. How can you measure the ENOB of an ADC using FFT method?
7.2. Describe how SNR is computed through FFT.
7.3. How can you compute ENOB using SNR?
7.4. Define the differential nonlinearity (DNL) and integral nonlinearity (INL) for any particular values for the gain and the offset of an ADC.
7.5. How can sinusoidal histogram method be used in the measurement of an ADC?
7.6. How can you determine the instantaneous ENOB of an ADC using Haar wavelet transform?
7.7. What is the process of measuring the instantaneous DNL of an ADC?
7.8. Describe the method by which we can measure the instantaneous ENOB of an ADC using the Daubechies-4 wavelet transform.
7.9. Describe the method by which we can estimate DNL of an ADC using the Daubechies-4 wavelet transform.
7.10. Determining the ENOB and DNL measurements of an ADC using a DWT can be described by way of an implementable flowchart. Draw the flowchart.

References

Akujuobi, C.M. (1998). *Implementation of wavelet-based solutions to signal processing applications*, Proc. DSP World & ICSPAT, Toronto, Canada, September 13–16, 1998.
Akujuobi, C.M. (2000). *Quantitative performance image analysis using different wavelet filter taps*. Proc. ICSPAT, Dallas, Texas, October 12–16, 2000.
Akujuobi, C. M., & Awada, E. (2008). Wavelet-based ADC testing automation using LabVIEW. *International Review of Electrical Engineering Transaction, 3*, 922–930.
Akujuobi, C.M., & Hu, L. (2002). A novel parametric test methods for communication system mixed signal circuit using discrete wavelet transform, *IASTED International Conference on Communication, Internet, and Information Technology*, U.S. Virgin Island, pp. 132–135, November 18–20, 2002.
Akujuobi, C. M., Awada, E., Sadiku, M., & Warsame, A. (2007). Wavelet-based differential nonlinearity testing of mixed signal system ADCs. *IEEE SoutheastCon*, 76–81. March 2007.
Awada, E. A., & Akujuobi, C. M. (2018a). A wavelet-based ADC/DAC differential nonlinearity measurement analysis. *Journal of Engineering and Applied Sciences, 13*(16), 398–405. ISSN: 1816-949X © Medwell Journals, 2018.
Awada, E. A., & Akujuobi, C. M. (2018b). ADC testing algorithm for ENOB by wavelet transform using Labview measurements and MATLAB simulations. *Journal of Engineering and Applied Sciences, 13*(16), 6668–6679. ISSN: 1816-949X © Medwell Journals, 2018.
Burns, M., & Roberts, G. W. (2004). *An introduction to mixed-signal IC test and measurement* (p. 2004). Oxford University Press.
Marshall, R. O., & Akujuobi, C. M. (2002). *On the use of wavelet transform in testing for the DNL of ADCs* (pp. 25–28). IEEE Conference on Circuits and Systems.
Yamaguchi, T., & Soma, M. (1997). *Dynamic testing of ADCs using wavelet transform* (pp. 379–388). IEEE International Test Conference. Nov. 1997.

Chapter 8
Wavelet-Based Static Test of ADCs

> *"Reflect upon your blessings, of which every man has plenty,*
> *not on your past misfortunes, of which all men have some."*
> Charles Dickens.

8.1 Introduction

We define the static test of an ADC as the test when the input signal to the ADC under test is slowly varied. The output results are the same as for a constant signal being input. The differential nonlinearity (DNL) error, integral nonlinearity error (INL), gain error (GE), and offset error (OE) are the important static error parameters in the testing process of ADCs. Traditionally, the maximum values of the integral nonlinearity (INL) and differential nonlinearity (DNL) are used as measures of evaluating the nonlinearities of analog-to-digital converters (ADCs). Because the DNL represents the difference between the step size of each code and the step size corresponding to a least significant bit (LSB) change, it is capable of detecting a defect that exists locally in a specific code.

However, even if it is possible to statistically estimate the maximum value of the DNL, it is difficult to establish cause-and-effect information, such as identifying the location of circuit deviations that are causing a problem (Akujuobi & Hu, 2002). In addition, the DNL cannot be used to guide the cell design to higher quality. Yamaguchi (1997) proposed a new method, which is based on wavelet transform, for identifying the location of circuit deviations that are responsible for the ADC problems. This new technique measures the energy of nonlinearity, thus enabling the analysis of causes of defects by using the magnitude of this energy measure. This chapter explores and discusses this wavelet-based technique for ADC mixed signal testing with implementation examples. We are using the wavelet idea in the discussions in this chapter (Yamaguchi, 1997; Gandelli & Ragaini, 1996).

In the case of the gain error and offset error in analog-to-digital converters, they are relevant in applications requiring matched converters such as simultaneous sampling, interleaving, and I/Q signal processing, where there is a need to match relative gain and offset between the individual converters (Burns & Roberts, 2004).

© Springer Nature Switzerland AG 2022
C. M. Akujuobi, *Wavelets and Wavelet Transform Systems and Their Applications*,
https://doi.org/10.1007/978-3-030-87528-2_8

These two errors are compensated by providing appropriate biasing. They invariably do not affect the electronic system in which the converters are used.

In Chap. 2, we covered different types of wavelets and, in Chap. 4, we covered wavelet transforms. A wavelet is a mathematical function used to divide a given function or continuous-time signal into different scale components. It is possible to assign a frequency range to each scale component. Each scale component can then be studied with a resolution that matches its scale. A wavelet transform is the representation of a function by wavelets.

In this chapter, we cover the use of wavelets and the wavelet transforms, for identifying the location of circuit deviation that may cause problems for ADCs. This method measures the energy of the nonlinearity, thus enabling the analysis of causes of defects by using the magnitude of the energy measure. We explore and show in this chapter that wavelet transform plays a key role in obtaining better static testing results of ADCs. We also compare the results of two types of commonly used wavelets, the Haar wavelet and Daubechies wavelet in the static testing of ADCs.

8.2   Static Testing of ADCs by Transfer Curve

In this section, we cover the testing of ADCs using the idea of transfer curve.

8.2.1   ADC Transfer Curve

To accurately measure the nonlinearity of the amplitude level in the input-output characteristics of an ADC, it is necessary to use an amplitude-to-output-code representation (Akujuobi et al., 2007; Akujuobi & Awada, 2008; Awada & Akujuobi, 2018a, 2018b). This two-dimensional amplitude-to-output-code representation is known as the transfer curve. The transfer curve of an ADC is in the form of a staircase (Kollar & Blair, 2005; Plassche, 1994). Figure 8.1 shows the transfer curve of a 4-bit ADC. The dashed line (red) represents the simulated transfer curve of a particular converter, while the solid line (blue) represents its ideal transfer curve.

8.2.2   Testing for ADC Errors Using the DNL

The DNL of an ADC represents the difference between two analog signal upper bounds. That is, the DNL of an ADC is defined as shown in Eq. (8.1).

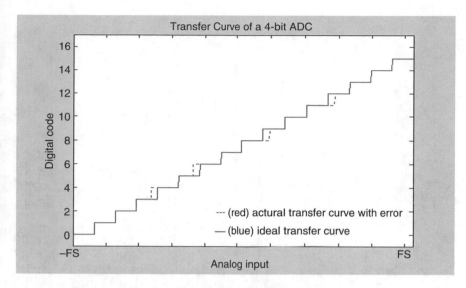

Fig. 8.1 Transfer curve of a 4-bit ADC with gain error

$$DNL = A_{input}(Q_{m+1}) - A_{input}(Q_m) - 1 [LSB] \qquad (8.1)$$

where:

Q_{m+1} and Q_m are adjacent digital output code
$A_{input}(Q_n)$ is the upper bound of the analog input signal corresponding to the digital output code Q_n.

If some DNLs are not zero, that means some differences between adjacent transition amplitudes are not the same. Then the DNL curve is not a straight line, which shows the existence of gain error. Figure 8.2 shows the DNL as computed from the transfer curve shown in Fig. 8.1. If the DNL is zero, that means all the differences between adjacent transition amplitudes are the same and correspond to a step size of 1 LSB, like an ideal transfer curve of an ADC. Hence, if offset error exists in an ADC whose transfer curve is shown in Fig. 8.3, then its DNL graph shows a straight line as shown in Fig. 8.4. Therefore, we cannot know whether or not the offset error exists in an ADC by looking at its DNL graph.

We have seen that the DNL could detect a defect, which exists locally at a specific code. By virtue of this capability, the DNL provides a rigorous test for nonlinearity. However, it is possible to measure the effectiveness of a pipelined ADC digital error correction circuit by the drop in the DNL and INL number of bits. Neither the DNL nor INL gives the energy; it is difficult to perform design optimization of such an ADC system. In the next section, the idea of using the wavelet transform to test the ADC is discussed.

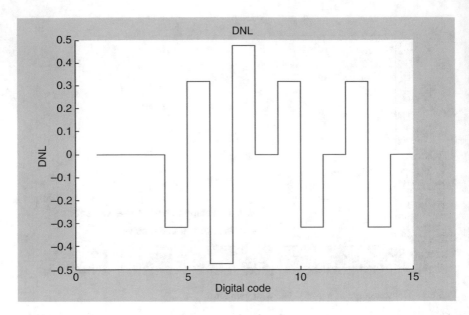

Fig. 8.2 DNL of a 4-bit ADC with gain error

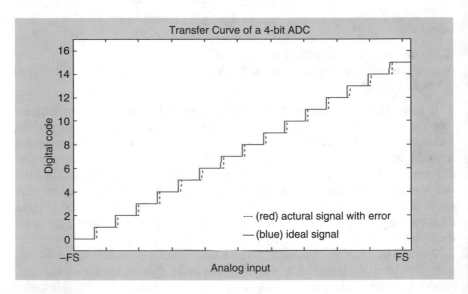

Fig. 8.3 Transfer curve of a 4-bit ADC with offset error

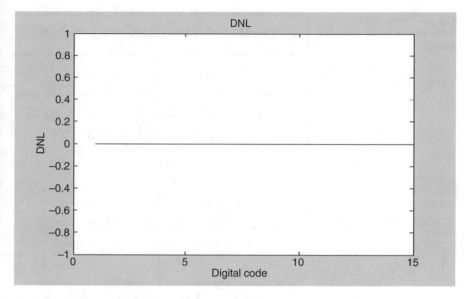

Fig. 8.4 DNL of a 4-bit ADC with offset error

8.3   Wavelet Transform-Based Static Testing of an ADC

The wavelet transform is used to analyze and reconstruct the original signal and utilize this transform to evaluate ADC transfer characteristics, that is, more specifically to evaluate the amplitude-to-output-code representation of an ADC. The gain error and offset error are found in the multiresolution figures. The procedure is illustrated in the flowchart shown in Fig. 8.5.

8.3.1   The ADC Static Testing Method and Choice of Wavelet Transforms

In the wavelet transforms, it is possible to use a variety of different types of wavelets (Strang, 1989; Young, 1993). For example, two popular wavelets are the Haar wavelet, which is discontinuous and symmetric about zero, and the Daubechies wavelet, which is a continuous and asymmetric about zero. In order to find an optimal wavelet transform for evaluating the staircase transfer curve of an ideal ADC, the ADC transfer curve is first decomposed into its wavelet components. Figures 8.6 and 8.7 show the results of the decomposition of an ideal 2-bit ADC into its Daubechies and Haar wavelet components. In Fig. 8.6, the energy of signals up to the scale level 10 can be clearly seen, but in Fig. 8.7, there is energy only up to the scale level 2. In addition, it can be seen from Fig. 8.7 that the transfer curve of an ideal ADC has finite energy and that its upper bound scale level, above which its

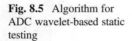

Fig. 8.5 Algorithm for ADC wavelet-based static testing

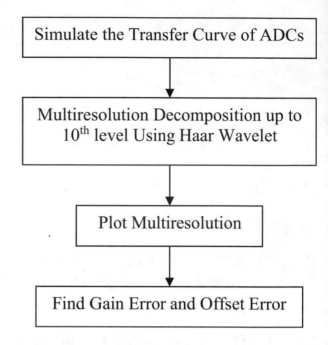

energy vanishes, corresponds to the number of ADC bits B. That is, the energy of an ideal B-bit ADC is limited to a scale level of B.

Using this idea, it is possible to measure a new noise-to-signal ratio (NSR), defined as the ratio of energy within scale level B (i.e., the signal scale band) in a B-bit ideal ADC to the energy in scale levels ($B + 1$) and above (i.e., the noise scale band). The NSR can be calculated using Eq. (8.2).

$$NSR = \frac{\sum_{a=B+!}^{MAX} \sum_{\tau} (DWT_x(\tau, a) h_{a,\tau}(t))^2}{\sum_{a=!}^{B} \sum_{\tau} (DWT_x(\tau, a) h_{a,\tau}(t))^2} \qquad (8.2)$$

The gain error and offset error can also be calculated. If the gain is larger than the ideal gain, the transfer curves are reduced along the amplitude axis toward the center. Similarly, if the gain is smaller than the ideal gain, the transfer curves are enlarged along the amplitude axis toward plus/minus full-scale extremities. Figure 8.8 shows the transfer curve of the 2-bit ADC with gain error. Offset error causes a parallel translation of the entire transfer curve. Figure 8.9 shows the transfer curve of a 2-bit ADC with offset error. The simulation of the ADC static testing measurement using MATLAB and DWT is discussed in Sect. 8.3.2.

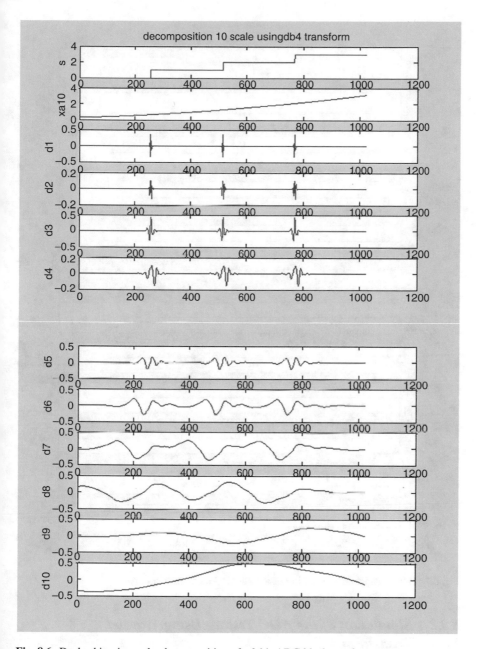

Fig. 8.6   Daubechies-4 wavelet decomposition of a 2-bit ADC ideal transfer curve

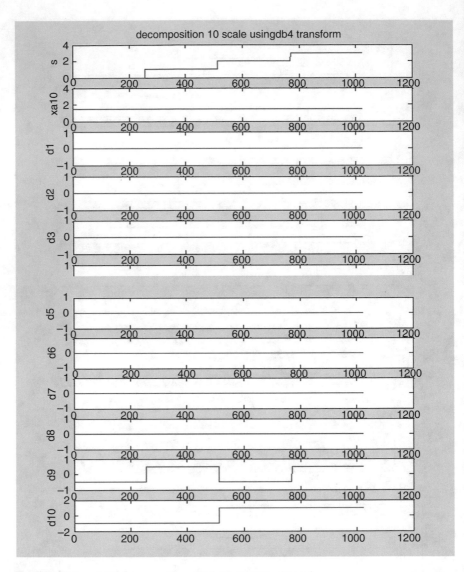

Fig. 8.7 Haar wavelet decomposition of a 2-bit ADC ideal transfer curve

8.3.2   Simulation of the ADC Testing Using Wavelet Transform

By the idea of finite energy in Haar wavelet decomposition, it is possible to find the gain error and offset error of ADCs as discussed in Sect. 8.3.1. When the ADC transfer curve with gain error is decomposed into its Haar wavelet components, the

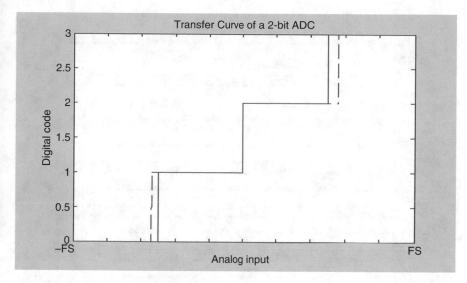

Fig. 8.8 Transfer curve of a 2-bit ADC with gain error

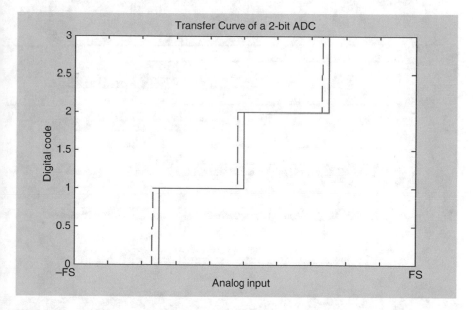

Fig. 8.9 Transfer curve of a 2-bit ADC with offset error

distortion energy due to the ADC gain error falls into the noise scale band. Figure 8.10 shows this result.

Similarly, the energy falls into the noise scale band for transition amplitudes of all codes, as shown in Fig. 8.11. The above result can also be seen in other cases with bigger bit numbers. Here, let us give another two examples with 4-bit and 8-bit

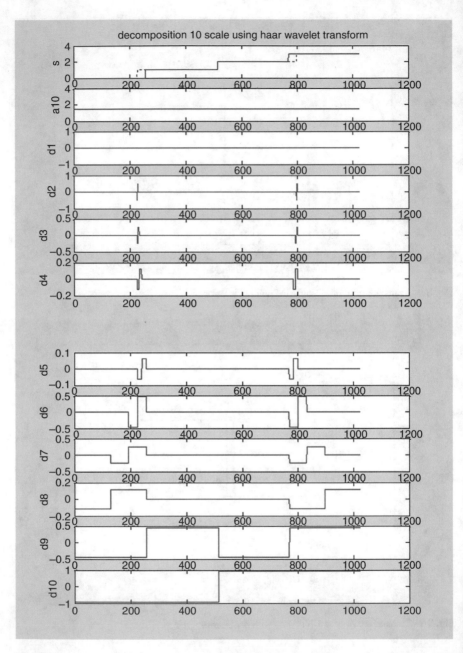

Fig. 8.10 Haar wavelet decomposition of a 2-bit ADC with gain error

ADCs. Figures 8.12 and 8.13 show the transfer curve of a 4-bit ADC with gain error and offset error, respectively. The Haar wavelet decompositions of the 4-bit ADC with gain error and offset error are shown in Figs. 8.14 and 8.15, respectively.

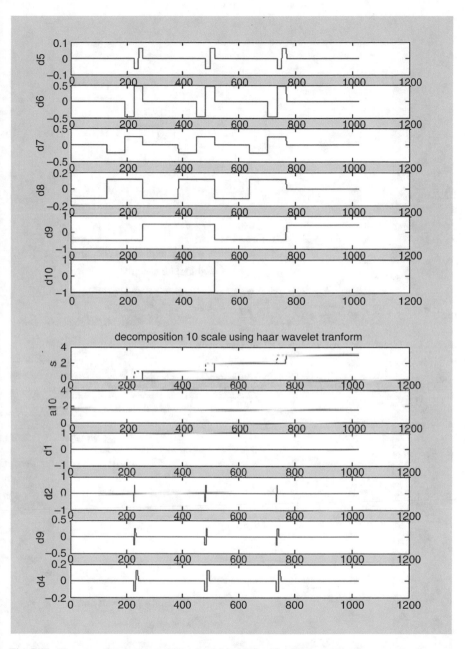

Fig. 8.11 Haar wavelet decomposition of a 2-bit ADC with offset error

Similarly, we can obtain the results for 8-bit ADCs. The results are shown in Figs. 8.16, 8.17, 8.18 and 8.19. From these figures, we get the same conclusion, i.e., the location of the ADC gain error can be found by locating the distortion of the

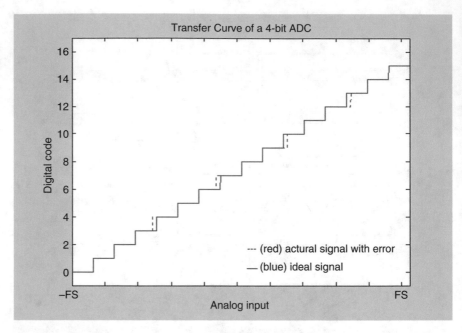

Fig. 8.12 Transfer curve of a 4-bit ADC with gain error

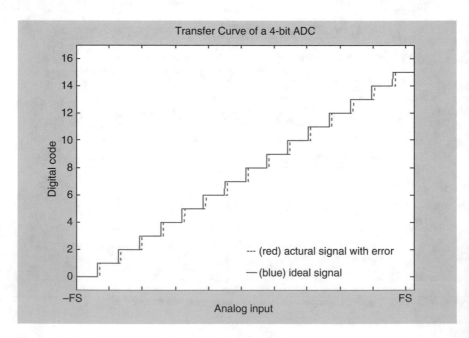

Fig. 8.13 Transfer curve of a 4-bit ADC with offset error

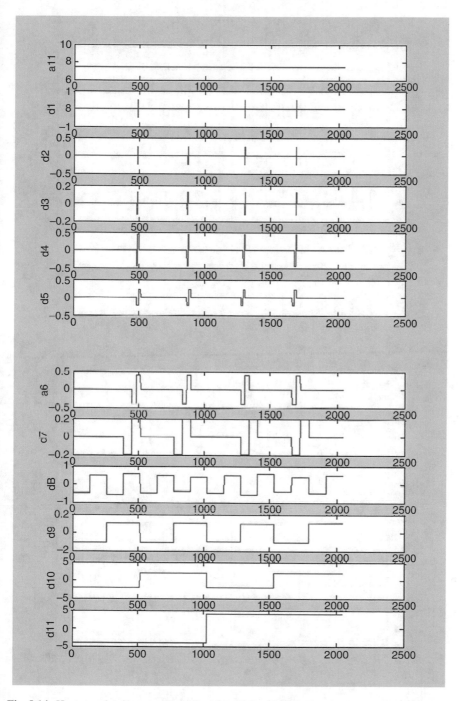

Fig. 8.14 Haar wavelet decomposition of a 4-bit ADC with gain error

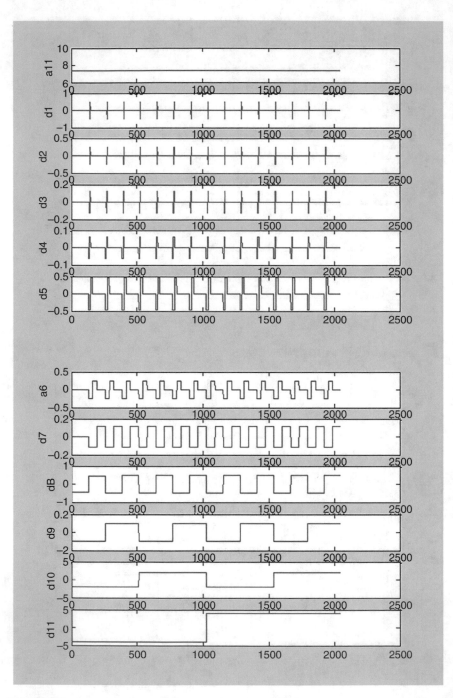

Fig. 8.15 Haar wavelet decomposition of a 4-bit ADC with offset error

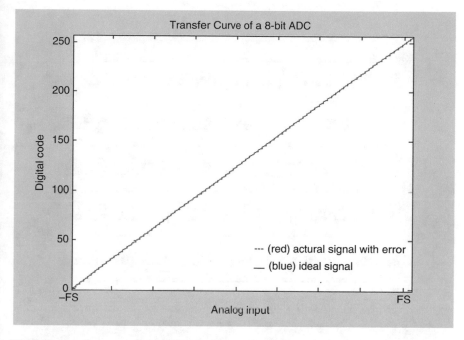

Fig. 8.16 Transfer curve of an 8-bit ADC with gain error

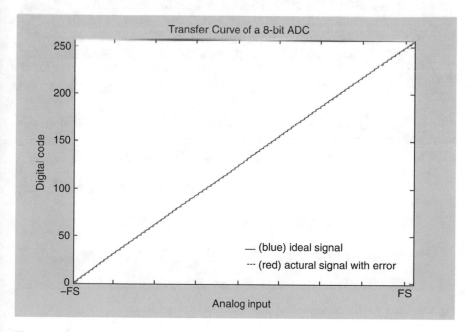

Fig. 8.17 Transfer curve of an 8-bit ADC with offset error

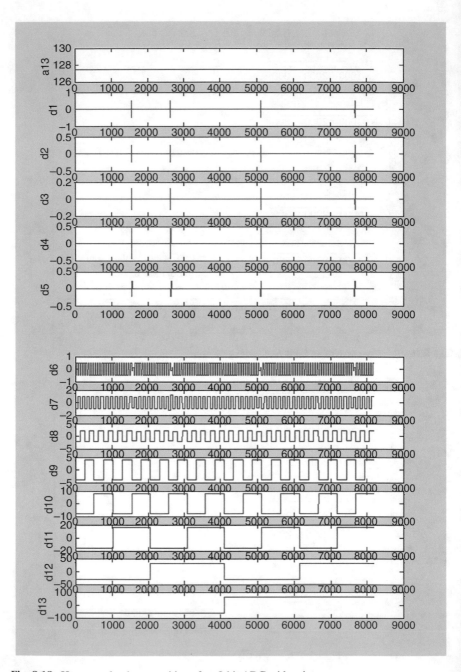

Fig. 8.18 Haar wavelet decomposition of an 8-bit ADC with gain error

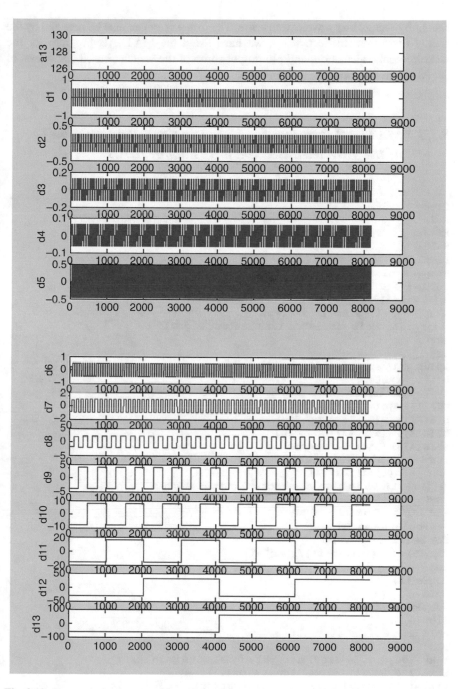

Fig. 8.19 Haar wavelet decomposition of an 8-bit ADC with offset error

energy by using Haar wavelet transform. The offset error can also be detected by the energy distortion. In Sect. 8.3.3, we have listed the MATLAB program for the measurement, calculation, and plotting of some of the wavelet-based static tests of the ADCs in this chapter.

8.3.3   MATLAB Program for Measuring and Plotting some of the Wavelet-Based Static Testing of ADCs

```
Clear
n=2;
FS=5;
N=10;
samples=2^N;
step=samples/2^n;
for i=1:1:2^n;
 b(i)=(i-1)*step;
 for j=((i-1)*step+1):i*step;
 l(j)=i-1; %x is simulated ideal output signal
 end
end
%plus gain error or offset error
b(i+1)=i*step;
a=[2,4]; %change some points to plus some error
b(a(1))=(a(1)-1)*step-20;
b(a(2))=(a(2)-1)*step-20;
%b(a(3))=(a(3)-1)*step-20;
%b(a(4))=(a(4)-1)*step+20;
for i=1:1:2^n; %simulate actual output signal with error
 for j=(b(i)+1):b(i+1);
 s(j)=i-1; %s is simulated actual output signal
 end
end
%calculate the DNL
for i=1:1:(2^n-1);
 DNL(i)=((b(i+1)-b(i))/step)-1;
end
figure (1)
hold off
subplot(2,1,1);
plot(s,'r:');
hold on
plot(l);
text(300,0.7,'-- (red) actural signal with gain error');
text(300,0.2,'__ (blue) ideal signal');
title(['Transfer Curve of a ',num2str(n),'-bit ADC']);
axis([0 2^N -1 2^n+1]);
ylabel('Digital code');
```

```
xlabel('Analog input');
ls=length(l);
[Cs,Ls]=wavedec(s,N,'haar'); %10 scale decomposition using Haar
ca10=appcoef(Cs,Ls,'haar',N); %extract 10 level low pass coefficents
for i=1:1:length(ca10)
 NSR2=ca10(i).^2;
end
for i=(N-n+1):1:N
 cd=detcoef(Cs,Ls,i);
 x=cd.*cd;
 y=0;
 for j=1:1:length(x);
 y=x(j)+y;
 end
 NSR2=NSR2+y;
end
NSR1=0;
for i=1:1:(N-n)
 cd=detcoef(Cs,Ls,i);
 x=cd.*cd;
 y=0;
 for j=1:1:length(x);
 y=x(j)+y;
 end
 NSR1=NSR1+y;
 end
end
NSR=10*log10(NSR1/NSR2);
hold off
subplot(2,1,2);
i=1:1:2^n-1;
stairs(i,DNL);
title('DNL');
xlabel(['Digital code NSR=',num2str(NSR),'dB']);
ylabel('DNL');

figure(2)
hold off
subplot(6,1,1);plot(s,'r:');
hold on
plot(l);
title(['decomposition',num2str(N),' scale using haar wavelet
transform']);
ylabel('s');

aN=wrcoef('a',Cs,Ls,'Haar',N);
hold off
subplot(6,1,2);plot(aN);ylabel(['a',num2str(N)]);
for i=1:4
 decmp_s=wrcoef('d',Cs,Ls,'Haar',i);
 subplot(6,1,i+2);
 hold off
 plot(decmp_s);
```

```
ylabel(['d',num2str(i)]);
end

figure(3)
hold off

for i=5:N
decmp_s=wrcoef('d',Cs,Ls,'Haar',i);
subplot(6,1,i-4);
hold off
plot(decmp_s);
ylabel(['d',num2str(i)]);

end
```

Summary

1. Traditionally, the maximum values of the integral nonlinearity (INL) and differential nonlinearity (DNL) are used as measures of evaluating the nonlinearities of ADCs.
2. The Haar and Daubechies wavelets are used for testing gain error and offset error of an ADC.
3. For an ideal transfer curve, the energy only exists up to the bit scale level in the Haar decomposition, and the energy exists at every scale level in the Daubechies decomposition.
4. This property mentioned in number 3 makes it possible to test the gain error and the offset error of the ADCs by using Haar wavelet transform.
5. It also suggests that Daubechies wavelet is not good in this testing method.
6. It is possible to measure the effectiveness of a pipelined ADC digital error correction circuit by the drop in the DNL and INL number of bits.
7. If the gain is larger than the ideal gain, the transfer curves are reduced along the amplitude axis toward the center.
8. If the gain is smaller than the ideal gain, the transfer curves are enlarged along the amplitude axis toward plus/minus full-scale extremities.

Review Questions

8.1 The DNL represents the difference between the step size of each code and the step size corresponding to a least significant bit (LSB) change.

   (a) True.
   (b) False.

8.2 The maximum values of the integral nonlinearity (INL) and differential nonlinearity (DNL) are not used traditionally as measures of evaluating the nonlinearities of ADCs.

(a) True.
(b) False.

8.3 To accurately measure the nonlinearity of the amplitude level in the input-output characteristics of an ADC, it is necessary to use an amplitude-to-output-code representation.

(a) True.
(b) False.

8.4 The transfer curve of an ADC is not in the form of a staircase.

(a) True.
(b) False.

8.5 The two commonly used wavelets discussed in this chapter for the static testing of an ADC are:

(a) Haar wavelet and Coiflet wavelet.
(b) Daubechies and Symlet wavelet.
(c) Haar and Daubechies wavelet.
(d) None of the above.

8.6 If some DNLs are not zero, that means some differences between adjacent transition amplitudes are the same.

(a) True.
(b) False.

8.7 Offset error causes a parallel translation of the entire transfer curve.

(a) True.
(b) False.

8.8 The DNL could detect a defect, which exists locally at a specific code.

(a) True.
(b) False.

8.9 By the idea of finite energy in Haar wavelet decomposition, it is possible to find the gain error and offset error of ADCs.

(a) True.
(b) False.

8.10 When the ADC transfer curve with gain error is decomposed into its Haar wavelet components, the distortion energy due to the ADC gain error does not fall into the noise scale band.

(a) True.
(b) False.

Answers: 8.1a, 8.2b, 8.3a, 8.4a, 8.5c, 8.6b, 8.7a, 8.8a, 8.9a, 8.10b.

Problems

8.11 Why is DNL capable of detecting a defect that exists locally in a specific code of and ADC?

8.12 Why is it important to use wavelet transforms in the static testing of ADCs?

8.13 Why is the use an amplitude-to-output-code representation necessary? What is your understanding of what a transfer curve of an ADC is? What about the staircase of an ADC?

8.14 What does the DNL of an ADC represent?

8.15 How can you define the DNL of an ADC?

8.16 What does it mean in some DNL measurements when some of the values are not zero?

8.17 What does it mean if the DNL is zero?

8.18 Why is the DNL measurement important?

8.19 Why is the use of wavelet transform method important in the static testing of an ADC?

8.20 What are the two most common types of wavelets used in the static testing of an ADC and why?

8.21 What is the significance of measuring the noise-to-signal ratio (NSR) using wavelets in static ADC testing?

8.22 What are the relationships between the gain error, offset error, and the transfer curves in the static measurement of an ADC?

References

Akujuobi, C. M., & Awada, E. (2008). Wavelet-based ADC testing automation using LabVIEW. *International Review of Electrical Engineering Transaction, 3*, 922–930.

Akujuobi, C.M., & Hu, L. (2002) A novel parametric test methods for communication system mixed signal circuit using discrete wavelet transform. *IASTED International Conference on Communication, Internet, and Information Technology*, U.S. Virgin Island, pp. 132–135.

Akujuobi, C. M., Awada, E., Sadiku, M., & Warsame, A. (2007). Wavelet-based differential nonlinearity testing of mixed signal system ADCs. *IEEE SoutheastCon*, 76–81.

Awada, E. A., & Akujuobi, C. M. (2018a). A wavelet-based ADC/DAC differential nonlinearity measurement analysis. *Journal of Engineering and Applied Sciences, 13*(16), 398–405. ISSN: 1816-949X © Medwell Journals, 2018.

Awada, E. A., & Akujuobi, C. M. (2018b). ADC testing algorithm for ENOB by wavelet transform using LabView measurements and MATLAB simulations. *Journal of Engineering and Applied Sciences, 13*(16), 6668–6679. ISSN: 1816-949X © Medwell Journals, 2018.

Burns, M., & Roberts, G. W. (2004). *An introduction to mixed-signal IC test and measurement.* Oxford University Press.

Gandelli, A., & Ragaini, E. (1996). ADC transfer function analysis by means of a mixed wavelet-Walsh transform. *IEEE Instrumentation and Measurement Technology Conference*, 1314–1318.

Kollar, I., & Blair, J. (2005). Improved determination of the best fitting sine wave in ADC testing. *IEEE Transactions on Instrumentation and Measurement, 54*, 1978–1983.

Plassche, R. (1994). *Integrated analog-to-digital and digital-to-analog converters.* Kluwer Academic Publishers.

Strang, G. (1989). Wavelets and dilation equations: A brief introduction. *SIAM Review, 31*, 614–627.

Yamaguchi, T. (1997). Static testing of ADCs using wavelet transforms. *IEEE Test Symposium*, 188–193.

Young, R.K. (1993). , "Wavelet theory and its application", Kluwer Academic Publishers, 1993.

Chapter 9
Mixed Signal Systems Testing Automation Using Discrete Wavelet Transform-Based Techniques

"Even if you are on the right track, you will get run over if you just sit there."

Will Rogers

9.1 Introduction

In Chap. 6, we covered the test point selection using wavelet transform for digital-to-analog converters (DACs). In Chap. 7, we covered the wavelet-based dynamic testing of ADCs. In Chap. 8, we covered the wavelet-based static testing of analog-to-digital converters (ADCs). We discussed the conventional testing methods such as fast Fourier transform (FFT) and sine wave histogram used to estimate for the mean square value of the quantization error sequence of effective number of bits (ENOB), dynamic nonlinearity (DNL), and integral nonlinearity (INL), not instantaneous values. The mean square value of the quantization error sequence causes any deviations occurring due to noise and/or bit(s) failures summed into the total errors. However, since wavelet transform is very sensitive to any changes of the signal through singularity and could present the fault information, we can use the discrete wavelet transform (DWT) algorithms in the estimation of worst-case ENOB and instantaneous DNL and INL.

In this chapter, we cover how we can enhance the testing of the mixed signal converters, we call systems, with the goal of developing real-time, automated diagnostic tool for analysis using testing algorithms based on discrete wavelet transform. The wavelet-based automated algorithms developed characterize both ADC and DAC parameters such as effective number of bits and differential nonlinearity by analyzing their output signal through filter banks and decimation processes. This testing technique introduces a major reduction in number of tested samples (codes) and significant decrease in testing duration based on the multiresolution property of the wavelet transform.

In addition, developing a universal testing graphical user interface (GUI) for both ADC and DAC is another aspect of the goal of the techniques covered in this chapter. The designed GUI is based on the powerful automation and data acquisition tools of LabVIEW to assist testing engineers with faster and easier testing setup. In

© Springer Nature Switzerland AG 2022
C. M. Akujuobi, *Wavelets and Wavelet Transform Systems and Their Applications*,
https://doi.org/10.1007/978-3-030-87528-2_9

this chapter, we demonstrate how to develop algorithms for other parameters in testing mixed signal converters such as integral nonlinearity (INL). The developed DWT-based automated algorithm is intended to cut down on the cost of resources by reducing test hardware and utilizing shelf test pieces of equipment. Finally, we explore several wavelets to determine suitability for this type of testing and compare with fast Fourier transform (FFT) techniques.

This DWT-based automated evaluation technique can meet the exponential growing demands for highly sophisticated converters with powerful, high-speed, high-resolution, and high internal complexity by shortening testing time and simplifying evaluation processes. Through simulation and actual testing, wavelets seem to show improvements in testing with very satisfactory results. In this chapter, we show some of the results based on real-time automated testing and the results compared with the other traditional testing techniques such as fast Fourier transform, histogram, and the device under test specifications. The testing results of the various N-bit ADCs (10–14 bits) and DACs (10–14 bits) demonstrate very promising automation testing algorithms for future devices.

The overall idea is to demonstrate, in this chapter objectively, how a wavelet-based automation process can enhance the testing of mixed signal systems such as ADCs and DACs in the following ways:

- Be able to estimate static parameters, such as INL and DNL, and dynamic parameters such as effective number of bits (ENOB), based on wavelet transform applications.
- Be able to compare with the conventional testing techniques and device specifications for result validations.
- Reduce the number of computed codes using wavelet-based testing in contrast with conventional techniques per each parameter.
- Shorting testing duration using wavelet transform versus conventional techniques.
- Reduce testing costs and testing processes by reducing the final computed codes.
- Reduce the cost of actual testing by consolidating test hardware.
- Implement algorithms into a fully controlled and automated testing system.
- Develop complete testing algorithms and testing GUI for both ADC and DAC parameters.
- Be able to layout the path for mobility testing pieces of equipment (CompactRIO) and onboard testing.

9.2   Noise and Quantization Error

In the testing of ADC parameters, it is more desirable to estimate for the worst-case ENOB and instantaneous magnitude of linearity deviation than the mean square error, which corresponds to obtaining the maximum value of quantization noise (Akujuobi et al., 2007; Nowak & Baraniuk, 1999; Yamaguchi & Soma, 1997).

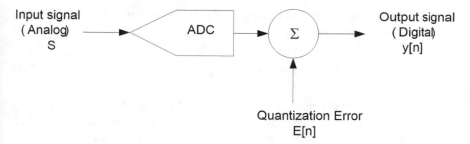

Fig. 9.1 Noise summation into ADC output digitized signal

Ideally, ADC deviation is negligible (zero). However, realistically, ADC performance produces quantization error that distorts the output signal $\widehat{X}[n]$ as shown in
Fig. 9.1. Therefore, ADC output can be viewed as shown in Eq. (9.1).

$$\widehat{X}[n] = S + e[n] \qquad (9.1)$$

where:

S is the original value
$e[n]$ is quantization error.

The DAC ideal analog output has a negligible deviation from digital input data
and 0 offset voltage (Baker, 2003; Burns & Roberts, 2004; Awada et al., 2010;
Awada & Akujuobi, 2018a, b). However, in a practical environment, DAC analog
output $\bar{x}(t)$ contains errors that distort the output signal due to noise, distortion error,
and heat, i.e.,

$$\bar{x}(t) = x(t) + e[n] \qquad (9.2)$$

where:

$x(t)$ = original input digital signal
$e[n]$ = error value.

From Eqs. (9.1) and (9.2), the output signal is the summation of both input signal
and quantization error and noises. Quantization is directly related to the converter
number of bits through quantization step size (Δ) that determines the distance
between adjacent codes in LSB, i.e.,

$$\Delta = \frac{Voltage[2^n - 1] - Voltage[0]}{2^n - 1} \qquad (9.3)$$

where:

n is converter number of bits,

However, the maximum instantaneous value of the quantization error is half of the quantization step Δ, i.e.,

$$-\frac{\Delta}{2} \leq e[n] \leq \frac{\Delta}{2} \tag{9.4}$$

Thus, the modulated output signal carries the information describing any failure of the device. Through the DWT analysis and the device step Δ as defined in Eq. 9.3, failures can easily be determined as described in Sects. 9.3, 9.4 and 9.5.

9.3 Worst-Case Effective Number of Bits (ENOB)

The process is to first capture the device under test (DUT) output data $\widehat{X}[n]$ and apply them via the DWT algorithms, using the wavelet-based multiresolution techniques. The combination of the instantaneous low- and high-frequency components is produced. While low-frequency components s_n are stationary over a period of time (approximation coefficients), high-frequency components d_n provide the detail coefficients as shown in Eqs. (9.5) and (9.6), respectively.

$$
\begin{bmatrix}
\vdots \\
s_{n-1,-1} \\
s_{n-1,0} \\
s_{n-1,1} \\
\vdots
\end{bmatrix}
=
\begin{bmatrix}
\cdots & \cdots & \cdots & & & \\
\cdots & \tilde{h}_{-1} & \tilde{h}_0 & \tilde{h}_1 & \tilde{h}_2 & \cdots \\
& \cdots & \tilde{h}_{-1} & \tilde{h}_0 & \tilde{h}_1 & \tilde{h}_2 & \cdots \\
& & \cdots & \tilde{h}_{-1} & \tilde{h}_0 & \tilde{h}_1 & \tilde{h}_2 & \cdots \\
& & & \cdots & \cdots & \cdots
\end{bmatrix}
\begin{bmatrix}
\vdots \\
s_{n,-1} \\
s_{n,0} \\
s_{n,1} \\
\vdots
\end{bmatrix}
\tag{9.5}
$$

$$
\begin{bmatrix}
\vdots \\
s_{n-1,-1} \\
d_{n-1,-1} \\
s_{n-1,0} \\
d_{n-1,0} \\
s_{n-1,1} \\
d_{n-1,1} \\
\vdots
\end{bmatrix}
=
\begin{bmatrix}
\cdots & \cdots & \cdots & & & \\
\cdots & \tilde{h}_{-1} & \tilde{h}_0 & \tilde{h}_1 & \tilde{h}_2 & \cdots \\
\cdots & \tilde{g}_{-1} & \tilde{g}_0 & \tilde{g}_1 & \tilde{g}_2 & \cdots \\
& \cdots & \tilde{h}_{-1} & \tilde{h}_0 & \tilde{h}_1 & \tilde{h}_2 & \cdots \\
& \cdots & \tilde{g}_{-1} & \tilde{g}_0 & \tilde{g}_1 & \tilde{g}_2 & \cdots \\
& & \cdots & \tilde{h}_{-1} & \tilde{h}_0 & \tilde{h}_1 & \tilde{h}_2 & \cdots \\
& & \cdots & \tilde{g}_{-1} & \tilde{g}_0 & \tilde{g}_1 & \tilde{g}_2 & \cdots \\
& & & \cdots & \cdots & \cdots \\
& & & \cdots & \cdots & \cdots
\end{bmatrix}
\begin{bmatrix}
\vdots \\
\vdots \\
\vdots \\
s_{n,-1} \\
s_{n,0} \\
s_{n,1} \\
\vdots \\
\vdots
\end{bmatrix}
\tag{9.6}
$$

From the high-pass filter, we obtain the detail coefficients as shown in Eq. (9.7).

$$(\ldots d_{n-1,-1}d_{n-1,0}, d_{n-1,1}, d_{n-1,2}, d_{n-1,3}, d_{n-1,4}, d_{n-1,5}, d_{n-1,6}, d_{n-1,7}, \ldots) \quad (9.7)$$

We downsample by 2 by taking the odd values in Eq. (9.7). The remaining transfer data equate to half the original output data as shown in Eq. (9.8).

$$(\ldots d_{n-1,-1}, d_{n-1,1}, d_{n-1,3}, d_{n-1,5}, d_{n-1,7}, \ldots) \quad (9.8)$$

However, to determine the DUT effective number of bits, as in Adamo et al. (2003) and Awada et al. (2010), the largest components of the DWT high-pass coefficient at first decomposition define the dynamic range (DR) that is used to estimate worst-case ENOB. DR can be computed from Eq. (9.9).

$$DR \equiv -20 \log_{10} \left[\frac{1}{\sqrt{2}} \left(\frac{\Delta}{2} \right) \right] = -20 \log_{10} \left[\left(\frac{1}{2^{\widehat{B}-0.5}} \right) \right] [dB] \quad (9.9)$$

where:

DR is the dynamic range.

$\widehat{B}$ is the effective number of bits (ENOB)

Δ is the quantization step size.

By rearranging Eq. (9.9), the worst-case instantaneous ENOB can be directly computed from Eq. (9.10).

$$\widehat{B} = \frac{DR}{20 \log_{10}(2)} - 0.5 [\text{bit}] \quad (9.10)$$

9.4 Instantaneous Differential Nonlinearity (DNL)

During the process of the estimation of instantaneous individual deviation codes, otherwise called the DNL, the captured converter digitized output data are analyzed through DWT as done in the ENOB analysis. However, unlike the ENOB, in DNL instantaneous estimation, the DUT outputs are decomposed twice as shown in Fig. 9.2 to refine the output data (codes) and magnifying error deviation.

The collected detail coefficients in the first DWT multiresolution are used again for a second level DWT multiresolution by repeating Eqs. (9.5) and (9.6), and downsample twice resulting to Eq. (9.11).

$$(\ldots d_{n-1,-1}, d_{n-1,1}, d_{n-1,3}, d_{n-1,5}, d_{n-1,7}, \ldots) \quad (9.11)$$

To compute the instantaneous DNL using DWT coefficients, high-pass coefficients at the second level of the DWT multiresolution are used (Akujuobi et al.,

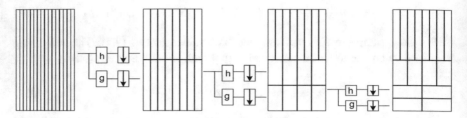

Fig. 9.2 Data decomposition process

2007; Yamaguchi & Soma, 1997). Since the DNL is defined as the maximum differences between adjacent codes($V_{j+1} - V_j$) (Mendoncaa & Silva, 2001; Akujuobi & Hu, 2002, 2003; Akujuobi et al., 2007), in the case of DWT, the instantaneous magnitudes are used in place of the original codes (codes in time-frequency domain are different version of the same signal). By applying DWT coefficients into DNL estimation, maximum instantaneous DNL can be computed as shown in Eq. (9.12).

$$DNL(n) = \frac{\max\left\{\left||d_{n-1,j}| - |d_{n-1,j+1}|\right|\right\}}{\Delta_{ideal}} - 1 \qquad (9.12)$$

where:

Δ_{ideal} is the ideal LSB

9.5   Instantaneous Integral Nonlinearity (INL)

The DNL defines the difference between actual step width and ideal value of 1 least significant bit (LSB) for individual monotonicity (Hoeschele, 1994); however, in the case of the INL error, it describes the overall deviation in LSB of an actual transfer function from straight line (Sunter & Nagi, 1997; Baker, 2003; Burns & Roberts, 2004). Therefore, in the wavelet transform-based testing technique, INL curve is determined based on the same data used in the instantaneous DNL estimation.

To compute the instantaneous INL using DWT coefficients, the high-pass coefficients at the second stage of the DWT multiresolution are used. As in the case of instantaneous DNL, the instantaneous magnitudes are used in place of the original codes because the codes in time-frequency domain are different versions of the same signal for INL estimation.

We apply the DWT coefficients into the INL estimation. The INL curve is determined by subtracting the reference calculated straight line based on instantaneous magnitudes from the converter actual curve of the instantaneous magnitudes. It is normalized to the DUT Δ_{ideal}. The worst-case maximum INL can be computed as shown in Eq. (9.13).

$$INL(n) = \frac{\max\left\{\left|\left|d_n\right| - \left|d_{ref(n)}\right|\right|\right\}}{\Delta_{ideal}} \tag{9.13}$$

where:

Δ_{ideal} is ideal LSB
d_n is instantaneous magnitudes
$d_{ref(n)}$ is the corresponding straight-line magnitudes.

In Sect. 9.6, we discuss how the automated DWT can be implemented using LabVIEW software.

9.6 Automation Testing Setup with LabVIEW and DWT

In this section, we illustrate how to develop a fully automated wavelet-based testing algorithm for faster testing measurements and better characterizations using LabVIEW software. Appendix E shows the automation testing process operations

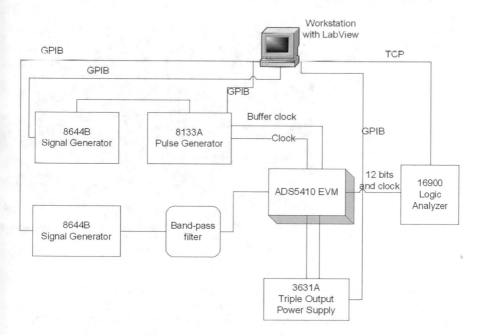

Fig. 9.3 ADC testing setup diagram

Fig. 9.4 ADC testing setup

Fig. 9.5 ADS5410 EVM
testing board

manual for mixed signal systems using DWT. Figure 9.3 shows the diagrammatic
representation of the testing setup. Figure 9.4 shows the actual arrangement of the
pieces of equipment that are used in the testing of ADCs. The ADS5410 EVM
shown in Fig. 9.5 is a 12-bit ADC EVM board that serves as the DUT in Figs. 9.3

and 9.4. However, a complete understanding of all testing instruments, methodology, and hardware setup (instrument interface and connectivity) is highly required to obtain proper results. Due to the testing setup complexity, costly test equipment, and the motivation of developing a simplified complete testing system for ADCs and DACs with less cost and higher mobility, we can setup a testing automation process based on National Instruments PXI Express chassis and controllers that incredibly reduces test costs by saving valuable test development time and hardware.

In general, we can have a measurement system that uses off the shelf hardware and software components to improve testing quality of mixed signal ADCs and DACs. Using LabVIEW testing automation and control, testing GUI can be developed and integrated into a PXI system offering signal generation, data acquisition, and clock synchronization for complete real-time converter testing. The use of National Instruments (NI) PXI in conjunction with LabVIEW developed GUI, for data presentation and feature reporting, makes the automation system especially suited for testing application enhancements of mixed signal converters. The process is as described starting from Sect. 9.7.

9.7   Testing Automation Programming Process

The starting of the automation process is as shown in Fig. 9.6 testing setup, utilizing the advantage of LabVIEW testing measurement and automation. The measurement automation system is based on the National Instruments PXI components capable of performing a variety of measurements including integral nonlinearity (INL), differential nonlinearity (DNL), and signal-to-noise and distortion (SINAD) to characterizing effective number of bits (ENOB).

The measuring station also consists of a programmable power supply, multimeter, 100 MHz oscilloscope, 400 MHz digitizing oscilloscope, and external function/ arbitrary waveform generator. External instruments are used in conjunction with the PXI to validate converter characteristics. In addition, a group of ADCs and DACs are used to validate testing algorithms and verify testing results. As stated in Sect.

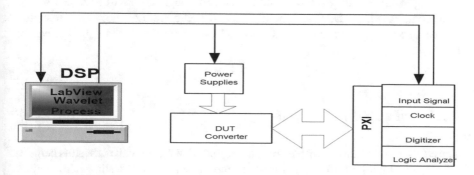

Fig. 9.6 Hardware setup

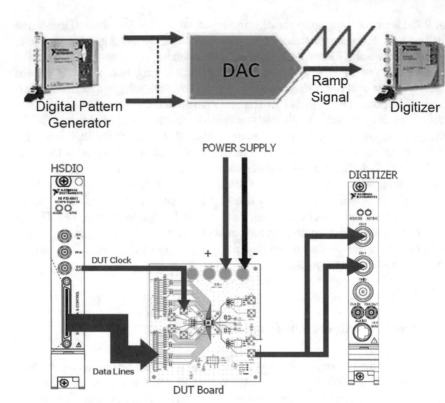

Fig. 9.7 DAC testing pieces of equipment and layout

9.6, Appendix E shows the automation testing process operations manual for ADCs and DACs using DWT.

We use the National Instruments pieces of equipment to set the proper testing parameters for the DAC as shown in Fig. 9.7. Testing begins with supplying digital ramp signal using National Instruments NI PXI-6552 pattern generator. The output analog signal is captured and digitized using NI PXI-5922 fixable resolution digitizer with capability up to 24 bits. Clock source, through NI PXI-6552, is used to synchronize the DUT DAC and the digitizer NI PXI-5922. Low noise power supply can be also used to power the DAC testing board and provide the digital power of bits triggering.

9.7.1   NI PXI-1042 Chassis

The NI PXI-1042 Chassis with its rugged PC-based platform with eight slots helps in offering a high-performance, low-cost measurement and testing automation. Through interconnect electrical bus PCI card, the PXI is connected directly to

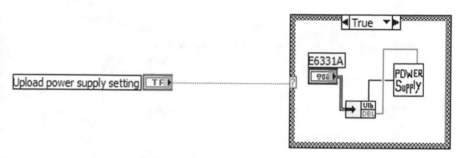

Fig. 9.8 LabVIEW codes for power supply

external PC for direct interaction with users. The PXI also adds mechanical, electrical, and software features that define complete systems for test and measurement, data acquisition, and equipment synchronization.

9.7.2   Power Supplies

The triple output power supply (3631A) provides a 3.3 V supply to the analog input of the DAC EVM and a 1.8 V supply for the digital input of the board to drive the digital circuits. Power supply is also controlled by LabVIEW through special VI as shown in Fig. 9.8.

9.7.3   HSDIO Card PXI-6552: Arbitrary Digital Waveform Generator

The HSDIO Card PXI-6552 is an arbitrary digital waveform generator with high-speed digital stimulus-response instrument that is capable of producing a variety of digital waveforms. It features up to 20 channels with programmable voltage levels and clock cycle. In addition, it is easy to set up and control by LabVIEW as shown in Fig. 9.9.

9.7.4   Scope Card PXI-5922

The PXI-5922 is a digitizer (delta-sigma ADC) that has flexible resolution and can sample anywhere from 24 bits at 500 kS/s to 16 bits at 15 MS/s rather than having a fixed resolution for all sample rates with integrated anti-alias protection for all sampling rates. With onboard memory of 8 MB, oversampling rate and sampling frequency should not exceed the memory limit. In addition, this digitizer has an

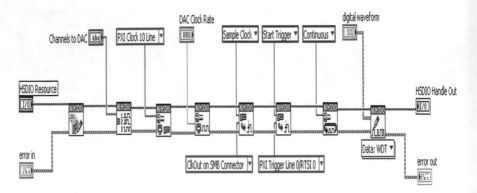

Fig. 9.9 Pattern generator setup and control VI

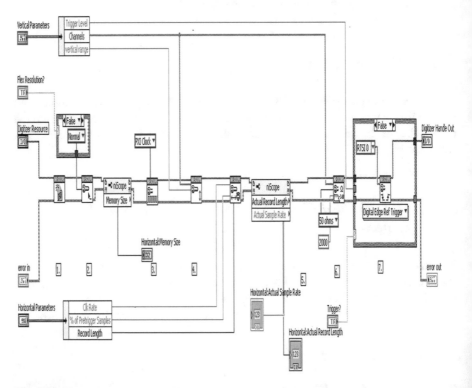

Fig. 9.10 Scope setup and control VIs

extremely low noise density and low spurious-free dynamic range (SFDR). The PXI-5922 is controlled by LabVIEW VIs as shown in Fig. 9.10.

In this automation design, to avoid the dead time, where noise is collected before the actual ramp signal is acquired, we generate and acquire about three cycles of

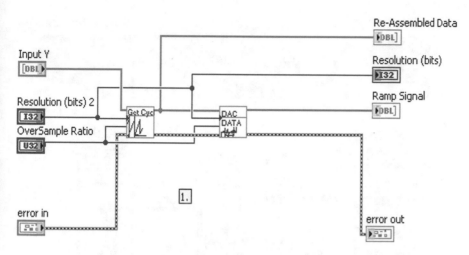

Fig. 9.11 VI of extracting one complete waveform for analysis

ramp signal. By locating the first maximum index of the derived ramp cycle (using derivative analysis), the starting of the second cycle of ramp signal can be located by the immediate minimum index and complete signal from maximum index to mini-mum index extracted for further data analysis as shown in Fig. 9.11. The isolated chunks of data, which represent the complete ramp signal, consist of data for each DAC Code. We can by removing 10% at the beginning and 10% from the end of the data chunk; any ringing issues in the data can be minimized.

9.8   ADC Testing Setup and LabVIEW VIs

We use the National Instruments pieces of equipment in testing the ADC and then set the proper testing parameters. The pieces of hardware that can be used are arbitrary analog waveform generator PXI-5421, digital waveform analyzers PXI-6541, sam-pling clock, and low noise power supply as shown in Fig. 9.12.

9.8.1   HSDIO Card PXI-6552

We use the HSDIO Logic Analyzer with high-speed digital stimulus-response instrument that is capable of capturing and storing digital waveforms in testing the

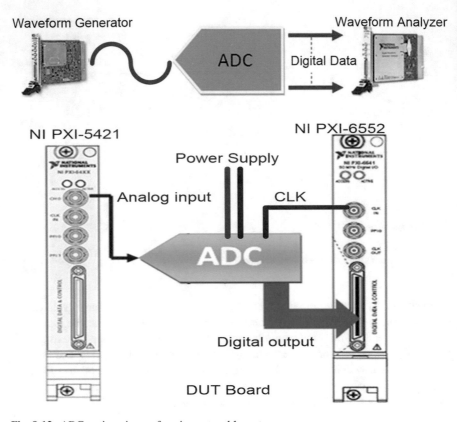

Fig. 9.12   ADC testing pieces of equipment and layout

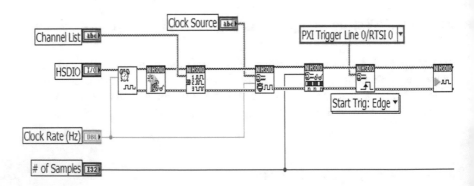

Fig. 9.13   Waveform capture VIs

ADCs. The hardware setup and control is accomplished with the aid of the LabVIEW applications as shown in Fig. 9.13.

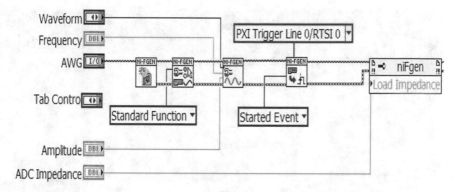

Fig. 9.14 Waveform generator VI

9.8.2   Waveform Generator NI PXI-5421

The PXI-5421, an arbitrary waveform generator, is used to produce stimulus sine waveform (analog) for sampling purposes. With built-in high-resolution (16 bit) DAC and up to 512 MB of onboard memory, the PXI-5421 is capable of producing clean analog waveform with significant spurious-free dynamic range. The selection of a particular waveform, frequency, amplitude, and impedance can be controlled through the LabVIEW drivers' sub VIs as illustrated in Fig. 9.14.

9.9   The GUI (Graphic User Interface)

In testing mixed signal converters (ADCs and DACs), the challenges have always been the automation of the testing process and combining all the key resources. For example, in the industry testing, analog group for testing ADCs are separated from DAC testing group in many design and manufacturing industries. We can therefore put more emphasis on the development of a complete testing algorithm for both ADCs and DACs. It is important in the development of a graphical user interface (GUI) for the DWT-based automatized testing of mixed signal systems to be one that can accommodate both the ADCs and DACs capable of assisting test engineers for faster and easier processes.

The GUI should be intuitively obvious to the user; keeping that in mind, the designer should think like the user and should not force the user to think like the software. Therefore, there are few tips that should be considered and followed throughout the development of a GUI. The tips are as follows:

- Creating a logical layout.
- Using colors to separate functions.
- Using images for easier indications.

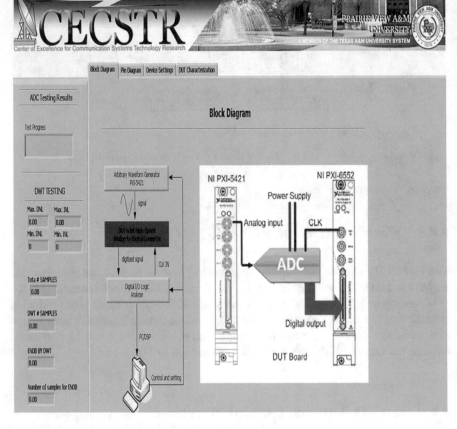

Fig. 9.15 GUI for ADC DWT-based testing methodology and guidance

- Being consistent with the designs and layout.
- Separating inputs from outputs.
- Providing useful error messages.
- Providing setup wizards.
- Using animation/progress bars.

It is important to avoid making mistakes. We successfully can do this by making sure we do not have the following:

- Complicated setup.
- Confusing layout.
- Hidden essential options.
- Useless error messages.
- Too many options.

We develop a complete testing GUI to test the ADCs and DACs following the guidelines itemized in this section. For example, Figs. 9.15 and 9.16 illustrate a

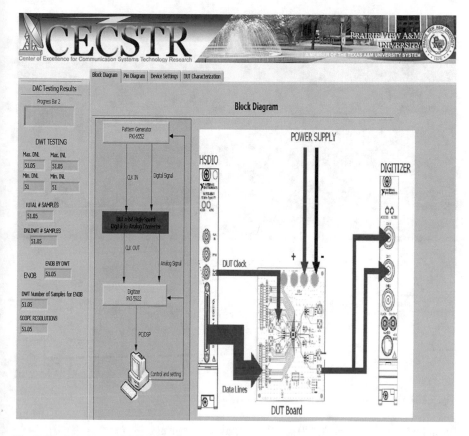

Fig. 9.16   GUI for DAC DWT-based testing methodology and guidance

developed GUI that is used in the testing of DWT-based methodology for both ADCs and DACs, respectively. This can give the users a complete idea of the testing procedures. In addition, a complete device under test (DUT) hock-up list illustration, detailed steps on parameter setup, and testing result can all be integrated into the developed GUI to assist the user or the test engineer in implementing a fast and accurate mixed signal systems automation testing.

9.10   Implementation of an Automated DWT-Based Algorithm for the Testing of ADCS

The implementation of the automation technique is done to enhance the testing process of analog-to-digital converters (ADCs) and digital-to-analog converters (DACs). The automated wavelet-based testing algorithms are implemented with

Fig. 9.17 Actual ADC testing boards used in the implementation

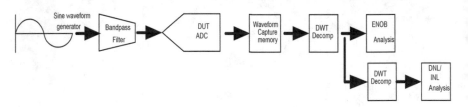

Fig. 9.18 Model setup for automated ADC wavelet test

the aid of the development of LabVIEW testing graphical user interface (GUI). As an example, a DWT can be implemented to test high-speed 14-bit ADC (ADS5423), 12-bit ADC (ADS5410), and 10-bit (TLC876). Their EVM testing boards are as shown in Fig. 9.17. These are high-resolution ADCs. The model setup for automated ADC wavelet test is as shown in Fig. 9.18. The testing procedures are constructed based on single-tone sine wave. One of the key elements in this process is to automate and enhance the testing process of the ADC characterization code by integrating the discrete wavelet algorithm, LabVIEW automation, and PXI hardware setup into one complete setup as shown in Fig. 9.19. Doing so allows the enhancement of testing process with higher accuracy results and less cost and time consumption. Therefore, LabVIEW program codes (sub-VIs) are developed for faster and easier testing process.

Different input frequency signals are used in obtaining results for testing purposes in this characterization. In addition, amplitude can be adjusted to check for any changes or effect on the testing result. To obtain a large scale of discrete wavelet testing results, several wavelet algorithms such as Haar, db2, db4, bior3-1, and Coif1 can be used. The results are compared with each other along with the results obtained by the conventional testing methods to determine the best testing technique for ADC characterization codes.

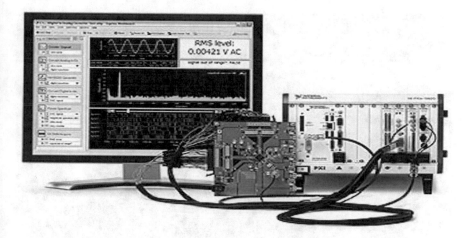

Fig. 9.19   Laboratory setup for ADC automated testing with DWT

Fig. 9.20   Test parameters at 10 MHz, 10-bit ADC using DWT

Intensive testing is performed using several parameters such as multiple stimulus frequency and waveforms, different clock frequency, and several devices with a range of bit numbers. For example, in testing 10-bit ADC (TLC876), Fig. 9.20 illustrates testing parameter setup of stimulus 10-MHz sine waveform with

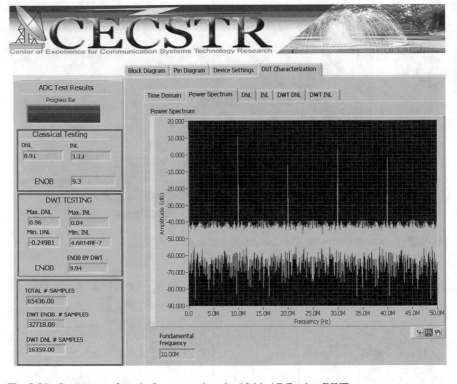

Fig. 9.21 Output waveform in frequency domain, 10-bit ADC using DWT

amplitude 0.8 V, 100-MHz clock frequency, and choice of a particular wavelet family. The 12-bit and 14-bit ADCs testing setup at 10 MHz are shown in Figs. F1 and F2, respectively, in Appendix F. In Fig. 9.21, an actual 10-MHz sine waveform is digitized by 10-bit ADC. The digitized signal is transformed in frequency domain.

Instantaneous DNL measurements and analysis for 10 bit at 10 MHz are illustrated in Fig. 9.22. The 12-bit and 14-bit ADCs DNL measurements at 10 MHz are shown in Appendix F, Figs. F3 and F4. Instantaneous INL measurements and analysis at 10 MHz are illustrated in Fig. 9.23.

9.11 A Comparative Tabular Summary of the Automated ADC Testing Using DWTs

We summarize in a tabular format some of the key testing results using different wavelets and comparing them with the conventional techniques such as FFT. Tables 9.1 to 9.6 summarize the testing results for ADC effective number of bits (ENOB).

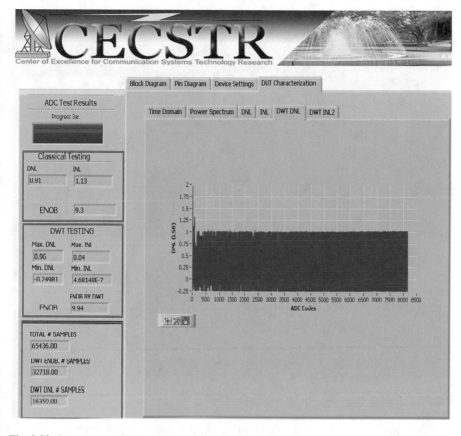

Fig. 9.22   Instantaneous DNL, 10-bit ADC using DWT

In Table 9.1, for 10 bit ADC, ENOB based on FFT algorithm measurements is in the range of 9.3 bits for 10 bits; however, in wavelet transform analysis, db2 and Coif1 are the closest to perfect performance of 10 bits.

In Table 9.2, for a 12-bit ADC, ENOB based on FFT algorithm measurements is in the range of 10.5 bits for 12 bits. However, in wavelet transform analysis, db2 and Coif1 were closest to perfect performance of 12 bits.

However, in Table 9.3, for 14-bit ADC, ENOB based on FFT algorithm measurements is in the range of 11.5 bits for 14 bits. However, in wavelet transform analysis, db2 and Coif1 are closest to perfect performance of 14 bits. This illustrates the effect of noise summation over decreasing size LSB separation as we increase number of bits. In addition, the same observations are seen at 20-MHz testing.

In testing for DNL, we compare the wavelet-based test algorithms with both the device specification shown in Appendix G and conventional testing techniques approved by industry.

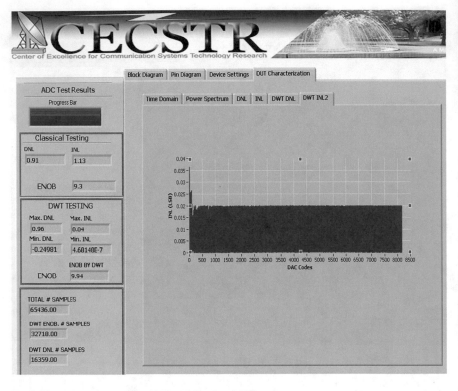

Fig. 9.23 Instantaneous INL, 10-bit ADC using DWT

Table 9.1 ENOB results for 10-bit ADC testing at 10 MHz sine wave using DWTs

Sampling frequencies	FFT	db4	db2	Haar	Bior3-1	Coif1
100 MHz	9.27	10.6	10.38	10.57	11.56	9.92
50 MHz	9.31	10.65	10.29	10.61	11.45	9.86
40 MHz	9.3	10.51	10.27	10.55	11.52	9.9

Table 9.2 ENOB result for 12-bit ADC testing at 10 MHz sine wave using DWTs

Sampling frequencies	FFT	db4	db2	Haar	Bior3-1	Coif1
100 MHz	10.57	12.6	12.08	12.57	13.51	11.94
50 MHz	10.57	12.56	11.91	12.44	13.5	11.89
40 MHz	10.56	12.62	11.96	12.28	13.66	11.64

Table 9.3 ENOB result for 14-bit ADC testing at 10 MHz sine wave using DWTs

Sampling frequencies	FFT	db4	db2	Haar	Bior3-1	Coif1
100 MHz	11.4	14.1	13.6	14.25	14.93	13.8
50 MHz	11.51	14.24	13.57	14.2	14.86	13.45
40 MHz	11.5	14.05	13.58	14.38	14.95	13.74

Table 9.4 DNL testing result for 12-bit ADC at 10 MHz sine wave using DWTs

Sampling frequencies	ADS5410 Spec.	Conventional histogram testing	db4	db2	Haar	Bior3-1	Coif1
100 MHz	−0.9 to 1	0.86	0.48	0.67	0.5	0.03	0.65
50 MHz	−0.9 to 1	0.85	0.46	0.69	0.52	0.01	0.57
40 MHz	−0.9 to 1	0.85	0.48	0.64	0.61	−0.03	0.61

Table 9.5 INL testing result for 10-bit ADC at 10 MHz sine wave using DWTs

Sampling frequencies	TLC876 Spec.	Conventional histogram testing	db4	db2	Haar	Bior3-1	Coif1
100 MHz	±1.5	1.11	0.13	0.14	0.16	0.03	0.14
50 MHz	±1.5	1.15	0.12	0.15	0.18	0.05	0.14
40 MHz	±1.5	1.13	0.15	0.15	0.2	0.04	0.16

In Table 9.4, for 12-bit ADC, the DNL based on histogram is in the range of 0.8 LSB. The device specification allows a range from −0.9 to 1 LSB. In implementing the wavelet transform analysis, the Haar and Coif1 are the closest to conventional measurements and with the device specification. Table 9.5 shows the results of the ADC INL testing using histogram and DWT techniques.

In the INL automation testing, we compare the wavelet test algorithms with both the device specifications in Appendix G and conventional testing techniques approved by industry. Table 9.5 shows the INL results for 10-bit ADC. The INL based on straight-line method was in the range of 1.1 LSB. The device specification allows a range from −1.5 to 1.5 LSB. However, in the DWT analysis, Haar, dbN, and Coif1 are the closest to convention measurements and with the device specification as shown in Appendix G.

9.12   Implementation of an Automated DWT-Based Algorithm for the Testing of DACs

In the real-time, (actual) DWT-based automation testing, DAC analog output signal can be directly digitized using an instrument with higher-precision ADC to sample the DAC analog output. The digitized output data can be analyzed using FFT and DWT to measure DAC parameters such as ENOB, DNL, and INL and determine a better method of testing. The testing procedures as shown in Fig. 9.24 can be implemented based on ramp-up digital input pattern for INL and DNL static testing and sine wave for ENOB dynamic testing.

As with the DWT-based automated testing of ADC discussed in Sect. 9.10, the DWT-based automated DAC testing algorithm and data acquisition control can be

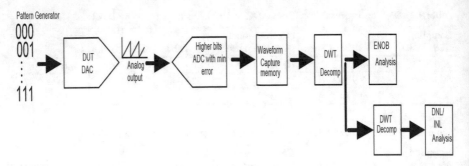

Fig. 9.24 Model setup for automated DAC wavelet test

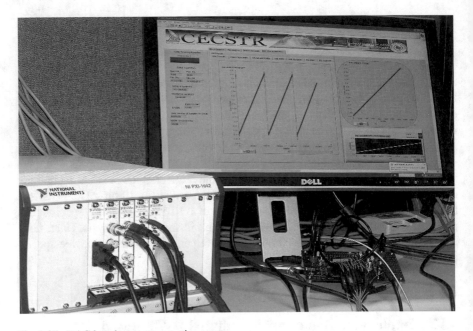

Fig. 9.25 DAC bench prototype testing setup

implemented using LabVIEW. Figure 9.25 shows a DAC bench prototype testing setup. The testing setup consists of the following:

- National Instruments PXI-1042 that is capable of generating:

  Various waveforms (analog and digital).
  Built-in clock synchronization.
  High-resolution digitizer.
  Built-in logic analyzer.

Fig. 9.26   Actual DAC testing boards

- Device under test DAC.
- PC terminal for testing engineering interfaces and interpolation of testing results.

The validation of the results can be implemented by using numerous Texas Instruments (TI) DAC boards such as the 10-bit DAC2900, 12-bit DAC 2902, and 14-bit DAC 2904 as shown in Fig. 9.26, while Fig. 9.27 shows the test parameter setup using the 10-bit DAC with the automated results based on DWT. In Figs. 9.28 and 9.29, the DAC output analog signal is digitized and presented in time domain for both sinusoid and ramp-up waveform. The 12-bit output DAC test parameter setup is as shown in Appendix H1. Figures 9.30 and 9.31 show an illustration of computing and analysis of the conventional DNL and instantaneous DNL, respectively. For the 12-bit and 14-bit DACs, the DNL measurements are shown in Appendix H, Figs. H2 and H3 for the 12-bit DAC and Figs. H4 and H5 for the 14-bit DAC. The specification sheet of Appendix I helps to validate the results as especially with the DAC 10-bit converter used in testing.

In addition, for the INL computation and analysis, the conventional INL and the instantaneous INL automated testing results using DWT for the 10-bit DAC are shown in Figs. 9.32 and 9.33, respectively. Tables 9.6, 9.7 and 9.8 show the testing results for the DAC effective number of bits (ENOB) for the 10, 12, and 14 bits, respectively, at various frequencies.

In Table 9.6, for 10-bit DAC, ENOB based on FFT algorithm measurements is in the range of 8.7 bits for 10 bits; however, in wavelet transform analysis; db2 and Coif1 are closest to perfect performance of 10 bits.

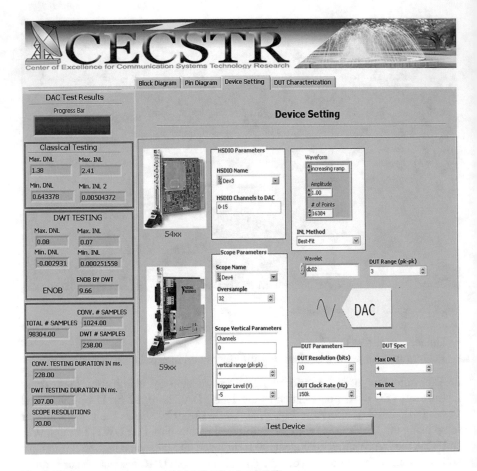

Fig. 9.27 Test parameter setup, 10-bit DAC using DWT

In Table 9.7, for 12-bit DAC, ENOB based on FFT algorithm measurements is in the range of 10.5 bits for 12 bits. However, in wavelet transform analysis, db2, Haar, and Coif1 are closest to perfect performance of 12 bits.

In Table 9.8, for 14-bit DAC, ENOB based on FFT algorithm was in the range of 12.1 bits for 14 bits. However, in wavelet transform analysis, db2 and Coif1 were closest to perfect performance of 14 bits.

9.13 Cost Analysis in Terms of Test Duration Reduction

In the testing automation process discussed in this chapter for the mixed signal systems (ADCs and DACs), the wavelet transforms have shown significant reduction in the number of sampling data. This fact leads to huge simplification in data

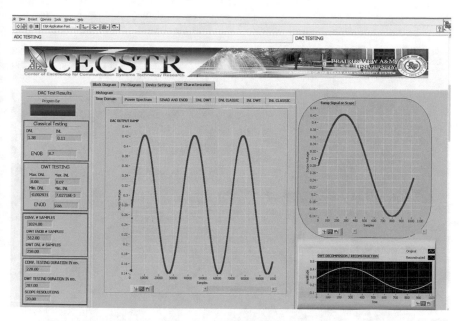

Fig. 9.28 Sine wave output waveform, 10-bit DAC using DWT

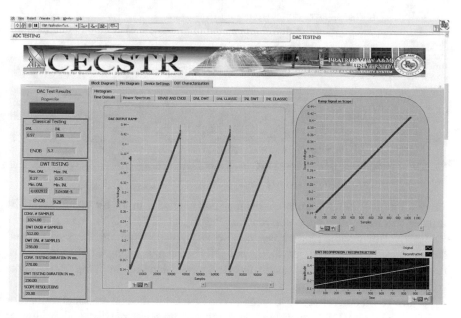

Fig. 9.29 Ramp-up output waveform, 10-bit DAC using DWT

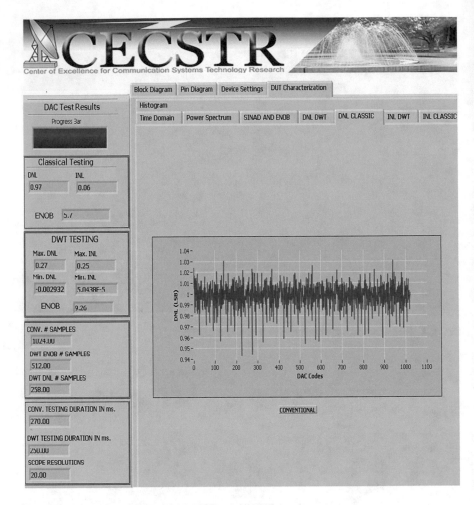

Fig. 9.30 Conventional DNL, 10-bit DAC using DWT

compilation and computation complexity. For example, in 10-bit converter, 1024 samples can be used for the conventional testing of ENOB, DNL, and INL independently. However, 512 samples can be used in determining the ENOB, and 256 samples can be used for DNL and INL estimation using DWT algorithms. This significant improvement results from testing the duration dropping from 228 ms to 200 ms at the rate of about 10%.

For the 12-bit converter, 4096 samples can be used in the conventional testing for ENOB, DNL, and INL separately. In using the wavelet transform algorithm, it reduced the number of collected data samples to 2048 in ENOB and 1024 samples

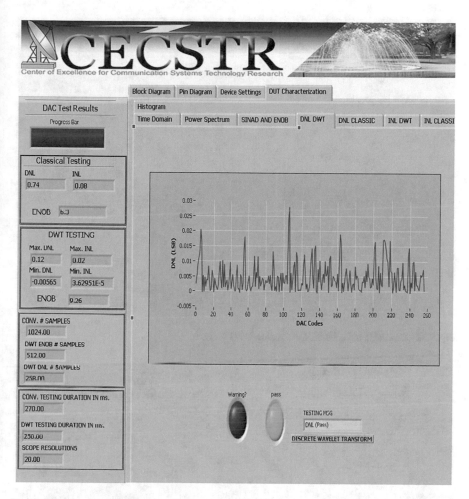

Fig. 9.31 Instantaneous DNL, 10-bit DAC using DWT

for DNL and INL estimation. In addition, we noticed that the test duration could drop from 460 ms to 275 ms at the rate of 40%. For the 14-bit converter, 16,384 samples can be used in the conventional testing of ENOB, INL, and DNL. However, 8192 samples can be used for the ENOB, and 4096 samples can be used for the INL and DNL using the DWT testing algorithm. In addition, it is possible to observe that the test duration can drop from 850 ms to 340 ms at the rate of 60%.

Summary

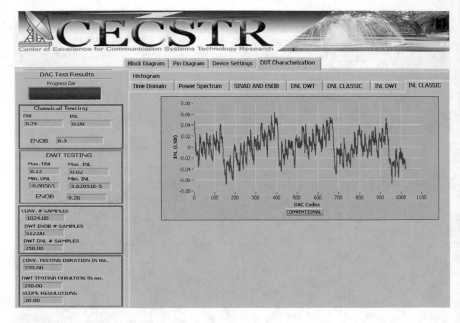

Fig. 9.32 Classical INL, 10-bit DAC using DWT

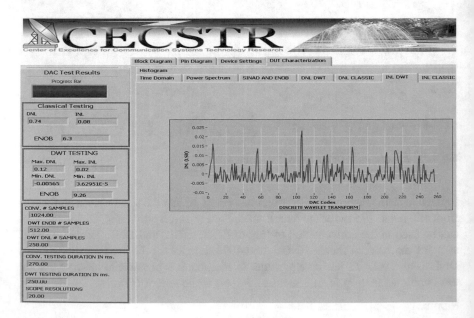

Fig. 9.33 Instantaneous INL, 10-bit DAC using DWT

1. The DWT-based automated evaluation technique can meet the exponential growing demands for highly sophisticated converters with powerful, high-

Table 9.6   10-bit DAC ENOB at various frequencies and analysis techniques

Fclock	Conventional testing	db4	db2	Haar	Bior3-1	Coif1
100 KHz	8.7	11.5	9.46	9.26	8.86	10.01
150 KHz	8.7	11.5	9.66	9.34	8.9	10.1
200 KHz	8.5	11.5	9.45	9.24	7.85	9.96

Table 9.7   The 12-bit DAC ENOB at various frequencies and analysis techniques using DWT

Fclock	Conventional testing	db4	db2	Haar	Bior3-1	Coif1
100 KHz	10.57	12.78	12.06	11.18	9.7	12.1
150 KHz	10.57	12.56	12	11	9.6	12.2
200 KHz	10.56	12.82	11.94	11.18	9.7	12.79

Table 9.8   The 14-bit DAC ENOB at various frequencies and analysis techniques using DWT

Fclock	Classical testing	db4	db2	Haar	Bior3-1	Coif1
100 KHz	12.5	13.67	13.77	12.64	11.27	13.98
150 KHz	12.1	13.57	13.73	12.62	11.22	13.26
200 KHz	12.1	13.14	13.65	12.61	11.17	13.69

speed, high-resolution, and high internal complexity by shortening testing time and simplifying evaluation processes.

2. In ADC parameter testing, it is more desirable to estimate for the worst-case ENOB and instantaneous magnitude of linearity deviation than the mean square error, which corresponds to obtaining the maximum value of quantization noise.

3. The DAC ideal analog output has a negligible deviation from digital input data and 0 offset voltage.

4. The maximum instantaneous value of the quantization error is half of the quantization stepΔ.

5. To determine the DUT effective number of bits, the largest components of the DWT high-pass coefficient at first decomposition define the dynamic range (DR) that is used to estimate worst-case ENOB.

6. To compute the instantaneous DNL using DWT coefficients, high-pass coefficients at the second level of the DWT multiresolution are used.

7. To compute the instantaneous INL using DWT coefficients, the high-pass coefficients at the second stage of the DWT multiresolution area are used.

8. Due to testing setup complexity, costly test equipment, and the motivation of developing a simplified complete testing system for ADCs and DACs with less cost and higher mobility, we can setup a testing automation process based on National Instruments PXI Express chassis and controllers that incredibly reduces test costs by saving valuable test development time and hardware.

9. The implementation of the automation technique is done to enhance the testing process of analog-to-digital converters (ADCs) and digital-to-analog converters (DACs).

10. The automated wavelet-based testing algorithms can be implemented with the aid of the development of LabVIEW testing graphical user interface (GUI).
11. In the real-time, (actual) DWT-based automation testing, DAC analog output signal can be directly digitized using an instrument with higher-precision ADC to sample the DAC analog output.
12. The GUI should be intuitively obvious to the user; keeping that in mind, the designer should think like the user and should not force the user to think like the software.

Review Questions

9.1. A wavelet-based automation process can enhance the testing of mixed signal systems such as ADCs and DACs.

(a) True
(b) False

9.2. Quantization is not directly related to the converter number of bits through quantization step size (Δ) that determines the distance between adjacent codes in LSB.

(a) True
(b) False

9.3. ADC performance produces quantization error that distorts the output signal.

(a) True
(b) False

9.4. The modulated output signal in an ADC carries the information describing any failure of the device.

(a) True
(b) False

9.5. During the process of the estimation of instantaneous individual deviation codes, otherwise called the DNL, the captured converter digitized output data are analyzed through DWT as done in the ENOB analysis.

(a) True
(b) False

9.6. The use of National Instruments (NI) PXI in conjunction with LabVIEW developed GUI, for data presentation and feature reporting, makes the automation system especially suited for testing application enhancements of mixed signal converters.

(a) True
(b) False

9.7. The implementation of the automation technique is not done to enhance the testing process of analog-to-digital converters (ADCs) and digital-to-analog converters (DACs).

(a) True
(b) False

9.8. The automated wavelet-based testing algorithms cannot be implemented with the aid of the development of LabVIEW testing graphical user interface (GUI).

(a) True
(b) False

9.9. Cutting down in the test duration cannot translate into cost savings in terms of work hours.

(a) True
(b) False

9.10. The wavelet transforms have shown significant reduction in the number of sampling data.

(a) True
(b) False

Answers: 9.1a, 9.2b, 9.3a, 9.4a, 9.5a, 9.6a, 9.7b, 9.8b, 9.9b, 9.10a.

Problems

9.1. In what demonstrable ways can a wavelet-based automation process enhance the testing of mixed signal systems such as ADCs and DACs?

9.2. What are the desirable and the realistic issues that a test engineer must consider in ADC parameter testing?

9.3. What are the ideal and the practical environment issues that a test engineer must consider in DAC parameter testing?

9.4. What is the process of computing the worst-case ENOB using DWT?

9.5. What is the process of estimating the instantaneous DNL using DWT?

9.6. What is the process of computing the instantaneous INL using DWT?

9.7. How can the automation process for mixed signal systems such as ADCs and DACs be set up using DWTs?

9.8. What are the key challenges in the testing of mixed signal converters with DWTs especially with the GUI?

9.9. What are few tips that should be considered and followed throughout the development of a GUI for the testing of mixed signal systems using DWTs?

9.10. What are some of the important mistakes that you must avoid in the testing of mixed signal systems using DWT automation process?

9.11. Describe the implementation of an automated DWT-based algorithm for the testing of ADCs and why?

9.12. Describe the implementation of an automated DWT-based algorithm for the testing of DACs and why?

9.13. How can you validate the DWT-based automation process of the DAC mixed signal systems?
9.14. How can you justify the cost savings as a result of the automation process using DWTs?

References

Adamo, F., Attivissimo, F., Giaquinto, N., & Trotta, A. (2003). A/D converters nonlinearity measurement and correction by frequency analysis and dither. *IEEE Transaction Instrumentation and Measurement, 1*, 1200–1205.

Akujuobi, C. M., & Hu, L. (2002). A novel parametric test methods for communication system mixed signal circuit using discrete wavelet transform. In *IASTED international conference on communication, internet, and information technology* (pp. 132–135). U.S. Virgin Island.

Akujuobi, C. M., & Hu, L. (2003). Implementation of the Wavelet Transform-based Technique for Static Testing of Mixed Signal System. *IASTED International Conference on Modeling and Simulation*, 56–59.

Akujuobi, C. M., Awada, E., Sadiku, M., & Warsame, A. (2007). Wavelet-based differential nonlinearity testing of mixed signal system ADCs. *IEEE SoutheastCon*, 76–81.

Awada, E. A., & Akujuobi, C. M. (2018a). A wavelet-based ADC/DAC differential nonlinearity measurement analysis. *Journal of Engineering and Applied Sciences, 13*(16), 398–405. ISSN: 1816-949X © Medwell Journals.

Awada, E. A., & Akujuobi, C. M. (2018b). ADC testing algorithm for ENOB by wavelet transform Using LabView measurements and MATLAB simulations. *Journal of Engineering and Applied Sciences, 13*(16), 6668–6679. ISSN: 1816-949X © Medwell Journals.

Awada, E., Akujuobi, C. M., Matthew, N. O., & Sadiku, M. N. O. (2010). A reduced-coded linearity test for DAC using wavelet analysis. *International Journal of Engineering Research & Innovation, 2*(1), 69–76.

Baker, M. (2003). *Demystifying mixed signal test methods*. Newnes, Elsevier Science.

Burns, M., & Roberts, G. W. (2004). *An introduction to mixed-signal IC test and measurement*. Oxford University Press.

Hoeschele, D. (1994). *Analog-to-digital and digital-to-analog conversion techniques*. Wiley.

Mendonçaa, H., & Silvaa, J. (2001). ADC testing using joint Time–frequency analysis. *Elsevier Science Direct, 23*, 129–135.

Nowak, R. D., & Baraniuk, R. (1999). Wavelet-based transformations for nonlinear signal processing. *IEEE Transaction on Signal Processing, 47*, 489–499.

Sunter, S., & Nagi, N. (1997). Simplified polynomial-fitting algorithm for DAC and ADC BIST. *IEEE International Test Conference*, 389–395.

Yamaguchi, T., & Soma, M. (1997). Dynamic testing of ADCs using wavelet transform. *IEEE International Test Conference*, 379–388.

Part III
Wavelets and Wavelet Transform
Application to Compression

Chapter 10
Wavelet-Based Compression Using Nonorthogonal and Orthogonally Compensated W-Matrices

"We cannot truly face life until we face the fact that it will be taken away from us."

Billy Graham.

10.1 Introduction

In this chapter, we discuss the W-transform multiresolution wavelet analysis for nonorthogonal and orthogonally compensated image compression. Kwong and Tang first introduced this idea in the early 1990s (Kwong, 1994; Kwong & Tang, 1994). In the traditional wavelet transforms, orthogonality is a condition required for perfect reconstruction to occur (Akansu & Haddad, 1992; Daubechies, 1989; 1998, 1990; Mallat, 1989; Vetterli & Herley, 1992). The W-transform is constructed by relaxing the orthogonality conditions required in the traditional wavelet transforms (Kwong & Tang, 1994). One of the advantages of the W-transform over some of the classical wavelet transforms is that it allows for the efficient handling of images of any sizes without the usual restriction to multiples of a power of two. The W-transform is an efficient tool for image compression since it does not extend the output image (Akujuobi & Lian, 2005; Akujuobi & Kwong, 1995). Particularly useful for image compression is a special nonorthogonal W-transform called the quadratic spline transform. It has symmetrical properties and the scaling functions are smooth.

However, there are cases when orthogonality may be required in order to improve image compression and reconstruction performance. This is especially true when dealing with images that have high oscillating properties and images that are degraded by noise. Orthogonality can be achieved by using a process known as orthogonal compensation. In this chapter, we discuss, demonstrate, and compare the effects of nonorthogonal and orthogonally compensated compression on images. The effects of nonorthogonal and orthogonally compensated W-transform compression using the quadratic spline and Daubechies wavelet coefficients are also compared.

We can achieve compression by first implementing the nonorthogonal or the orthogonally compensated W-transform by decomposing the image into four sub-images. The various matrix levels are quantized using a specific quantization

© Springer Nature Switzerland AG 2022
C. M. Akujuobi, *Wavelets and Wavelet Transform Systems and Their Applications*,
https://doi.org/10.1007/978-3-030-87528-2_10

ratio. Alternatively, small components of the output matrices can be discarded based on a specified threshold value. Then the image is reconstructed. Different statistical measures of compression quality such as compression rate (CR) and bits per pixel (bpp), mean square error (MSE), signal-to-noise ratio (SNR), entropy, and histogram are used. Images of varying sizes, intensities, textures, and backgrounds are used.

In Sect. 10.2 of this chapter, the orthogonality condition is discussed, while the W-transform and W-matrix are discussed in Sect. 10.3. The orthogonality compensation process and compression algorithms for the two types of W-transform are discussed in Sects. 10.4 and 10.5, respectively. The discussion of some simulation example results is in Sect. 10.6. The performance evaluations are discussed in Sect. 10.7.

10.2   Orthogonality Condition

Orthogonality conditions are generally desirable for good compression and reconstruction performance, when dealing with noise-degraded images and with images that have high oscillating properties. As an example, if we let x be an infinite image input and h and g are a class of finite impulse response (FIR) filters. The composite operation of filtering and decimation can be represented by the filter matrices $\mathbf{H}$ and $\mathbf{G}$ as shown in Eqs. (10.1) and (10.2), respectively.

$$\mathbf{H} = \begin{bmatrix} h_{N-1}\ h_{N-2} \neq\ h_1\ h_0 & & & \\ & h_{N-1}\ h_{N-2} \neq h_1\ h_0 & & \\ & & - & \\ & & & h_{N-1}\ h_{N-2} \neq h_1\ h_0 \end{bmatrix} \tag{10.1}$$

and

$$\mathbf{G} = \begin{bmatrix} g_{N-1}\ g_{N-2} \neq g_1\ g_0 & & & \\ & g_{N-1}\ g_{N-2} \neq\ g_1\ g_0 & & \\ & & - & \\ & & & g_{N-1}\ g_{N-2} \neq g_1\ g_0 \end{bmatrix} \tag{10.2}$$

where $\mathbf{H}$ and $\mathbf{G}$ are infinite in length. The rows of $\mathbf{H}$ and $\mathbf{G}$ are successively shifted over by two because of the decimation operation. If h and g are assumed to form an orthonormal set, then as shown in Eqs. (10.3) and (10.4),

$$\mathbf{H}\mathbf{H}^{\mathrm{T}} = \boldsymbol{I} \tag{10.3}$$

and

$$\mathbf{GG}^{\mathrm{T}} = \mathbf{I}, \tag{10.4}$$

where $\mathbf{I}$ is the identity matrix and the superscript $\mathbf{T}$ denotes transposition. We also have

$$\mathbf{GH}^{\mathrm{T}} = \mathbf{0}. \tag{10.5}$$

When the above conditions are satisfied, $\mathbf{H}$ and $\mathbf{G}$ are then orthogonal to each other; and they span two disjoint signal spaces. Furthermore, the inverse of $\mathbf{H}$ and $\mathbf{G}$ are $\mathbf{H}^{-1} = \mathbf{H}^{\mathrm{T}}$ and $\mathbf{G}^{-1} = \mathbf{G}^{\mathrm{T}}$. $\mathbf{H}x$ and $\mathbf{G}x$ give the projections of x onto the subspaces spanned by $\mathbf{H}$ and $\mathbf{G}$, respectively. Since the projections of x are onto two orthogonal subspaces, we have Eq. (10.6).

$$V_{-1} = V_0 \bigoplus W_0, \tag{10.6}$$

where V_{-1} is the original signal space, V_0 is the subspace spanned by $\mathbf{H}$, and W_0 is the subspace spanned by $\mathbf{G}$. This is equivalent to filtering x at the first level of a filter bank. By iterating this procedure for subsequent levels, we obtain Eqs. (10.7) to (10.9).

$$V_{j-1} = Vj \bigoplus W_j, j = 0, 1 \tag{10.7}$$

$$V_j \subset V_{j-1}, \qquad j = 0, 1, \tag{10.8}$$

and

$$V_{-1} = W_0 \bigoplus W_1 \bigoplus \tag{10.9}$$

where j is the number of decomposition levels. This decomposition of the signal space.

V_{-1} is demonstrated to be an orthogonal multiresolution signal decomposition for discrete sequences (Vetterli & Herley, 1992). In addition, it leads to an orthogonal wavelet decomposition (Mallat, 1989).

10.3   W-Transform and W-Matrix

The W-transform treats the signals as finite but does not constrain the length to be a power of two. Note that, although we do require the filter coefficients to form an orthogonal basis, we do not disregard the possibility. Hence, the W-transform leads to a possibly nonorthogonal multiresolution signal decomposition (Kwong & Tang, 1994).

Let us reconsider the filter matrices in Eqs. (10.1) and (10.2). For illustration purposes, let the length of the filters h and g be $N = 4$ and let the signal, x, be of even finite length. Form the array as in Eq. (10.10).

$$
\begin{bmatrix}
h_3 \, h_2 \, h_1 \, h_0 & & & & \\
& h_3 \, h_2 h_1 \, h_0 & & & \\
& & h_3 \, h_2 \quad h_1 \, h_0 & & \\
& & - & & \\
& & & h_3 \, h_2 \quad h_1 \, h_0 & \\
& & & & h_3 \, h_2 \quad h_1 \, h_0
\end{bmatrix}
\tag{10.10}
$$

The coefficients h_0 and h_3 fall outside the square matrix because of the finite extension of the input. To include these coefficients in the matrix, we add them back to the nearest neighborhood that is retained. Thus, we obtain the following matrix as shown in Eq. (10.11).

$$
\mathbf{H}
\begin{bmatrix}
h_3 + h_2 \, h_1 \, h_0 & & & \\
& h_3 \, h_2 \, h_1 \, h_0 & & \\
& & h_3 \, h_2 \, h_1 \, h_0 & \\
& & \div & \\
& & & h_3 \, h_2 \, h_1 \, h_0 \\
& & & h_3 \, h_2 \, h_1 + h_0
\end{bmatrix}
\tag{10.11}
$$

The $\mathbf{G}$ matrix is constructed in a similar fashion. Next, we interleave the rows of the $\mathbf{H}$ and $\mathbf{G}$ matrices to obtain Eq. (10.12).

$$
\begin{bmatrix}
h_3 + h_2 \, h_1 \, h_0 & & & \\
g_3 + g_2 g_1 \, g_0 & & & \\
& h_3 \, h_2 \quad h_1 \, h_0 & & \\
& g_3 \, g_2 \quad g_1 \, g_0 & & \\
& & h_3 \, h_2 \quad h_1 \, h_0 & \\
& & g_3 \, g_2 \quad g_1 \, g_0 & \\
& & \div & \\
& & & h_3 \, h_2 \quad h_1 \, h_0 \\
& & & g_3 \, g_2 \quad g_1 \, g_0 \\
& & & h_3 \, h_2 \quad h_1 + h_0 \\
& & & g_3 \, g_2 \quad g_1 + g_0
\end{bmatrix}
\tag{10.12}
$$

This is defined to be the even-sized **W**-matrix. For odd-length signals, the odd-sized **W**-matrix is given as shown in Eq. (10.13).

$$
\mathbf{W} =
\begin{bmatrix}
h_3 + h_2\ h_1\ h_0 & & & & & & & \\
g_3 + g_2 g_1\ g_0 & & & & & & & \\
& h_3\ h_2 & h_1\ h_0 & & & & & \\
& g_3\ g_2 & g_1\ g_0 & & & & & \\
& & h_3\ h_2 & h_1\ h_0 & & & & \\
& & g_3\ g_2 & g_1\ g_0 & & & & \\
& & & \div & & & & \\
& & & & h_3\ h_2 & h_1\ h_0 & & \\
& & & & g_3\ g_2 & g_1\ g_0 & & \\
& & & & & h_3\ h_2 & h_1 & h_0 \\
& & & & & g_3\ g_2 & g_1 & g_0 \\
& & & & & & h_3\ h_2 + h_1 + h_0
\end{bmatrix}
\tag{10.13}
$$

The W-transform of the image is given in Eq. (10.14).

$$
y = \mathbf{W}x, \quad x = \mathbf{W}^{-1}y
\tag{10.14}
$$

A pair of vectors containing, respectively, the odd and even components of y is formed as a result of the W-transform.

$$
y1 = [y_1, y_3,]', \mathbf{y2} = [y_2, y4,]
\tag{10.15}
$$

We denote the columns of the matrix W^{-1} as

$$
\left[\widehat{g}_1, \widehat{h}_1, \widehat{g}_2, \widehat{h}_2 \ldots \right]
\tag{10.16}
$$

The equivalent form of the second part of Eq. (10.14) can be shown in Eq. (10.17).

$$
x = \left(y_{11}\widehat{g}_1 + y_{12}\widehat{g}_2 \ldots \right) + \left(y_{21}\widehat{h}_1 + y_{22}\widehat{h}_2 \ldots \right).
\tag{10.17}
$$

We can interpret Eq. (10.17) as the x decomposition along the subspaces **G** and **H**, spanned by $\hat{g}$- and $\hat{h}$- vectors, respectively. Observe that for an odd-sized **W**-matrix, the resulting low-pass signal will contain one more sample than the high-pass signal. In either case, the length of the output is always equal to the length of the input. In general, $\mathbf{W}^{-1} \neq \mathbf{W}^{\mathrm{T}}$ and the decomposition are not orthogonal. However, for image compression purposes, only the inverse of the transform is necessary. Note that for orthogonal matrices, if the condition number of the matrix is large, then small data impurities may be present in the transformed signal. It turns out that the **W**-matrices have moderate condition numbers (Kwong & Tang, 1994). In the 2D case, we can assume separability and apply the 1D W-transform to the rows and then to the columns of the image.

Kwong and Tang provided a theorem for generating the coefficients of general **W**-matrices. A particularly useful example is to use $[h_3, h_2, h_1, h_0] = [-1, 3, -3, 1]$ and.

$[g_3, g_2, g_1, g_0] = [-1, 3, 3, -1]$. For this choice, the coefficients exhibit compact support of length 4 and symmetry, and the associated scaling function is relatively smooth. Also the corresponding wavelet has vanishing moments up to order 2. An advantage of relaxing the orthogonality constraint is that linear phase filters can be used in the analysis, thereby ensuring no nonlinear phase distortions. However, orthogonality may be required in the wavelet transformation of signals with high oscillating properties or signals with noise degradation. In such cases, a process known as orthogonal compensation (OC) is used with the W-transform.

10.4   The Orthogonality Compensation Process

As discussed in Sect. 10.3, the W-transform leads to a nonorthogonal wavelet transform. In those situations where orthogonality may lead to better performance, the process of orthogonal compensation is used with the W-transform. Let the vector to be discarded be v. The vector v is decomposed into a linear combination of vectors (g_i) and an error vector that is orthogonal to G as shown in Eq. (10.18).

$$v = (a_1(\widehat{g}_1) + a_2(\widehat{g}_2) + \ldots) + \mathbf{e}. \tag{10.18}$$

We determine a_i and add to the corresponding y_{li}. This ensures that the actual part that is discarded is e, which is orthogonal to G. We take the inner products of v with each $(\widehat{g}_i)$.

10.5   Compression Algorithms for the Nonorthogonal
and Orthogonally Compensated Cases

The compression algorithms applied to the images are slightly different for both the nonorthogonal and orthogonally compensated cases. They are as follows:

- **Nonorthogonal Case.**

  - Implement a W-transform decomposition on the image.
  - Discard small components of the output matrices based on a threshold value or quantize the detail submatrices based some quantization ratios.
  - Reconstruct the compressed image.

- **Orthogonal Compensated Case.**

  - Implement an orthogonally compensated W-transform decomposition on the image in both the x and y directions.
  - Quantize the various matrix levels using a specified quantization ratio.
  - Reconstruct the compressed image.

10.6 Simulation Examples

In this section, we explore some examples to illustrate the compression idea of the wavelet-based W-transform and W-matrices. The nonorthogonal and the orthogonally compensated W-transforms are implemented in the decomposition of the images. We can decompose the images in all cases up to three levels. As many as 12 different types of images are in the examples used. These images vary in sizes, intensities, textures, and backgrounds. The quadratic spline *(kw)* and Daubechies *(kwdau)* wavelet coefficients that can be used in the nonorthogonal and the orthogonally compensated image compression analysis illustration examples are $kw = [0.25, 0.75, 0.75, 0.25]$ and $kwdau = [0.4830, 0.8365, 0.2241, -0.1294, 0.2679, -3.7321]$, respectively. The quantization ratios *(r3)* used in all cases are $r3 = [32.0, 16.0, 8.0]$ for levels 3, 2, and 1, respectively. MATLAB© is used in the implementation of all the simulations.

10.7 Performance Evaluation for the Simulation Examples

In the evaluation of performances of different wavelets and algorithms, many statistical measures of image quality can be used (Akujuobi et al., 1993a, 1993b), as discussed in this section. We can use MSE, SNR, entropy, histogram, compression ratio (CR), and bits per pixel (bpp). In all examples, the measurements are confined to 8 bits per pixel (bpp) grayscale images. The actual formulas used for SNR, MSE, entropy and histogram, compression ratio (CR), and bits per pixel (bpp) are shown in Eqs. (10.19)–(10.24), respectively. The SNR is determined by Eq. (10.19).

$$SNR\ (dB) = 10 \log_{10} \left[\frac{\sum_{i=1}^{N} \sum_{j=1}^{N} (x_{ij})^2}{\sum_{i=1}^{N} \sum_{j=1}^{N} (x_{ij} - \widehat{x}_{ij})^2} \right] \tag{10.19}$$

The MSE is determined by Eq. (10.20).

$$MSE = \left[\frac{\sum_{i=1}^{N} \sum_{j=1}^{N} (x_{ij} - \widehat{x}_{ij})^2}{N^2} \right] \tag{10.20}$$

where x and $\widehat{x}$ in Eqs. (10.19) and (10.20) are the original and the reconstructed images, respectively. The entropy of the images before and after compression is determined by Eq. (10.21).

$$H(x) = - \sum_{i=1}^{N} (P_i \log P_i) \tag{10.21}$$

where Pi is the relative probability of the pixels. Pi can be calculated based on the law of large numbers and the relative frequency interpretation of probability, that is, the probability that a certain pixel (level of gray) occurred divided by the total number of pixels. This can be verified by the fact that

$$\sum_{i=1}^{N} (P_i) = 1 \tag{10.22}$$

Histogram is defined as a representation of the number of occurrences of each i (i = 1 to N) image pixel value. Compression ratio (CR) is determined as shown in Eq. (10.23).

$$CR = \frac{\text{Number of bits in the Original Image}}{\text{Number of bits in the Compressed Image}} \tag{10.23}$$

The bit rate per pixel (bpp) is determined as shown in Eq. (10.24).

$$\text{Bit Rate}\ (bpp) = \frac{8 \text{ bits per pixel}}{CR} \tag{10.24}$$

10.8   The Simulation Example Results and Discussions

The MSE and SNR computations and their differences for the nonorthogonal and orthogonally compensated W-transform using 12 different images are shown in Tables 10.1 and 10.2, respectively. The entropy values of the 12 different images

Table 10.1 W-transform nonorthogonal and orthogonal compensated MSE and their differences for 12 different images

mages	Wavelet	Orthogonal	Nonorthogonal	Difference	Quantization ratio
Food	Kw	134.05	207.76	73.71	32,16,8
	Kwdau	103.04	103.54	0.50	32,16,8
Pepper	Kw	21.81	31.88	10.07	32,16,8
	Kwdau	21.49	22.5422	1.05	32,16,8
Durer	Kw	43.47	62.95	19.47	32,16,8
	Kwdau	39.85	40.12	0.27	32,16,
Clown	Kw	37.07	51.94	14.87	32,16,8
	Kwdau	31.50	31.65	0.15	32,16,8
Earth	Kw	13.55	21.42	7.87	32,16,8
	Kwdau	12.40	12.52	0.12	32,16,8
Porsche	Kw	29.90	42.54	12.64	32,16,8
	Kwdau	25.72	26.72	1.00	32,16,8
Spine	Kw	1.15	1.86	0.72	32,16,8
	Kwdau	0.94	0.94	0.00	32,16,8
Flower	Kw	85.65	127.19	41.54	32,16,8
	Kwdau	71.06	71.50	0.44	32,16,8
Finger	Kw	30.78	45.51	14.73	32,16,8
	Kwdau	31.34	31.60	0.26	32,16,8
Acc car	Kw	32.82	46.75	13.92	32,16,8
	Kwdau	27.51	27.60	0.10	32,16,8
Boat	Kw	26.02	36.88	10.86	32,16,8
	Kwdau	24.70	25.90	1.20	32,16,8
Aerial view	Kw	58.67	83.31	24.64	32,16,8
	Kwdau	56.50	57.67	1.14	32,16,8

for nonorthogonal and orthogonally compensated W-transform compression are shown in Table 10.3. The compression ratios for the nonorthogonal and orthogonally compensated (OC) W-transform compression using 12 different images are shown in Table 10.4. Their differences are also shown in Table 10.4. The original 8-bit 256 × 256 pepper and 256 × 256 race car images are shown in Figure 10.1a and d, respectively.

Figure 10.1b is the histogram of the original pepper image before compression, while Figure 10.1c shows the entropy histogram of the original pepper image and its entropy value of 7.589. Figure 10.2a and b shows the reconstructed pepper image after nonorthogonal and orthogonally compensated W-transform compression, respectively, using the quadratic spline wavelet coefficients (*kw*). Figure 10.2c and d shows the respective histograms of Figure 10.2a and b.

Figure 10.3a and b shows the reconstructed pepper image after nonorthogonal and orthogonally compensated W-transform compression, respectively, using the Daubechies wavelet coefficients (*kwdau*). Figure 10.3c and d shows the respective histograms of Figure 10.3a and b. The entropy histograms of the reconstructed

Table 10.2 W-transform nonorthogonal and orthogonally compensated compression SNR and their differences for 12 different images

Images	Wavelet	Orthogonal	Nonorthogonal	Difference	Quantization ratio
Food	Kw	20.56	15.65	1.91	32,16,8
	Kwdau	21.70	21.68	1.91	32,16,8
Pepper	Kw	29.01	18.65	1.91	32,16,8
	Kwdau	29.08	28.86	0.22	32,16,8
Durer	Kw	20.57	18.96	1.61	32,16,8
	Kwdau	20.95	20.92	0.03	32,16,
Clown	Kw	15.52	14.06	1.46	32,16,8
	Kwdau	16.23	16.21	0.02	32,16,8
Earth	Kw	18.33	16.34	1.99	32,16,8
	Kwdau	18.71	18.67	0.04	32,16,8
Porsche	Kw	26.01	24.47	1.54	32,16,8
	Kwdau	26.72	26.49	0.23	32,16,8
Spine	Kw	26.86	24.76	2.10	32,16,8
	Kwdau	27.71	22.71	0.00	32,16,8
Flower	Kw	23.96	22.25	1.71	32,16,8
	Kwdau	24.77	24.75	0.02	32,16,8
Finger	Kw	24.28	22.58	1.70	32,16,8
	Kwdau	24.20	24.17	0.03	32,16,8
Race car	Kw	27.47	25.94	1.53	32,16,8
	Kwdau	28.24	28.23	0.01	32,16,8
Boat	Kw	28.51	27.00	1.51	32,16,8
	Kwdau	28.73	28.52	0.21	32,16,8
Aerial view	Kw	25.76	24.23	1.53	32,16,8
	Kwdau	25.92	25.83	0.09	32,16,8

pepper image after the nonorthogonal and the orthogonally compensated W-transform compression are shown in Figure 10.4a and b, respectively, using the quadratic spline wavelet coefficients (kw). The entropy histograms of the reconstructed pepper image after the nonorthogonal and the orthogonally compensated W-transform compression are shown in Figure 10.4c and d, respectively, using the Daubechies wavelet coefficients ($kwdau$). Fig. 10.5a and 10.5b shows the difference between the original and nonorthogonal and OC reconstructed compressed pepper images using the quadratic spline wavelet (kw), while Fig. 10.5c and 10.5d, respectively, shows the differences using the Daubechies wavelet coefficients (kwdau).

The MSE for kw is higher than the kwdau in both the OC and nonorthogonal cases. The SNR for kw is slightly lower than the SNR for $kwdau$ in both the OC and nonorthogonal cases. There should be no surprises in these MSE and SNR results for kw and $kwdau$. The scaling function of kw is smoother than that of $kwdau$. Because of that, kw is more suitable to approximate smooth functions, but $kwdau$ is better for functions that oscillates very fast, especially when there is a lot of noise. For images,

Table 10.3 Entropy for nonorthogonal and orthogonal compensated W-transform for 12 different images

Images	Wavelet	Original	Orthogonal	Nonorthogonal	Quantization ratio
Food	Kw	5.030	7.781	7.950	32,16,8
	Kwdau	5.030	7.672	7.764	32,16,8
Pepper	Kw	7.589	7.618	7.623	32,16,8
	Kwdau	7.5897	7.610	7.607	32,16,8
Durer	Kw	6.930	6.985	7.007	32,16,8
	Kwdau	6.930	6.982	6.982	32,16,
Clown	Kw	5.126	6.058	6.148	32,16,8
	Kwdau	5.126	5.994	5.990	32,16,8
Earth	Kw	5.229	5.552	5.690	32,16,8
	Kwdau	5.229	5.509	5.506	32,16,8
Porsche	Kw	7.140	7.547	7.536	32,16,8
	Kwdau	7.140	7.435	7.436	32,16,8
Spine	Kw	4.094	4.229	4.296	32,16,8
	Kwdau	4.094	4.185	6.74	32,16,8
Flower	Kw	6.425	7.794	4.296	32,16,8
	Kwdau	6.425	7.772	6.695	32,16,8
Finger	Kw	6.192	6.953	6.985	32,16,8
	Kwdau	6.192	6.928	6.930	32,16,8
Race car	Kw	7.111	7.491	7.499	32,16,8
	Kwdau	7.111	7.318	7.318	32,16,8
Boat	Kw	6.970	7.017	7.032	32,16,8
	Kwdau	6.970	6.994	7.032	32,16,8
Aerial view	Kw	7.312	7.356	7.379	32,16,8
	Kwdau	7.312	7.355	7.356	32,16,8

the data usually have much noise, so *kw* will filter out quite a bit the noise, while *kwdau* can preserve some of the noise (perhaps distort some of the smoother part). Hence, for *kw*, the error between the original and the reconstructed compressed image may be larger (since it contains the noise that has been filtered out) than the corresponding original and the reconstructed compressed image for *kwdau*. The simulation example results, therefore, show much larger MSE and lower SNR for *kw*. Does this mean that *kw* is not as good as *kwdau*? The answer is obviously *no*, as can be demonstrated in the values of the CRs shown in Table 10.4 and in the values of the bit rates, shown in Table 10.5. MSE and SNR may not be the proper measures of quality. Nevertheless, it is commonly used in the elevation of compression techniques, and it does provide some measure of relative performance.

The entropy values for *kw* are larger than the entropy values for *kwdau* in both types of W-transform as shown in Table 10.3 and Figure 10.4a, b, c, and d, respectively. For the histograms, because MATLAB© processes vectors, therefore, when it gets a 2D input, it calculates the histogram for each new row in the array independently and then superimposes them on top of each other. The "lines" you see

Table 10.4 W-transform nonorthogonal and orthogonal compensated compression ratios and their differences for 12 different images

Images	Wavelet	Orthogonal	Nonorthogonal	Difference	Quantization ratio
Food	Kw	6.92	5.02	1.90	32,16,8
	Kwdau	8.04	8.03	0.02	32,16,8
Pepper	Kw	11.67	11.41	0.25	32,16,8
	Kwdau	9.08	9.08	0.00	32,16,8
Durer	Kw	9.10	8.49	0.61	32,16,8
	Kwdau	8.02	8.02	0.00	32,16,
Clown	Kw	18.85	16.07	2.78	32,16,8
	Kwdau	15.06	15.08	−0.03	32,16,8
Earth	Kw	26.25	23.28	2.96	32,16,8
	Kwdau	20.95	20.95	0.00	32,16,8
Porsche	Kw	10.61	10.37	0.24	32,16,8
	Kwdau	8.84	8.85	−0.01	32,16,8
Spine	Kw	54.90	53.53	1.37	32,16,8
	Kwdau	36.13	36.13	0.00	32,16,8
Flower	Kw	6.24	5.23	1.00	32,16,8
	Kwdau	6.25	6.24	0.01	32,16,8
Finger	Kw	10.39	10.29	0.10	32,16,8
	Kwdau	8.22	8.22	0.00	32,16,8
Race car	Kw	8.73	8.57	0.16	32,16,8
	Kwdau	7.28	7.28	0.00	32,16,8
Boat	Kw	11.22	11.32	−0.10	32,16,8
	Kwdau	8.77	8.79	−0.02	32,16,8
Aerial view	Kw	6.03	5.89	0.14	32,16,8
	Kwdau	5.32	5.32	0.00	32,16,8

on the histograms result from the different superimposed histograms. As shown in Fig. 10.5a and 10.5b, *kw* produces a more explicit contour detail of the difference images than kwdau shown in Fig. 10.5c and 10.5d.

Table 10.5 shows the bit rates (bpp) and their differences for the nonorthogonal and orthogonally compensated W-transform compression analysis using 12 different images. The original 8-*bit* 512 × 512 fingerprint and 256 × 256 aerial view images are shown in Figure 10.6a and b, respectively. The reconstructed fingerprint and aerial view compressed images using the nonorthogonal W-transform quadratic spline wavelet coefficients (*kw*) are shown in Figure 10.6c and d, respectively. For the aerial view compressed image, the CR is 5.89, while that of fingerprint is 10.29. The reconstructed fingerprint and aerial view compressed images using the orthogonally compensated W-transform quadratic spline wavelet coefficients are shown in Fig. 10.7a and 10.7b, respectively. Figure 10.7c and 10.7d is the corresponding reconstructed images using the Daubechies wavelet coefficients (*kwdau*). Their CRs are 8.22 and 5.32 respectively. The reconstructed images using the orthogonally compensated W-transform, for *kwdau*, are shown in Fig. 10.8a and 10.8b. The

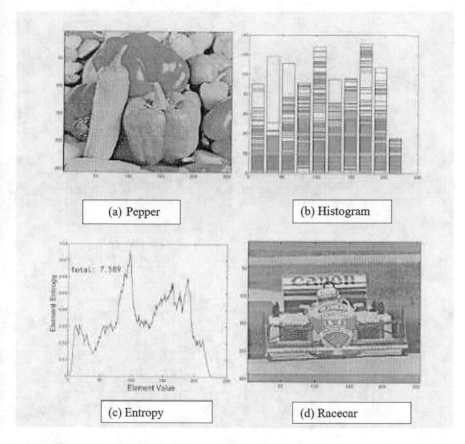

Fig. 10.1 (**a**) Original 8-bit 256 × 256 pepper image; (**b**) histogram of the original pepper image; (**c**) entropy histogram of the original pepper image; (**d**) original 8-bit 256 x 256 race car image

difference between the original and the reconstructed nonorthogonal and orthogonally compensated compressed aerial view image for *kw* is shown in Fig. 10.8c and 10.8d, respectively. At the bottom of Figs. 10.5c and 10.5d, 10.6, and 10.7a and 10.7b are the total nonzero (nz) values after quantization for levels 3, 2, and 1, respectively.

It should be observed that in over 80 percent of the images used in the simulation examples from both the OC and the nonorthogonal cases, the CRs for the *kw* were higher than the CRs for *kwdau*. In addition, the bit rates for *kw* were lower than the bit rates for the kwdau in both the OC and nonorthogonal cases in over 80 percent of the images. Differences exist between the OC and nonorthogonal W-transform applications to compression. The CRs are much lower in the nonorthogonal case than in the OC case, whether we use *kw* or *kwdau*, although there were cases where noticeable changes were not observed when *kwdau* is used. The bit rates are much lower using *kw* than *kwdau* in the nonorthogonal and OC cases.

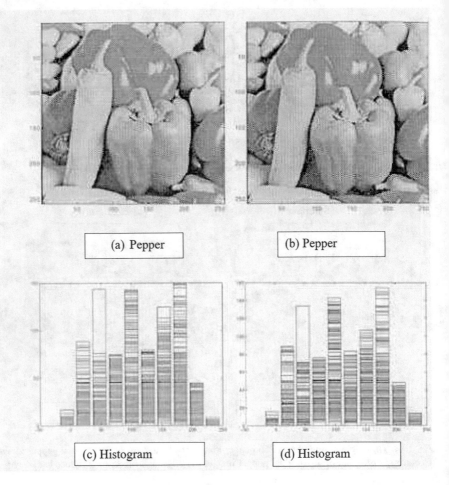

(a) Pepper

(b) Pepper

(c) Histogram

(d) Histogram

Fig. 10.2 (**a**) Reconstructed pepper image after compression using nonorthogonal W-transform for kw; (**b**) reconstructed pepper image after compression using orthogonal compensated W-transform for kw; (**c**) histogram of reconstructed pepper image after compression using nonorthogonal W-transform for kw; (**d**) histogram of reconstructed pepper image after compression using orthogonal compensated W-transform for kw

Summary

1. The W-transform is constructed by relaxing the orthogonality conditions required in the traditional wavelet transforms.
2. The W-transform is an efficient tool for image compression since it does not extend the output image.

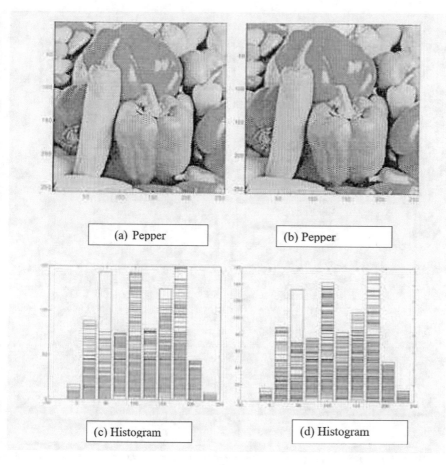

Fig. 10.3 (**a**) Reconstructed pepper image after compression using nonorthogonal W-transform for kwdau; (**b**) reconstructed pepper image after compression using orthogonal compensated W-transform for kwdau; (**c**) histogram of reconstructed pepper image after compression using nonorthogonal W-transform for kwdau; (**d**) histogram of reconstructed pepper image after compression using orthogonal compensated W-transform for kwdau

3. One of the advantages of the W-transform over some of the classical wavelet transforms is that it allows for the efficient handling of images of any sizes without the usual restriction to multiples of a power of two.
4. Orthogonality conditions are generally desirable for good compression and reconstruction performance, when dealing with noise-degraded images and with images that have high oscillating properties.
5. The W-transform treats the signals as finite but does not constrain the length to be a power of two.
6. The bit rates are much lower using *kw* than *kwdau* in the nonorthogonal and OC cases.

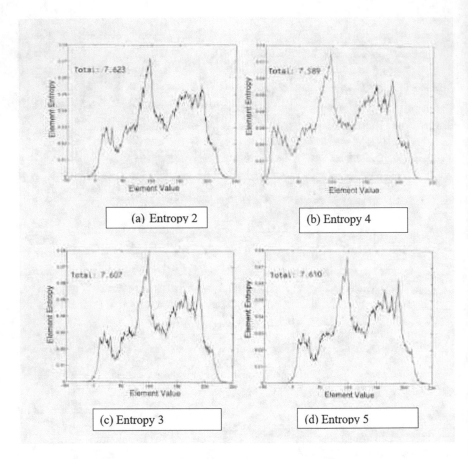

Fig. 10.4 (**a**) Entropy histogram of reconstructed pepper image after compression using nonorthogonal W-transform for kw; (**b**) entropy histogram of reconstructed pepper image after compression using orthogonal compensated W-transform for kw; (**c**) entropy histogram of reconstructed pepper image after compression using nonorthogonal W-transform for kwdau; (**d**) entropy histogram of reconstructed pepper image after orthogonal compensated compression for kwdau

Review Questions

10.1 The W-transform can be constructed by relaxing the orthogonality conditions required in the traditional wavelet transforms.

   (a) True.
   (b) False.

10.2 The W-transform is an efficient tool for image compression since it does extend the output image.

   (a) True.
   (b) False.

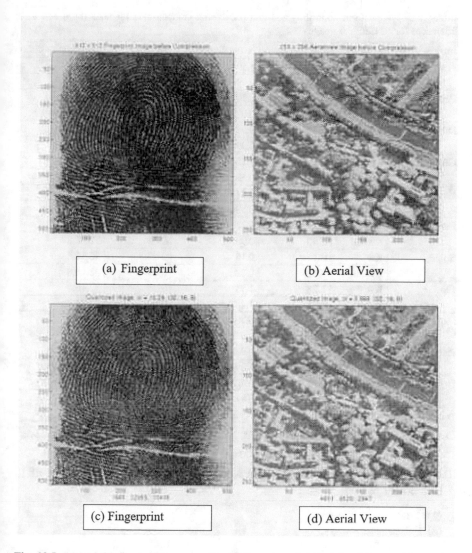

(a) Fingerprint

(b) Aerial View

(c) Fingerprint

(d) Aerial View

Fig. 10.5 (a) and (b), *kw* produces a more explicit contour detail of the difference images than *kwdau* shown in (c) and (d) of the fingerprint and aerial images

10.3 Particularly useful for image compression is a special nonorthogonal W-transform called the quadratic spline transform because:

   (a) It has symmetrical properties.
   (b) The scaling functions are smooth.
   (c) It has symmetrical properties and the scaling functions are smooth.
   (d) None of the above.

Table 10.5 W-transform nonorthogonal and orthogonal compensated bit rates (bpp) and their differences for 12 different images

Images	Wavelet	Orthogonal	Nonorthogonal	Difference	Quantization ratio
Food	Kw	1.15	1.59	0.44	32,16,8
	Kwdau	0.99	0.99	0.00	32,16,8
Pepper	Kw	0.68	0.70	0.02	32,16,8
	Kwdau	0.88	0.88	0.00	32,16,8
Durer	Kw	0.88	0.94	0.06	32,16,8
	Kwdau	0.99	0.99	0.00	32,16,
Clown	Kw	0.49	0.42	0.07	32,16,8
	Kwdau	0.53	0.53	0.00	32,16,8
Earth	Kw	0.30	0.34	0.04	32,16,8
	Kwdau	0.32	0.38	0.06	32,16,8
Porsche	Kw	0.75	0.77	0.02	32,16,8
	Kwdau	0.90	0.90	0.00	32,16,8
Spine	Kw	0.15	0.15	0.00	32,16,8
	Kwdau	0.22	0.22	0.00	32,16,8
Flower	Kw	1.28	1.53	0.25	32,16,8
	Kwdau	1.28	1.28	0.00	32,16,8
Finger	Kw	0.76	0.78	0.02	32,16,8
	Kwdau	0.97	0.97	0.00	32,16,8
Race car	Kw	0.91	0.93	0.02	32,16,8
	Kwdau	1.09	1.09	0.00	32,16,8
Boat	Kw	0.71	0.70	0.01	32,16,8
	Kwdau	0.91	0.91	0.00	32,16,8
Aerial view	Kw	1.32	1.35	0.03	32,16,8
	Kwdau	1.50	1.50	0.00	32,16,8

10.4 The W-transform treats the signals as finite but does not constrain the length to be a power of two.

(a) True.
(b) False.

10.5 An advantage of relaxing the orthogonality constraint is that linear phase filters cannot be used in the analysis, thereby ensuring no nonlinear phase distortions.

(a) True.
(b) False.

10.6 The compression algorithms applied to the images are slightly not different for both the nonorthogonal and orthogonally compensated cases.

(a) True.
(b) False.

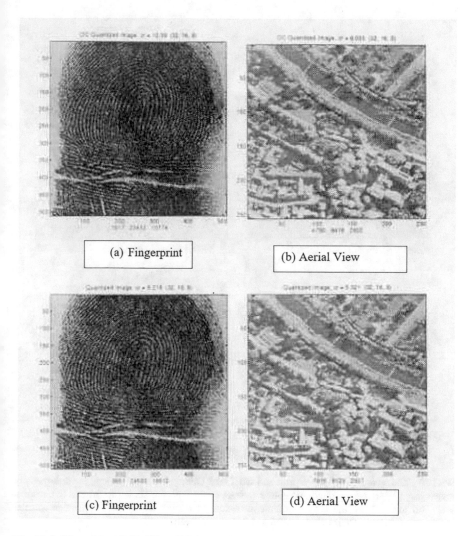

(a) Fingerprint

(b) Aerial View

(c) Fingerprint

(d) Aerial View

Fig. 10.6 The original 8-*bit* 512×512 fingerprint and 256×256 aerial view images are shown in (a) and (b), respectively. The reconstructed fingerprint and aerial view compressed images using the nonorthogonal W-transform quadratic spline wavelet coefficients (*kw*) are shown in (c) and (d), respectively. For the aerial view compressed image, the CR is 5.89, while that of fingerprint is 10.29

10.7 The MSE for *kw* is higher than the *kwdau* in both the OC and nonorthogonal cases.

(a) True.
(b) False.

10.8 The SNR for *kw* is not slightly lower than the SNR for *kwdau* in both the OC and nonorthogonal cases.

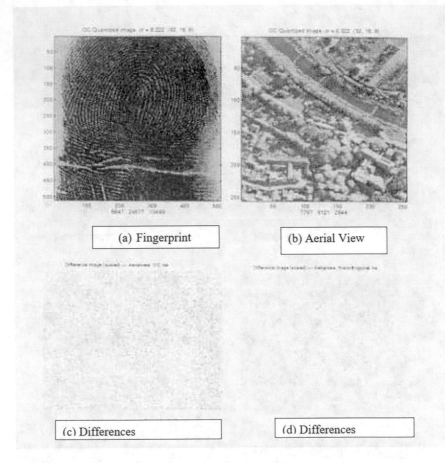

(a) Fingerprint

(b) Aerial View

(c) Differences

(d) Differences

Fig. 10.7 The reconstructed fingerprint and aerial view compressed images using the orthogonally compensated W-transform quadratic spline wavelet coefficients are shown in (**a**) and (**b**), respectively. (**c**) and (**d**) are the corresponding reconstructed images using the Daubechies wavelet coefficients (*kwdau*). Their CRs are 8.22 and 5.32 respectively

(a) True.

(b) False.

10.9 The bit rates are much lower using *kw* than *kwdau* in the nonorthogonal and OC cases.

(a) True.

(b) False.

10.10 The compression ratios for the *kw* are higher than the CRs for *kwdau*.

(a) True.

(b) False.

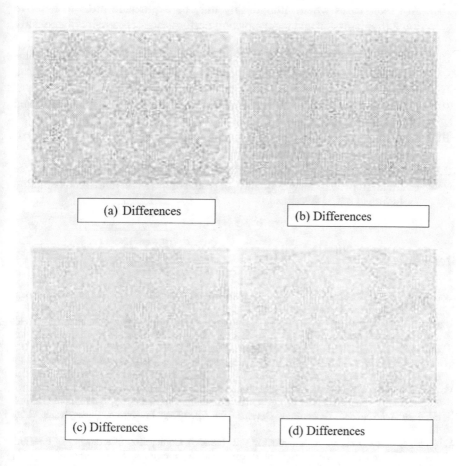

Fig. 10.8 (**a**) Difference between original and reconstructed nonorthogonal compressed pepper image kw; (**b**) difference between original and reconstructed orthogonally compensated compressed pepper image for kw; (**c**) difference between original and reconstructed nonorthogonal compressed pepper image for kwdau; (**d**) difference between original and reconstructed orthogonally compensated compressed pepper image for kwdau

Answers: 10.1a, 10.2b, 10.3c, 10.4a, 10.5a, 10.6a, 10.7b, 10.8b, 10.9a, 10.10a.

Problems
10.1. What is one of the differences between the W-transform and the traditional wavelet transform techniques?
10.2. What is one of the advantages of the W-transform over some of the classical wavelet transforms?
10.3. Why is W-transform an efficient tool for image compression?
10.4. Why is the quadratic spline wavelet transform particularly useful for image compression?

10.5. Are there cases when orthogonality may be required in order to improve image compression and reconstruction performance? How can orthogonality be achieved in this case?

10.6. Describe briefly how compensation can be achieved in the orthogonal compensated and nonorthogonal compression of images using the W-transform.

10.7. Determine the W-transform of a signal x using W-matrix.

10.8. State the algorithm for the nonorthogonal case of the W-transform-based compensation process.

10.9. State the algorithm for the orthogonal compensated case of the W-transform-based compensation process.

10.10. Name some of the statistical measures that can be used in the performance evaluation of the image quality of the compression and reconstruction techniques.

References

Akansu, A. N., & Haddad, R. A. (1992). *Multiresolution signal decomposition: Transforms, subbands, wavelets*. Academic Press.

Akujuobi, C. M. & Kwong, M. K. (1995). Nonorthogonal and orthogonally compensated wavelet analysis: an application to image compression using the W-transforms. *Mathematics and Computer Science Division Preprint*, Argonne National Laboratory, 1995.

Akujuobi, C. M., & Lian, J. (2005). Image compression using nonorthogonal and orthogonally compensated W-Matrices. *Chinese Journal of Engineering Mathematics, 22*(5), 914–928.

Akujuobi, C. M., Parikh, V., & Baraniecki, A. Z. (1993a). Performance evaluation of multiresolution image analysis using wavelet transform. *Proceedings of International Association of Science and Technology Development (IASTED)*, Pittsburgh, Pennsylvania, May 10–12.

Akujuobi, C. M., Parikh, V., & Baraniecki, A. Z. (1993b). Comparison of efficiency of time and frequency domain implementation of discrete wavelet transform. *Proceedings of IASTED*, Pittsburgh, Pennsylvania, May 10–12.

Daubechies, I. (1989). Orthonormal bases of with finite support-connection with discrete filters. In J. M. Combes, A. Grossmann, & Tchamitchian (Eds.), *Wavelets: Time frequency methods and phase space, proceedings of the 1987 international workshop on wavelets and applications, Marseille, France, December 14–18* (pp. 36–66). Springer.

Daubechies, I. (1990). The wavelet transform, time-frequency localization and signal analysis. *IEEE Transactions on Information Theory, 36*(5), 961–1005.

Daubechies, I. (1998). Orthonormal bases of compactly supported wavelets. *Communications on Pure and Applied Mathematics, 41*, 969–996.

Kwong, M. K. (1994). MATLAB implementation of W-matrix multiresolution analysis. *Mathematics and Computer Science Division Preprint*, MCS-P462–0894, Argonne National Laboratory.

Kwong, M. K. & Tang, P. T. P. (1994) W-matrices Multiresolution Analysis. *Mathematics and Computer Science Division Preprint*, MCS-P449–0794, Argonne National Laboratory.

Mallat, S. G. (1989). A theory for multiresolution signal decomposition: The wavelet representation. *IEEE Transactions on Pattern Analysis and Machine Intelligence, 11*(7), 674–693.

Vetterli, M., & Herley, C. (1992). Wavelets and filter banks: Theory and design. *IEEE Transactions on Signal Processing, 40*(9), 2207–2232.

Chapter 11
Wavelet Application to Image and Data Compression

"We shall draw from the heart of suffering itself the means of inspiration and survival."

Sir Winston Churchill

11.1 Introduction

The compression of images and data of all kinds has become essential more than ever in the twenty-first century where information is expected at a much more faster rate in key application areas such as transmission of information in a communication system considering the limitations of the channel capacity and the storage issues in databases.

The science or art of coding digital images efficiently to reduce the number of bits needed to represent an image is known as image compression.

Compression of images and data is required to reduce the storage and transmission costs while at the same time trying to maintain good image and data quality. The compression of image data is achieved by taking advantages of the redundancies in the data image. The redundancies can be temporal, spectral, or spatial. Temporal redundancy is caused because of the correlation of different frames in a sequence of images such as videoconferencing applications or in broadcast images. The spectral redundancy is caused because of correlation between different color planes. The spatial redundancy is caused because of the correlation between neighboring pixels.

In this chapter, we discuss wavelet-based application to image and data compression, need for compression, its principles, and classes of compression and various algorithm of image compression. This chapter gives a recipe for selecting one of the popular image compression algorithms based on wavelet, JPEG/DCT, VQ, and fractal approaches. We discuss the advantages and disadvantages of these algorithms for compressing grayscale images. We discuss the wavelet-based EZWT, SPIHT, EBCOT, WDR, and ASWDR algorithms and the usefulness of the wavelet-based compression coding schemes (Rawat et al., 2015; Taubman, 2000; Walker & Nguyen, 2001; Raja & Suruliandi, 2011; Kumar & Chaudhary, 2013).

© Springer Nature Switzerland AG 2022
C. M. Akujuobi, *Wavelets and Wavelet Transform Systems and Their Applications*,
https://doi.org/10.1007/978-3-030-87528-2_11

11.2 Compression Ideas

The idea behind the compression of images is that since most of these images have characteristics that are common of having pixels that are neighbors and are correlated, they therefore contain information that are redundant. It becomes necessary then to find a representation of the image that may not be correlated. In the compression of images, there are two important components. They are irrelevancy reduction component and redundancy reduction component. Omitting that part of the signal that may not be noticeable at the signal-receiving end, especially, by the human visual system (HVS) is known as irrelevancy reduction. When you attempt to remove duplication from the source of the signal/image/video, that is known as redundancy reduction. Information redundancy r can be defined as represented in Eq. (11.1).

$$r = b - H_e \qquad (11.1)$$

where b is the smallest number of bits with which the image quantization levels can be represented and H_e are the extra bits not needed to represent the information. The three types of redundancy are irrelevant information redundancy, spatial and temporal redundancy, and finally coding redundancy.

11.2.1 Irrelevant Information Redundancy

The human visual system (HVS) normally does ignore information that may not make sense in the recognition of the information (image) of interest. Therefore, that extra information becomes irrelevant to the HVS because without it, you can still recognize the image, which means that it is redundant. In image compression, the key is reducing all of the redundant/irrelevant information leaving only the relevant information. This means the reduction in the number of bits that can be used to represent the image. It also means the removal of most of the spectral and spatial redundancies.

11.2.2 Spatial and Temporal Redundancy

Image pixels are highly correlated and, because of that, you find lots of information that are duplications. As an example, in video sequences, information is duplicated in temporally correlated pixels. Another example is elements that are duplicated within a structure, such as pixels in a still image and bit patterns in a file. Exploiting spatial redundancy is how compression is performed. Spatial (or intra-frame) compression takes place on each individual frame of the video, compressing the pixel information as though it were a still image. Temporal (or inter-frame) compression happens over a series of frames and takes advantage of areas of the image that remain unchanged from frame to frame, throwing out data that are repeated pixels.

Temporal compression relies on the placement of key frames interspersed throughout the frames sequence. The key frames are used as masters against which the following frames (called delta frames) are compared. It is recommended that a key frame be placed once every 3–10 s. Videos without a lot of motion, such as talking head clips, take the best advantage of temporal compression. Videos with pans and other motions are compressed less efficiently.

11.2.3   Coding Redundancy

Using a system of symbols such as bits, letters, and numbers to represent information is known as coding the information. Coding redundancy is associated with the representation of information. The information is represented in the form of codes. As an example, every piece of coded information has code symbols called codeword. The length of the codeword is the number of code symbols associated with that particular information. In most of the grayscale images, 8-bit codes are used to represent the intensities of the images. If the gray levels of an image are coded in a way that uses more code symbols than necessary to represent each of the gray levels, then the resulting image is said to contain coding redundancy.

11.3   Justification for Compression

The principle idea of compression is to efficiently reduce the number of bits needed to represent an image and data. Is it really important that we compress images and why? To answer these questions, let us assume that we have a 512×512 pixel, 8-bits-per-pixel, gray-level image. It will require about 2,097,152 bits for storage and approximately about 4 min to transmit the image over a communication system. In the case of a color image of 512×512 pixel multimedia data of 8-bits-per-pixel, it will require about 786 Kbytes for storage, 6.29 M bits per image as its transmission bandwidth, and about 3 min and 39 s as the transmission time. It is obvious that the storage or even the transmission of a few of the images would definitely pose a problem, especially in the twenty-first century where information is needed at a much faster rate. There are many such examples that could be used to justify the reasons for the compression of images. Other examples may include facsimile transmission of graphic documents over telephone lines; medical image data compression, archival storage, and broadcast compression are just some of the few potential application areas that may need image data compression. If these images were not compressed, storage and transmission of information will suffer very immensely. However, there are different types of compression techniques as discussed the next section.

11.4   The Different Modes of Compression

There are different types of compression moes. These different modes can be classified as lossless compression, lossy compression, predictive compression, and transform coding compression.

11.4.1   Lossless Compression Mode

In lossless compression techniques, there is no loss of information. One of the characteristics of the losslessly compressed image data is that the original image data can be recovered exactly from the compressed data. This kind of compression technique is generally used for applications that cannot tolerate any difference between the original and reconstructed data. As an example, text compression is an important area for lossless compression. This is because it is important that the reconstruction of the text be identical to the original text, as very small differences can result in statements with very different meanings. Just consider the sentences "John not dead" and "John now dead." The same thing can be said for certain kinds of bank records and for computer files.

When it is absolutely necessary that data of any kind are to be processed or "enhanced" later to yield more information, it is important that the integrity be preserved. As an example, suppose we compressed an x-ray (medical image) of a sick patient in a lossy fashion; and the difference between the reconstruction and the original was visually undetectable in this era of telemedicine. If this image was later enhanced, the previously undetectable differences may cause the appearance of artifacts that could seriously mislead the medical doctor. The price to pay for this kind of mishap may be a human life; therefore, it makes sense to be very careful about using a compression scheme that generates a reconstruction that is different from the original.

It is very true that there may be many situations that require compression where we want the reconstruction to be identical to the original. It is also very true that there may be many occasions in which it is possible to relax this requirement in order to get more compression. In these occasions, we can look to lossy compression techniques.

11.4.2 Lossy Compression Mode

In lossy compression techniques, there is some loss of information. One of the characteristics of the lossy compressed image data is that generally it cannot be recovered or reconstructed exactly as the original image. The advantage of this kind of compression is that we can generally obtain much higher compression ratios than is possible with lossless compression. There are many applications where this lack of exact reconstruction is not a problem. As an example, when transmitting or storing speech, the exact value of each sample of the speech is not necessary. Depending on the quality required of the reconstructed speech, varying amounts of loss of information about the value of each sample can be acceptable. If the quality of the reconstructed speech is to be similar to that heard on the telephone, a significant loss of information can be acceptable. However, if the reconstructed speech needs to be of the quality heard on a compact disc, the amount of information loss that can be acceptable may be much lower.

In a similar way, when viewing a reconstruction of a video sequence, the fact that the reconstruction is different from the original is generally not important as long as the differences do not result in annoying artifacts. Thus, video is generally compressed using lossy compression. Once we have developed a data compression scheme, we need to be able to measure its performance. Because of the number of different areas of application, different terms have been developed to describe and measure the performance.

11.4.3 Predictive Compression Mode

The predictive compression technique is a spatial domain technique because it operates on the pixel values directly. It predicts a data point by extrapolating previous data points. The number of previous data points that are used to formulate a predicted value is known as the order of the prediction. In applications with a time variant forcing function, a sample is correlated with only a fixed number of previous data points, and thus, a higher order prediction may not necessarily be better. The prediction of a sample should, in general, be based upon the mathematical descriptor that best describes the variable being sampled. Generally, this is a very hard task, especially, when the forcing function is not known.

Predictive compressions use image information redundancy (correlation of data) to construct an estimate $\sim f(i, j)$ of the gray-level value of an image element (i,j) from values of gray levels in the neighborhood of (i, j). In image parts where data are not correlated, the estimate $\sim f$ will not match the original value. The differences between estimates and reality, which may be expected to be relatively small in absolute terms, are coded and transmitted together with the prediction model parameters of which the whole set now represents compressed image data. The gray value at the location

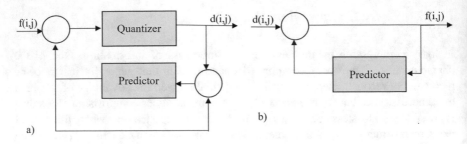

Fig. 11.1 Differential pulse code modulation. (**a**) Compression, (**b**) reconstruction

(i, j) is reconstructed from a computed estimate $\sim\!f(i,j)$ and the stored difference $d(i,j)$ as shown in Eq. (11.2).

$$d(i,j) = \dot{f}(i,j) - f(i,j) \tag{11.2}$$

One particular example of predictive coding is the differential pulse code modulation (DPCM). The quantizer is used in the quantization of the image gray-level values of an image element. The predictor is used in predicting the data points (Fig. 11.1).

11.4.4 Transform Coding Compression Mode

The transform coding compression technique is typically lossless and can be reversed easily on its own but is used to enable better and more targeted quantization. This then results in a lower-quality copy of the original input. It is a type of data compression for "natural" data like audio signals or photographic images. In transform coding, knowledge of the application is used to choose information to discard, thereby lowering its bandwidth. The remaining information can then be compressed through a variety of methods. When the output is decoded, the result may not be identical to the original input, but is expected to be close enough for the purpose of the application. The transform coding compression techniques provide greater image data compression compared to predictive techniques; however, it does this at the expense of greater computation.

11.5   The Different Compression Techniques

In the compression of images, there are different types of compression techniques that can be used. The techniques have their different characteristics. Some of the most prominent techniques are JPEG/DCT, VQ, fractal, and wavelet compression techniques.

11.5.1   JPEG/DCT Compression Technique

The Joint Photographic Experts Group (JPEG) as a committee created the JPEG standard and other still picture coding standards. The word "Joint" is as a result of the fact that other members, such as ISO, TC97, WG8, and CCITT SGVIII, joined the group. The JPEG standard was developed in the late 1980s. They examined several transform coding techniques and finally selected the discrete cosine transform (DCT) because it was by far the most efficient practical compression technique in their decision. The JPEG standard was published in 1992. The DCT, which was initially proposed by Nasir Ahmed in 1972, expresses a finite sequence of data points in terms of a sum of cosine functions oscillating at different frequencies. It is a widely used transformation technique in signal processing and data compression. It is used in most digital media, including digital images such as JPEG where small high-frequency components can be discarded. However, the JPEG2000 standard that replaced JPEG uses wavelet transform instead of DCT in its image decomposition. The wavelet-based compression is discussed in Sect. 11.5.4.

11.5.2   Vector Quantization (VQ) Compression Technique

Vector quantization (VQ) is an effective and important compression technique with high compression efficiency and widely used in many multimedia applications. The goal is to reduce the bit rate to minimize communication channel capacity or digital storage memory requirements while maintaining the necessary fidelity of the data. VQ compression is a fixed-length algorithm for image block coding. A vector quantizer is composed of two operations. The first is the encoder, and the second is the decoder. The encoder takes an input vector and outputs the index of the codeword that offers the lowest distortion. In this case, the lowest distortion is found by evaluating the Euclidean distance between the input vector and each codeword in the codebook. Once the closest codeword is found, the index of that codeword is sent through a channel. This channel could be a computer storage, communications channel, and many others. When the encoder receives the index of the codeword, it replaces the index with the associated codeword. Figure 11.2 shows a block diagram of the operation of the encoder and decoder.

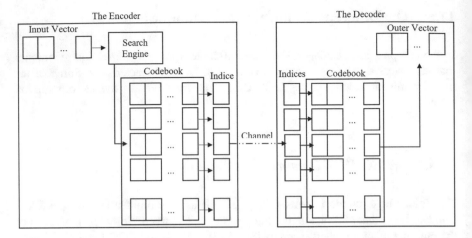

Fig. 11.2 The encoder and decoder in a vector quantizer. Given an input vector, the closest codeword is found and the index of the codeword is sent through the channel. The decoder receives the index of the codeword and outputs the codeword. (Source: http://mqasem.net/vectorquantization/vq.html)

11.5.3   Fractal Image Compression Technique

Image compression using fractals is a relatively useful technique. It is one of the lossy image compression techniques by Michael Barnsley in the late 1980s and early 1990s. The technique, however, is computationally very demanding. Fractals are used in image processing techniques such as color separation, edge detection, spectrum analysis, and texture-variation analysis. The image is divided into segments. These segments are looked up in a library of fractals. The library does not contain literal fractals that would require astronomical amounts of storage. Instead, the library contains relatively compact sets of numbers, called IFS codes, that will reproduce the corresponding fractals. IFS codes are discussed in Sect. 5.4. Once all the segments are looked up in the library, and their IFS codes found, the original digitized image can then be discarded and the codes kept, thus achieving a very high image compression ratio.

11.5.4   Wavelet Image Compression Technique

We discussed wavelets and wavelet transforms in Chaps. 2 and 4, respectively. One of the best image compression techniques is using wavelet transform. Wavelet theory uses a two-dimensional expansion set to characterize and give a time-frequency localization of a one-dimensional signal. Founded on the same principles of Fourier theory, the wavelet transform calculates inner products of a signal with a set of basis functions to find coefficients that represent the signal. Since this is a linear system, the signal can be reconstructed by a weighted sum of the basis functions. In contrast to the one-dimensional Fourier basis localized in only frequency, the wavelet basis is two-dimensional—localized in both frequency and time. A signal's energy, therefore, is usually well represented by just a few wavelet expansion coefficients. The wavelet system characteristics are as follows:

- Normally generated from a single scaling function or wavelet by scaling and translation.
- It satisfies multiresolution conditions, which means that if the basic functions are made half as wide and translated in steps half as wide, they will represent a larger class of signals exactly or give a better approximation of any signal.
- Always has the efficient calculation of the discrete wavelet transform (DWT), which means that the lower-resolution coefficients can be calculated from the higher-resolution coefficients by a tree-structured algorithm known as a filter bank.

The DWT of a signal $f(t)$ can be represented as shown in Eq. (11.3).

$$f(t) = \sum_{jk} a_{j,k} 2^{j/2} \Psi\left(2^j t - k\right) = \sum_{jk} a_{j,k} \Psi_{j,k}(t) \tag{11.3}$$

where the two-dimensional set of coefficients $a_{j,k}$ is the DWT of $f(t)$.

Wavelet analysis produces several important benefits, particularly for image compression. First, an unconditional basis causes the size of the expansion coefficients to drop off with j and k for many signals. Since wavelet expansion also allows a more accurate local description and separation of signal characteristics, the DWT is very efficient for compression. Second, an infinity of different wavelets creates a flexibility to design wavelets to fit individual applications.

Example 11.1

Given an $M \times N$ image, we decompose it using wavelet transform. Then filter each row and decimate it by 2, that is, take every other sample. This results to a two $N \times M/2$ images. Take each column, filter it, and subsample, that is, decimate by 2. Then we get four $N/2 \times M/2$ images. Among these four sub-images, the one resulting from the low-pass filtering of rows and columns is referred to as the **LL** image, the one that obtained low-pass filtering of the rows and high-pass filtering of the columns is referred to as the **LH** image, and the other two are **HL** and **HH**. Figures 4.2 and 4.6 in Chap. 4 illustrate this example. This first process in the analysis/decomposition is called level 1 decomposition. Depending on how many levels of decomposition that is needed, the process can be extended to levels 2, 3, 4, and more. The size of the image determines how much compression that is needed.

Solution 11.1

The MATLAB code that can be used in the decomposition of a 2D image is shown as MATLAB Program 11.1.

Original Image

Decomposed Level 1

Decomposed Level 2

Reconstructed Level 1

Reconstructed Level 2

Fig. 11.3 Two-level decomposition and reconstruction of an image

MATLAB Program 11.1.

```
% 2-D, 2 Level Haar Wavelet Decomposition
clear
clc
close all
im = imread('cameraman.tif'); % Load the image of a cameraman.
subplot(2,2,1), imshow(im); title('Original Image');
% Obtain the 2-D Haar transform using the |'integer'| flag.
[a,h,v,d]=haart2(im,1,'integer');
ff = uint8([a,h;v,d]);
subplot(2,2,2), imshow(ff); title('Decomposed Image Level 1');
[a2,h2,v2,d2]=haart2(ff,1,'integer');
hh = uint8([a2,h2;v2,d2]);
subplot(2,2,3), imshow(hh); title('Decomposed Image Level 2')
disp('The Decomposition Filter Coefficients:');
[LoD,HiD] = wfilters('haar')
```

Example 11.2

The decomposed image of Example 11.1 is reconstructed as shown in Example 11.2. The reconstructed image using Haar wavelets matched the input image perfectly, which can be seen looking at the image decomposition and reconstruction in Fig. 11.3. It is interesting to see that the image is more precise while being decomposed. It may be due to the code used with the integer function. However, the reconstruction of the image proved to be a success.

Solution 11.2

The MATLAB code used in the reconstruction of the 2D image is shown as MATLAB Program 11.2.

MATLAB Program 11.2.

```
% 2-D, 2 Level Haar Wavelet Reconstruction
clear
clc
close all
im = imread('cameraman.tif'); % Load the image of a cameraman.
subplot(2,3,1), imshow(im); title('Original Image');
% Obtain the 2-D Haar transform using the |'integer'| flag.
[a,h,v,d]=haart2(im,1,'integer');
ff = uint8([a,h;v,d]);
subplot(2,3,2), imshow(ff); title('Decomposed Level 1');
[a2,h2,v2,d2]=haart2(ff,1,'integer');
hh = uint8([a2,h2;v2,d2]);
subplot(2,3,3), imshow(hh); title('Decomposed Level 2')
% Reconstruct the image using the inverse 2-D Haar transform
xrec = ihaart2(a2,h2,v2,d2,'integer');
subplot(2,3,4),imshow(xrec,[]); title('Reconstructed Level 1');
xrec2 = ihaart2(a,h,v,d,'integer');
subplot(2,3,5),imshow(xrec2,[]); title('Reconstructed Level 2');
disp('The Reconstruction Filter Coefficients:');
[LoR,HiR] = wfilters('haar')
```

11.6   The Compression and Decompression of an Image/Data Using Wavelet Transform

The wavelet-based compression of an image starts with the image as a source input being sent to the wavelet transform block as shown in Fig. 11.4. The wavelet transformation is discussed briefly in Example 11.1. Its output is passed to the quantizer and to the encoder and finally the compressed image is born. The quantizer is used for the reduction of the precision of the floating point values of the wavelet transform. These floating points are typically either 32-bit or 63-bit floating point numbers. To use less bits in the compressed transform, which is necessary if

Fig. 11.4 Wavelet-based compression of an image

Fig. 11.5 Wavelet-based decompression of an image

compression of 8 bpp or 12 bpp images is to be achieved, these transform values must be expressed with less bits for each value. This leads to rounding error. These approximate, quantized, wavelet transforms produce approximations to the images when an inverse transform is performed, thus creating the error inherent in lossy compression. The encoder does the conversion of the image data from one format to another helping to reduce the amount of space necessary to hold the information and sometimes to change it to a compatible format. In the case of the wavelet-based decompression, it starts with the compressed image as a source input sent to the decoder and then to the approximate wavelet, transform, and inverse wavelet transform blocks as shown in Fig. 11.5. In wavelet-based compression, different algorithms such as the embedded zero wavelet trees (EZWT), the set partitioning in hierarchical trees (SPIHT), the embedded block coding with optimized truncation (EBCOT), the wavelet difference reduction (WDR) algorithm, and the adaptively scanned wavelet difference reduction (ASWDR) algorithm can be applied. We discuss these different wavelet-based compression algorithms in the next section.

11.7   The EZWT Algorithm

The embedded zero wavelet trees (EZWT) is one of the first techniques, which gave image compression a new direction that was superior to the JPEG standard. The original work started with work of Shapiro in the early 1990s. The EZWT algorithm was one of the first algorithms to show the full power of wavelet-based image compression. EZWT is a lossy and zero tree-based image compression scheme. The idea is to use the statistical properties of the trees in order to efficiently code the locations of the significant coefficients. Because a majority of the coefficients are zero, the spatial locations of the significant coefficients make up a large portion of the total size of a typical compressed image. A coefficient just like a tree is considered significant if its magnitude or magnitudes of a node and all its descendants in the case of a tree are above a particular threshold. By starting with a threshold, which is close to the maximum coefficient magnitudes, and iteratively decreasing the threshold, it is possible to create a compressed representation of an image, which progressively adds finer detail. Due to the structure of the trees, it is very likely that if a coefficient in a particular frequency band is insignificant, then all its descendants, the spatially related higher-frequency band coefficients, will also be insignificant.

EZWT uses four symbols to represent (a) a zero tree root, (b) an isolated zero (a coefficient which is insignificant but which has significant descendants), (c) a significant positive coefficient, and (d) a significant negative coefficient. Two binary bits may thus represent the symbols. The compression algorithm consists of a number of iterations through a dominant pass and a subordinate pass; the threshold is updated (reduced by a factor of two) after each iteration. The dominant pass encodes the significance of the coefficients, which have not yet been found significant in earlier iterations, by scanning the trees and emitting one of the four symbols. The children of a coefficient are only scanned if the coefficient was found to be significant, or if the coefficient was an isolated zero. The subordinate pass emits 1 bit, the most significant bit of each coefficient not so far emitted, for each coefficient, which has been found significant in the previous significance passes. The subordinate pass is therefore similar to bit-plane coding. In addition to producing a fully embedded bit stream, EZWT consistently produces compression results that are competitive with virtually all known compression algorithms.

11.8 The SPIHT Algorithm

The set partitioning in hierarchical trees (SPIHT) belongs to the next generation of wavelet encoders, employing more sophisticated coding (Dodla et al., 2013). It is identical for the encoder and decoder. In fact, SPIHT exploits the properties of the wavelet-transformed images to increase its efficiency. Its key advantages are the fact that it can provide good image quality with high peak signal-to-noise ratio (PSNR) and it is the best method for progressive image transmission. The image is first decomposed into four subbands. The decomposition process is repeated until it reaches the final scale. Each decomposition consists of one low-frequency subband with three high-frequency subbands. The example of the subband decomposition process can be seen in Chap. 4, Fig. 4.2. The SPIHT algorithm is the extension and efficient implementation of the EZWT algorithm. It is represented as shown in Eq. (11.4).

$$In(T) = \{1, \, max_{(i,j) \in T}\{|C_{i,j}|\} \geq 2^n \qquad\qquad 0, \text{otherwise}\}$$

$$(11.4)$$

where:

$n(T)$ is the importance of a set of coordinate T
$C_{i,j}$ is the coefficient value at each coordinate (i, j).

The complete SPIHT algorithm in three steps such as sorting, refinement, and quantization does compression. The SPIHT algorithm encodes the image data using three lists such as List of Insignificant Pixels (LIP), List of Significant Pixels (LSP), and List of Insignificant Set (LIS). LIP contains the individual coefficients having the magnitudes smaller than the threshold values. LSP is the set of pixels having magnitudes greater than the threshold value of the important pixels. LIS contains the overall wavelet coefficients defined in tree structure having magnitudes smaller than the threshold values. This is very much like a parent-child relationship.

The largest coefficient in the spatial orientation tree obtained by a maximum number of bits is n_{max}, and it is represented as shown in Eq. (11.5).

$$n_{max} = \left[log_2 \left(max_{i,j} \{ |C_{i,j}| \} \right) \right] \tag{11.5}$$

In the sorting process, all the pixels in the LIP list undergo a verification process whether they are important, and then the pixels and the coordinates with the coefficients in all the three lists are tested using Eq. (11.5). One coefficient only may be found as important SPIHT, a set partitioning in hierarchical trees algorithm 267, as an example, and it will be eliminated from the subsets and then inserted into the LSP, or it will be inserted into the LIP. In the refinement process, the *nth* most significant bit (MSB) of the coefficient in the LSP is the final output. The value of *n* is decreased; here again sorting with refinement is applied until *n = 0*. The decrement of *n* by 1 before going to the next step is the process of the quantization idea. Since the SPIHT algorithm controlled the bit rate exactly and the execution can be terminated at any time. Once the encoding process is over, then the decoding process is applied.

11.9   The EBCOT Algorithm

The embedded block coding with optimized truncation (EBCOT) algorithm exhibits state-of-the-art compression performance while producing a bit stream with a rich feature set, including resolution and signal-to-noise (SNR) scalability together with a random-access property schemes (Rawat et al., 2015; Taubman, 2000). It is an advanced technique for image compression. The EBCOT algorithm uses a wavelet transform to generate the subband samples which are to be quantized and coded, where the usual dyadic decomposition structure is typical, but other "packet" decompositions are also supported and occasionally preferable. The EBCOT algorithm has modest complexity and is extremely well suited to applications involving remote browsing of large compressed images. It lends itself to explicit optimization with respect to mean-square-error (MSE) as well as more realistic psychovisual metrics, capable of modeling the spatially varying visual masking phenomenon.

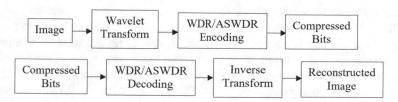

Fig. 11.6   WDR/ASWDR compression and decompression system

11.10   The WDR Algorithm

Wavelets applied to compression have shown a good adaptability to a wide range of data while being of reasonable complexity. The wavelet difference reduction (WDR) algorithm is a method for efficient embedded image coding (Raja & Suruliandi, 2011). This method retains all of the important features like low complexity, region of interest, embeddedness, and progressive SNR. In the WDR algorithm, the wavelet transform is applied first to the image. Then the bit plane-based WDR encoding algorithm for the wavelet coefficients is carried out. The advantages of the WDR algorithm are that it produces an embedded bit stream, thereby facilitating progressive transmission over small bandwidth channels and/or enabling multiresolution searching and processing algorithms. It also helps to encode the precise indices for significant transform values, thereby allowing for the region of interest (ROI) capability and for image processing operations on compressed image files. Figure 11.6 shows the diagram of the WDR and ASWDR compression and decompression systems. The ASWDR is discussed in Sect. 11.11.

11.11   The ASWDR Algorithm

In the early 2000s, Walker and Nguyen came up with the idea of the adaptively scanned wavelet difference reduction (ASWDR) algorithm (Walker & Nguyen, 2000; 2001). The work of Raja and Suruliandi (2011) continued with this same idea with different wavelet codecs. ASWDR is one of the most important image compression algorithms. It is said to be adaptively scanned because of the fact that the algorithm can modify the scanning order used in order to achieve better performance. The algorithm produces an embedded bit stream with region of interest capability. The ASWDR is a generalization of the compression method called the wavelet difference reduction (WDR). While the WDR method employs a fixed ordering of the positions of wavelet coefficients, the ASWDR method employs a varying order, which aims to adapt itself to specific image features. ASWDR algorithm aims to improve the subjective perceptual qualities of compressed images and improves the results of objective measures. Figure 11.6 shows the diagram of the WDR and ASWDR compression and decompression system, respectively. The WDR is discussed in Sect. 11.10.

The ASWDR algorithm adapts the scanning so as to predict locations of new significant values making the reduced binary expansion of the number of steps to be empty. If a prediction is correct, then the output specifying that location will just be the sign of the new significant value the reduced binary expansion of the number of steps will be empty. It means, therefore, that a good prediction scheme will significantly reduce the coding output of WDR. The scanning of the ASWDR dynamically adapts to the location of edge details in an image, and this enhances the resolution of these edges in ASWDR compressed images.

Steps a–g for performing the ASWDR algorithm on a grayscale image are as follows as implemented by Walker and Nguyen (2000) and Walker and Nguyen (2001):

(a) *Do a wavelet transform of the image.*
(b) *Initialize a scanning order for the transformed image.*
(c) *Determine the initial threshold, T, in such a way that at least one transform value has magnitude less than or equal to T and all transform values have magnitudes less than 2 T.*
(d) *Perform the significance pass process. This is done by recording the positions for new significant values: new indices m for which |x[m]| is greater than or equal to the present threshold. Then encode these new significant indices using difference reduction.*
(e) *Perform the refinement pass process. This is done by recording the refinement bits for the significant transform values determined using larger threshold values.*
(f) *Perform the new scan order. This is done by running through the significant values at level j in the wavelet transform.*
(g) *Divide the present threshold by 2. Steps d–f should be repeated until either a bit budget is exhausted or a distortion metric is satisfied.*

11.12   Usefulness of Wavelet-Based Compression

There have been many schemes developed over the years that have been used in the compression of images and data. Some of these schemes are the JPEG compression based on DCT, VQ, fractal compression, and many others. Each of these has always had their shortcomings. The fact that the input of the image has to be "blocked" results to correlations across the block boundaries not being eliminated. What you end up with are "blocking artifacts" that are noticeable, especially, at low bit rates. However, in many cases, the lapped orthogonal transforms (LOT) have been used to try to solve the problem by using smoothly overlapping blocks. The result is just minimal solutions in the reduction of the blocking effects in LOT compressed images. Even at that, it is done at the expense of having an increased computational complexity of using such algorithms. It therefore does not justify the replacement of DCT-based algorithms by LOT-based algorithms.

The wavelet-based compression schemes have gained very wide acceptability in the signal processing community. This is because the wavelet-based compression schemes have outperformed other coding compression coding schemes like the DCT and many others. In the wavelet-based compression application, there is no need to block the input image, and its basis functions have variable length. The wavelet coding schemes avoid blocking artifacts at higher compression rates. It facilitates progressive transmission of images and is more robust under transmission and in decoding errors. The wavelet-based compressing coding schemes have much better matching in the human visual system (HVS) characteristics. They are especially suitable for applications where scalability and tolerable degradation are important because of their inherent multiresolution nature. In addition, the multiresolution character of the wavelet decomposition leads to superior energy compaction and perceptual quality of the decompressed image.

Summary

1. Compression of images and data is required to reduce the storage and transmission costs while at the same time trying to maintain good image and data quality.
2. The compression of image data is achieved by taking advantages of the redundancies in the data image.
3. The idea behind the compression of images is that since most of these images have characteristics that are common of having pixels that are neighbors and are correlated, they therefore contain information that are redundant. It becomes necessary then to find a representation of the image that may not be correlated.
4. In image compression, the key is reducing all of the redundant/irrelevant information leaving only the relevant information.
5. One of the characteristics of the losslessly compressed image data is that the original image data can be recovered exactly from the compressed data.
6. One of the characteristics of the lossy compressed image data is that generally it cannot be recovered or reconstructed exactly as the original image.
7. The predictive compression technique is a spatial domain technique because it operates on the pixel values directly.
8. The transform coding compression technique is typically lossless and can be reversed easily on its own but is used to enable better and more targeted quantization.
9. The idea of the EZWT algorithm is to use the statistical properties of the trees in order to efficiently code the locations of the significant coefficients.
10. The set partitioning in hierarchical trees (SPIHT) belongs to the next generation of wavelet encoders, employing more sophisticated coding.
11. The embedded block coding with optimized truncation (EBCOT) algorithm exhibits state-of-the-art compression performance while producing a bit stream with a rich feature set, including resolution and signal-to-noise (SNR) scalability together with a random-access property.
12. The wavelet difference reduction (WDR) algorithm is a method for efficient embedded image coding.

13. The ASWDR algorithm adapts the scanning to predict locations of new significant values.
14. In the wavelet-based compression application, there is no need to block the input image, and its basis functions have variable length.

Review Questions

11.1. The compression of image data is achieved by taking advantages of the redundancies in the data image.

   (a) True.
   (b) False.

11.2. The major important components in image compression are:

   (a) Irrelevancy reduction component and additive reduction component.
   (b) Irrelevancy reduction component and redundancy reduction component.
   (c) Redundancy reduction component and additive reduction component.

11.3. Image pixels are highly correlated and, because of that, you find lots of information that are duplications.

   (a) True.
   (b) False.

11.4. The different types of compression techniques are:

   (a) Predictive compression and transform coding compression.
   (b) Lossless compression, lossy compression.
   (c) Lossless compression, lossy compression, predictive compression, and transform coding compression.

11.5. JPEG/DCT, VQ, fractal, and wavelet are not compression algorithms.

   (a) True.
   (b) False.

11.6. In wavelet-based compression, you can apply EZWT, SPIHT, EBCOT, WDR, and ASWDR algorithms to aid and enhance the compression.

   (a) True.
   (b) False.

11.7. EZWT is not a lossy and zero tree-based image compression scheme.

   (a) True.
   (b) False.

11.8. SPIHT exploits the properties of the wavelet-transformed images to increase its efficiency.

   (a) True.
   (b) False.

11.9. ASWDR is a generalization of the compression method called the wavelet difference reduction (WDR).

(a) True.
(b) False.

11.10. In the wavelet-based compression application, there is a need to block the input image, and its basis functions have no variable length.

(a) True.
(b) False.

Answers: 11.1a, 11.2b, 11.3a, 11.4c, 11.5b, 11.6a, 11.7b, 11.8a, 11.9a, 11.10b.

Problems
11.1. What are the three types of redundancy?
11.2. Describe irrelevant information redundancy.
11.3. Describe spatial and temporal redundancy.
11.4. Describe coding redundancy?
11.5. Why is compression important?
11.6. What are the different types of compression techniques?
11.7. What is your understanding of a lossless compression and why is it important?
11.8. What is your understanding of a lossy compression and why is it important?
11.9. What is predictive compression?
11.10. Describe transform coding compression?
11.11. Name some of the different types of compression algorithms.
11.12. Describe the JPEG/DCT compression algorithm.
11.13. Define vector quantization system and how is it used as a compression algorithm.
11.14. How does the fractal image compression work?
11.15. Describe briefly the wavelet image compression technique.
11.16. Write a MATLAB program for a two-level image decomposition using the Haar wavelet for image compression. Use any 2D image of choice.
11.17. Describe briefly the EZWT algorithm.
11.18. How is the SPIHT algorithm used in a wavelet-based compression?
11.19. How does the EBCOT algorithm function in the wavelet-based image compression?
11.20. Describe the WDR algorithm.
11.21. Describe the ASWDR algorithm.
11.22. What are the steps for performing the ASWDR algorithm on a grayscale image?
11.23. Why is wavelet-based compression useful?

References

Dodla, S., SolmonRaju, Y. D., & Murali Mohan, M. K. V. (2013). Image compression using wavelet and SPIHT encoding scheme. *International Journal of Engineering Trends and Technology (IJETT), 4*(9).

Kumar, T., & Chaudhary, D. (2013). Compression study between 'Ezw', Spiht, Stw, Wdr, Aswdr, and Spiht_3d. *International Journal of Scientific & Engineering Research, 4*(10).

Raja, S. P., & Suruliandi, A. (2011). Image compression using WDR & ASWDR techniques with different wavelet codecs. *ACEEE International Journal on Information Technology, 01*(02), 23–26.

Rawat, P., Rawat, A., & Chamoli, S. (2015). Analysis and comparison of EZW, SPIHT and EBCOT coding schemes with reduced execution time. *International Journal of Computer Applications, 130*(2), 24–29.

Taubman, D. S. (2000). High performance scalable image compression with EBCOT. *IEEE Transaction Image Processing, 9*, 1158–1170.

Walker, J. S., & Nguyen, T. O. (2000). Adaptive scanning methods for wavelet difference reduction in Lossy image compression. *Proceedings of IEEE International Conference on Image Processing, 3*, 182–185.

Walker, J. S., & Nguyen, T. Q. (2001). Chapter 6: Wavelet-based image compression. In K. R. Rao et al. (Eds.), *The transform and data compression handbook*. CRC Press LLC.

Chapter 12
Application of Wavelets to Video Compression

"To succeed. . .you need to find something to hold on to, something to motivate you, something to inspire you."

Tony Dorsett.

12.1 Introduction

The application of wavelet-based image and video compression techniques has gained momentum over the past several decades (Sang, 2013; Yang & Wang, 2013; Adam et al., 2007; Hsiang & Woods, 2004; Wang et al., 2014; He & Dong, 2003). The Joint Photographic Experts Group (JPEG) in March of 2000 released JPEG2000. This JPEG2000 is a wavelet-based image compression standard, and it is possible that it may replace the original JPEG standard. Video is one of those real-time media whose transport places exceptional demands on the capacity of existing network infrastructure, so that compression is essential for virtually any digital video communications. Since channel capacity must be utilized effectively in any communication system, it is desirable for the compressed representation to contain embedded subsets corresponding to successively higher bit rates and reconstructed video qualities. Therefore, in the video arena, wavelet-based techniques can be applied in the compression of video signals (Yang and Wang, 2013). This chapter discusses the application of wavelets and wavelet transforms in the compression of video signals. We discuss some of the key issues that hamper better video compression quality, what metrics can be used in the determination of the image quality, and how to mitigate against those impairments.

The idea of using a pre-processing algorithm based on the wavelet transform for the removal of noise in images prior to video compression is discussed. The intelligent removal of noise reduces the entropy of the original signal, aiding in compressibility. The wavelet-based denoising method can computationally help speed up to at least an order of magnitude than previously established image denoising methods and a higher peak signal-to-noise ratio (PSNR). A video compression denoising algorithm, which eliminates both intra- and inter-frame noise, is discussed in this chapter. The inter-frame noise removal technique estimates the amount of motion in the image sequence. Using motion and noise level estimates, a

C. M. Akujuobi, *Wavelets and Wavelet Transform Systems and Their Applications*, https://doi.org/10.1007/978-3-030-87528-2_12

video compression denoising technique can be established which may be robust to various levels of noise corruption and various levels of motion (Adam et al., 2007). We discuss a virtual object video compression method. Object-based compression methods have come to the forefront with the adoption of the MPEG-4 (Moving Picture Experts Group) standard. Object-based compression methods promise higher compression ratios without further cost in reconstructed quality. In this chapter, we discuss 3D video compression and different types of image and video standards.

Example 12.1: Why Video Compression I

First, let us consider a 100-gigabyte disc (100×109 bytes), which can store about 1 to 4 hours of high-quality video. The raw video contains vast amount of data. Sending the video through a communication channel should require very high bandwidth and storage capacity. Without compression, the communication of the video information will take lots amount of time and a very high cost. A 20 Mb/s of high-definition television (HDTV) channel bandwidth requires compression by a factor of 70.

Secondly, in considering the idea of video compressed and uncompressed images, let us consider a video image that has 640×480 resolution, 8-bit color, and 24 frames per second (fps) on *compressed video* containing 307.2 Kbytes per image (frame), 7.37 Mbytes per second, 442 Mbytes per minute, and 26.5 GB per hour. Consider in comparison another video that has 640×480 resolution, 24-bit colors, and 30 fps *uncompressed video* containing 921.6 Kbytes per image (frame), 27.6 Mbytes per second, 1.66 GB per minute, and 99.5 GB per hour. The advantages are obvious between the compressed and uncompressed video images. We therefore can understand the importance in the compression of video signals.

Example 12.2: Why Video Compression II

Imagine having a frame of a Super 35 format motion picture that may be digitized (via telecine) to a 3112 lines by 4096 pels/color, 10 bits/color image. This results to 1 second of the movie taking approximately 1 GB of memory. Consider a typical progressive scan (noninterlaced) HDTV sequence that may have 720 lines and 1280 pixels with 8 bits per luminance and chroma channels. The data rate corresponding to a frame rate of 60 frames/sec is $720 \times 1280 \times 3 \times 60 = 165$ Mbytes/second. It therefore becomes very necessary for the image sequences to be significantly compressed for efficient storage and transmission as well as for efficient data transfer among various components of a video system.

12.2   Video Compression Quality and the Metrics

With the wide application of the wireless video services, such as video conference and video on demand, there is an increasing importance of assessing the video quality accurately in real time. In spite of improvements in video compression and transmission techniques, defects or impairments are often introduced along the several stages of a communication system. It is important to accurately estimate the visual quality of the received video in real time, especially for wireless applications. The most accurate quality metric is to measure video quality directly through performing psychophysical experiments with human subjects (subjective quality metric). Unfortunately, these psychophysical experiments are expensive and time-consuming. On the other hand, the simple objective quality metric peak signal-to-noise ratio (PSNR) is not accurate, i.e., sometimes, it does not match the properties of the human visual system (HVS).

For example, a high PSNR does not necessarily mean a good subjective video compression quality. Many researchers are actively doing research in this area and trying to design an accurate objective video compression quality metric. Full reference system or no reference system is another aspect that can be considered when designing video compression quality metric. In real application, at the receiver side, it is hard to get the original video sequence, so no-reference system is preferred. Despite the fact that so much effort has been put in this area, it is hard to find key standards on video compression quality metrics. In this chapter, we discuss some mitigation possibilities. However, it is important to understand the factors that normally decrease video compression qualities. Most of the time, a video signal to be transmitted through network must be compressed first to save bandwidth; thus, two groups of errors arise. The first group contains compression errors, while the second group is made up of transmission errors.

12.3 Video Compression Errors

In most instances, a video signal to be transmitted through any network must be compressed first to save bandwidth. In the process of compressing the information, compression error may occur (Wang et al., 2014). This includes blocking artifacts, blurriness, and motion estimation errors. We discuss these types of errors in Sects. 12.3.1, 12.3.2 and 12.3.3.

12.3.1 Blocking Artifacts

The current existing video coding standards, such as MPEG-2 and H.264, can provide superior video compression performance for wide range applications. These coding standards are DCT-based coding systems that perform block-based operations, such as block-based DCT and block-based motion estimation and motion compensation (MEMC). The well-known consequence of block-based coding is blocking artifact, as seen in Fig. 12.1. Although H.264 uses an in-loop deblocking filter to help prevent the blocking artifacts, MPEG codecs do suffer from these kinds of blockiness in the coded video sequences, especially in low bit rate. In MPEG

(a) Original Image

(b) MPEG Encoded Image with Blocking Artifacts

Fig. 12.1 Blocking artifacts

codecs, standard 8×8 block size is used; therefore, this kind of blockiness distortion is relatively easy to detect. Most existing metrics focus on the block boundaries. Cross correlation is used to align grid to the edge map of a frame. It is assumed that the largest correlation value corresponds to the actual location of the *block boundaries*.

12.3.2   Blurriness

Blurriness is caused by low-pass processing on the video signal. Essentially, all lossy video coding algorithms compress video signals by discarding certain amount of high-frequency (detail) information while retaining low-frequency information. It is because human eyes are more sensitive to low-frequency information. This operation is similar to low-pass processing on the processed signal, causing blurriness errors, as shown in Fig. 12.2. Most of the existing blurriness metrics are based on the fact that blurring makes the edges larger or less sharp. Blurriness is determined by measuring the width of edges in the frame.

(a) Original Image   (b) Lowpass Image

Fig. 12.2 Blurrines

12.3.3 Motion Estimation Errors

Video signals are different from still images, because video signals not only display spatial information but also provide temporal information. Video quality depends on spatial (edges, etc.) as well as on temporal (motion, etc.) features of the video sequence. The blocking artifacts and blurriness mentioned in Sects. 12.3.1 and 12.3.2, respectively, are the two errors that are impairments on video spatial information, which can be measured using quality estimation methods designed for still images. With temporal information, the motion characteristics are used to characterize a video sequence; hence it can be used as a metric for measuring video compressed quality. Different motion features are used to analyze regions with different movements. For example, zero motion vectors imply still regions. The motion vectors also allow the detection of local movements or character of global movements. All these features can help to measure the perceptual quality reduction in the temporal direction.

12.4 Transmission Errors: Packet Loss

The transmission error contributes to the issues of video compression signal quality. It is part of the digital video streams that are transmitted as a series of independent macroblocks. In a heavy traffic network environment, some blocks may arrive out of order, while some blocks may not arrive at all. The result of these packet losses is a corrupted image that is awfully distracting to a human viewer, or even a previous image persisting for several frames. It is a challenge to design a metric to measure

(a) Original Image   (b) Image with Packet Loss

Fig. 12.3 Packet loss

this kind of error, since it is very hard to separate packet loss errors from object motion in a given frame. Both of these may cause sudden numerical changes in luminosity in some regions in a frame. In addition, the current standards, such as MPEG-2, define different types of video frames: I-frame, P-frame, etc. When coded in bit stream, different types of frames have different sizes. Packet loss in different frames has different influence on the perceptual quality. It is possible to evaluate these packet loss errors with the help of the network parameters. Figure 12.3 shows the original image data along with the image with packet loss.

12.5   Justification for Wavelet-Based Video Compression

2D and 3D wavelet transforms are in the compression of motion video. Wavelet-based compression is very low in complexity to compress entire images, and therefore does not suffer from the boundary artifacts seen in DCT-based techniques. They usually do not interfere with spatial scalability requirements and allow a high degree of quality scalability, which, in many cases, results in the possibility of optimally truncating the coded bit stream at arbitrary points (bit stream embedding). In the internetworking of the various video compression standards and video services, one of the important things needed is compatibility. It is very necessary in the sense that no matter what the original communication channel bandwidth or video format is, the arbitrary decoder should always be able to decode and display the incoming compressed video signal. Therefore, it presents the idea of a multiresolution systems technique, whereby the decoding and displaying of the received compressed video are independent of the following:

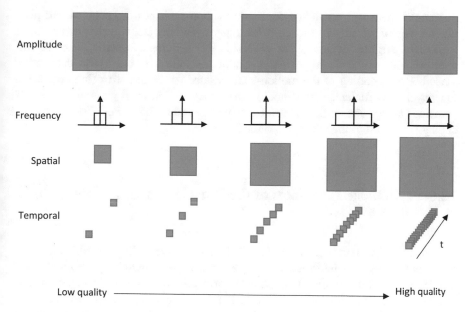

Fig. 12.4 The four forms of scalability

- The frame rate or the number of frames per second.
- The visual quality of the decoded frames.
- The spatial frequencies contained in the frames.
- The spatial resolution of the frames or fields in the sequence.

Because of the increasing demand for multimedia applications such as video streaming on the Internet or the support of decoders with differing complexities, a high level of scalability is required for the encoded data (Ohm et al., 2004). Full spatial, temporal, and signal-to-noise (SNR) scalability should, therefore, be provided by video compression coding schemes. To achieve spatial scalability, the discrete wavelet transform (DWT) is successfully applied in image and video compression. It is used in the still image-coding standard JPEG2000, hence the justification for the application of wavelets and wavelet transform numerous techniques to video compression.

The four forms of flexible decoding and displaying are spatial, frequency, amplitude, and temporal scalability respectively as shown in Fig. 12.4. The combination of the amplitude and frequency scalability is commonly called "signal-to-noise (SNR) scalability." Scalable Video Coding (SVC) differs from the traditional single point approaches mainly because it allows to encode in a unique bit stream several working points corresponding to different quality, picture size, and frame rate. Wavelet-based SVC (WSVC) motion-compensated technique idea can be applied in the current state-of-the-art SVC. In the wavelet-based spatially Scalable Video Coding schemes with in-band prediction, the wavelet decomposition of each frame of the video data is succeeded by the computation of a blockwise motion-compensated prediction for the individual subbands. To achieve half-pixel motion vector accuracy, the shift-invariant overcomplete wavelet transform is exploited. Higher accuracy is obtained by additional interpolation of the overcomplete subbands to maintain spatial scalability. A blockwise decision is made between coding of the motion-compensated prediction error (inter mode) or the wavelet coefficients (intra mode). The concept of in-band prediction is easily combined with a conventional intra-frame wavelet coder. The JPEG2000 arithmetic coding is applied to the resulting prediction error.

12.6   The Basic Principles of the Wavelet-Based Technique to SVC

A typical SVC system is shown in Fig. 12.5. It shows the coding of a video signal at an original Common Intermediate Format or Common Interchange Format (CIF) also known as Full Common Intermediate Format (FCIF) resolution and a frame rate of 30 frames per second (fps). CIF is a standardized format for the picture resolution, frame rate, color space, and color subsampling of digital video sequences used in video teleconferencing systems. In the example shown in Fig. 12.5, the highest operating point and decoding quality level correspond to a bit rate of 2 Mbps associated with the data at the original spatiotemporal resolution. A stream portion is extracted from the originally coded stream by a functional block called extractor.

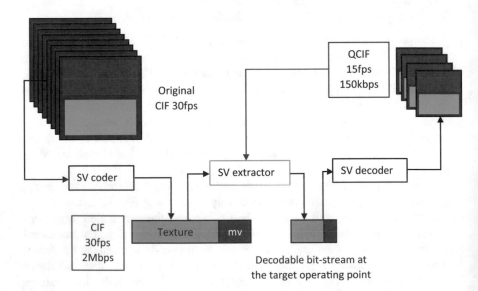

Fig. 12.5 SVC rationale. A unique bit stream is produced by the coder. Decodable substream can be obtained with simple extractions. (Source: https://d3i71xaburhd42.cloudfront.net/03fb7f51 59403e0919629d4159e51bd0011d8af6/1-Figure1-1.png)

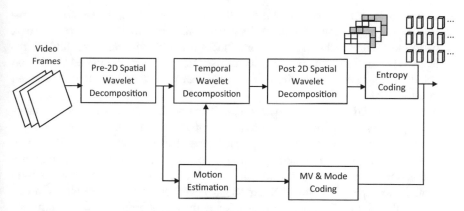

Fig. 12.6 Diagram of the general framework of the video wavelet

The extractor is arranged between the coder and the decoder. The extractor receives the information related to the desired working point, a lower spatial resolution Quarter CIF (QCIF), a lower frame rate of 15 fps, and a lower bit rate quality of 150 kilobits per second (kbps) and extracts a decodable bit stream matching or almost matching the working point specification. One of the main differences between an SVC system and a transcoding solution is the low complexity of the extractor, which does not require coding/decoding operations and typically consists of simple parsing operations on the coded bit stream.

The general framework of the video wavelet compression system (VWCS) is as shown in Fig. 12.6. It has a pre- and post spatial decomposition wavelet-based SVC (WSVC) configurations that can be implemented. The various components of the WSVC are the pre-2D spatial wavelet decomposition, temporal wavelet decomposition, motion estimation, motion video (MV), and mode coding and entropy coding.

- *Pre-2D Spatial Wavelet Decomposition:* All the necessary spatial pre-processing work required for the entire system is done in this block. It helps to remove not only the spatial domain noise but also noise in the temporal domain. Using the wavelet transform, we remove both spatial and temporal noise providing a higher compression gain with 3D wavelet compression. This necessitates the WSVC configurations to be implemented similar to t + 2D and 2D + t architectures, where "t" stands for temporal and "2D" for spatial. The wavelet-based coding and decoding (codec) processing is often referred to as a t + 2D scheme because the temporal wavelet transform is followed by the 2D spatial wavelet transform. In the 2D + t scheme, the spatial wavelet transform is applied before the temporal transform. There is also a 2D + t + 2D scheme reference to the order that the temporal and spatial wavelet transforms are performed. The most popular codec is the t + 2D architecture, where the motion-compensated temporal filtering (MCTF) precedes the 2D spatial discrete wavelet transform (DWT). It offers higher video compression efficiency than the other two schemes.

- *Temporal Wavelet Decomposition:* In this module, the frame-wise motion compensation (or motion alignment) wavelet transform on a lifting structure is performed. The block-based temporal model is adopted for temporal filter alignment. The motion vectors are directly used for the motion-aligned prediction (MAP) lifting step, while the motion field can be inverted for motion-aligned update (MAU).
- *Post-2D Spatial Wavelet Decomposition*: The post-2D spatial wavelet decomposition process allows for a simple syntax configuration for the post-spatial transforms on each frame or previously transformed subband. In the post-2D spatial wavelet decomposition, it helps to remove not only the post-spatial domain noise but also noise in the temporal domain. It is similar to what is done in the pre-2D spatial decomposition. The different spatial decompositions are used for different temporal subbands for possible adaptation to the signal properties of the temporal subbands.
- *Motion Estimation:* In this module, the macroblock size is scaled according to the decimation ratio in support of the spatial scalability. Interpolation accuracy is also scaled. Therefore, full pel (full pixel) estimation on 64×64 macroblocks at 4CIF resolution can be used at quarter-pel precision on a 16×16 macroblock basis for QCIF reconstructions. Hence, the motion is estimated at full resolution and kept coherent across scales by simply scaling the motion field partition and the motion vector values.
- *The Motion Video (MV) and Mode Coding (MC):* There has to be a way in which each block can select a forward, backward, or bidirectional motion model. This requires a mode selection information. The mode selection also determines the block that are aspect ratio and informs about the possibility of determining current MV values formal ready estimated ones. The motion vector modes and values are estimated using rate-constrained Lagrangian optimizations and coded with variable length and predictive coding, similarly to what happens in H.264/AVC (where AVC is advanced video coding). In the case of the MC filtering, a connected/disconnected flag is included to the motion information on units of 4×4 pixels.
- *Entropy Coding:* In the entropy module, after the spatiotemporal modules, the coefficients are coded with a 3D (spatiotemporal) extension of the embedded block coding with optimized truncation (EBCOT) of the embedded bit streams. EBCOT algorithm is called 3D EBCOT. Each spatiotemporal subband is divided into 3D blocks, which are coded independently. For each block, fractional bit plane coding and spatiotemporal context-based arithmetic coding are used.

12.7 Wavelet-Based Three-Dimensional Video Compression

In this section, we discuss the video compression of information using wavelet-based 3D techniques (Lewis & Knowles, 1993). The wavelet-based compression techniques provide some of the benefits of object-based compression methods without the difficulties of true object-based compression. An object-based wavelet compression algorithm, called virtual object compression, can be implemented to support high-quality, low bit rate video. Virtual object compression separates the portion of the video that exhibits motion from the portion of the video that is stationary. The stationary video portion is then grouped as the background, and the portion of the video, which exhibits motion, is grouped as the virtual object. After separation, both background and virtual object are coded independently by means of 2D wavelet compression and 3D wavelet compression, respectively.

There are two separate processing areas in object-based video compression. Object extraction is the method of separating different objects in a video image sequence, and the compression of those objects is a method of compressing arbitrarily shaped objects. In the virtual object video compression method, the wavelet transform is used for both object extraction and compression.

When the wavelet transform is applied in the temporal domain, the motion of objects is detected by large coefficient values. Therefore, the wavelet transform is used in the identification and extraction of the moving objects prior to object-based compression. Virtual object-based compression uses the non-decimated wavelet transform in the temporal domain in the separation of objects and stationary background. Virtual object compression also restricts the shape of the virtual object to be rectangular. This restriction enables the use of 3D wavelet video compression method for the compression of the virtual object. In addition, with a rectangular object restriction, the location and shape of the object can be completely defined with only two sets of spatial coordinates (the starting horizontal and vertical locations of the virtual object and the width and height of the virtual object), virtually eliminating shape coding overhead.

Figure 12.7 shows the wavelet-based 3D video compression system. The input to the system is a group of frames (GoFs) into the first 2D wavelet transform block, which does the spatial transformation of each of the frames of the video image sequence into the wavelet domain. The wavelet transform techniques have already been discussed in Chap. 4 of this textbook. The type of wavelet that can be used is

Group of Frames (GoF)

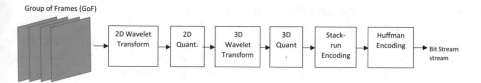

Fig. 12.7 Wavelet-based 3D video compression

the biorthogonal type of wavelet, or 5/3 wavelet. The biorthogonal wavelet is given as shown in Eq. 12.1.

$$
\begin{aligned}
h[\cdot] &= \{-1/8, 1/4, 3/4, 1/4, -1/8\} \\
g[\cdot] &= \{1/2, -1, 1/2\}
\end{aligned}
\tag{12.1}
$$

where h[·] and g [·] are the low-pass and high-pass filters, respectively.

As an example, let us consider the biorthogonal 5/3 wavelet kernel of Eq. (12.1). Specifically, let $x_{2k}[m, n]$ and $x_{2k+1}[m, n]$ denote even and odd frame subsequences, respectively, from the original video sequence. Ignoring motion, the 5/3 transform may be implemented by the following lifting steps.

$$
h_k[m, n] = x_{2k+1}[m, n] - \tfrac{1}{2}(x_{2k}[m, n] + x_{2k+2}[m, n])
\tag{12.2}
$$

$$
g_k[m, n] = x_{2k}[m, n] + \tfrac{1}{4}(h_{k-1}[m, n] + h_k[m, n])
\tag{12.3}
$$

Let $W_{k1, k2}$ denote the motion-compensated mapping from frame k_1 onto the coordinate system of frame k_2. As before, no particular motion model is assumed here. We modify the lifting steps of Eqs. (12.2) and (12.3) as shown in Eqs. (12.4) and (12.5), respectively.

$$
h_k[m, n] = x_{2k+1}[m, n] - \tfrac{1}{2}(W_{2k,2k+1}(x_{2k})[m, n] + W_{2k+2,2k+1}(x_{2k+2})[m, n])
\tag{12.4}
$$

$$
g_k[m, n] = x_{2k}[m, n] + \tfrac{1}{4}(W_{2k-1,2k}(h_{k-1})[m, n] + W_{2k+1,2k}(h_k)[m, n])
\tag{12.5}
$$

This shows from Eqs. (12.4) and (12.5) that when there is no motion, this sequence of lifting steps reduces to the 5/3 wavelet transform (up to a scale factor). It is also worth noting that each high-pass subband frame is essentially the residual from a bidirectional motion-compensated prediction of the relevant odd indexed original video frame.

This type of biorthogonal wavelet can be used because it can give the best overall quality for a given compression ratio for a wavelet which produces only integer coefficients. The integer wavelet coefficients have great benefits as they help to reduce both the computational complexity and memory requirement. The spatially wavelet transformed coefficients are then uniformly quantized in the 2D quantization block. This helps the coefficients to be represented with no more precision than is necessary to obtain the desired reconstruction quality.

The frames in the GoF are transformed in the temporal domain to exploit the intra-frame redundancy after they have been spatially transformed and quantized, *hence the idea of 3D wavelet transformation.* The temporal domain transformation has the capability of allowing for greater compression, given that the frames in the GoF are similar. Following the same process as in the 2D wavelet transformation and quantization, the 3D wavelet coefficients are also quantized as shown in Fig. 12.6 once they are obtained. There is specific reasoning to why two quantization processes are necessary. This is because the statistical properties of the horizontal and vertical dimensions in a video signal are similar to each other but differ from the time dimension. Hence, a different quantization step applied to the spatial and temporal domains is reasonable. The quantization step leads to artifact generation in signal reconstruction. However, the artifacts that appear from quantization of the 2D wavelet coefficients and the 3D wavelet coefficients are perceptibly vastly different.

The quantization of spatial domain wavelet coefficients leads to blurring and softening of the video signal, while the quantization of the 3D wavelet coefficients leads to "trails" of moving objects from frame to frame. Therefore, to mitigate the differing types of artifacts generated from the wavelet transformation in the two domains, two quantization step sizes are necessary. The formulations of the 2D and 3D wavelet transform may not be consistent with the traditional symmetric wavelet transformation of a three-dimensional signal. In the symmetric case, each dimension is transformed at a certain multiresolution level, and the lowest subband is then processed further for the next level. In this type of formulation, however, the wavelet transform is applied in the spatial domain through all subbands, and only afterward is it applied in the temporal domain. *This is what may be referred to as the decoupled 3D wavelet transform, and it may be the preferred wavelet transformation method for video compression.* Owing to the localization and multiresolution properties of the three-dimensional (3D) discrete wavelet, transform (DWT), 3D wavelet-based SVC offers high coding performance with three types of scalabilities in a flexible manner, namely, temporal, spatial, and quality scalability.

Figure 12.7 shows the difference between the 2D wavelet transform and 3D wavelet transform (both symmetric and decoupled) when viewing the differing sizes and shapes of the various subbands that are calculated. Figure 12.7 gives the size and shapes of each of the subbands calculated by the various wavelet transforms. In Fig. 12.7, the 2D wavelet transform does not apply temporal domain processing; thus there are no segmentation lines crossing the temporal domain separating different subbands. There are only segmentation lines crossing the horizontal and vertical dimensions. There exists a greater number of subbands generated by the decoupled 3D wavelet transform than in the symmetric 3D wavelet transform, allowing for greater frequency analysis in both the spatial and temporal domains. Each subband generated by the 3D wavelet transform is a three-dimensional bandpass signal representing the original signal.

The quantized 3D video data serves as input to the stack-run encoder shown in Fig. 12.7 where a simple form of video data compression is done in which a stream of data is given the input, for example, "AAAAABBBBBCCCCCC," and the output comes out as a sequence of counts of consecutive data values in a row with a compressed output of "5A4B6C." This may be stacks of data. This type of data compression is lossless, meaning that when decompressed, all of the original data will be recovered when decoded. Its simplicity in both the encoding (compression) and decoding (decompression) is one of the most attractive features of the algorithm. Then Huffman encoding is performed by replacing each letter in the string with its

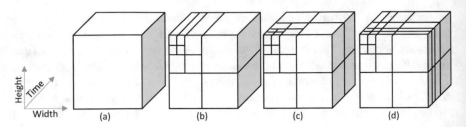

Fig. 12.8 Three-dimensional (3D) video transformation: (**a**) size and shape of the original 3D video signal, (**b**) 2D wavelet transform, (**c**) symmetric 3D wavelet transform, (**d**) decoupled 3D wavelet transform

binary code, thereby producing the compressed video bit stream. In Fig. 12.8, we have the three-dimensional (3D) video transformation (a) showing size and shape of the original 3D video signal, (b) 2D wavelet transform, (c) symmetric 3D wavelet transform, and (d) decoupled 3D wavelet transform.

12.8 The Basic Image and Video Compression Standards

Image and video standardization of compression algorithms helps reduce the cost of codecs, triggers product development and industrial growth, and enables the compatibility among products from different manufacturers. A good example is the standards for facsimile. The basic image and video compression standards are the Joint Pictures Experts Group (JPEG), JPEG2000, the H.26X, and the MPEG-X.

12.8.1 The Joint Pictures Experts Group (JPEG)

The JPEG committee for digital images developed the JPEG standard in the 1980s. It operates on 8×8 or 16×16 blocks of image data. The images compressed by JPEG are segmented into processing blocks called macroblocks. The JPEG uses the discrete cosine transform (DCT) techniques in compressing each macroblock separately, then quantizing the resultant coefficients, run-length encoding, and finally coding with a variable length entropy coder. The block-based encoder facilitates simplicity, computational speed, and a modest memory requirement.

12.8.2 The Joint Pictures Experts Group (JPEG) 2000

The JPEG2000 came into existence around March 2000. It is based on wavelet transform, which is far more superior to the DCT that the original JPEG was based on. This allows the user to specify the size of the processing block because small block sizes reduce the memory requirement while large block sizes improve compression gain and reconstructed image quality. After transformation, the coefficients are quantized and encoded as in the JPEG standard.

12.8.3   The H.261 Video Compression Standard

The H.261 is a video compression standard that was first ratified in November 1988 by the International Telecommunication Union Telecommunication (ITU-T) standard sector. It was the first video coding standard that was useful in practical terms. The H.261 video compression standard was originally designed for transmission over ISDN lines on which data rates are multiples of 64 kbit/sec. The compression algorithm involves block-based DCT transformation as in JPEG but also inter-frame prediction and motion compensation (MC) for temporal domain compression. Temporal domain compression starts with an initial frame, the intra-frame (or I-frame). Compression is achieved by creating a predicted (P) frame by subtracting the motion-compensated current frame from the closest reconstructed I-frame.

The design goals of the H.261 video standard were that it was intended for video telephony. In the case of low delay, each frame is coded as it arrives and only needs a small bit stream buffer on output to smooth to the constant bit rate (CBR) which adds a little delay. In the case of the CBR, it only sends a small number of intra-coded blocks in each frame, so data rate variation is only a function of video content. It adjusts the quantization based on occupancy of the bit stream buffer.

In the non-design goals, the H.261 is not intended for recording and playback. It has no way to seek backward or forward because you do not normally encode any frames with entirely intra-coded blocks. However, it could do this, but would not give the CBR flow needed for ISDN usage. For limited robustness to bit errors, the errors cause corruption (incorrect Huffman decoding of rest of group of blocks (GOB)). It can possibly be detected by hitting an illegal state in decoder. It stops decoding, the search for the next GOB. It starts decoding again. The intra-blocks always recover the damage slowly over next few seconds.

The input image format is usually the Common Intermediate Format (CIF). This is mainly used for high bit rate application. The Quarter CIF (QCIF) input is one quarter of CIF and it is mainly used for low bit rate application. The H.261 has mainly two compression modes, namely, intra mode that is similar to JPEG still image compression and is based on block-by-block DCT coding technique, and the inter mode, which is a temporal prediction that is employed with or without motion compensation. Then the inter-frame prediction error is DCT encoded.

12.8.4   The H.263 Video Compression Standard

The H.263 is a video compression standard originally designed as a low bit rate compressed format for videoconferencing. The ITU-T Video Coding Experts Group (VCEG) standardized it in 1995. The H.263 is based on discrete cosine transform (DCT) video compression. The standard is similar to H.261 but provides more advanced techniques such as half-pixel precision MC, whereas H.261 uses full pixel precision MC. It was primarily used as a starting point for the development of MPEG (that is optimized for higher data rates). MPEG-4 part 2 is similar to H.263.

12.8.5   The H.264 Video Compression Standard

The H.264 video compression standard provides high-quality encoding and decoding for streaming video applications in real time, at rates ranging from one quarter to half the file size of previous video formats. The file size that can be achieved is three times smaller than those achieved with MPEG-2 video standards. It produces good-quality images achieving both high and low ratios with better picture quality than MPEG-2, MPEG-4, or H.263. The MPEG-4 part 10 and H.264 video standards are technically identical. It is twice more efficient than MPEG-4. It is easy to integrate and covers a wide range of image. The disadvantages of the H.264 video compression standard are that it requires more time to coding and licensing agreements are complicated.

12.8.6   The MPEG-1 Video Compression Standard

MPEG-1 is a standard for lossy compression of video and audio. The MPEG-1 standard is published as ISO/IEC 11172 – Information technology—Coding of moving pictures and associated audio for digital storage media at up to about 1.5 Mbit/s. ISO/IEC JTC 1 is a joint technical committee of the International Organization for Standardization (ISO) and the International Electrotechnical Commission (IEC). Its purpose is to develop, maintain, and promote standards in the fields of information technology (IT) and information and communications technology (ICT).

It was founded on February 23, 1947, and headquartered in Geneva, Switzerland, and works in 164 countries. MPEG-1 is designed to compress VHS-quality raw digital video, CD audio, and metadata such as subtitles and closed captioning without excessive quality loss, making video CDs, digital cable and satellite TV, and digital audio broadcasting (DAB) possible as well as allowing for easy file sharing of moving pictures for such uses as reference, archiving, and transcription.

MPEG-1 is used in a large number of products and technologies, most notably the MP3 audio format.

12.8.7   The MPEG-2 Video Compression Standard

The MPEG-2 video compression standard was finalized in 1994. It extends MPEG-1 video with tools for the efficient compression of interlaced pictures, as required for standard-definition (SD) and high-definition (HD) video. The compression factor of MPEG-2 video is comparable to MPEG-1 video. Depending on the video content, a compression factor in the range between 20 and 60 can be achieved. It is used to reduce spatial redundancy within a picture. It uses a discrete cosine transform (DCT) techniques, based on spatial blocks with 8 by 8 samples each. For using 8×8 blocks, each picture is subdivided into slices, that is, horizontal rows of consecutive so-called macroblocks, whereby each macroblock contains 16 luminance pixels from 16 lines, as well as the spatially corresponding color samples. Macroblocks are adjacent to each other and do not overlap.

12.8.8   The MPEG-4 Video Compression Standard

MPEG-4 video compression standard was finalized as MPEG-4 part 2 in 1998. It was not intended to be as a successor of MPEG-2 video. Instead, it aims on very low bitrates for use on the Internet, at quality levels required by MPEG-2 target applications. The MPEG-4 video is slightly 10–20% more efficient than MPEG-2 video. The advanced video coding (AVC) standard, published as MPEG-4 part 10 in ISO/IEC, as H.264 in ITU-T was finalized in 2003 and is the successor of MPEG-2 video, typically providing an improvement of the compression factor by two or more compared to MPEG-2 video. AVC provides a compression factor in the range between 40 and 120, depending on the video content. MPEG-4 offers higher compression ratio, also beneficial for digital video composition, manipulation, indexing, and retrieval. It is an entirely new standard for composing media objects to create desirable audiovisual scenes, multiplexing and synchronizing the bit streams for these media data entities so that they can be transmitted with guaranteed quality of service (QoS). It also offers interaction with the audiovisual scene at the receiving end providing a toolbox of advanced coding modules and algorithms for audio and video compressions.

12.8.9   The MPEG-7 Video Compression Standard

MPEG-7 is a multimedia content description standard (MCDI). It was standardized in ISO/IEC 15938 as MCDI following the successful development of the MPEG-1, MPEG-2, and MPEG-4 standards. The main objective of the MPEG-7 is to serve the need of audiovisual content-based retrieval or audiovisual object retrieval in applications such as digital libraries. It is also applicable to any multimedia applications involving the generation (content creation) and usage (content consumption) of multimedia data. MPEG-7 became an international standard in September 2001 with the formal name Multimedia Content Description Interface. It supports a variety of multimedia applications. Its data may include still pictures, graphics, 3D models, audio, speech, video, and composition information on how to combine these elements. These MPEG-7 data elements are represented in textual format, or binary format, or both.

Summary
1. In the video arena, wavelet-based technique is applied in the compression of video signals.
2. In spite of improvements in video compression and transmission techniques, defects or impairments are often introduced along the several stages of a communication system.
3. Most of the time, a video signal to be transmitted through network must be compressed first to save bandwidth; thus, two groups of errors arise. The first group contains compression errors, while the second group is made up of transmission errors.
4. In the process of compressing information, compression error may occur. This includes blocking artifacts, blurriness, and motion estimation errors.
5. The transmission error contributes to the issues of video compression signal quality.
6. Wavelet-based compression is very low in complexity to compress entire images, and therefore does not suffer from the boundary artifacts seen in DCT-based techniques.
7. The four forms of flexible decoding and displaying are spatial, frequency, amplitude, and temporal scalability respectively.
8. The wavelet-based compression techniques provide some of the benefits of object-based compression methods without the difficulties of true object-based compression.
9. Image and video standardization of compression algorithms helps reduce the cost of codecs, triggers product development and industrial growth, and enables the compatibility among products from different manufacturers.
10. ISO/IEC JTC 1 is a joint technical committee of the International Organization for Standardization (ISO) and the International Electrotechnical Commission (IEC).

Review Questions

12.1 Video is not one of those real-time media whose transport places exceptional demands on the capacity of existing network infrastructure, so that compression is essential for virtually any digital video communications.

(a) True
(b) False.

12.2 Wavelet-based techniques can be applied in the compression of video signals.

(a) True
(b) False

12.3 With the wide application of the wireless video services, such as video conference and video on demand, there is an increasing importance of assessing the video quality accurately in real time.

(a) True
(b) False

12.4 In most instances, a video signal to be transmitted through any network must be compressed first to save bandwidth.

(a) True
(b) False

12.5 In the process of compressing information, compression error may occur. This includes:

(a) Blocking artifacts errors only
(b) Blurriness errors only
(c) Motion estimation errors
(d) Blocking artifacts, blurriness, and motion estimation errors

12.6 The two groups of errors that arise in video compression are:

(a) Compression and decompression errors
(b) Transmission and compression errors
(c) Compression and transmission errors

12.7 CIF is a standardized format for the picture resolution, frame rate, color space, and color subsampling of digital video sequences used in video teleconferencing systems.

(a) True
(b) False

12.8 The frames in the GoF are transformed in the temporal domain to exploit the intra-frame redundancy after they have been spatially transformed and quantized.

(a) True

(b) False

12.9 MPEG-4 video compression standard was finalized as MPEG-4 part 2 in 1998. It is not intended as a successor of MPEG-2 video. Instead, it aims on very low bitrates for use on the Internet, at quality levels required by MPEG-2 target applications.

(a) True
(b) False

12.10 Image and video standardizations of compression algorithms do not help reduce the cost of codecs, nor trigger product development and industrial growth, and do not enable the compatibility among products from different manufacturers.

(a) True
(b) False

Answers: 12.1b, 12.2a, 12.3a, 12.4a, 12.5d, 12.6c, 12.7a, 12.8a, 12.9a, 12.10b.

Problems

12.1. (a) Why is video compression necessary? (b) Give an example.
12.2. (a) Why is video compression quality important? (b) What type of metric can be used to measure the quality of compressed video data image?
12.3. What are some of the factors that normally decrease video compression qualities?
12.4. Name what constitute compression errors in video compression.
12.5. (a) Describe blocking artifacts in a video compression system. (b) What type of metric can be used to measure the quality of the video compressed image?
12.6. (a) What causes blurriness in a video compression system? (b) What type of metric can be used to measure the quality of the video compressed image?
12.7. (a) Describe the motion estimation errors? (b) What metric can you use to measure quality of the video image in a motion state?
12.8. (a) Describe a transmission error in a video compression system? (b) How can you measure the quality of a video image because of transmission error?
12.9. Why is wavelet-based video compression important?
12.10. Name the four forms of flexible decoding and displaying.
12.11. What are the basic principles of the wavelet-based techniques to Scalable Video Coding (SVC)?
12.12. Describe the general framework of the video wavelet compression system (VWCS)?
12.13. Describe the pre-2D spatial wavelet decomposition of the video wavelet compression system (VWCS)?
12.14. Describe the pretemporal wavelet decomposition of the video wavelet compression system (VWCS)?

12.15. Describe the post-2D spatial wavelet decomposition of the video wavelet compression system (VWCS)?

12.16. Describe the motion estimation module of the video wavelet compression system (VWCS)?

12.17. Describe the motion video (MV) and mode coding (MC) module of the video wavelet compression system (VWCS)?

12.18. Describe the entropy-coding module of the video wavelet compression system (VWCS).

12.19. Why is the 3D wavelet-based video compression technique necessary?

12.20. Describe the 3D wavelet-based video compression system.

12.21. Why are video compression standards important?

12.22. Describe the JPEG standard.

12.23. Describe the JPEG2000 standard.

12.24. Describe the H.261 video compression standard.

12.25. Describe the H.263 video compression standard.

12.26. Describe the H.264 video compression standard.

12.27. Describe the MPEG-1 video compression standard.

12.28. Describe the MPEG-2 video compression standard.

12.29. Describe the MPEG-4 video compression standard.

12.30. Describe the MPEG-7 video compression standard.

References

Adam, N., Signoroni, A., & Leonardi, R. (2007). State-of-the-art and trends in scalable video compression with wavelet-based approaches. *IEEE Transactions on Circuits and Systems for Video Technology, 17*(9), 1238–1255.

He, C., & Dong, J. (2003). Optimal 3-D coefficient tree structure for 3-D wavelet video coding. *IEEE Transactions on Circuits and Systems for Video Technology, 13*(10), 961–972.

Hsiang, S. T., & Woods, J. W. (2004). Embedded video coding using invertible motion compensated 3-D Subband/Wavelet filter bank. *Signal Processing: Image Communication, 16*, 705–724.

Lewis, A. S., & Knowles, G. (1993). Video Compression Using 3D Wavelet Transforms. *Electronic Letters, 26*(6), 396–398.

Ohm, J.-R., van der Schaar, M., & Woods, J. W. (2004). Interframe wavelet coding – Motion picture representation for universal scalability. *Signal Processing: Image Communications, 19*(9), 877–908.

Sang, L. (2013). Wavelet transform based video compression algorithm and its application in video transmission. *Computer Modelling & New Technologies, 17*(5B), 60–63.

Wang, D., Zhang, L., Klepko, R., & Vincent, A. (2014). *A wavelet-based video codec and its performance. https://www.researchgate.net/publication/228575408*, May 2014.

Yang, A., & Wang, C. (2013). Video compression based on wavelet transform and DBMA with motion compensation. *Journal of Communications, 8*(7), 440–448.

Part IV
Wavelets and Wavelet Transforms to Medical Application

Chapter 13
Wavelet Application to an Electrocardiogram (ECG) Medical Signal

"And in the end, it's not the years in your life that count. It's the life in your years."

Abraham Lincoln

13.1 Introduction

In the field of medicine and in many other areas, wavelets have been applied to solving various problems. In this chapter, we discuss the application of wavelets to the processing of an electrocardiogram (ECG) medical signal. The electrocardiogram (ECG or EKG) is a recording test of the electrical activities of the heart that is broadly used for the diagnosis of heart diseases. It is used to evaluate the heart performance and determine any problems that might occur (National Heart, Lung, and Blood Institute, 2011). Wavelet-based techniques have been used in converting the data from magnetic resonance imaging (MRI) machines, mammograms, and other medical equipment giving medical doctors a much needed clearer pictures of the internal parts of the body and a chance to detect disease and other problems earlier before they become major issues. This new technique seems to be displacing the conventional analytical methods. In this chapter, specifically we discuss how we can process ECG signals by removing noise, thereby making it much easier to detect anomalies. We use different kinds of wavelets such as the Haar, Daubechies, and Symlet types of wavelets. We use performance statistical metrics to evaluate which type of wavelet does a better job in removing noise from the ECG signals. We start in Sect. 13.2 in describing what an ECG medical signal is and how it can be represented.

13.2 Description of an ECG Signal

The electrocardiogram (ECG) is a test for recording the electrical activities of the heart that is broadly used for the diagnosis of heart diseases and to evaluate the heart performance and determine any problems that might occur. ECG is a nonlinear,

© Springer Nature Switzerland AG 2022
C. M. Akujuobi, *Wavelets and Wavelet Transform Systems and Their Applications*,
https://doi.org/10.1007/978-3-030-87528-2_13

Fig. 13.1 Ideal ECG signal. (Source: http://www. intechopen.com/books/ adaptive-filtering- applications/adaptive-noise- removal-of-ecg-signal- based-on-ensemble- empirical-mode- decomposition)

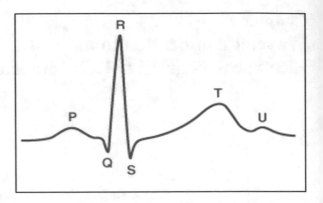

non-stationary signal, which is distorted by different types of noises, e.g., baseline drift, power line interference, electrode contact noise, and electromyogram (EMG) interference (Limaya & Deshmukh, 2016; Tareen, 2008). The ECG signal is characterized by six peaks and valleys, which are traditionally labeled as P, Q, R, S, T, and U, as shown in Fig. 13.1 (Zhidong et al., 2011). ECG signal noise removal is the process of isolating the essential signal component from unwanted signals to obtain a noise-free ECG signal that provides more diagnosis that is accurate. The earlier method of ECG signal analysis was established on time domain only (Singh & Kaur, 2013; Limaya & Deshmukh, 2016). Other methods included using digital filters (Mbachu et al., 2011).

However, this is not enough to analyze all the characteristics of the ECG signal. The frequency components of a signal are necessary. To achieve this, fast Fourier transform (FFT) method is used. Nevertheless, the restriction of the FFT is that the method cannot give information concerning the exact location of frequency components. Hence, the use of the short-term Fourier transform (STFT) is a solution. However, the main restriction of the STFT is that the tuning of time-frequency is not an optimum solution. Therefore, we decided to experiment what we felt should be a more appropriate method to solve this problem. The wavelet transform seems to be the most effective method for this purpose (Akujuobi & Baraniecki, 1994; Akujuobi et al., 2001; El-Dahshan, 2010).

Some of the methods used for noise removal of the ECG signals are adaptive filtering (Zhidong et al., 2011), FIR filtering (Rani et al., 2011), and wavelet transform (El-Dahshan, 2010; Tikkaren, 1999; Banerjee et al., 2012). Adaptive filtering is the most commonly used method for noise removal. Wavelet transform is the recent approach with different types of wavelet families. Wavelet transform methods provide the best performance for non-stationary signals that make them appropriate for ECG signal analysis. This chapter compares different wavelets, namely, the Daubechies-4, Haar, and Symlet types of wavelets, and then determines which type gives the better performance.

In the ECG signal, an extra wave can be seen at the end of the T-wave, and this is called the U-wave. This may be due to the repolarization of the papillary muscles.

The P-wave represents the activation of the upper chambers of the heart, the atria, while the QRS complex and the T-wave represent the excitation of the ventricles or the lower chamber of the heart. The interval between the S-wave and the T-wave is called the ST segment. The horizontal segment preceding the P-wave is designated as the baseline or the isopotential line. The discrete wavelet transform (DWT) is applied to the ECG signal to remove noise, thereby isolating the essential signal component from unwanted signals to obtain a noise-free ECG signal that provides diagnosis that is more accurate.

13.3 Discrete Wavelet Transform Application to ECG Signals

In this section, we start by examining the first levels of decomposition and reconstruction of the ECG signals and then extend that to the other levels for better enhancements.

13.3.1 One-Level DWT Decomposition and Reconstruction of the ECG Process

Given an original ECG signal (the signal with the noise) as shown in Fig. 13.2, we can apply a high-pass filter and a low-pass filter on the samples based on the DWT convolution techniques. We apply the high-pass filter in order to preserve the high-frequency components, and we apply the low-pass filter in order to preserve the low-frequency components. Then we apply downsampling on the high-frequency components to get the detail coefficients and apply downsampling on the low-frequency components to get the approximation coefficients. This process is called the wavelet-based decomposition process as shown in Fig. 13.3. To reconstruct the signal, the inverse is applied. The signal is reproduced by upsampling the detail coefficients and then applying a high-pass filter to get the high-frequency components, and we also can do the upsampling for the approximation coefficients and then apply a low-pass filter to get the low-frequency components. After that, we do the summation between the high-frequency and low-frequency components in order to get the original signal or the reconstructed signal. This process is called the wavelet-based reconstruction process as shown in Fig. 13.4. This process is extended to five levels of decomposition and reconstruction with actual ECG signals.

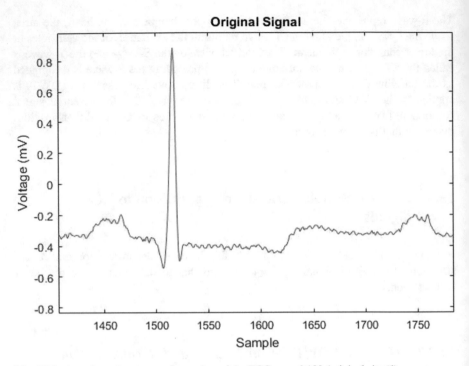

Fig. 13.2 An enlarged segment of a portion of the ECG record 100 (original signal)

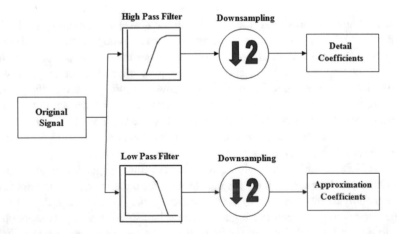

Fig. 13.3 Discrete wavelet transform (DWT) (decomposition)

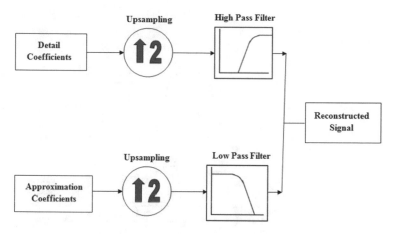

Fig. 13.4 Inverse discrete wavelet transform (IDWT) (reconstruction)

13.3.2 Extension to Five Levels of Decomposition and Reconstruction of the ECG Signals

In Sect. 13.3.1, we discussed how we could apply the ECG signal to one-level decomposition and reconstruction. In this section, we extend that process to a five-level process using actual ECG signals as input. The input ECG signals used are collected from the MIT-BIH Arrhythmia Database. Several samples can be collected. In this example, samples numbering from 100 to 124 were collected. The ECG signals can then be analyzed (decomposed) and synthesized (reconstructed) up to five levels using different types of wavelets. In this example, the Haar, Daubechies, and Symlet wavelets are used. Figures 13.5 and 13.6, respectively, show the five-level decomposition and reconstruction processes used for the ECG signal processing. We use the MATLAB Wavelet Toolbox for the ECG processing (MATLAB, 2010) or the newest version. The idea of using these wavelets is to remove noise from the ECG signal, thereby enhancing the quality of the medical information. We compared the noise-removed ECG signal from the original ECG signal using signal-to-noise ratio (SNR), peak signal-to-noise ratio (PSNR), mean squared error (MSE), and maximum squared error (MAXERR).

Similar to the one-level decomposition, when the next level of decomposition is performed, the process will still result in two sub-signals. The approximation of the previous level will fall into the details and approximation sub-signals in the next level. In this procedure, the denoising is done using five-level wavelet-based decomposition technique.

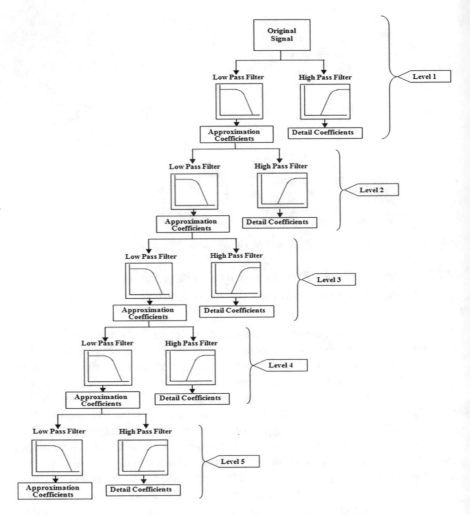

Fig. 13.5 Discrete wavelet transform (DWT) of the ECG signals for five levels

13.3.3   The DWT Noise Removal Technique Using ECG Signals

The DWT is applied to the ECG signal via five levels of the decomposition process. The Haar, Daubechies-4, and Symlet types of wavelets can be used. A threshold value is determined through a loop to find the value where the minimum error can be obtained between the detailed coefficients of thresholded noisy signal and the original. The overall technique is divided into seven steps.

Step 1: Eliminate baseline drift.

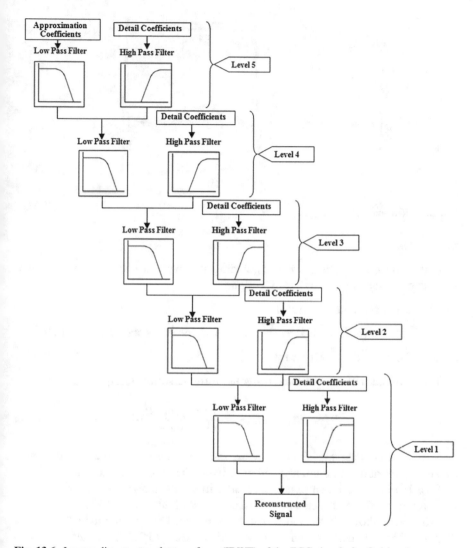

Fig. 13.6 Inverse discrete wavelet transform (IDWT) of the ECG signals for five levels

The baseline drift is efficiently removed by smoothing the signal using a moving average filter than span the moving average, and this elimination is done by using the smooth function in MATLAB.

Step 2: Apply wavelet transform.

Perform a multi-level wavelet analysis using the Haar, Daubechies-4, and Symlet types of wavelets. This results in the low- and high-pass filter coefficients (approximation and detail coefficients), thereby removing noise (the high-frequency component) of the ECG signal using wavelet transform.

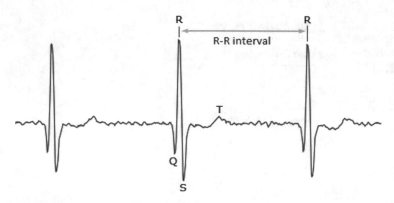

Fig. 13.7 R-R interval

Step 3: Thresholding.

Determine the threshold value by finding the average between the maximum value and the mean value of the ECG signal.

Step 4: Calculate the *R-R* interval.

The *R-R* interval is the interval between the R-peaks as shown in Fig. 13.7.

Step 5: Calculate the heartbeat rate.

The heartbeat rate is calculated using the formula shown in Eq. (13.1).

$$Heart\ Beat\ Rate = \frac{60 * Sampling\ Rate}{R - R\ interval} \tag{13.1}$$

Step 6: Calculate the signal-to-noise ratio, peak signal-to-noise ratio, mean squared error, and maximum squared error using the MATLAB built-in functions.

Step 7: Plot the original signal, baseline drift elimination, main ECG signal after noise removal, and *R*-detected signal using the plot function in MATLAB.

Figure 13.8 shows the flowchart that we used in the calculation and in the removal of noise from an ECG signal as described in Sect. 13.3.3 from step 1 to step 7, respectively.

13.4   Metrics for Performance Evaluation

The metrics for the evaluation of the performance of these different types of wavelets on the ECG signal noise removal are discussed in this section. The metrics are classified as the statistical parameters for the performance evaluations of the effects of the different types of wavelets on the ECG signal. They are signal-to-noise ratio (SNR), peak signal-to-noise ratio (PSNR), mean squared error (MSE), and

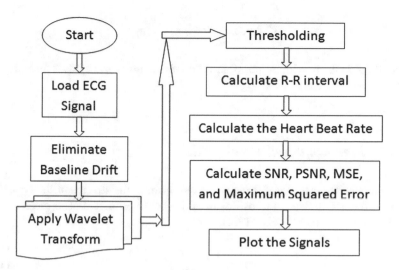

Fig. 13.8 Flowchart of the ECG signal denoising

maximum squared error (MAXERR). They are used as a basis for the comparison between the original signal (the signal with the noise) and the noise-removed signal in order to determine which type of wavelet gives the best performance. An overview of the statistical parameters is illustrated as follows.

13.4.1 Signal-to-Noise Ratio (SNR)

The signal-to-noise ratio (SNR) is a measure used to compare the desired signal to the noise. SNR is the ratio between the signal power and the noise power, often expressed in decibels. The signal-to-noise ratio is given as shown in Eq. (13.2).

$$SNR_{dB} = 10\log_{10}\left(\frac{P_{signal}}{P_{noise}}\right) \tag{13.2}$$

where P_{signal} is the power of the signal and P_{noise} is the power of the noise.

13.4.2 Peak Signal-to-Noise Ratio (PSNR)

Peak signal-to-noise ratio (PSNR) is a term for the ratio between the maximum possible power of a signal and the power of noise. It is used to measure the efficiency of the wavelet. The peak signal-to-noise ratio is given as shown in Eq. (13.3).

$$PSNR_{dB} = 10 \log_{10} \left(\frac{MAX^2}{MSE} \right) \tag{13.3}$$

where MAX is the maximum possible value of the signal and MSE is mean squared error.

13.4.3   Mean Squared Error (MSE)

The mean squared error (MSE) measures the mean of the squares of the errors, that is, the difference between the estimator and what is estimated. The mean squared error is given as shown in Eq. (13.4).

$$MSE = \frac{1}{n} \sum_{i=1}^{n} \left(\widehat{Y}_i - Y_i \right)^2 \tag{13.4}$$

where n represents the number of samples and $\left(\widehat{Y}_i - Y_i \right)^2$ is the square of the errors.

13.4.4   Maximum Squared Error (MAXERR)

The maximum squared error (MAXERR) is the maximum absolute squared deviation of the data from the approximation. The maximum squared error is given as shown in Eq. (13.5).

$$MAXERR = Max \left(\left| \widehat{Y}_i - Y_i \right|^2 \right) \tag{13.5}$$

where $\left| \widehat{Y}_i - Y_i \right|^2$ is the absolute squared deviation of the data from the approximation.

The performance of the noise removal technique and the wavelet type is determined by calculating the ECG signal performance metrics. These metrics determine the quality of the noise removal technique and the wavelet type that gives the best results. We discuss in Sect. 13.5 the obtained results from the simulation using the technique discussed in Sects. 13.3 and 13.4.

13.5   Evaluations of the Performance Measures

The input ECG signal is as seen in Fig. 13.9, which is a sample of the ECG record 100 (original signal) out of a total of 100–124 samples we used from the MIT-BIH Arrhythmia Database (MIT-BIH Arrhythmia Database [Online], 2016). Figure 13.10 shows an enlarged segment of a portion of the ECG record 100 (original signal). Figures 13.11, 13.12 and 13.13 show the enlarged segment of a portion of the ECG record 100 ECG after denoising using the Haar, Symlet, and Daubechies wavelets, respectively. Tables 13.1, 13.2 and 13.3 show the data list of the ECG records, the SNR, PSNR, MSE, MASERR, BPM (beats per minute), and exact BPM for the Haar, Symlet, and Daubechies-4 wavelets, respectively. All of the wavelets seem to have performed well with the Daubechies-4 wavelet performing the best.

Summary
1. The electrocardiogram (ECG) is a test for recording the electrical activities of the heart that is broadly used for the diagnosis of heart diseases and to evaluate the heart performance and determine any problems that might occur.
2. ECG is a nonlinear, non-stationary signal, which is distorted by different types of noises, e.g., baseline drift, power line interference, electrode contact noise, and electromyogram (EMG) interference.

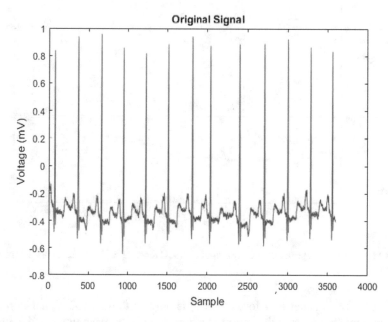

Fig. 13.9 ECG record 100 (original signal). (Source: The MIT-BIH Arrhythmia Database [Online] (2016)

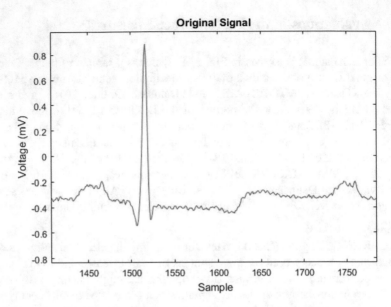

Fig. 13.10   An enlarged segment of a portion of the ECG record 100 (original signal)

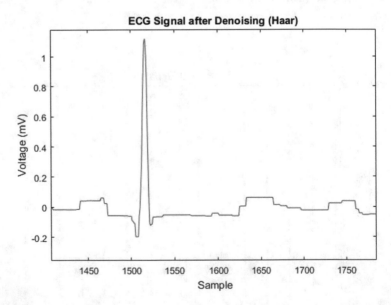

Fig. 13.11   An enlarged segment of a portion of the ECG record 100 ECG after denoising using the Haar wavelet

3. The discrete wavelet transform (DWT) is applied to the ECG signal to remove noise, thereby isolating the essential signal component from unwanted signals to obtain a noise-free ECG signal that provides diagnosis that is more accurate.

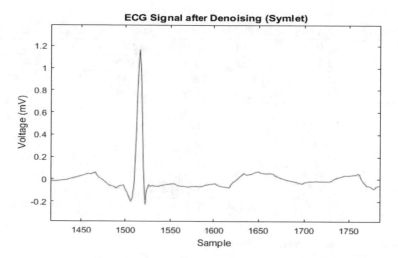

Fig. 13.12 An enlarged segment of a portion of the ECG record 100 after noise removal of ECG signal using the Symlet DWT

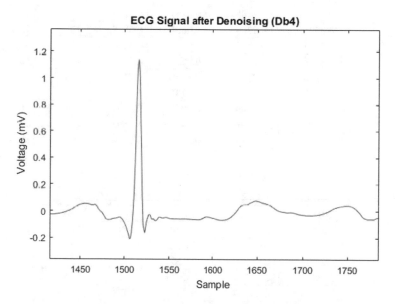

Fig. 13.13 An enlarged segment of a portion of the ECG record 100 ECG after denoising using the Daubechies wavelet

4. The overall technique for the DWT-based processing of an ECG signal can be divided into seven steps.
5. The performance of the noise removal technique and the wavelet type is determined by calculating the ECG signal performance metrics of the wavelets.

Table 13.1 Evaluations using the Haar wavelet

ECG record	SNR (db)	PSNR (db)	MSE	MAXERR	BPM	Exact BPM
100	19.1618	57.9591	0.1040	0.4387	74	75
101	19.5360	58.0765	0.1013	0.5289	67	68
102	19.1409	61.5310	0.0457	0.3796	112	118
103	23.3518	60.3162	0.0605	0.3785	70	72
105	21.7054	60.1751	0.0625	0.4433	83	84
106	20.4780	60.9351	0.0524	0.5985	80	76
108	15.6740	57.9347	0.1046	0.9246	113	116
112	18.0194	49.0270	0.8136	1.1311	86	89
114	22.9716	69.0777	0.0080	0.5323	53	56
124	25.5502	50.0102	0.6487	1.1890	49	52
Mean	20.5589	58.50426	0.2001	0.65445	78.7	80.6

Table 13.2 Evaluations using the Symlet wavelet

ECG record	SNR (dB)	PSNR (dB)	MSE	MAXERR	BPM	Exact BPM
100	19.6528	57.9599	0.1040	0.4332	74	75
101	20.4557	58.0777	0.1012	0.4960	67	68
102	20.6524	61.5345	0.0457	0.3814	113	118
103	24.8695	60.3209	0.0604	0.3805	70	72
105	23.3164	60.1778	0.0624	0.4570	83	84
106	21.8878	60.9362	0.0524	0.6359	74	76
108	16.7258	57.9369	0.1046	0.9462	113	116
112	19.4080	49.0272	0.8135	1.1278	86	89
114	24.1727	69.0369	0.0081	0.5332	53	56
124	26.7486	50.0099	0.6488	1.1693	49	52
Mean	21.789	58.5018	0.20011	0.65605	78.2	80.6

Table 13.3 Evaluations using the Daubechies-4 wavelet

ECG record	SNR (dB)	PSNR (dB)	MSE	MAXERR	BPM	Exact BPM
100	20.0215	57.9607	0.1040	0.4256	75	75
101	20.8425	58.0781	0.1012	0.4982	68	68
102	20.9180	61.5307	0.0457	0.3798	121	118
103	25.6822	60.3238	0.0604	0.3764	71	72
105	23.9864	60.1762	0.0624	0.4682	85	84
106	22.4385	60.9287	0.0525	0.6358	75	76
108	17.0721	57.9376	0.1045	0.9401	115	116
112	19.8880	49.0271	0.8135	1.1349	88	89
114	24.5679	69.0551	0.0081	0.5228	54	56
124	27.3165	50.0098	0.6488	1.1638	50	52
Mean	22.2733	58.50278	0.2001	0.65436	80.2	80.6

6. All of the wavelets used seem to have performed well with the Daubechies-4 wavelet performing the best.

Review Questions

13.1 Wavelets such as the Haar, Daubechies, and Symlet can be used in the processing of ECG signals.

   (a) True
   (b) False

13.2 The electrocardiogram (ECG) is a test for recording the electrical activities of the heart that is broadly used for the diagnosis of heart diseases and to evaluate the heart performance and, in addition, determine any problems that might occur.

   (a) True
   (b) False

13.3 ECG is not a nonlinear, non-stationary signal, which is distorted by different types of noises, for example, baseline drift, power line interference, electrode contact noise, and electromyogram (EMG) interference.

   (a) True
   (b) False

13.4 The discrete wavelet transform (DWT) is applied to the ECG signal to remove what?

   (a) Baseline drift only.
   (b) Power line interference only.
   (c) Electrode contact noise only.
   (d) EMG interference only.
   (e) Any type of noise in general.

13.5 The process of having the summation between the high-frequency and low-frequency components in order to get the original signal is called what?

   (a) The decomposition process.
   (b) The analysis process.
   (c) The reconstruction process.
   (d) None of the above.

13.6 The process of applying the downsampling on the high-frequency components to get the detail coefficients and applying downsampling on the low-frequency components to get the approximation coefficients is called what?

   (a) The synthesis process.
   (b) The decomposition process.
   (c) The reconstruction process.
   (d) None of the above.

13.7 The performance of the noise removal technique and the wavelet type is determined by calculating the ECG signal performance metrics of the wavelets.

(a) True
(b) False

13.8 The signal-to-noise ratio (SNR) is not a measure used to compare the desired signal to the noise.

(a) True
(b) False

13.9 Peak signal-to-noise ratio (PSNR) is a term for the ratio between the maximum possible power of a signal and the power of noise, and it is used to measure the efficiency of the wavelet.

(a) True
(b) False

13.10 The overall technique for the DWT-based processing of an ECG signal can be divided into seven steps.

(a) True
(b) False

Answers: 13.1a, 13.2a, 13.3b, 13.4e, 13.5c, 13.6b, 13.7a, 13.8b, 13.9a, 13.10a.

Problems

13.1 What is an electrocardiogram (ECG)?
13.2 Describe the ECG signal.
13.3 Name some of the different types of noises that may distort the ECG signal.
13.4 Describe the application of the decomposition of the first-level DWT to the ECG signal.
13.5 Describe the application of the reconstruction of the first-level DWT to the ECG signal.
13.6 What are the necessary seven steps using DWT noise removal technique for ECG signals?
13.7 Draw flowchart that can be used in the calculation and in the removal of noise from an ECG signal.
13.8 What are the four different statistical parameters you can use in the performance evaluation of the ECG signals using DWT?
13.9 Describe signal-to-noise ratio (SNR) as it relates to the ECG signal.
13.10 Describe peak signal-to-noise ratio (PSNR) as it relates to the ECG signal.
13.11 Describe mean squared error (MSE) as it relates to the ECG signal.
13.12 Describe the maximum squared error (MAXERR) as it relates to the ECG signal.

References

Akujuobi, C. M., & Baraniecki, A. Z. (1994). "A comparative analysis of wavelets and fractals," an invited book chapter to the 2D and 3D Digital Signal Processing Techniques and Applications, Control and Dynamic Systems Series, ed. C. T. Leondes, Vol. 67, *Academic Press, Inc.*, San Diego, pp. 143–197.

Akujuobi, C. M., Odejide, O., & Fudge, G. (2001). Development of wavelet based signal detection and measurement algorithm. In *Proc. of ASEE 6th Global Colloq. On Engr. Edu.* Oct. 1–4.

Banerjee, S., Gupta, R., & Mitra, M. (2012). Delineation of ECG characteristic features using multiresolution wavelet analysis method. *Measurement, 45*(3), 474–487.

El-Dahshan, E. (2010). Genetic algorithm and wavelet hybrid scheme for ECG signal denoising. *Telecommunication Systems, 46*, 209–215.

Limaya, H., & Deshmukh. (2016). ECG noise sources and various noise removal techniques: A survey. *International Journal of Application or Innovation in Engineering & Management (IJAIEM), 5*(2), 86–92.

MATLAB. (2010). http://www.mathworks.com/help/wavelet/index.html

Mbachu, C. B., Onoh, G. N., Idigo, V. E., Ifeagwu, E. N., & Nnebe, S. U. (2011). Processing ECG signal with Kaiser window-based FIR digital filters. *International Journal of Engineering, Science and Technology, 3*(8), 6775–6783. August 2011.

MIT-BIH Arrhythmia Database [Online]. (2016). Available at: http://physionet.org/physiobank/database/mitdb/

National Heart, Lung, and Blood Institute. (July 2011). "What Is Arrhythmia?" Retrieved 7 March 2015. http://www.nhlbi.nih.gov, https://en.wikipedia.org/wiki/Cardiac_arrhythmia

Rani, S., Kaur, A., & Ubhi, J. S. (2011). Comparative study of FIR and IIR filters for the removal of baseline noises from ECG signal. *International Journal of Computer Science and Information Technologies, 2*(3), 2011.

Singh, G., & Kaur, R. (2013). Removal of EMG interference from electrocardiogram using back propagation. *International Journal of Innovative Research in Computer and Communication Engineering, 1*(6), 1300–1305.

Tareen, S. G. (2008). Removal of power line interference and other single frequency tones from signals. In *M.Sc., Computer Science and Electronics*. Malardalen University.

Tikkanen, P. E. (1999). Nonlinear wavelet and wavelet packet denoising of electrocardiogram signal. *Biological Cybernetics, 80*(4), 259–267.

Zhidong, Z., Yi, L., & Qing, L. (2011). Adaptive noise removal of ECG signal based on ensemble empirical mode decomposition. In L. Dr Garcia (Ed.), *Adaptive filtering applications*. InTech, Hangzhou Dianzi University.

Part V
Wavelet and Wavelet Transform Application to Segmentation

Chapter 14
Application of Wavelets to Image Segmentation

"The only man who makes no mistakes is the man who never does anything."

Theodore Roosevelt.

14.1 Introduction

Segmentation is the process of dividing an image into different regions. It is also the process of partitioning a digital image into multiple regions which we can call sets of pixels where the pixels in each region have similar attributes. In the process of reducing images to information, one of the most used techniques is segmentation. It helps in the process of dividing an image into regions that may correspond to units that are structural in the scene or even in the distinguishing objects of interest. In the segmentation of an original image, we want to have an arrangement of intensities and be able to represent them as one intensity. We also want to make sure that the user or possibly a machine can be allowed to easily determine which pixels belong to which object within the image. The traditional approaches to image segmentation include pixel-based and region-based or edge-based segmentation processes. Texture analysis plays a significant role in many tasks such as scene classification, shape determination, or image processing. Wavelet transform has been used in the segmentation of images (Dandare & Kant, 2014; Gavlasova et al. 2006a, b; Arivazhagan & Ganesan, 2003; Mostafa & Gharib, 2001). In this chapter, we describe the technique of wavelet transform application for feature extraction segmentation associated with individual image pixels. For the image decomposition and feature extraction segmentation, we use the Haar wavelet transform as an example in this chapter. In addition, we discuss the thresholding ideas as well for different image segmentation processes and techniques.

© Springer Nature Switzerland AG 2022
C. M. Akujuobi, *Wavelets and Wavelet Transform Systems and Their Applications*,
https://doi.org/10.1007/978-3-030-87528-2_14

14.2   Image Segmentation Idea

Image segmentation is usually an initial and important step in a series of processes aimed at overall image understanding. Texture analysis is one of the important techniques in the segmentation idea with many tasks such as scene classification and shape determination. The applications of image segmentation include the following:

- Object identification in a scene for object-based measurements such as shape and size.
- Object identification in a moving scene for object-based MPEG4 video compression.
- Object identification, which are at different distances from a sensor using depth measurements from a laser range finder enabling path planning for a mobile robot.

Image segmentation processes are of different types. They include thresholding techniques, region-based techniques, watershed techniques, k-means clustering, template matching, contour-based techniques, and wavelet-based segmentation techniques.

14.3   Thresholding Technique

Most kinds of measurement or even the understanding of a scene or an image lies as a prerequisite, in the selection of the features within that scene or image. The key is the idea of thresholding. Traditionally, thresholding can be accomplished to define a range of brightness values in the original image. Then, select the pixels within this range as belonging to the foreground, and reject all of the other pixels to the background. Such an image is then displayed usually as a binary or two-level image, using any colors to distinguish the regions. This operation is called thresholding. Thresholding is usually the first step in any segmentation approach (Yazid & Arof, 2013). It is the simplest technique based on the variation of intensity between the object pixels and the background pixels. In this section, we discuss the technique for partitioning images directly into regions based on intensity values and properties of the values. One way to extract the object from the background is to select threshold T. Any point (r, s) in the image at which $f(r, s) > T$ is called the object point; otherwise, it will be called background. A parameter θ called the brightness threshold is chosen and applied to an image $A[r, s]$ as shown in Eq. (14.1). That alone is not yet segmentation; the result is just a binary image. The result must further be worked out by joining pixels together into one or more regions or areas.

Fig. 14.1 Example of thresholding similarity approach

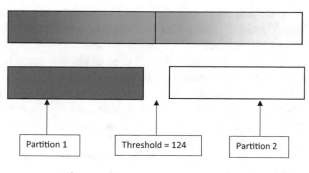

Fig. 14.2 Example of thresholding histogram approach

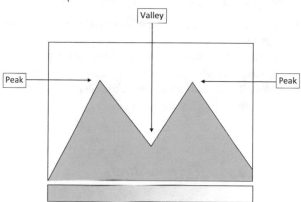

If $\Lambda[r, \sigma] - > 0$

$A[g, h] = \text{object} = 1$

$$\text{else,} \quad A[g, h] = \text{background} = 0 \qquad (14.1)$$

This version of the algorithm assumes that we are interested in light objects on a dark background. Within the thresholding technique idea, there is what is called the "similarity approach." In this case, you pick a threshold T, which as shown in Fig. 14.1 is 124. Therefore, based on the criteria set up in Eq. (14.1), and as shown in Fig. 14.1, pixels below the threshold get a new intensity A called partition 1. The pixels above the threshold get new intensity B called partition 2. Another thresholding technique is called the "histogram thresholding technique." In this case, the histogram of the data can be determined. Figure 14.2 shows a diagrammatic example of what the real plotted data may look like. The two peaks of the data signify the separation of the most prevalent intensities of the data image. The valley signifies the threshold value showing the separation of the two intensities. The valley may be called the "threshold" of the image needed to segment the pixel intensities.

Thresholding is one of the most important techniques in image segmentation. In this technique, pixels that are alike in grayscale (or in some other feature) are grouped together.

Often an image histogram is used to determine the best setting for the threshold (s). Some images (such as scanned text) tend to be bimodal, and a single threshold is suitable. Other images may have multiple modes, and multiple thresholds may be helpful. In general, multi-level thresholding is less reliable than single-level thresholding, mostly because it is very difficult to determine thresholds that, adequately, separate objects of interest.

There are three types of thresholding, which are local thresholding, adaptive thresholding, and global thresholding, respectively.

- Local Thresholding, where threshold *(T)* depends only on $f(r, s)$, that is, only on gray-level values.
- Adaptive Thresholding, where threshold depends on both $f(r, s)$ and $p(r, s)$ which are each different gray-level values.
- Global Thresholding, the threshold depends on the spatial coordinates r and s and it is called dynamic threshold.

14.3.1 Local Thresholding

The local threshold statistically examines the intensity values of the local neighbor of each pixel. The statistic, which is most appropriate, depends largely on the input image. Simple and fast functions include the mean of the local intensity distribution. It also includes the median value, or the mean of the minimum and maximum values.

14.3.2 Adaptive Thresholding

In adaptive thresholding, the threshold is changed dynamically over the image (Yazid & Arof, 2013). This tends to be the more type of the sophisticated version of the thresholding. It can accommodate changing light conditions in the image, for example, those occurring because of a strong illumination gradients or shadows. Adaptive thresholding takes typically a grayscale or color image as input and, in the simplest implementation, outputs a binary image representing the segmentation. For each pixel in the image, a threshold has to be calculated. If the pixel value is below the threshold, it is set to the background value; otherwise, it assumes the foreground value. There are two main approaches to finding the threshold. They are (i) Chow and Kaneko approach and (ii) local thresholding.

- **Chow and Kaneko:** Chow and Kaneko tend to divide an image into an array of overlapping sub-images and then find the optimum threshold for each sub-image by investigating its histogram (Chow & Kaneko, 1972). The threshold for each single pixel is found by interpolating the results of the sub-images.
- The local thresholding we have already discussed in Sect. 14.3.1.

14.3.3   Global Thresholding

The global thresholding technique requires homogeneous object and background intensities. Serious segmentation error may occur if that condition is not met. The global thresholding is based on the histogram of an image. We partition the image histogram using a single global threshold. It turns out that the success of this technique very strongly depends on how well the histogram can be partitioned. The basic global threshold, T, is calculated as follows:

(a) Select an initial estimate for T (typically the average gray level in the image).
(b) Segment the image using T to produce two groups of pixels: G_1 consisting of pixels with gray levels $>T$ and G_2 consisting pixels with gray levels $\leq T$.
(c) Compute the average gray levels of pixels in G_1 to give μ_1 and G_2 to give μ_2.
(d) Compute a new threshold value.
(e) Repeat steps (b)–(d) until the difference in T in successive iterations is less than a predefined limit T_∞.

This algorithm seems to work very well for finding thresholds when the histogram is suitable.

14.4   Implementation of the Region-Based Technique

In considering the region-based segmentation technique, it is implemented by breaking a large image into simpler and smaller region of common identities (Kaur & Singh, 2011). Region-based segmentation is totally based on common identities in whole regions like same gray level. We discuss segmentation techniques that are based on finding the region directly. The basic formulation is as follows:

$$\cup^n_{i=1} R_i = R \tag{14.2}$$

$$R_i \text{ is a connected region} : i = 1, 2, 3, \ldots n \tag{14.3}$$

$$R_i \cap R_1 = \varnothing \, (for \, all \, i \, and \, j) : i \neq j \tag{14.4}$$

$$P(R_i) = TRUE \, for \, i = 1, 2, 3, \ldots \tag{14.5}$$

$$P\left(R_i \cup R_j\right) = FALSE \, for \, i \neq j. \tag{14.6}$$

In finding the region directly, certain conditions must be followed. They are as follows:

- First point as shown in Eq. (14.2) indicates that segmentation must be complete, that is, every pixel must be in a region.
- Equation (14.3) indicates that any point in a region must be connected to some predefined sense/identities.
- Equation (14.4) shows that the region must be disjoint.
- Equation (14.5) deals with the property that must be satisfied by pixel in region it falls.
- Equation (14.6) indicates that R_i and R_j are different to each other.

14.4.1 Growing the Region

The basic idea of region growing is to collect pixels with similar properties to form a region. In growing the region, we select a seed point and start integrating until a small region of the same pattern is completely integrated. It is possible that more than one seed point can be there in the natural problem and then to treat them all simultaneously. It is also possible that sometimes cluster can be formed and then the pixel nearest to the centroid can be treated as a seed point. We have to use the connectivity idea in this process; otherwise, descriptors can mislead results. Another problem is the formulation of stopping rule. The region growing should stop when no more pixel satisfy the condition, but sometimes it keeps adopting other pixels on the basis of size and the average gray-level matching of region to their gray level. As an example, region growing can successfully be used in weld failures.

14.4.2 Region Splitting and Merging

The region splitting and merging is an alternative technique that can be used in image segmentation (Mostafa & Gharib, 2001). It takes an image and then subdivides it into arbitrary disjointed regions. The arbitrary regions can then be split and merged in order to satisfy the condition necessary for segmentation. The image is assumed homogeneous, and if it is not, it can be divided into four parts. The splitting process stops when the whole image breaks down in small homogeneous regions. If the original image is square $N \times N$, thereby having dimensions that are powers of 2 ($N = 2n$). All the regions are produced, but the splitting algorithm are squares having dimensions $M \times M$, where M is a power of 2 as well ($M = 2m, M < = n$).

The splitting technique is explained better in the quad tree technique as shown in Figs. 14.3 and 14.4, respectively, similar to the multiresolution technique in Mallat

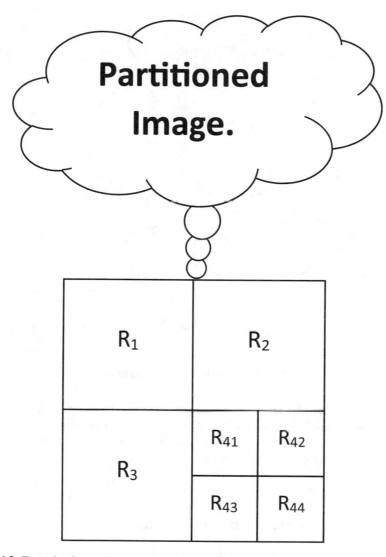

Fig. 14.3 Example of a quad tree technique in region splitting and merging

(1989) and Akansu and Haddad (1992). It is also similar to the decomposition and reconstruction processes with wavelet-based techniques discussed in Chap. 4. Splitting and merging has an algorithm that does splitting and merging simultaneously. The algorithm is stated as follows, where R is the region of interest:

- $(P(R) = \text{False})$: If R is nonhomogeneous, then split it into four sub-regions.
- $(P(R_i \cup R_j) = \text{TRUE})$: R_i and R_j are two adjacent regions, and as they are homogeneous, they can be merged.
- This process continues until there is no more possibility of splitting and merging.

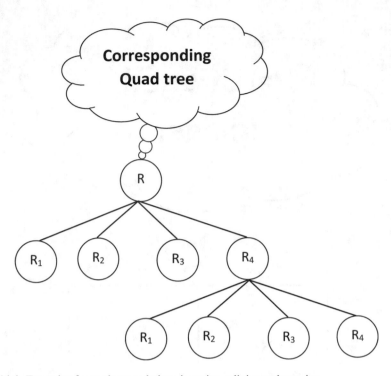

Fig. 14.4 Example of a quad tree technique in region splitting and merging

It should be noted that splitting and merging is more effective when done together simultaneously than using only splitting algorithms. The application can be extended to 3D images and magnetic resonance imaging (MRI). An example of region splitting and merging is as shown in Fig. 14.5.

14.5   Watershed Segmentation Technique

The watershed segmentation technique produces more stable segmentation and continuous segmentation boundaries. It provides a simple framework for incorporating knowledge-based constraints. Images can be visualized in 3D. We have two spatial coordinates with the gray levels. Some gray images can be classified as topological images. The topographical interpretation involves points belonging to regional minimum, catchment basins/watershed, and divide lines/watershed lines. The main objective is to find out the watershed lines. Once that is done, then punch the regional minimum, and flood the entire topography at uniform rate from below. As an example, a dam is built to prevent the rising water of different catchment basins to merge on. Eventually only the tops of the dams are visible above the water line. The dam boundaries correspond to the divide lines of the watershed.

Fig. 14.5 An example of region splitting and merging

The topographical view of a mountain can show that the height of the mountain is proportional to the grayscale value of the original image of the mountain. Since water levels can be rising, however, to prevent water spilling out of the structure, the entire topography has to be enclosed by the dams of height greater than the mountains. The heights of mountain can be represented to be the highest possible gray-level value in the image.

Dam construction is based on binary images, which are members of a 2D integer space Z^2. The overall objective of building a dam is to stop water spilling across basin. If we let M_1 and M_2 to be sets of coordinates of points of two regional minima, the coordinates of the points in the catchment basin associated with the minima in the flooding level are $C_n(M_1)$ and $C_n(M_2)$. $C[n]$ is the union of these sets. Let q denote the connected component that can be formed by dilation from flooding stage $(n-1)$ to stage n. The dilation of the connected components by the structuring elements in the dam can be subjected to two conditions: (i) Dilation has to be constrained to q and (ii) the center of the structuring elements can be located only at the points of q during dilation. Dilation cannot be performed on the set of points that may cause the sets being dilated to merge. We have to do the following: (a) by satisfying every point during dilation and (b) that it did not apply to any point during the dilation process.

14.5.1 Watershed Segmentation Algorithm

In developing the watershed segmentation algorithm, we let M_1, M_2, M_3. . . .M_n be the set of coordinates of points in the regional minima of the image $g(x, y)$. Let $C(M_i)$ be the coordinates of points of the catchment basin associated with regional minima M_i.

$$T[n] = \{(s,t)|g(s,t) < n\} \tag{14.7}$$

where *T[n]* is the set of points in $g(x, y)$ which are lying below the plane $g(x, y) = n$; n, stage of flooding, which varies from min + 1 to max+1; min, minimum gray-level value; and max, maximum gray-level value.

Let $C_n(M_1)$ be the set of points in the catchment basin associated with M_1 that are flooded at stage n.

$$C_n(M_1) = (M_1) \cap [n]. \tag{14.8}$$

$$C_n(M_i) = 1 \text{ at location}(x,y) \text{ if } (x,y) \in C(M_i). \tag{14.9}$$

$$(x, y) \in T[n], \textit{otherwise it is 0}$$

Let $C[n]$ be the union of flooded catchment basin portion at the stage n.
where $C[n] = \cup_{i=1}^{R} Cn(M_1)$ *and* $C[Max + 1] = \cup_{i=1}^{R} C(M_i)$.

What happens is that $C[n]$ keeps on increasing the level of flooding and during this process $C_n(M_i)$ and T[n] either increase or remain constant. The algorithm initializes at C[min + 1] = T[min +1] and then proceeds recursively assuming that at step n, $C[n-1]$ has been constructed. If we let Q be set of connected components in T[n], then for each of the connected components q $\in$ Q[n], there are three possibilities, namely:

(i) $q \cap C[n-1]$ is empty.
(ii) $q \cap C[n-1]$ contains one connected component of $C[n-1]$.
(iii) $q \cap C[n-1]$ contains more than one connected component of $C[n-1]$.

The interpretation is that (i) occurs when a new minima is encountered; in this case, *q* is added to set C[n-1] to form C[n]. Then, (ii) occurs when *q* lies within a catchment basin of some regional minima, in that case. Finally, (iii) occurs when ridge between two catchment basins is hit and further flooding can cause the waters from two basins to merge, so a dam must be built within *q*.

14.5.2   *Gradient of the Image*

In the gradient of an image, the image characterized by small variations in the gray levels has small gradient values, so watershed segmentation can be applied on the gradient of the image rather than the actual image. The regional minima of catchment basins correlate nicely with the small value of the gradients corresponding to the objects of interest.

14.6   K-Means Clustering

The term clustering refers to the method of division of data vectors into small amount of groups. The K-means clustering is a clustering algorithm (Kaur & Singh, 2011). It approximates an NP (non-deterministic polynomial-time)-hard combinatorial optimization problem. It is unsupervised. The K stands for the number of clusters, and it is based on a user inputting to the algorithm. From a set of data or observations, we have K-means to classify them into k clusters. The algorithm is iterative in nature. If we let $X_1, X_2, \ldots .X_n$ to be data points or vectors or observations, each observation can be assigned to one and only one cluster. C_i denotes cluster number for i^{th} observation. The dissimilarities can be measured using the Euclidean distance metric. The K-means minimizes within cluster point scatter as shown in Eq. (14.10).

$$W(C) = 1/2 \sum\nolimits_{k-1}^{K} \sum(i) = k \sum C(j) = k)\|x_i - x_j\|^2$$
$$= \sum\nolimits_{k=1}^{K} N_k \sum C(i) = k)\|x_i - m_k\|^2 \qquad (14.10)$$

where:

$m_k = $ *mean vector of the* k^{th} cluster
$N_k = $ number *of observations in* the kth cluster.

It should be noted that the K-means result is always noisy. The cluster algorithm can be stated as follows:

- For a given assignment C, we compute the cluster means m_k shown in Eq. (14.11).

$$m_k = \frac{\sum_{i:C(i)=k} x_i}{N_k}, k = 1 \ldots k \qquad (14.11)$$

- For a current set of cluster means, assign each observation as shown in Eq. (14.12).

$$C(i) = \arg \min \|x_i - m_k\|^2, i = 1, \ldots \ldots N \qquad (14.12)$$

- Iterate both steps above until convergence is attained.

In the K-means clustering segmentation technique, K-means converges, but finds local minima of the cost function. It works only for numerical observation. Normally, it requires fine-tuning when working on image segmentation, because there is no imposed spatial coherency in K-means algorithm. It works as a starting point for sophisticated image segmentation algorithm.

14.7   Template Matching Segmentation Technique

The template matching segmentation technique is used for classifying objects. The technique compares portions of images against one another. Sample image can be used for recognizing the similar objects in source image. If the standard deviation of the template image when compared to the source image is small enough, the template matching may be used. The templates are most often used to identify printed characters, numbers, and other simple objects. The template matching consists of three main factors:

- Definition and method.
- Bi-level image.
- Gray-level image.

14.7.1   The Definition and Method Template Matching

In the definition and method template matching, as shown in Fig. 14.6, the matching process moves the template image in all possible positions in a larger source image and then computes a numerical index that indicates how well the template matches the image in that position. The actual match is done on a pixel-by-pixel basis. The correlation process is a measure of the degree to which two variables agree, not necessarily in actual value but in general behavior. The two variables are the corresponding pixel values in two images, template and source.

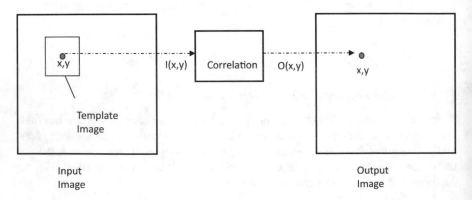

Fig. 14.6 The template matching destination and method idea

14.7.2   The bi-Level Image Template Matching

In the bi-level template matching, the template is normally a small image that is usually a bi-level image. Then an attempt is made to find the template in a source image with a yes or no option answer matching the template image.

14.7.3   Gray-Level Image Template Matching

In the gray-level template matching technique, when used on a gray-level image, it is unreasonable to expect a perfect match of the gray levels. Instead of a yes or no match at each pixel, the differences in the levels should be used. The formula for the correlation (r) calculation is as shown in Eq. (14.13).

$$r = \frac{\sum_{i=0}^{N-1}(x_i - \bar{x}) \cdot (y_i - \bar{y})}{\sqrt{\sum_{i=0}^{N-1}(x_i - \bar{x})^2 \cdot \sum_{i=0}^{N-1}(y_i - \bar{y})^2}} \tag{14.13}$$

where:

x is the template gray-level image,
$\bar{x}$ is the average gray level in the template image
y is the average gray level in the source image,
N is the number of pixels in the section image (section image = template image size) columns*rows.

The value of correlation r is between -1 and $+1$, with larger values representing a stronger relationship between the two images. Correlation is a computational intensive statistical parameter.

Example 14.1
Let the template matching image size be 67 × 55. Let the source image size be 155 × 214. We assume the template image to be inside the source image. Let the correlation (search) matrix size to be 88 × 159, that is, ((155–67) × (214–55)). The computation count then becomes 88 × 159 × 67 × 55 = 51,560,520.

14.8   Contour-Based Segmentation Technique

In the contour-based segmentation technique, the active contour models are also known as deformable models, dynamic contours, or snakes. The term "snake" is used to refer to active contour models. The segmentation of real images, such as medical images, is very complicated as they consist of noise and different artifacts. The performance of active contour models is effective in real images. This approach is immune to noise, boundaries, and gap present in the image. These are also effective in 3D and dynamic data.

The snake is the concept of computations of minimal energy paths. It follows the region growing process discussed in Sect. 14.4. The initially contour is placed like a seed pixel inside the region of interest. The contour, which consists of small curves like the elastic band, grows or shrinks to fit in the region of interest. The computational algorithm can be as follows where the points of the snake are represented as a spline curve with a set of control points. In specifying all the points, it makes the algorithm computationally intense. The snake is represented in parametric form as $x = X(s)$ and $y = Y(s)$. Then the curvature can be calculated as shown in Eq. (14.14).

$$E_{length} = \sum_{i=1}^{n} \left(\frac{\partial x_i}{\partial s}\right)^2 + \left(\frac{\partial y_i}{\partial s}\right)^2 \qquad (14.14)$$

Forces or energy functions move the contour. The forces of the energy function are derived from the image itself. There are two types of snakes, and they are internal energy and external energy.

14.8.1   Internal Energy

The internal energy depends on intrinsic properties like boundary length or curvature. The minimization of these factors leads to the shrinkage of the contour.

14.8.2   External Energy

The external energy is derived from the image structures, and it determines the relationship between the snake and the image. The external energy depends on factors such as image structure and user supplied constraints. These constraints also control the external force. However, snakes do have problems such as the following:

- Minimizing snakes is a problem, which is an iterative process.
- The points of a snake are evaluated iteratively.
- The parameters are adjusted so energy is reduced.
- Iteration continues till it settles to the minimum of E_{total}.
- If the chosen contour is good, snake converges very quickly.

We can calculate E_{total} and E_{length} using Eqs. (14.15) and (14.16), respectively.

$$E_{\text{total}} = \alpha E_{\text{length}} + \beta E_{\text{curvature}} - \gamma E_{\text{image}} \qquad (14.15)$$

$$E_{\text{length}} = \text{total length of snake} \qquad (14.16)$$

where:

$$E_{\text{curvature}} = \text{its curvature}$$

$$E_{\text{image}} = \text{counteracting or opossing force}$$

α, β, and γ are used to control the energy functions of the snake.

It must be noted that the main advantages of snake is that, very often, it provides closed coherent areas.

14.9   Wavelet-Based Segmentation Technique

In Fourier analysis, which is the traditional signal processing tool, a signal is broken into harmonics of various frequencies, whereas wavelet analysis consists of the breaking up of a signal into shifted and scaled versions of the mother wavelet (Akansu & Haddad, 1992). It has superior analyzing performance and multiresolution properties, so it is suitable for image analysis. Wavelets are functions generated from a single function by its dilations and translations. The wavelet-related properties discussed in Chap. 4 make it suitable for the segmentation application. The wavelet advantage is that it is a mathematical tool that has hierarchically decomposition functions in the frequency domain by preserving the spatial domain (Gavlasova et al. 2006a, b). In using wavelets, we can produce an image pyramid that is capable of representing the entropy levels for each frequency. This particular property can therefore be exploited to segment objects in noisy images based on frequency response in various frequency bands, separating them from the background and from other objects. In wavelet-based image segmentation process, we use the same image decomposition and reconstruction processes discussed in Chap. 4. In the wavelet transform segmentation application, we use the Haar wavelet as an example in the image decomposition and feature extraction processes. The algorithm of the Haar wavelet image decomposition includes image feature-based segmentation.

14.9.1   Image Feature Extraction

In image segmentation, texture plays an important role (Arivazhagan & Ganesan, 2003). Image textures are characterized by the spatial distribution of gray levels in a neighborhood. An image region has a constant texture if a set of its local properties in that region is constant, slowly changing or approximately periodic. Texture analysis is one of the most important techniques used in wavelet-based feature extraction. There are three primary issues in texture analysis, which are classification, segmentation, and shape recovery from texture. The analysis of texture requires the identification of proper attributes or features that differentiate the textures of the image. We can perform texture segmentation by performing co-occurrence matrix features we can call *Contrast* and *Energy* of size N × N derived from discrete wavelet transform overlapping but adjacent to sub-images $C_{i,j}$ of size 4 × 4, both horizontally and vertically. The algorithm of the image feature extraction involves the following steps:

Step 1: Decomposition, using one-level DWT with the Haar transform, of each sub-image $C_{i,j}$ of size 4 × 4.

Step 2: Computation of the co-occurrence matrix features energy and contrast given in Eqs. (14.17) and (14.18), respectively, from the detail coefficients, obtained from each sub-image $C_{i,j}$.

Step 3: Forming new feature matrices.

$$\text{Energy} = \sum_{(i,j=1)}^{N} C^2_{i,j} \tag{14.17}$$

$$\text{Contrast} = \sum_{i,j=1}^{N} (i-j)^2 C(i,j) \tag{14.18}$$

14.9.2   Pixel Differences

A new matrix with the differences is obtained once the computation of the co-occurrence matrix features is completed. This is done by performing the calculation between the value-by-value pixels of the features in both the horizontal and vertical directions. After this process, we can begin to see the segmentation band being formed across the texture boundaries.

14.9.3 Circular Averaging Filtering

It is very possible that as we see the segmentation band being formed across the texture boundaries, we can also begin to notice some artifacts or some spurious noisy spots. These spurious elements can be removed by applying a circular averaging filter. We first create the filter with suitable radius and then apply to the segmented image to minimize and try to eliminate noise or artifacts on the image.

14.9.4 Thresholding

In this process, we can use the thresholding technique to effect segmentation on the image. We have discussed the thresholding technique in Sect. 14.3 of this chapter. Thresholding helps in situations where we want to thin thick boundaries on the line of one pixel thickness. Not only can we use the thresholding technique, along with the wavelet transform, but we can also use one of the other techniques such as region-based technique, watershed techniques, k-means clustering, template matching, and contour-based technique as we already discussed in Sects. 14.3, 14.4, 14.5, 14.6, 14.7 and 14.8.

Summary
1. Segmentation helps in the process of dividing an image into regions that may correspond to units that are structural in the scene or even in the distinguishing objects of interest.
2. Texture analysis plays a significant role in many tasks such as scene classification, shape determination, or image processing.
3. Image segmentation processes are of different types. They include thresholding technique, region-based technique, watershed techniques, k-means clustering, template matching, contour-based technique, and wavelet-based segmentation technique.
4. Thresholding is one of the most important techniques in image segmentation. In this technique, pixels that are alike in grayscale (or in some other feature) are grouped together.
5. The template matching consists of three main factors: definition and method, bi-level image, and gray-level image.
6. Forces or energy functions move the contour. The forces of the energy function are derived from the image itself. There are two types of snakes, and they are internal energy and external energy.
7. The basic idea of region growing is to collect pixels with similar properties to form a region.
8. In using wavelets, we can produce an image pyramid that is capable of representing the entropy levels for each frequency. This particular property can therefore be exploited to segment objects in noisy images based on frequency

response in various frequency bands, separating them from the background and from other objects.

Review Questions

14.1. In the segmentation of an original image, we want to have an arrangement of intensities and be able to represent them as one intensity.

(a) True
(b) False

14.2. The application of image segmentation does not include identifying objects in a scene for object-based measurements such as size and shape.

(a) True
(b) False

14.3. Thresholding is not one of the most important techniques in image segmentation.

(a) True
(b) False

14.4. The global thresholding is based on the histogram of an image.

(a) True
(b) False

14.5. The region-based segmentation technique is done by breaking a large image into simpler and smaller region of common identities.

(a) True
(b) False

14.6. The region splitting and merging is not an alternative technique that can be used in image segmentation.

(a) True
(b) False

14.7. The template matching consists of these three main factors.

(a) Definition and method, bi-level image, and color level image
(b) Method, bi-level image, and gray-level image
(c) Definition and method, tri-level image, and gray-level image
(d) Definition and method, bi-level image, and gray-level image

14.8. An image region has a constant texture if a set of its local properties in that region is constant, slowly changing, or approximately periodic.

(a) True
(b) False

14.9. In using wavelets, we cannot produce an image pyramid that is capable of representing the entropy levels for each frequency.

(a) True
(b) False

14.10. The analysis of texture does not require the identification of proper attributes or features that differentiate the textures of the image.

(a) True
(b) False

Answers: 14.1a, 14.2b, 14.3b, 14.4a, 14.5a, 14.6b, 14.7d, 14.8a, 14.9b, 14.10b.

Problems

14.1. (a) Why is image segmentation important? (b) What are some of the applications of image segmentation?
14.2. What are the different types of image segmentation processes?
14.3. What is image segmentation thresholding?
14.4. What are the three types of thresholding?
14.5. Describe local thresholding.
14.6. Describe adaptive thresholding.
14.7. What are the two main approaches to finding the threshold in adaptive thresholding?
14.8. Describe the global thresholding technique.
14.9. Describe the region-based segmentation technique.
14.10. Describe the idea of growing the region in the segmentation process.
14.11. Describe the region splitting and merging technique.
14.12. State the splitting and merging algorithm.
14.13. Describe the watershed segmentation technique.
14.14. Describe the watershed segmentation algorithm.
14.15. Describe the gradient of the image in the segmentation technique.
14.16. Describe the K-means clustering technique in image segmentation.
14.17. Describe the template matching segmentation technique.
14.18. What are the three main factors of template matching?
14.19. What is definition and method template matching?
14.20. Describe the bi-level image template matching.
14.21. Describe the gray-level image template matching.
14.22. Given a template image matching size of 77×65 and a source image size of 165×224 and assuming the template image to be inside the source image, calculate the computation count.
14.23. Describe the contour-based segmentation technique.
14.24. What are the two types of snakes?
14.25. What is meant by internal energy in the image segmentation technique?
14.26. What is meant by external energy in the image segmentation technique?
14.27. What is the advantage in using the wavelet-based image segmentation?

14.28. What is the role that image feature extraction plays in using the wavelet-based image segmentation?

References

Akansu, A. N., & Haddad, R. A. (1992). *Multiresolution signal decomposition, transforms, subbands, wavelets*. Academic Press.

Arivazhagan, S., & Ganesan, L. (2003). Texture SegmentationUsing wavelet transform. *Pattern Recognition Letters., 24*(16), 3197–3203.

Chow, C. K., & Kaneko, T. (1972). Automatic boundary detection of the left ventricle from cineangiograms. *Computers and Biomedical, 5*, 388–410.

Dandare, S. N., & Kant, N. N. (2014). Automatic image segmentation using wavelet transform based on normalized Grapg cut. *International Journal of Modern Engineering Research (IJMER), 4*(6).

Gavlasova, A., Prochazka, A., & Mudrova, M. (2006a). *Wavelet based image segmentation. https://www.researchgate.net/publication/228453305*, Retrieved May 2021, 2006.

Gavlasova, A., Prochazka, A., & Mudrova, M. (2006b). Wavelet use for image classification. In *15th international conference on process control*. Pleso.

Kaur, G., & Singh, B. (2011). Intensity based image segmentation using wavelet analysis and clustering techniques. *Indian Journal of Computer Science and Engineering (IJCSE), 2*(3), 379–384.

Mallat, S. (1989). Multifrequency Channel decomposition of images and wavelets models. *IEEE trans, acoustic speech and. Signal Processing, 37*(12), 2091–2110.

Mostafa, M.G., Gharib, T.F. (2001). Medical image segmentation using a wavelet-based multiresolution EM algorithm. *IEEE International conference on industrial electronics technology & automation.*

Yazid, H., & Arof, H. (2013). Gradient based adaptive thresholding. *Journal Vision Communication Image Research, 24*, 926–936.

Chapter 15
Hybrid Wavelet- and Fractal-Based Segmentation

*Our imagination is the only limit to what we can hope to have
in the future.*

– Charles F. Kettering

15.1 Introduction

In Chap. 14, we discussed image segmentation using wavelets. In this chapter, we
compare the quality of segmentation of images for a wavelet-based segmentation
with a fractal-based segmentation. We compute the error probability of
misclassification. We explore various wavelets and wavelet filter lengths and use
different images to see image effects. In Chap. 5, Sect. 5.6.5, we indicated that
wavelets and fractals can be applied to the problem of texture image segmentation.
Wavelets and fractals are similar in so many ways (Akujuobi & Baraniecki, 1994,
1992a, 1992b). We also discussed the similarity between wavelets and fractals in
Chap. 5. In this chapter, we develop a wavelet- and fractal-based model for the
segmentation of images. We compare the different approaches using as a criterion
the number of iterations required for segmentation and error probability of
misclassification. We also use the developed segmentation model to study the effects
of different wavelet filter lengths on image segmentation. We compare the effects of
the different wavelet filter lengths using as a criterion the number of iterations
required for segmentation and error probability of misclassification. The
Daubechies-4, Daubechies-12, Daubechies-18, and biorthogonal 18 wavelets are
used and are discussed in this chapter. In the next section, we discuss the wavelet-
and fractal-based segmentation model.

© Springer Nature Switzerland AG 2022   329
C. M. Akujuobi, *Wavelets and Wavelet Transform Systems and Their Applications*,
https://doi.org/10.1007/978-3-030-87528-2_15

15.2   The Hybrid Wavelet- and Fractal-Based Segmentation Model

The hybrid wavelet- and fractal-based segmentation model is shown in Fig. 15.1. It is developed using an approach that is based on the wavelet and fractal analysis processes as described in Sects. 15.3 and 15.4, respectively, and a classification algorithm for segmentation that is described in Sect. 15.6. The wavelet- and fractal-based segmentation is developed using an iterative process. This technique is based on the computation of the mean and covariance for each window of choice. The means of the windows are then used for the segmentation by computing new means for each region and then comparing with the given mean. This process iteratively continues until a minimum distance is achieved. Both the fractal and wavelet analysis data share a common segmentation scheme in the segmentation model shown in Fig. 15.1.

The concept of determining the local fractal dimension is used in computing the fractal analysis image data for segmentation. This is done in the fractal analysis block of the segmentation model. The fractal analysis of the image helps to characterize the local fractal behavior at each image point. The wavelet-based technique is applied to the input image by using the idea of feature extraction through a compactly supported two-dimensional wavelet of the neighborhood window. This is done in the wavelet analysis block of the segmentation model. Such analysis decomposes the textural information into orientation-sensitive multiple channels. The information in each channel is then reduced to a discriminant feature by the appropriate statistics. Each of the major blocks in the segmentation model is described in full starting with the wavelet block in the next section.

HYBRID WAVELET AND FRACTAL-BASED SEGMENTATION MODEL

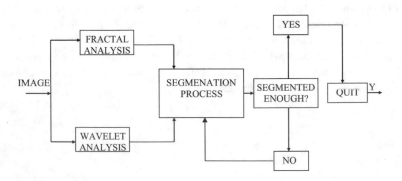

Y: OPTIMUM SEGMENTED OUTPUT IMAGE

Fig. 15.1 Wavelet- and fractal-based segmentation model

15.3 Computation of Wavelet-Based Analysis Image Data for the Hybrid Segmentation Process

The wavelet-based analysis block in Fig. 15.1 is discussed in this section. The block is used in computing the wavelet analysis image data for segmentation. For each pixel, a window of pixels centered at the original is obtained. This window is decomposed into seven "channels" using a selected wavelet analysis. These channels are then used to compute an "energy vector" composed of the variance of each channel. The energy vector is then turned into a dimensional pixel value by a transform w that averages the variances.

15.3.1 Wavelet-Based Analysis Image Data Computation Process for Segmentation

In the wavelet-based analysis block of the segmentation model, a two-stage local wavelet analysis is applied to every pixel neighborhood supported by window size w in order to get a feature which could compactly represent the statistical property of a region. This leads to seven channel images in this particular case. These seven channels are obtained as a result of combining the three output channels of the first stage and the four output channels of the second stage of the decomposition process as shown in Fig. 15.2. The window size is important because for textural images, a single pixel has no information about the texture type. To know the texture type, the region around the pixel within a given window must be known (Chang & Kuo, 1992; Teshome, 1991; Pentland, 1984). We found the decomposition process and the local energy function proposed by Teshome (1991) and Perez-Lopez and Sood (1994) to be appropriate for this segmentation process. Thus, we use X^j to represent the textural windowed image at location (k, l) with resolution (projection level) j; the separate channel images are given as shown in Fig. 15.3. The decomposition process is as represented in Eqs. (15.1, 15.2, 15.3 and 15.4), respectively.

$$Y^{i,1} = HLX^{i,1} \tag{15.1}$$

$$Y^{i,2} = LHX^{i-1} \tag{15.2}$$

$$Y^{i,3} = HHX^{i-1} \tag{15.3}$$

$$X^2 = LLX^1 \tag{15.4}$$

where:

$i = 1, 2.$
$L =$ low-pass filter operator.
$H =$ high-pass filter operator.

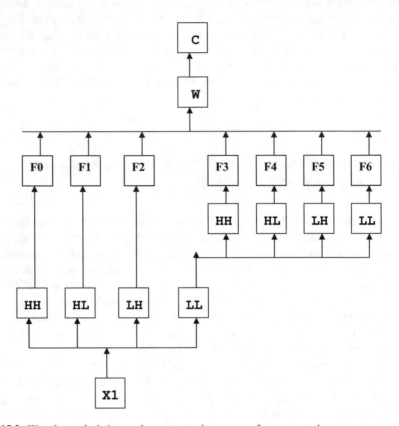

Fig. 15.2 Wavelet analysis image data computation process for segmentation

Fig. 15.3 Two-stage wavelet transform decomposition process resulting in seven image controls

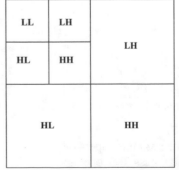

MULTIRESOLUTION DECOMPOSITION REPRESENTATION

L=LOWPASS

H=HIGHPASS

The overall strength of each channel is measured using variance. This is because variance has a direct association to the energy content of the channel and sample variance gives good estimation. Channel energy feature vector representation is defined after two levels of decomposition as shown in Eqs. (15.5, 15.6, 15.7, 15.8, 15.9, 15.10, 15.11, and 15.12), respectively.

$$e = \langle e_0, e_1, e_2, e_3, \ldots \ldots e_6, \rangle^t \tag{15.5}$$

where:

$$e_0 = \frac{1}{N_1^2} \sum_{m,n} \left(x_{m,n}^2 - \bar{x}^2 \right)^2 \tag{15.6}$$

$$e_1 = \frac{1}{N_1^2} \sum_{m,n} \left(y_{m,n}^{2,1} - \bar{y}^{2,1} \right)^2 \tag{15.7}$$

$$e_2 = \frac{1}{N_1^2} \sum_{m,n} \left(y_{m,n}^{2,2} - \bar{y}^{2,2} \right)^2 \tag{15.8}$$

$$e_3 = \frac{1}{N_2^2} \sum_{m,n} \left(y_{m,n}^{2,2} - \bar{y}^{2,2} \right)^2 \tag{15.9}$$

$$e_4 = \frac{1}{N_2^2} \sum_{m,n} \left(y_{m,n}^{2,2} \quad \bar{y}^{2,1} \right)^2 \tag{15.10}$$

$$e_5 = \frac{1}{N_2^2} \sum_{m,n} \left(y_{m,n}^{2,1} - y^{2,-2} \right)^2 \tag{15.11}$$

$$e_6 = \frac{1}{N_2^2} \sum_{m,n} \left(y_{m,n}^{2,2} - \bar{y}^{1,1} \right)^2 \tag{15.12}$$

In this particular case, $x_{m,n}^l$ and $y_{m,n}^{l,k}$ are gray levels in images X^l and $Y^{l,k}$, respectively, and $N_1 = N/2$ and $N_2 = N/4$ where N is the size of the local window. Fig. 15.2 shows the entire wavelet-based segmentation process. Each of the channel outputs is processed by feature extractors F_0 to F_6 which reduces the total image neighborhood to a feature vector of dimension seven as shown in Eq. (15.5). The computed seven element feature vectors for each pixel are combined into a single value that is used as a dimensional feature input as shown in Figs. 15.4 and 15.5 for the classification stage discussed in Sect. 15.6. This feature input is computed by averaging the energy maps over all m element feature vectors. A classifier C is used to determine the class where the pixel belongs. The classification algorithm is discussed in Sect. 15.6.

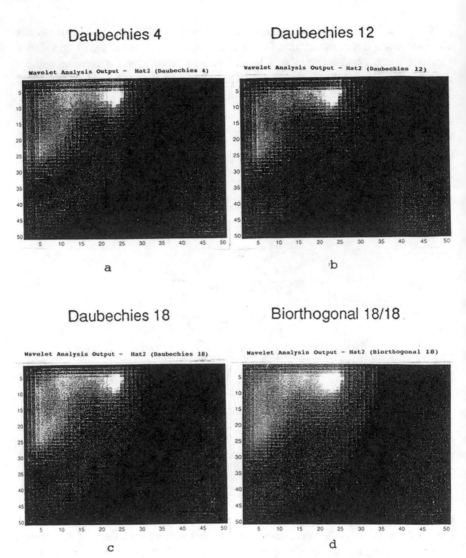

Fig. 15.4 Input (wavelet-based) analysis image for the classification stage using the 50×50 Hat2 image. (**a**) Daubechies-4, (**b**) Daubechies-12, (**c**) Daubechies-18, and (**d**) biorthogonal 18

15.3.2   Algorithm for the Computation of the Wavelet-Based Analysis Image Data for Segmentation

The algorithm for the computation of the wavelet-based analysis image data for segmentation is as follows.

1. Determine the window size. In this case, we can use a 9 x 9 window size as shown in Fig. 15.6.

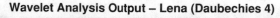

Wavelet Analysis Output – Lena (Daubechies 4)

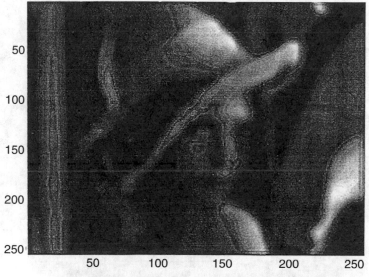

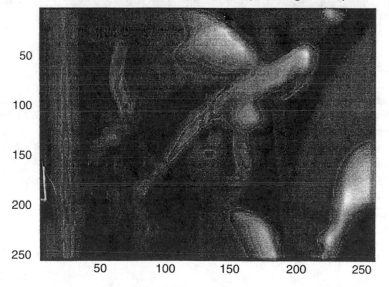

Fig. 15.5 Input (wavelet-based) analysis for the classification stage using the 256×256 Lena image. (**a**) Daubechies-4 and (**b**) biorthogonal 18

2. Choose a wavelet type. The Daubechies-4, Daubechies-12, Daubechies-18, and biorthogonal 18 wavelets can be used.
3. Perform a two-stage decomposition on each pixel on the image data array. This results to a seven-element feature vector.

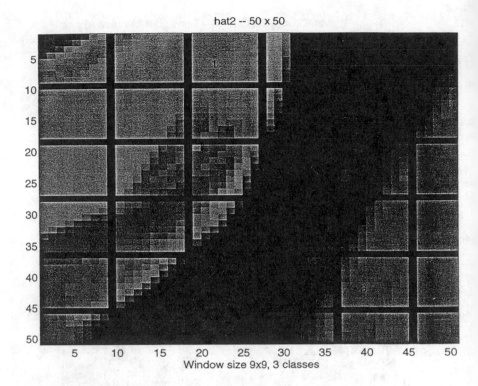

Fig. 15.6 A 9 × 9 window size division using the 50 × 50 Hat2 image

4. Generate the variance energy vector.
5. Determine one energy map data for the image. This is done by averaging the energy maps over all **m** element feature vectors.
6. Determine the absolute value of the image energy data. This makes the magnitude of all data to have real output.
7. Scale the image data output to be in the grayscale range of 0–255.
8. Save the image data to a file ("filename").
9. Call the classification (segmentation) algorithm.
10. End.

15.4   Computation of Fractal-Based Analysis Image Data for the Hybrid Segmentation Process

The fractal-based analysis block in Fig. 15.1 is discussed in this section. The block is used in computing the fractal-based analysis image data for segmentation.

15.4.1 The Fractal-Based Analysis Image Data Computation Process

The fractal-based analysis process used in the segmentation model is discussed. The analysis is done using a fractal dimension technique to create a fractal map. We use power spectral method in determining the fractal dimension of textured images on a pixel-by-pixel basis using a 9-by-9 window size. The power spectral decay exponent β is computed using a log-log linear regression, that is, the slope of $log(P(f))$ versus $log(f)$. In the case of 2D image ($E = 2$), the radial Fourier transform is given as shown in Eq. 15.13.

$$F(r) \propto r^{\frac{-\beta}{2}} = r^{-H-1} \tag{15.13}$$

with the power spectral density given by

$$P(r) = |F(r)|^2 \propto r^{-2H-2} \tag{15.14}$$

We use Eq. (15.14) to compute H since

$$\beta = 2H + E \tag{15.15}$$

and we compute the fractal dimension as

$$D = E - H + 1 \tag{15.16}$$

which is

$$D = 3 - H, (E = 2) \tag{15.17}$$

We can digitize the images and the fractal dimension computed for each 9 x 9 block of pixels by means of the Fourier technique. That is, the parameter H is estimated by means of a least-squares regression of the Fourier domain fractal definition onto the power spectrum of the block of pixels. Figure 15.7 shows a fractal dimension feature input for classification using the 256 × 256 Lena and the 50 × 50 Hat2 images.

15.4.2 Algorithm for the Computation of the Fractal-Based Analysis Image Data for Segmentation

The algorithm for the computation of the fractal-based analysis image data for segmentation is detailed in this section. The algorithm is as follows.

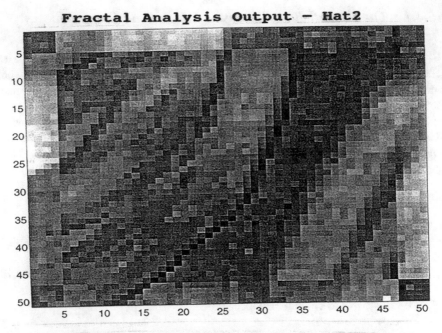

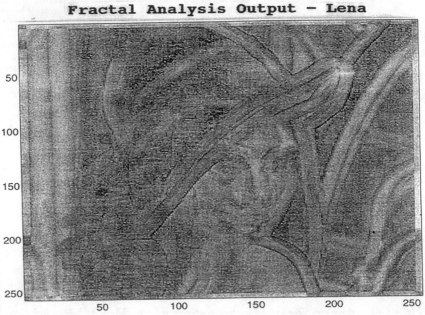

Fig. 15.7 Fractal dimension feature input for classification stage using 50×50 Hat2 and 256×256 Lena images

1. Determine the window size. We used 9 x 9 window size in this case as shown in Fig. 15.6.
2. Initialize the output data image.
3. Build frequency template based on window size, i.e., radial frequency squared, record squared radial frequencies, and record number of "hits."
4. Convert to frequency squared-power spectral density.
5. Process all pixels, not just the ones with full windows around them.
6. Calculate the fractal dimension of each pixel—using least squares.
7. Take absolute value to obtain real data.
8. Scale the data output to be in the grayscale range of 0–255.
9. Save the fractal-based analysis image data output to a file ("filename").
10. Call the classification (segmentation) algorithm.
11. End.

15.5 Formalizing the Notion of Segmentation

In this section, we formalize the definition of the segmentation problem. Let Z and R denote the set of integers and real numbers, respectively. Let region $R \subset Z^2$ and $r_i \subset Z^2$ such that $\cup_i r_i = R$ and $r_i \cap r_j = 0$, $i \neq j$. On each sub-region r_i, let a spatial point process $X_i(m, n)$ be defined. Therefore, the segmentation problem is to find

$$f : R \rightarrow \{1, 2, 3, \ldots, K\}$$

such that

$$f(m, n; X_1, X_2, X_3, \ldots, X_K) = i \leftrightarrow (m, n) \in r_i \qquad (15.8)$$

Image segmentation, the process of grouping image data into regions with similar features, is a component process in image understanding systems and also serves as a tool for image enhancement. The approach used in this chapter in developing the wavelet- and fractal-based segmentation model is to find discriminant features of each of these regions and make decisions based upon these features. Figure 15.8 shows three discriminant features for the 256×256 Lena image using Daubechies-4 wavelet for 10, 20, and 30 segmentation iterations, respectively. The discriminant function $f(m, n; X_1, X_2, X_3, \ldots, X_K)$ generally is not algebraic. It is expressed as a statistical decision algorithm. The processes $X_i(m, n)$ are assumed to be homogeneous (stationary under translation). It should be noted that this assumption is necessary in order to do any statistical inference about the region. In the next section, we discuss the segmentation process and algorithm.

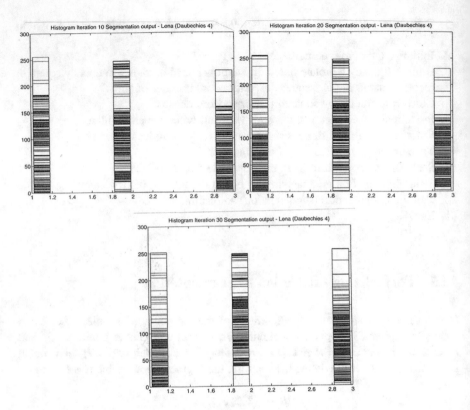

Fig. 15.8 Histogram of three discriminant features for the 256×256 Lena image using Daubechies-4 wavelet for 10, 20, and 30 segmentation iterations, respectively

15.6   The Segmentation Model Process

The segmentation model process is discussed in this section. The wavelet-based analysis is used in conjunction with the image intensity for multifeature image segmentation using the classification scheme described in Sect. 15.6.1. It should be noted that after getting the random sample, the moment method can be used to estimate mean and covariance. By the moment method, we mean a technique that matches moments between sample moments and texture distribution moments. Therefore, if $\widehat{e} = < e_o, \ldots\ldots, e_6 >$ represent the mean energy vector and $C = [c_{ij}]$ represent covariance matrix and if also $\widehat{e}$ and $\widehat{C}$ represent their estimates, then

$$\wedge e_i = \frac{1}{K} \sum_{k=1}^{k=K} e_{ki} \qquad (15.19)$$

$$\widehat{c_i} = \frac{1}{K-1} \sum_{k=1}^{k=K} (e_{ki} - \wedge e_i)(e_{kj} - \wedge e_j) \tag{15.20}$$

where:

e_{ki} is the i_{th} component of the K_{th} sample,
K is the total number of sample points chosen in the image texture surface

$$A(t+1) = \frac{[P(t)a(t) + P(t+1)a(t+1)]}{P(t) + P(t+1)} \tag{15.21}$$

Moment technique is given as shown in Eq. 15.21.
where:

a(t) is the current value of the parameter (mean or covariance),
a(t + 1) is the updated value of the parameter (mean or covariance),
P(t) is the value of the current prior probabilities.
P(t + 1) is the updated value of the prior probabilities.
A(t + 1) is the suggested update to be used in place of a(t + 1).

This "moment" idea is like an "update rule filter," in which the update rule filters (or smooths) the dynamics of the iteration. We update the values by weighted averaging of the current and updated estimates instead of simply replacing a parameter (mean or covariance) by its update. We find this idea to have favorable effect on the convergence performance. The relative weights of these two estimates are the corresponding prior probabilities. In the classification algorithm discussed in this chapter, we can compute the mean vector and the covariance matrix using this update rule. The prior probabilities however can be updated by simple replacement. The fractal dimension transformation can also be used in conjunction with the image intensity for multifeature image segmentation using the classification scheme described in Sect. 15.6.1.

15.6.1   Classification Theory Formulation

Let x_i represent a feature of different entities that are measurable. We choose x_1 as measured texture 1 and x_2 as measured texture 2. Let ω_i represent a class, or a "state of nature." We choose ω_1 and ω_2 to be class 1 and class 2, respectively. Supposing the underlying class is ω_1 (class 1), measurements of x_1 and x_2 classes can be expected to be small, whereas if the class being observed is ω_2 (class 2), it is expected that the values of ω_1 and ω_2 can be large on average or, at least, larger than those of ω_1, as shown in Fig. 15.9. The region where values of the features overlap is of significant importance. Classification errors are likely to occur in this region.

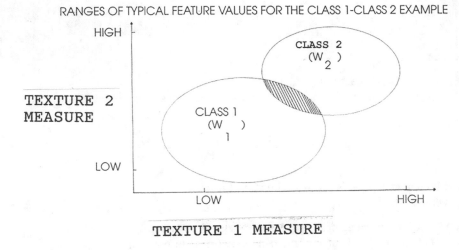

RANGES OF TYPICAL FEATURE VALUES FOR THE CLASS 1-CLASS 2 EXAMPLE

Fig. 15.9 Ranges of feature values that are typical for the "class 1 and class 2" example

Using a sample, the extracted features can be arranged as a *2 x 1* feature vector, $x^T = [x_1, x_2]$. Assume the class conditioned probability density function for the feature vector ($p(x_1, x_2/\omega_i)$ or simply $p(x/\omega_i)$ where $i = 1, 2$) are available. Assume also that something is known about the a priori likelihood of the occurrence of class ω_i or ω_2. Specifically, assume the a priori probabilities $P(\omega_i)$, $i = 1, 2$ are known. In the absence of this information, it is often reasonable to assume that $P(\omega_1) = P(\omega_2)$, that is, the a priori probabilities are equal. The classification problem therefore is as follows:

To determine the strategy for classifying samples, based on the measurement of x, such that classification error is minimized.

Bayes' theorem can be used in the development of this classification solution. Bayes' decision rule is considered an optimal solution in this case (Chang & Kuo, 1992). The a priori estimate of the probability of a certain class is converted to the a posteriori (or measurement conditioned) probability of a state through as shown in Eq. (15.22).

$$P(\omega_i|\underline{x}) = \frac{\left[p\left(\underline{x}|\omega_i \right) P(\omega_i) \right]}{p\left(\underline{x} \right)} \qquad (15.22)$$

where:

$$p(\underline{x}) = \sum_i p(\underline{x}|\omega_i) \qquad (15.23)$$

   Minimization of any reasonable choice of classification error requires that a given sample, x, is classified by choosing the class, ω_i, for which $P(\omega_i/x)$ is largest. This is also intuitively reasonable. In Eq. (15.22), it should be noted that the quantity $p(x)$ is common to all class-conditioned probabilities; therefore, it represents a scaling factor that may be eliminated. Thus, in the class 1 and class 2 example above, the classification algorithm is to decide using Eqs. (15.24) and (15.25), respectively,

$$\omega_1 \text{ if } p(\underline{x}|\omega_1)P(\omega_i) > p(\underline{x}|\omega_2)P(\omega_2) \tag{15.24}$$

or

$$\omega_2 \text{ if } p(\underline{x}|\omega_2)P(\omega_2) > p(\underline{x}|\omega_1)P(\omega_1) \tag{15.25}$$

It should be noted also that any monotonically nondecreasing function of $P(\omega_i/x)$ may be used for this test. Assuming a multidimensional Gaussian distribution, consider

$$p(\underline{x}) = (2\pi)^{\frac{-n}{2}}\left|\sum\right|^{\frac{-1}{2}} \exp\left[\frac{-1}{2}\left(\underline{x}-\underline{\mu}\right)^T \sum{}^{-1}\left(\underline{x}-\underline{\mu}\right)\right] \tag{15.26}$$

where:

x is $n \times 1$ with mean μ and covariance matrix Σ

$$v'(\underline{x}) = P(\underline{x}|\omega_i) \tag{15.27}$$

The class conditional density functions are assumed to be given by Eq. (15.26), and class dependence is through specific mean vectors and covariance matrices, that is, μ_i and Σ_i. We define a discriminant function for the ith class from Eq. (15.22) as shown in Eq. (15.27).

   With a given vector x, classification is based on finding the largest discriminant function. Assuming equal a priori probabilities, this means choosing the class for which $p(x|\omega_i)$ is largest. We already indicated above that any monotonically increasing function of $v_i(x)$ is also a valid discriminant function. The log function meets this requirement; that is, an alternative discriminant function is as shown in Eq. (15.28)

$$v'i(\underline{x}) = \log\{p(\underline{x}|\omega_i)\} \tag{15.28}$$

which, in the Gaussian case with equal covariance matrices (i.e., class dependence is only through the mean vectors), gives Eq. (15.29).

$$v'i(\underline{x}) = \frac{-1}{2}\left(\underline{x}-\underline{\mu_i}\right)^T \sum{}^{-1}\left(\underline{x}-\underline{\mu_i}\right) - \frac{n}{2}\log(2\pi) - \frac{1}{2}\log\left|\sum\right| \tag{15.29}$$

Notice that in the equal covariance matrix case, the second and third terms in Eq. (15.29) are constant biases and may be eliminated. Observe, however, that Σ influences classification through the first term, which is the squared distance of the feature vector from the ith mean vector, weighted by the inverse of the covariance matrix. In the case where $\Sigma = I$, a Euclidean distance norm results as shown in Eq. (15.30).

$$d_i^2 = \left(\underline{x} - \underline{\mu_i}\right)^T \sum^{-1} \left(\underline{x} - \underline{\mu_i}\right) \tag{15.30}$$

The factor in Eq. (15.29) is extremely significant. Given x, $v'_i(x)$ is largest when d_i^2 is smallest, this can be seen also as matching x against each of μ_i, to classify based on the best match. Assuming $\Sigma = I$ in Eq. (15.30), d_i^2 can be expanded as shown in Eq. (15.31).

$$\left\|\underline{x} - \mu_i\right\|^2 = \underline{x}^T \underline{x} - 2\underline{\mu_i}^T \underline{x} + \underline{\mu_i}^T \mu_i \tag{15.31}$$

15.6.2   The Segmentation (Classification) Model Algorithm

The classification (segmentation) algorithm used in the segmentation block of Fig. 15.1 is discussed in this section. The algorithm is as follows.

1. Determine the number of classes and window size iteratively, and then identify representative windows so prior probabilities can be calculated. The window size is 9 x 9 as shown in Fig. 15.6. The number of classes is dependent upon the number features in a particular image. We can use three classes for the hat2 and Lena images, respectively.
2. Calculate a priori probabilities, means, and covariances for a given class.
3. Generate log of class probabilities and covariances for pixel classification.
4. Determine the current pixel vector (pv) which is one pixel each from intensity and analysis data.
5. Calculate "distance" to each class iteratively, and determine the smallest distances for the classes using means and covariances.
6. Recalculate means, prior probabilities, and covariances based on pv.
7. Classification done for that particular iteration.
8. For each class, obtain the recommended parameter (mean or covariance) and updates using the "moment" function iteratively.
9. Save outputs after each iteration.
10. End.

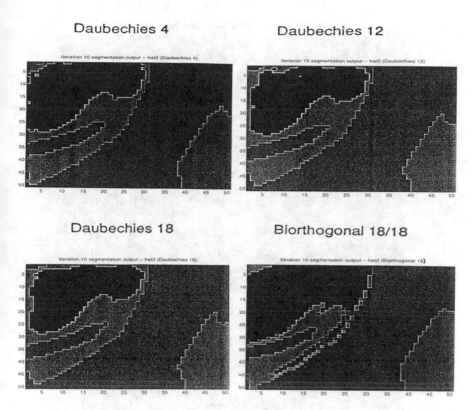

Fig. 15.10 Wavelet-based segmented 50 × 50 Hat2 image for ten iterations using Daubechies-4, Daubechies-12, Daubechies-18, and biorthogonal 18 wavelet filter length coefficients

Figures 15.10 and 15.11 show the wavelet-based segmented images for 50 × 50 Hat2 and 256 × 256 Lena images, respectively; while Fig. 15.12 shows the fractal-based segmented 50 × 50 Hat2 and 256 × 256 Lena images.

15.7   Example of Simulation Results and Discussions

The wavelet- and fractal-based segmentation model can be simulated using a MATLAB software on a PC-based system. The original 50 × 50 Hat2 image and original 256 × 256 Lena images used in the segmentation model are shown in Fig. 15.13. In order to compare the performances of both wavelet- and fractal-based segmentation processes, representative simulation runs can be conducted. Figure 15.14 shows the comparison of the fractal and Daubechies-4 wavelet segmented 50 × 50 Hat2 image after ten iterations. Figure 15.15 shows the comparison of the fractal, Daubechies-12, Daubechies-18, and biorthogonal 18 segmented 50 × 50 Hat2 image after ten iterations. Figure 15.16 is the comparison of the fractal,

Iteration 30 segmentation output - Lena (Daubechies 4)

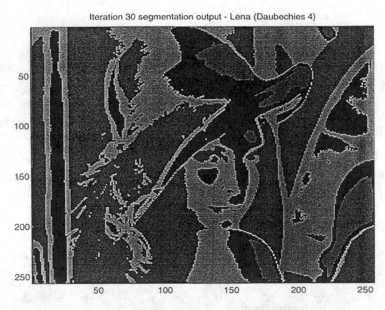

Iteration 30 segmentation output - Lena (Biorthogonal 18)

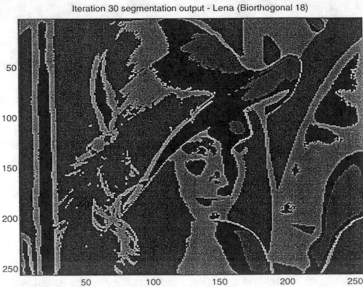

Fig. 15.11 Wavelet-based segmented 256×256 Lena image for 30 iterations using Daubechies-4 and biorthogonal 18 filter length coefficients

Daubechies-4, and biorthogonal 18 wavelet segmented 256×256 Lena image for 30 iterations. Equation 15.32 is used in the computation of error probabilities.

Table 15.1 is the fractal-based error probability of misclassification for the 256×256 Lena image after 30 iterations, and Table 15.2 is the fractal-based error

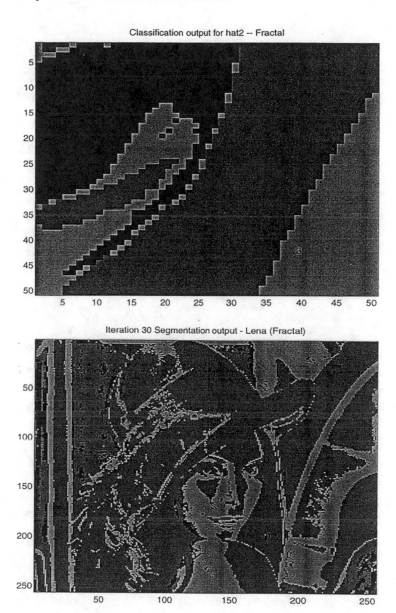

Fig. 15.12 Fractal-based segmented 50 × 50 Hat2 image after 10 iterations and 256 × 256 Lena image after 30 iterations

probability of misclassification for the 50 × 50 Hat2 image after 10 iterations. The total average error probability is 0.687 for the Lena image and 0.539 for the Hat2 image. Table 15.3 is the wavelet-based error probability of misclassification for the 256 × 256 Lena image (Daubechies-4 and biorthogonal 18 wavelets) after

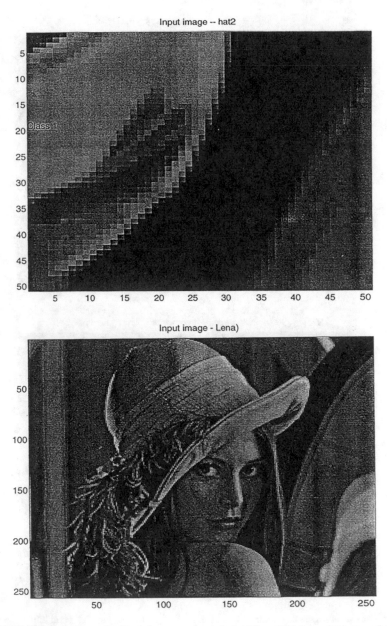

Fig. 15.13 Original 50 × 50 Hat2 image and original 256 × 256 Lena image used in the segmentation model

30 iterations. The total average error probability for Lena is 0.511 (Daubechies-4) and 0.513 (biorthogonal 18). Table 15.4 shows the wavelet-based error probability of misclassification for the 50 × 50 Hat2 image (Daubechies-4, Daubechies-12, Daubechies-18, and biorthogonal 18 wavelets) after ten iterations. The total average

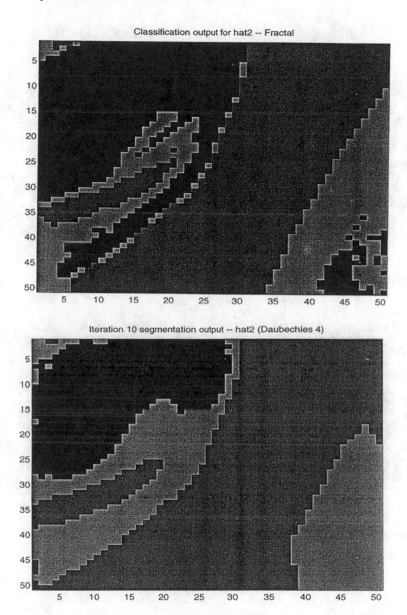

Fig. 15.14 Comparison of the fractal and Daubechies-4 wavelet segmented 50 × 50 Hat2 image after ten iterations

error probability for Hat2 image is 0.083 (Daubechies-4), 0.120 (Daubechies-12), 0.120 (Daubechies-18), and 0.129 (biorthogonal 18). The probability of error misclassification is computed as follows.

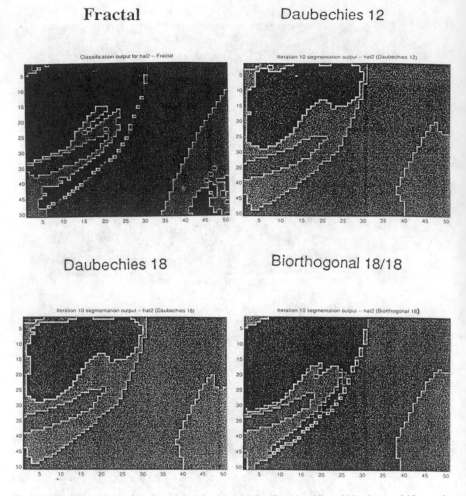

Fig. 15.15 Comparison of the fractal, Daubechies-12, Daubechies-18, and biorthogonal 18 wavelet segmented 50×50 Hat2 image after ten iterations

$$P_e = \frac{N_{c1} + N_{c2} + \ldots + N_{cn}}{N_{\text{total}}} \tag{15.32}$$

where:

N_{cn} is the number of points misclassified in class n
N_{total} is the total number of reference points.

We show in Fig. 15.17 the histogram representations of the fractal and wavelet dimension outputs for the 256×256 Lena image. The MATLAB histogram function (hist()) calculates the histogram for each row in a 2D array independently and then superimposes them on top of each other. The "lines" seen on the histogram

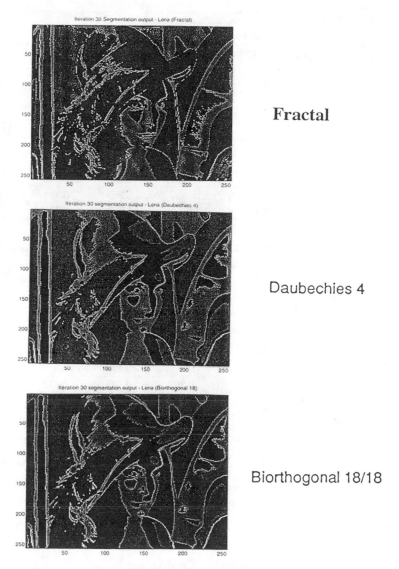

Fig. 15.16 Comparison of the fractal, Daubechies-4, and biorthogonal 18 wavelet segmented 256 × 256 Lena image after 30 iterations

outputs of Figs. 15.8 and 15.17, respectively, result from the different superimposed histograms. Figure 15.18 illustrates the class probabilities for the 50 × 50 Hat2 image after ten iterations for fractal, Daubechies-4, Daubechies-12, Daubechies-18, and biorthogonal 18 wavelet filter lengths.

Table 15.1 Fractal error probability of misclassification for 256 × 256 Lena image after 30 iterations

Error probabilities for Lena—fractal				
Class->	1	2	3	Total
Iteration				
1	0.243	0.297	0	0.54
2	0.333	0.213	0.33	0.877
3	0.237	0.003	0.003	0.243
4	0.263	0.11	0.333	0.707
5	0.283	0.33	0.33	0.943
6	0.263	0.02	0.003	0.287
7	0.29	0.003	0.003	0.297
8	0.267	0.03	0.333	0.63
9	0.237	0.263	0.333	0.833
10	0	0.333	0.333	0.667
11	0.233	0.33	0.333	0.897
12	0.333	0.327	0.003	0.663
13	0.26	0.333	0.003	0.597
14	0.017	0.257	0.333	0.607
15	0.26	0.003	0.003	0.267
16	0.323	0.32	0.333	0.977
17	0.273	0.333	0.297	0.903
18	0.317	0.33	0.333	0.98
19	0.257	0.33	0.327	0.913
20	0	0.333	0.333	0.667
21	0.203	0.333	0.333	0.87
22	0.237	0.333	0.33	0.9
23	0.237	0.33	0.33	0.897
24	0.313	0.333	0	0.647
25	0.28	0.33	0.003	0.613
26	0.127	0.333	0.333	0.793
27	0.247	0.333	0.003	0.583
28	0.06	0.29	0.333	0.683
29	0.243	0.003	0.22	0.467
30	0.323	0.01	0.333	0.667

15.8   Performance Complexity Evaluation

The computational complexity of the algorithms is measured using MATLAB's floating point operations (flops) and that can be performed in this case. For a 100 × 100 Hat2 image, it took 106,162,949 flops to compute the fractal dimension of the image. In the case of wavelet, it took 86,057,056 flops to compute the wavelet dimension of the same image using Daubechies-4 wavelet filter length coefficient.

Table 15.2 Fractal error probability of misclassification for 50 x 50 Hat2 image after ten iterations

Error probability for Hat2—fractal				
Class->	1	2	3	Total
Iteration				
1	0	0	0.057	0.057
2	0	0.33	0.333	0.663
3	0	0	0.002	0.002
4	0.333	0	0.333	0.667
5	0	0.333	0.333	0.667
6	0	0.333	0.333	0.667
7	0	0.333	0.333	0.667
8	0	0.333	0.333	0.667
9	0	0.333	0.333	0.667
10	0	0.333	0.333	0.667
Total average error probability = 0.539				

Summary

1. The hybrid model for wavelet- and fractal-based segmentation can be implemented successfully.
2. Image segmentation, the process of grouping image data into regions with similar features, is a component process in image understanding systems and also serves as a tool for image enhancement.
3. The wavelet- and fractal-based segmentations can be developed using an iterative process. This technique is based on the computation of the mean and covariance for each window of choice.
4. The concept of determining the local fractal dimension can be used in computing the fractal analysis image data for segmentation.
5. The results show the differences between a wavelet-based and a fractal-based segmented images.
6. The results further show the comparative analysis of different segmented images using different wavelets and wavelet filter lengths.
7. The wavelet-based segmentation process performed better than the fractal-based segmentation process.
8. The shorter wavelets used in the segmentation performed better than the longer wavelets.
9. The orthogonal wavelets—Daubechies-4, Daubechies-12, and Daubechies-18—performed better than the set of biorthogonal wavelet (Vetterli-Herley) biorthogonal 18.
10. The wavelet-based segmentation process needs fewer computations than the fractal-based segmentation process at least up to $O(N^2)$.

Table 15.3 Wavelet error probability of misclassification for 256×256 Lena image (Daubechies-4 and biorthogonal 18 wavelets) after 30 iterations

Error probability for Lena – Wavelet – Daubechies 4				Error Probability for Lena-Wavelet-Biothogonal 18			
Class -> 1	2	3	Total	Class -> 1	2	3	Total
Iteration				Iteration			
1 0.223	0.297	0	0.52	1 0.193	0.3	0	0.493
2 0.03	0.003	0.333	0.367	2 0.047	0.003	0.333	0.383
3 0.08	0.003	0.333	0.417	3 0.12	0.003	0.333	0.457
4 0.083	0.003	0.333	0.42	4 0.087	0.003	0.333	0.423
5 0.107	0.003	0.333	0.443	5 0.11	0.003	0.333	0.447
6 0.107	0.003	0.333	0.443	6 0.11	0.003	0.333	0.447
7 0.117	0.033	0.333	0.453	7 0.113	0.003	0.333	0.45
8 0.127	0.003	0.333	0.463	8 0.117	0.003	0.333	0.453
9 0.147	0.003	0.333	0.483	9 0.14	0.003	0.333	0.477
10 0.137	0.003	0.333	0.483	10 0.15	0.003	0.333	0.487
11 0.153	0.003	0.333	0.49	11 0.193	0.003	0.333	0.53
12 0.153	0.003	0.333	0.53	12 0.177	0.003	0.333	0.513
13 0.193	0.003	0.333	0.53	13 0.2	0.003	0.333	0.48
14 0.18	0.003	0.333	0.517	14 0.183	0.003	0.333	0.52
15 0.217	0.003	0.33	0.55	15 0.217	0.003	0.333	0.553
16 0.207	0	0.333	0.54	16 0.207	0.003	0.333	0.543
17 0.21	0.003	0.26	0.473	17 0.217	0.003	0.29	0.51
18 0.207	0	0.333	0.557	18 0.217	0.003	0.333	0.553
19 0.22	0.003	0.333	0.557	19 0.237	0.003	0.333	0.573
20 0.22	0	0.333	0.553	20 0.22	0.003	0.333	0.557
21 0.22	0.003	0.327	0.55	21 0.233	0.003	0.26	0.497
22 0.22	0.003	0.333	0.557	22 0.227	0.003	0.333	0.563
23 0.247	0.003	0.33	0.58	23 0.263	0.003	0.333	0.6
24 0.23	0.003	0.333	0.567	24 0.25	0.003	0.333	0.587
25 0.237	0.003	0.223	0.463	25 0.263	0.003	0.22	0.487
26 0.227	0.003	0.333	0.563	26 0.25	0.003	0.333	0.587
27 0.267	0.003	0.333	0.603	27 0.273	0.003	0.287	0.563
28 0.247	0.003	0.333	0.583	28 0.267	0.003	0.333	0.603
29 0.27	0.003	0.26	0.533	29 0.27	0.003	0.193	0.467
30 0.263	0.003	0.333	0.6	30 0.26	0.003	0.333	0.597

Review Questions

15.1 The hybrid wavelet- and fractal-based segmentation model can be developed using an approach that is based on the wavelet and fractal analysis processes.

(a) True.
(b) False.

15.2 The wavelet- and fractal-based segmentations can be developed using an iterative process.

Table 15.4 Wavelet error probability of misclassification for 50 × 50 Hat2 image (Daubechies-4, Daubechies-12, Daubechies-18, and biorthogonal 18 wavelets) after ten iterations

Daubechies-4					Daubechies-12				
Error probability for Hat2—Daubechies-4					Error probability for Hat2—Daubechies-12				
Class->	1	2	3	Total	Class->	1	2	3	Total
Iteration					Iteration				
1	0.012	0.015	0.008	0.035	1	0.018	0.012	0.008	0.038
2	0.032	0.005	0.005	0.042	2	0.072	0.005	0.007	0.083
3	0.06	0.005	0.007	0.072	3	0.092	0.003	0.012	0.107
4	0.07	0.005	0.023	0.098	4	0.098	0.005	0.038	0.142
5	0.072	0.003	0.013	0.088	5	0.095	0.002	0.03	0.127
6	0.068	0.005	0.035	0.108	6	0.092	0.002	0.05	0.143
7	0.068	0.002	0.032	0.102	7	0.093	0	0.045	0.138
8	0.052	0.002	0.047	0.1	8	0.08	0	0.058	0.138
9	0.052	0	0.05	0.102	9	0.082	0	0.067	0.148
10	0.035	0	0.052	0.087	10	0.068	0	0.067	0.135
Total average error probability = 0.120					Total average error probability = 0.120				
Daubechies-18					Biorthogonal 18/18				
Error probability for Hat2—Daubechies-18					Error probability for Hat2—biorthogonal 18/18				
Class->	1	2	3	Total	Class->	1	2	3	Total
Iteration					Iteration				
1	0.025	0.015	0.008	0.048	1	0.12	0.297	0	0.417
2	0.073	0.007	0.007	0.087	2	0.043	0.007	0.023	0.073
3	0.09	0.003	0.013	0.107	3	0.077	0.013	0.197	0.287
4	0.095	0.005	0.038	0.138	4	0.03	0.02	0.007	0.057
5	0.092	0.002	0.035	0.128	5	0.027	0.013	0.023	0.063
6	0.087	0.002	0.05	0.138	6	0.033	0.007	0.083	0.123
7	0.088	0	0.052	0.14	7	0.027	0.003	0.037	0.067
8	0.073	0	0.057	0.13	8	0.023	0.003	0.033	0.06
9	0.077	0	0.072	0.148	9	0.027	0	0.043	0.07
10	0.067	0	0.07	0.137	10	0.023	0	0.057	0.08
Total average error probability = 0.120					Total average error probability = 0.129				

(a) True.

(b) False.

15.3 The iterative technique is not based on the computation of the mean and covariance for each window of choice.

(a) True.

(b) False.

15.4 The concept of determining the local fractal dimension cannot be used in computing the fractal analysis image data for segmentation.

(a) True.

(b) False.

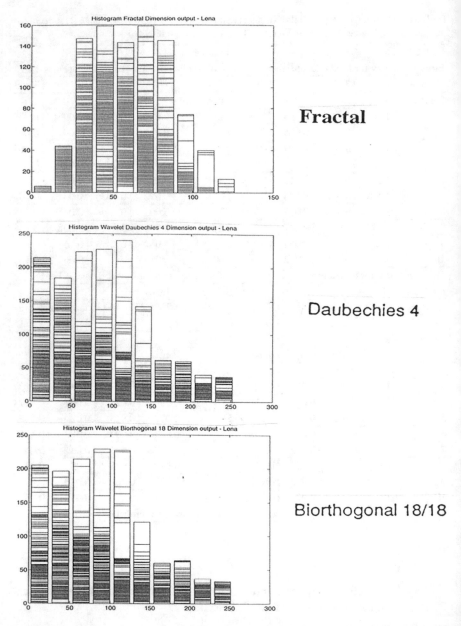

Fig. 15.17 Histogram representation of the fractal and wavelet dimension outputs for 256×256 Lena image. (**a**) Fractal, (**b**) Daubechies-4, and (**c**) biorthogonal 18

Fractal

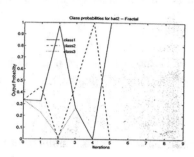

Daubechies 4

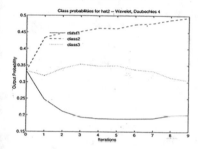

Daubechies 12

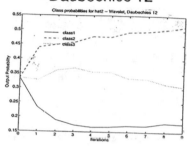

Daubechies 18

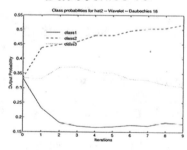

Biorthogonal 18/18

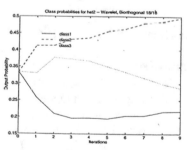

Fig. 15.18 Class probabilities for 50 × 50 Hat2 image after ten iterations. (**a**) Fractal, (**b**) Daubechies-4, (**c**) Daubechies-12, (**d**) Daubechies-18, and (**e**) biorthogonal 18

15.5 In the wavelet-based analysis block of the segmentation model, a two-stage local wavelet analysis is applied to every pixel neighborhood supported by window size **w** in order to get a feature which could compactly represent the statistical property of a region.

(a) True.
(b) False.

15.6 In the fractal-based analysis process segmentation model, the analysis is done using a fractal dimension technique to create a fractal map.

(a) True.
(b) False.

15.7 The power spectral method can be used in determining the fractal dimension of textured images on a pixel-by-pixel basis using any window size.

(a) True.
(b) False.

15.8 The fractal dimension transformation cannot be used in conjunction with the image intensity for multifeature image segmentation using the classification scheme.

(a) True.
(b) False.

15.9 The "moment" idea is like an "update rule filter," in which the update rule filters (or smooths) the dynamics of the iteration.

(a) True.
(b) False.

15.10 The region where values of the features overlap is of significant importance. Classification errors are likely to occur in this region.

(a) True.
(b) False.

Answers: 15.1a, 15.2a, 15.3b, 15.4b, 15.5a, 15.6a, 15.7a, 15.8b, 15.9a, 15.10a.

Problems

15.1 Describe in a general sense the idea of the hybrid wavelet- and fractal-based segmentation model.

15.2 Describe the general computation process for the wavelet-based analysis data for the hybrid segmentation process.

15.3 Describe the usage of the two-stage local wavelet analysis in the computation process for segmentation.

15.4 How can we represent the wavelet-based decomposition process for segmentation mathematically?

15.5 How can we represent the channel energy feature vector for the wavelet-based segmentation process?

15.6 Describe the algorithm for the computation of the wavelet-based analysis image data for segmentation.

15.7 Describe the fractal-based analysis image data computation process.

15.8 Describe the algorithm for the computation of the fractal-based analysis image data for segmentation.

15.9 Describe the moment technique for the segmentation of images.

15.10 Describe the segmentation classification model algorithm.

15.11 Describe one technique which can be used in the measurement of the performance evaluation of the complexity of the algorithms of the wavelet- and fractal-based segmentation of an image data.

References

Akujuobi, C. M., & Baraniecki, A. Z. (1992a). *Wavelets and fractals: Overview of their similarities based on application areas*. In Proceedings of the IEEE International symposium on time-frequency and time-scale analysis, Victoria, B.C., Canada, pp. 197–200.

Akujuobi, C. M., & Baraniecki, A. Z. (1992b). *Wavelets and fractals: A comparative study*. In Proceedings of the IEEE workshop on statistical signal and array processing, Victoria, B.C., Canada, pp. 42–45.

Akujuobi, C. M., & Baraniecki, A. Z. (1994). A comparative analysis of wavelets and fractals. In C. T. Leondes (Ed.), *2D and 3D digital signal processing techniques and applications* (Vol. 67, pp. 143–197). Academic Press.

Chang, T., & Kuo, C. C. J. (1992). *A wavelet transform approach to texture analysis*. In IEEE International conference on acoustic speech and signal processing, ICASSP-92, Vol. IV, pp. 661–664.

Pentland, A. P. (1984, November). *Fractal-based description of natural scenes*. In IEEE transactions on pattern analysis and machine intelligence, Vol. PAM 1–6, No. 6.

Perez-Lopez, K., & Sood, A. (1994, March). *Comparison of Subband features for automatic indexing of scientific image databases*. Center for Computational Statistics, George Mason University, Technical report no. 96.

Teshome, H. (1991). *Multichannel wavelets decomposition approach for texture segmentation*. Ph. D. thesis, Stevens Institute of Technology.

Part VI
Wavelet and Wavelet Transform Application to Cybersecurity Systems

Chapter 16
Wavelet-Based Application to Information Security

Things which matter most must never be at the mercy of things which matter least.

– Goethe

16.1 Introduction

The confidentiality, integrity, and availability (CIA) of electronic data from people with malicious intentions are guarded using the idea of information security (IS). This idea is responsible for handling the risk management encountered in transmitting information from point A to point B. Anything can be a risk to the CIA triad. We protect and maintain sensitive information so that without permission, it cannot be changed, altered, or transferred. For example, a message sent from point A to point B can be modified, if someone intercepts it during transmission, before it reaches the intended recipient. Good tools for cryptography can help mitigate this issue. Authentication processes are improved using digital signatures that guarantee that people are properly identified before access is granted to sensitive areas. The CIA triad is a concept that is a key information security necessity for the secure use, flow, and processing of data. Confidentiality, integrity, and availability are the three main objectives of information security. We will discuss these issues in this chapter.

New data delivery and processing networks are constantly being launched on the market. Every bit of information that communicates with these new tools must be protected properly without compromising or undermining the network's functions. Hackers can so quickly discover and exploit weaknesses that a whole system can be compromised before information security analysts even know a problem exists. Because these attacks are usually rendered by "zero-day bugs or flaws" in codes unknown to developers, they have become flash points of concerns to many security organizations.

Wavelets have immensely contributed to information security, and we will discuss their application in this chapter. Wavelets are deployed in various implementations of security schemes ranging from biometric identification to anomaly detection and cybersecurity and for both wired and wireless devices. The purpose of this chapter is to explore how wavelets can be applied to information security. We cover the applications of wavelets and wavelet transforms to information security, detection schemes, and methods. In addition, we discuss variance change and jump detections and finally cryptography and its different forms including elements of steganography. We start with information security schemes in Sect. 16.2.

16.2   Information Security Schemes

Every information security scheme that any individual or organization adopts must have to comply with regulations and as such must fall within the CIA triad. The CIA triad is developed to guide as a model for information security guideline within organizations. Confidentiality handles rules of access to ensure that only authorized persons have access to the right information. This is sometimes known as privacy. Integrity is the guarantee that a piece of information is accurate and trustworthy, while availability ensures reliable access to information by the appropriate individuals.

16.2.1   Confidentiality

Confidentiality ensures that sensitive information does not get into the hand of the wrong person while ensuring that the appropriate persons are equally not denied access to rightful data. Access should only be granted to those authorized to have them. It is also a common practice to classify data and create levels of access to them based on the amount of damage that can be caused should the information get into the hand of the wrong person. With this classification, restrictions that are more stringent could be placed on the information based on their categories. Sometimes, data security may involve special training for people that already have knowledge about the data. This kind of training is a threat on its own for the data requiring protection. The training is intended to familiarize the person that will be handling the data of the risks involved and how to mitigate against them. Deeper aspects of training can involve training on how to create strong passwords and password-related best practices and information about social engineering methods, so that they do not bend information, safekeeping rules, though they may have good intension, but the result will always have a disastrous ending.

A good example of confidentiality is encryption of banding or routing numbers when conducting online banking. User IDs and password as a standard procedure must be observed, but it is also becoming a norm that companies are now implementing a two-factor authentication. Other options of confidentiality being observed currently, depending on the class of information, include biometric verification and security tokens, key fobs, or soft token. *It is worth mentioning that all the mentioned schemes above like biometric verification and encryption can all be implemented efficiently using wavelet. Steganography will be another important wavelet application that could be helpful in information security.*

16.2.2   Integrity

Integrity involves the maintenance of data accuracy, consistency, and trustworthiness of data while the data lives. Integrity is achieved by deploying methods that ensure data is not altered by dubious or by unauthorized persons both while the data is at rest or in transit. We implement integrity by permissions and access control. Version control records changes to a file or set of files over time are known so that you can recall specific versions later. It is used to safeguard against erroneous changes or deleting the data by error both by authorized and unauthorized personnel. It is also necessary to put checks by adding some means to detect changes that may occur which are not human-caused like electromagnetic pulse or server crash. Backups may be necessary or redundant repositories. Error detection techniques such as checksum and cryptographic checksums for integrity verification may be handy and a good tool for data integrity.

16.2.3   Availability

Availability is best achieved through timely maintenance of all hardware and by conducting repairs immediately when the need arises and avoiding software conflict in your system. It is necessary also to be abreast of current system updates. Good communication bandwidths, with removal of possible bottlenecks, can enhance information availability. Other important practices worth adopting include redundancy, failover, "redundant array of inexpensive disks" or "redundant array of independent disks" (RAID), and high-availability clusters. Even distributed processing can prevent serious system failure. You do not want to find yourself in worst-case scenarios where you have no option but total system failure. Measures against data loss must include unforeseen unpredictable disasters like hurricane, flooding, etc. Arrangements for backup must be made in terms of different geographical location; probably isolated and fireproof materials should be considered. We cannot forget to mention the possibility of flood attacks like denial of service (DoS) and other cyberattacks capable of denying access to data or compromising the integrity of the data.

More challenges have arisen for information security experts with the flood of data being generated every second by data collection platforms and the security challenge of Internet of Things (IoT). We cannot delve into the complexity that these two platforms bring to table for information security experts. At all cost, data must be made available with integrity and confidentiality not compromised. Below are some of the best-recommended cybersecurity practices worth adopting for any organization.

- Biometric security implementation. While implementing regular known applications like facial recognition, consider other emerging methods like behavioral biometrics that analyze aspects like keystroke dynamics, mouse dynamics, and eye movement biometrics achieved using eye and gaze tracking devices.
- Implement hierarchical cybersecurity policy. Security is always best in layers. Unauthorized access to a level of information should put all data at risk.
- Deploy a risk-based technique to security. Thorough risk assessment is key. Never assume security. Auditing should be regular.
- Data backup. Failure and risk do not make announcement before they come. The best time to back up your data is when you have the data.
- Manage IoT security. Any device can be the gateway for data compromise. There was a demonstration of recent certain USB cables that can be used to introduce virus to your system and retrieve sensitive information. If any device has IP address or MAC address, then it is worth being looked after with the approach of security threat.
- Adopt the culture multi-factor authentication.
- You must have a secure means of handling passwords.
- Use the principle of least privilege. Not all can have the same level of access.
- Ensure to keep an eye on the privileged users. No trust when it comes to data security.
- Monitor third-party access to your data. A bridge of the third-party system may be intended for you. Monitor every inch of their processes on your platform.
- Be careful about phishing and ensure your workers are too.
- Raise employee awareness. Constant training, meetings, reminders, and so on must be adopted to keep employees away of possible attacks.

In majority of these cybersecurity practices, wavelets and wavelet transform techniques can be applied in the information security analysis as discussed in Sect. 16.3.

16.3   Wavelet Application Analysis to Information Security

Information security has become crucial due to the explosion in data and the amazing possibilities of what can be done with data. The worst is what hackers can do with data when they succeed to get hold of them. Wavelets have been applied in many information security schemes, particularly, in fields where sensitive and dynamic information flow is required. Wavelets have been maximally used in biometric verification for body parts, like iris eye, earlobe, palm, and facial recognition, and may be used in recent test of behavioral biometrics.

Wavelet has become a good analytical tool because of its adaptation to time-varying signal. Because it could analyze any dynamic signal in a time-frequency spectrum, it gives it a perfect advantage over many other analytical tools in varying signal analysis. This is why wavelets are good for anomaly detection as well. The scaling and dilating functions associated with wavelet give it a fitting value for any signal wavelength. More so, many wavelet forms mean there are options in the wavelet family and wavelet could be modified to adapt to almost any signal for analysis.

The beauty of wavelet plays out in its ability to dissect a signal into different frequencies, as already discussed in Chaps. 1 and 2, respectively, and hence a frequency of choice could be focused on for analytic investigations. This makes it possible for noise to be filtered off as many signals may be corrupted with different types of noise. The robustness of wavelets in noise filtering is perhaps one of their usages in applications where the information can easily be corrupted with noise like in telecommunications and wireless sensor networks.

Wavelets for years have dominated research work on cryptography and steganography, contributing immensely in the area of information assurance. Wavelet signal compression capability has made it very easy to use in the storage of vital data such as in the case of image data compression. This has enabled the security firms to keep pace in maintaining image databases compared to the enormous space images take. Without this wavelet-based algorithm application, it may be very costly to maintain databases of images of everyone with the flow of big data currently, and hence, many security schemes will never have been a possibility. We discuss in Sect. 16.4 some of the detection methods that are relevant to the wavelet application techniques to information security.

16.4 Detection Methods

Threat detection is a nightmare if detection occurs after attack. It is always better and safer to place detection ahead of Intrusion Prevention. Threat mitigation requires speed in all stages of threat occurrence. Security is an ongoing process. No individual or organization should assume security, as we know that being exposed to threat is imminent. Therefore, information security should be a top priority of every organization.

Security check must be put in place to detect threat in a timely manner so that attackers do not have enough time to gain full knowledge or access to every of the target network or data. The organization's security programs should be able to checkmate known threats, while the aim is to keep guard for the unknown threats that do the damages. Even known threats sometimes can get pass an organization's security programs. Some detection methods worth adopting for information security are discussed next. They are as follows:

- *Monitor and be at alert for spikes in signals and activities.*

    Spike activities could be a sign of threat. Activities like unusual modification or deleting of files, many wrong login trials, delay in between password letters, or so much noise or mixed signal in a packet transmission could be an indication of threat, and it should be investigated.

- *Monitor all access attempts, and search for anomalous ones.*

  Keep track of all access attempts, both successful and unsuccessful one. Find out the reason for the failed one. All access must be logged. Check for unusual frequencies of occurrence, and do not forget to check and verify all access after work hours. You must also check to ensure that all access accounts are within stipulated levels of permission.

- *Look for unusual VPN access to your network.*

  Unusual access to your organization network like abnormal speed of download and volume or access from an unusual location should be investigated to ascertain if it is real threat or not.

- *All privileged and service accounts must be monitored with utmost care.*

  Service and privileged accounts access information rarely for only privileges or activities other accounts cannot perform. Hence, any activity from any of these accounts should be monitored and logged, and any deviation from the company policy should be investigated to classify it as threat or not. Many networks have been compromised and privileged or even abused. Every account should be kept in check.

- *Check for abnormal access to sensitive data.*

  You should always double-check unusual access, quick scan, or any other unusual activity on sensitive data.

- *Monitor all shared account.*

  This happens to be one of the most common sources of data breach. All organizations shared must be monitored with great care and every form of abnormal activity quickly investigated.

- *All infrastructure resources of the organization must be monitored.*

  Activities around organization's infrastructure should be kept in perspective and in view which includes all events around data centers, DoS, and others. Suspicious activities should be reported and investigated.

- *Set intruder traps.*

  Sometimes intruder traps with honeypot targets may be the signal you need to know that there are unusual activities going on in the network.

- *Analyze user and attacker behavior analytics.*

  This is very necessary because it allows an organization to have fine knowledge of employee and user behavior. This lets you know where the employees live, possible login locations, type of data employees access, their access times and time spent on data, common downloads and the network bandwidths they use in accessing networks, and, further, the OS of the system the access is from. The analysis and log of this kind of behavior allow for easy spot of hackers.

In Sect. 16.5, we discuss some of the key detection schemes that are very useful in the protection of information. The survival of many organizations depends on how proactive they are in acknowledging and finding ways to mitigate against these possible detection schemes.

16.5 Detection Schemes

There are no definite ways on how threats could be detected and handled. However, there have been desperate means of threat handling of which Cisco is at the front. There are about five detection schemes, and we will attempt to explain them in this section.

i. *Endpoint Detection and Response (EDR).* This focuses on end node monitoring and responses. It holds record of endpoint behaviors for analysis and alerts the systems on anomaly detection. It monitors the endpoint network activities, processes, DLLs, registry settings, and file activities.

ii. *Network Traffic Analysis (NTA).* This technology channels its activities to analysis of network traffic in search of anomalous, malicious, and suspicious activities. This is one of the oldest and most successful schemes on threat detection. *Wavelet has a lot of application on programs deployed on this level.*

iii. *Malware Sandboxes.* This service detonates and analyzes files sent to it and usually deployed as an appliance, as cloud-based, or in some manner as hybrid configuration.

iv. *Cyber Threat Intelligence (CTI).* Threat detection in this technique is achieved by comparing internal security incidences with indicator of compromise (IoC) and known cyber foe tactics, techniques, and procedures (TTPs). This analysis is used to classify security incident as threat or not.

v. *Central Analytics and Management.* In this case, all incidences are analyzed at a center point. It allows for holistic approach to threat detection and response.

In every case of these schemes discussed in Sect. 16.5 for information security detection, numerous data are produced which must be analyzed to detect and prevent cyberattacks. In Sect. 16.6, we discuss the analysis of the data for information network security using wavelets and wavelet transforms.

16.6 Information Network Security Data Analysis Using Wavelet Transforms

Wavelet transforms discussed in Chap. 2 are applied in the information network security in the analysis and mining of time-serial data streams for the detection of anomalous in the information security events. Figure 16.1 shows the wavelet-based

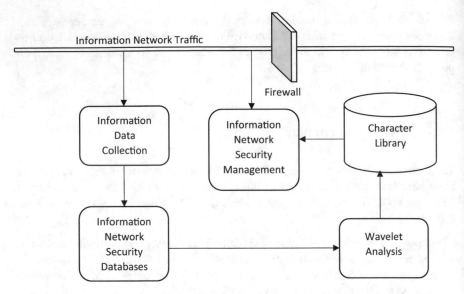

Fig. 16.1 Wavelet-based data analysis framework for information network security traffic. *Source*: Jian-ping et al. (2014)

data analysis framework for information network security traffic. The idea is to capture the original data using information network security tools. We then have to find the signatures of the information network security events included in the original data by wavelet analysis technology. Finally, according to the signatures of different information network security events, corresponding security policies can be adopted to automatically and intelligently prevent or stop these information network security events to enhance the information security protection capability of the whole network. The target in this case is to release the information network security administrators from massive alarm data and arduous tasks of security management with the advantages of data mining for a mass of data.

The framework for the wavelet analysis of information network security data has five segments. They are data collection, information network security database (INSD), denoised INSD via the wavelet analysis, character library, and network security policies.

16.6.1 Data Collection and Information Network Security Database (INSD) Segment

The data collection segment is used to capture all the packets passing through the network. Then it sends them to the information network security database (INSD) after simple classification by type of traffic. This process is enabled with the help of the network traffic tools that includes hardware and software such as the spectroscope, traffic data collection instrument, and tcpdump. The INSD is a database that is used to store the classified traffic data captured by the data collection segment.

16.6.2   The Wavelet Transform Application Including the Denoised Segment

The wavelet transform approach while decomposing the signal from the INSD performs the filtering (denoising) process at the same time. The denoising process takes care of the unwanted information (noise) leaving the needed information that will now be analyzed for any malicious attacking pattern for security purposes. This is the beauty of the wavelet transform techniques. It is also one of the advantages wavelet transform has over the Fourier transform. The thresholding mechanism is used in the noise removal process as well along with the wavelet analysis.

As an example, if the wavelet function in Eq. (16.1) is satisfied and is able to tolerate the condition in Eq. (16.2),

$$W_\Phi f(a,b) = f(t), \Phi_{a,b}(t) = \frac{1}{\sqrt{|a|}} \int_{-\infty}^{+\infty} f(t)\Phi\left(\frac{t-b}{a}\right) \tag{16.1}$$

$$\int_{-\infty}^{+\infty} \left(\frac{|\Phi'(\omega)|}{\omega}\right) < \infty \tag{16.2}$$

where $\Phi'(\omega)$ is the Fourier transform of the wavelet function. Then, we can reconstruct the original signals from the wavelet-decomposed signals using the threshold technique in the analysis.

The following are the necessary steps for the wavelet-based denoising algorithm:

1. *Signal Decomposition.* First, we choose a wavelet and choose a level N. Then, we compute the wavelet decomposition of the signal S at level N.
2. *Detail Coefficients' Threshold.* For each level j from 1 to N, select a threshold T, and apply soft thresholding to the detail coefficients.
3. *Signal Reconstruction.* Compute the wavelet-reconstruction signal based on the original approximation coefficients of level N and the modified detail coefficients of levels from 1 to N.

It is important to note that going through the reconstruction of the denoised signal, the INSD data will be more suitable for transmission on the Internet and have many other advantages such as being easier to explore any security issues in the transmitted information in the network of interest.

16.7 MATLAB Implementation of the Wavelet Transform-Based Analysis Algorithms

In this Sect. 16.7, we describe the MATLAB implementation of the wavelet-based approach as designed and implemented by Jian-ping et al. (2014).

- *Use the built-in MATLAB function WAVEDEC to analyze the INSD, which performs a multi-level 1D wavelet analysis of signal x using a specific wavelet given by "wname." The Haar and Daubechies type of wavelets can be used in this case as an example.*
- *[C, L] = WAVEDEC (X, N, 'wname') returns the wavelet decomposition of the signal X at level N, using "wname." The output decomposition structure contains the wavelet decomposition vector C and the vector L.*
- *The vector C is the set of DWT coefficients, which consists of the approximation and detail coefficients.*
- *For [C, L] = WAVEDEC(X, N, Lo_D, Hi_D), Lo_D is the decomposition low-pass filter, and Hi_D is the decomposition high-pass filter.*
- *The structure is organized as follows:*

  - $C = [app.coef.(N), detail. coef.(N), ... , det.coef.(1)]$
  - $L = [L(1), L(2), ... L(N + 2)]$
  - *Where, $L(1)$ = length of app.coef.(N)*
  - *$L(i)$ = length of det. coef.(N–i + 2) for i = 2, ..., N + 1*
  - *$L (N + 2)$ = length(X).*
  - *"app." represent approximation, and "det." detail.*

- *As INSD is very complex and numerous, to speed up the querying, we identify the indices first and then query using these indices.*
- *Approximately, between 10 and 15% of the DWT coefficients carry relevant information.*
- *We can use approximation coefficients (and a few detail coefficients) as indices.*
- Example 16.1 shows the (Jian-ping et al., 2014) MATLAB algorithm for getting indices—INDEX.
- The output $[C_\tau, L_\tau]$ are the desired indices, which help in getting the algorithm for fast querying from the INSD.
- Example 16.2 shows the (Jian-ping et al., 2014) MATLAB algorithm for the fast querying from the INSD—SEARCH.

Example 16.1: The MATLAB Algorithm for Getting the Indices—INDEX

```
(a) [C, L] : = WAVEDEC (X, N, 'wname');
(b) C_τ : = C; L : = 20% of the detail coefficients of X;
(c) While (length (L) >= 2)
   {X : = C;
   [C, L] : = WAVEDEC (X, N, 'wname');
   C_τ : = C; L : = 20% of the detail coefficients of X;
   }
```

Example 16.2: The MATLAB Algorithm for Fast Querying from INSD— SEARCH

```
(a) [C, L] : = WAVEDEC (Q, N, 'wname'); # where Q is the query
(b) Call INDEX to get indices (C_r, L_r)
(c) For each c in C
    {for each c_r in C_r
    {
    if ||c - c_r|| < ε; # ε is the given precision
      {Push c into C_out; }
    }
(d) For each 1 in L
    {for each 1 τ c_r in L_r
    {
        if ||1 - 1_r|| < ε;
        {Push c into L_out; }
    }
(e) [C_out, L_out] is the desired query.
```

16.8   Wavelet Transforms and Cryptography in Information Security

The idea of cryptography is to hide the data or the information from possible intruders. Figure 16.2 shows the block diagram of a cryptographic system. The information is transmitted in encrypted forms due to security reasons. At the receiving end, the transmitted information is reversed to get the original information. However, in the case for some reason the intruder gets access to the encrypted information, the technique that can be used to remedy such a case is to use a secret key. A secret key is generated using a password, which is assigned during the encryption process. Based on the key, the information is encrypted. Wavelets such as the Haar, Daubechies, and Symlet are some of the three wavelets that can be used in the encryption process. These wavelets are chosen for encryption based on the key generated at the initial step. The secret password and the encrypted information are shared with the intended receiver. This means that the intended receiver can decrypt

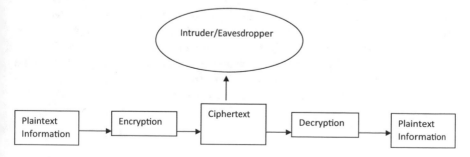

Fig. 16.2 Block diagram of a cryptographic system

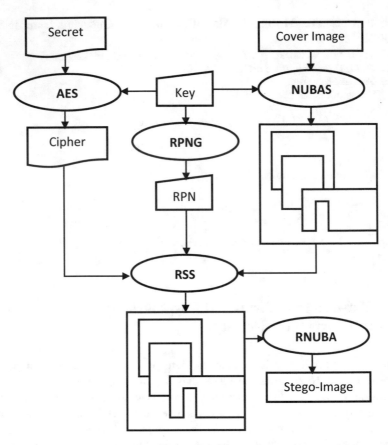

Fig. 16.3 Three layer securities for hiding secret information into the cover image. *Source:* Srinivasan et al. (2015). Use permission granted

the information using the same key. After applying the key, the same wavelet, which is used for encryption, is selected and applied to the encrypted information. The whole process is reversed and the original information is retrieved.

The information encryption, a video steganography method using Haar or Daubechies, or Symlet information wavelet (SIW) and least significant bits (LSB) are the schemes that are used to hide the data. Key generation-based encryption is used to encrypt the information using keys from the information. The secret information is encrypted using AES algorithm, Non-Uniform Block Adaptive Segmentation on Information (NUBASI), and Randomized Secret Sharing algorithm to hide the message in an image as shown in Fig. 16.3. The AES algorithm and the cryptographic algorithms (or ciphers) are discussed in Sects. 16.8.1 and 16.8.2, respectively.

16.8.1 AES Algorithm

The Advanced Encryption Standard (AES) is part of the three layer securities for hiding secret information into the cover image as shown in Fig. 16.3. The first layer has the cipher message produced by encrypting the secret message by the AES. Then the digital cover image is segmented by using the Non-Uniform Block Adaptive Segmentation on Information (NUBASI) discussed in Sect. 16.8.3. The third layer is the Randomized Secret Sharing (RSS) Section, which uses the same key, SEC_KEY, for embedding the secret message into the segmented information.

It was in 2001 that the National Institute of Standards and Technology (NIST) established the Advanced Encryption Standard (AES) algorithm specification for the encryption of electronic data. The AES is a block as shown in Fig. 16.3 with a block size of 128 bits, but three different key lengths: 128, 192, and 256 bits. The AES adopted by the US government superseded the Data Encryption Standard (DES) that was established in 1977. AES uses a symmetric key algorithm, which means that it uses the same key for both encrypting and decrypting the data. AES is included in the ISO/IEC 18033-3 standard. It is available in many different encryption packages and is the first (and only) publicly accessible cipher approved by the US National Security Agency (NSA) for top-secret information when used in an NSA-approved cryptographic module.

16.8.2 Description of the Ciphers

The algorithm used in encrypting the information for security is known as a *cipher* as shown in Fig. 16.3. AES is one of those algorithms for encryption of information. It is based on a design principle known as a substitution-permutation with an example discussed in Sect. 16.9.1 and is efficient in both software and hardware. It uses a fixed block size of 128 bits and a key size of 128, 192, or 256 bits. AES operates on a 4×4 column-major order array of bytes, termed the *state*. Most AES calculations are done in a particular finite field as shown in Example 16.3.

Example 16.3
Let us consider these 16 bytes, which have elements b_0, b_1, b_2, b_3, ... b_{15} represented as this two-dimensional array:

$$
\begin{array}{cccc}
b_0 & b_4 & b_8 & b_{12} \\
b_1 & b_5 & b_9 & b_{13} \\
b_2 & b_6 & b_{10} & b_{14} \\
b_3 & b_7 & b_{11} & b_{15}
\end{array}
$$

The key size used for an AES cipher specifies the number of transformation rounds that convert the input, called the information plaintext, into the final output, called the ciphertext. The numbers of rounds are as follows:

- 10 rounds for 128-bit keys
- 12 rounds for 192-bit keys
- 14 rounds for 256-bit keys

Each round consists of several processing steps, including one that depends on the encryption key itself. A set of reverse rounds are applied to transform ciphertext back into the original information plaintext using the same encryption key.

16.8.3 Non-Uniform Block Adaptive Segmentation on Information (NUBASI)

The Non-Uniform Block Adaptive Segmentation on Information (NUBASI) is an algorithm which produces the number of segments with different dimensions of an input cover information such as an image by accepting the same secret key, SEC_KEY, designed and implemented by Srinivasan et al. (2015). When you split an image into various sub-images, you refer to that process as the segmentation process. There are two types of segmentation depending on the size of the size of the sub-images. The two types are known as uniform and non-uniform segmentation processes. In the uniform segmentation, all the produced segments are having the same dimension. On the other hand, in the non-uniform segmentation, every segment has different dimensions. In this algorithm, a digital cover image with dimension "M x N" and a 128-bit key are taken as inputs and finally produce T numbers of non-uniform image segments. Here, the *key* plays a vital role in dividing the image. The (Srinivasan et al., 2015) algorithm and description are as illustrated in Example 16.4. The number T is calculated as T = floor (L/2) * (floor (L/2) +1).

Example 16.4: NUBASI Algorithm Implementation

   Algorithm NUBASI (Img, SEC_KEY[], Seg [])

   Input: Img → A digital cover image with dimension M x N.

   SEC_KEY →   A list containing sequence of characters with size L.

   Output: Seg → A list of segmented images with size T, each with different dimensions,
   *   where T is L/2 x L/2.*

   1. Find the Height and Width of the cover image Img, ImgW & ImgH
   *   ImgW ← Img.Width*
   *   ImgH ← Img.Height*

2. *Find the Length of the Key,* **KeyLen**
   KeyLen ← *Key.Size*

3. *Split the input Key list into two lists* **KeyV** *and* **KeyH**
   splitKey (Key[], KeyV[], KeyH[])

4. *Find the Length of the KeyV and KeyH,* **KeyLenH** *&* **KeyLenV**
   KeyLenH ← *KeyV.Size*
   KeyLenV ← *KeyH.Size*

5. *Find sum of all the keys in KeyV,* **SumKeyV**
   SumKeyV ← *getSum (KeyV[])*

6. *Find sum of all the keys in KeyH,* **SumKeyH**
   SumKeyH ← *getSum (KeyH[])*

7. *Calculate Partition Percentage,* **PPV** *and Pixel Length,* **PixV** *for each key in KeyV list*
   $for\ i ← 0,1,2,...,LenV\text{-}1\ do$
   $PPV_i = round ((KeyV_i\ /\ SumKeyV) *100)$
   $PixV_i = round ((PPV_i * ImgH)\ /\ 100$
   end for

8. *Calculate Partition Percentage,* **PPH** *and Pixel Length,* **PixH** *for each key in KeyH list*
   $for\ i ← 0,1,2,...,LenH\text{-}1\ do$
   $PPH_i - round ((KeyH_i\ /\ SumKeyH) *100)$
   $PixH_i = round ((PPH_i * ImgW)\ /\ 100$
   end for

9. *Perform segmentation over the cover image Img as* $X_1 ← X_2 ← 0$
   $T ← 0$
   $for\ i ← 0,1,2,..., LenH\text{-}1\ do$
   $X_2 ← X_1 + PixH_i$
   $for\ j ← 0,1,2,..., LenV\text{-}1\ do$
   $Y_2 ← Y_1 + PixV_j$
   $SegT ← getSubImage (Img, X_1, Y_1, X_2 - X_1, Y_2 - Y_1)$
   $T ← T + 1$
   end for
   $X_1 ← X_2$
   $Y_1 ← 0$
   /*Shift the list PixV left*/
   $t ← PixV_0$
   $for\ k ← 1,2,..., LenV\text{-}1\ do$
   $PixV_{k-1} ← PixV_k$
   end for
   $PixV_{LenV-1} ← t$
   end for

16.8.4   Randomized Secret Sharing (RSS)

The Randomized Secret Sharing (RSS) block shown in Fig. 16.3 is part of the third layer of the three layer securities for hiding secret information into the cover image as first mentioned in Sect. 16.8.1 of this Chap. 16. The non-uniform image segments, which are produced by the NUBASI algorithm, are numbered as Seg_0, Seg_1, and up to Seg_{T-1}. The cipher information, which is generated by the AES algorithm, is divided into T-1 numbers of blocks B_0, B_1, B_2, and up to B_{T-2} with equal size. A random number, RPN, selects the segments for each block of information. The random pattern number (RPN) shown in Fig. 16.3 is a number between 0 and 31. An algorithm random pattern number generator (RPNG) shown in Fig. 16.3 generates the numbers. The same 16-byte secret key, SEC_KEY, is used in generating the random pattern number. The order in which the segments are selected for embedding cipher information blocks is from 32 defined patterns. After embedding the message into the image segments, the Reverse Non-Uniform Block Adaptive Segmentation on Image (RNUBASI) algorithm is used in merging all the stego-image segments to produce the final stego-image. This algorithm is absolutely the reverse of NUBASI which is used in segmenting the information. Example 16.5 shows the (Srinivasan et al., 2015) RSS algorithm implementation process, while Example 16.6 shows the (Srinivasan et al., 2015) Embed algorithm implementation. The system uses 24-bit cover image for embedding secret information. Each pixel in the segmented image is composed of 3 bytes, each for representing the colors red, blue, and green, respectively. This algorithm uses one of the most popular steganographic techniques called least significant bit replacement.

Example 16.5: RSS Algorithm Implementation

```
Algorithm RSS ( Seg[], CMsg[], RPN, StegImg )

Input: Seg → A List of cover image segments, Seg₀, Seg₁, Seg₂, . . . , Seg_{T-1}

    CMsg → Cipher Message with size M
    RPN → A random pattern number from 0 to 31

Output: StegImg → A stego-image embedded with the secret message.

1. Split the CMsg into 'T-1' numbers of blocks, B₀, B₁, B₂, . . ., B_{T-2}

2. Embed the RPN into Seg₀
   Embed (RPN, Seg₀)

3. Embed the CMsg blocks into different image segments
   for each Blk in B₀, B₁, B₂, . . . , B_{T-1} do
     Choose a segment Seg_i based on RPN
     Embed (Blk, Seg_i)
   end for
```

4. *Merge all the cover image segments into a single image*
   End RSS

Example 16.6: Embed Algorithm Implementation

Algorithm Embed (Msg[], Img)

Input: Msg → A list containing sequence of bytes
   Img → A cover image with dimension R x C

Output: Img → A image embedded with Msg

Convert the Msg into sequence of bits, $MBit_0$, $MBit_1$, $MBit_2$, . . . , $MBit_N$
i ← 0
j ← 1

```
Repeat while i < N
  For r in 0, 1, 2, . . ., R
    For c in 0, 1, 2, . . . C
      getPixel (Img, r, c, P)
      getRGB (P, R, G, B)
       storeBit (B, 8-j, MBiti)
      storeBit (G, 8-j, MBiti+1)
      storeBit (R, 8-j, MBiti+2)
      setRGB (P, R, G, B)
      setPixel (Img, r, c, P)
       i ← i+3
    end for
  end for
  j ← j + 1
end while
```

In Sect. 16.9, we discuss the symmetric and asymmetric ideas of a cryptographic system.

16.9   Cryptographic Symmetric and Asymmetric Systems

Cryptography is derived from the Greek words for "secret writing." The field of cryptography deals with the technique for conveying information securely. When sensitive information is encrypted with a powerful algorithm, the information can be protected from eavesdroppers. The block diagram of a cryptographic system is shown in Fig. 16.2. The information in its original form (in intelligible form) is called *information plaintext*. The transmitter will encrypt the information plaintext in order to hide its meaning. This encrypted message is called *ciphertext*. The algorithm used in encrypting the message is known as a *cipher*. The process of converting from information plaintext to ciphertext is called *encryption*, while the reverse process is called *decryption*. Most cryptographic algorithms make use of a secret value called

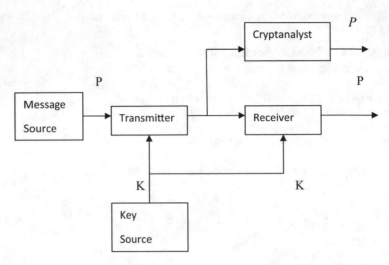

Fig. 16.4 Conventional (symmetric) cryptographic system

the *key*. The process of breaking ciphers when the key is unknown is called *cryptanalysis*.

Cryptographic algorithms (or ciphers) can be classified using several criteria, such as ciphers as either symmetric or asymmetric. In conventional cryptography, the sender and receiver of a message use the same secret key. This method is known as symmetric (or secret key) cryptography. The main challenge is getting the sender and receiver to agree on the secret key without anyone else finding out. The cryptography is asymmetric (or public key) if the sender and receiver use different keys. In their system, each person gets a pair of keys, one called the public key and the other called the private key. The public key is published, while the private key is kept secret. The two types of cryptosystems are shown in Figs. 16.4 and 16.5, respectively. The most important modern symmetric encryption algorithm is that contained in Advanced Encryption Standard (AES) algorithm discussed in Sect. 16.8.1. Public key cryptography provides advantages over symmetric cryptography. First, the problem of managing secret keys is greatly reduced. With a symmetric or conventional cipher, it is necessary to transfer secret keys to both communicating parties before secure communication can begin. As shown in Fig. 16.4, the key is distributed over a secure channel. Public keys may be transferred over a public channel. Second, public key cryptography provides digital signature. Authentication can be provided with public key cryptography. Digital signatures are set to replace handwritten signatures in many applications. Over the years, several encryption algorithms have been developed. Here, we will consider only the substitution ciphers, in Sect. 16.9.1.

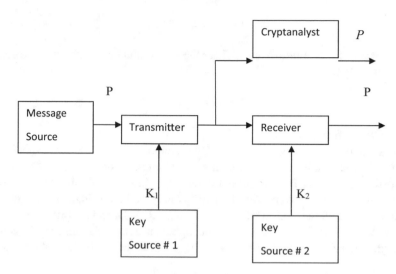

Fig. 16.5 Public key (asymmetric) cryptographic system

16.9.1   Substitution Permutation Cipher

Simple (monoalphabetic) substitution ciphers replace each letter in the information plaintext with another letter to produce the ciphertext. Although this class of ciphers are easy to implement and use, they are not difficult to break. An example is a substitution cipher known as *shifted alphabet*. One was used by Julius Caesar to send messages to his generals. Using Caesar cipher, we replace each letter in the alphabet with the letter three places further down the alphabet. The alphabet is wrapped around so that A follows Z. Thus, the plaintext letter A becomes D, B becomes E, ..., Z becomes C. By this method, we obtain

plain: a b c d e f g h i j k l m n o p q r s t u v w x y z

cipher: d e f g h i j k l m n o p q r s t u v w x y z a b c

For example,
plain:
COME
HOME
cipher:
FRPH
KRPH

If a numerical equivalent is assigned to each letter (a = 1, b = 2, ..., z = 26), then the encryption algorithm becomes

$$E(\lambda) = (\lambda + 3)\mathrm{mod}\,(26) \qquad\qquad (16.3)$$

where λ stands for an English letter. Since the shift could be any amount, the general Caesar algorithm is

$$E(\lambda) = (\lambda + k)\mathrm{mod}\,(26) \qquad\qquad (16.4)$$

where k varies from 1 to 25.

The ciphertext alphabet used in the Caesar cipher is orderly. Some substitution ciphers use a scrambled alphabet that has no apparent order. For example, A might be substituted with T, B with U, and C with Q. However, all simple substitution ciphers can be easily broken using frequency analysis. In the English language, E is the most frequent letter, followed by T, O, A, N, I, R, S, H, and so forth. Z is the least frequent letter. Polyalphabetic substitution ciphers use multiple alphabets to conceal the single letter frequency distribution of the plaintext letters in the ciphertext.

Example 16.3
We may use information reversal for encryption. In this case, the plaintext is written backward to produce the cipher. For example,

plaintext: you have to go home
cipher: emoh og ot evah uoy

Obviously, this can easily be decoded and the message is not secure. However, much stronger versions can be adopted to make it more difficult to decode.

Summary
1. The confidentiality, integrity, and availability (CIA) of electronic data from people with malicious intentions is guarded using the idea of information security (IS).
2. Confidentiality, integrity, and availability are the three main objectives of information security.
3. Wavelets are deployed in various implementations of security schemes ranging from biometric identification to anomaly detection and cybersecurity and for both wired and wireless devices.
4. Confidentiality handles rules of access to ensure that only authorized persons have access to the right information. This is sometimes known as privacy. Integrity is the guarantee that a piece of information is accurate and trustworthy, while availability ensures reliable access to information by the appropriate individuals.
5. Security must keep up with the explosion currently witnessed in big data, infrastructure, and IoT.
6. These growths are coming with accompanying vulnerabilities that must be addressed with the necessary urgency required.

7. The application of wavelet analysis in many information security applications and platforms has contributed to surmounting the growing challenge of information security as contained in the triad CIA.
8. The ability of wavelet to locally analyze dynamic signals in time-frequency domain is great and one of the reasons it will continue to be deployed in data and signal analysis seeing how dynamic data has become.
9. What you can do with wavelets and wavelet transforms in information security is dependent on your capability of understanding and knowing how to apply wavelet as described in this chapter.
10. The Non-Uniform Block Adaptive Segmentation on Information (NUBASI) is an algorithm which produces the number of segments with different dimensions of an input cover information such as an image by accepting the same secret key, SEC_KEY.
11. The Randomized Secret Sharing (RSS) block shown is part of the third layer of the three layer securities for hiding secret information into the cover image.
12. Cryptographic algorithms (or ciphers) can be classified using several criteria, such as ciphers as either symmetric or asymmetric.

Review Questions

16.1 The confidentiality, integrity, and availability (CIA) of electronic data from people with malicious intentions are guarded using the idea of information security (IS).

a. True
b. False

16.2 The three main objectives of information security are:

a. Confidentiality, reducibility, and availability
b. Cryptographic, integrity, and availability
c. Cryptographic, integrity, and adaptability
d. Confidentiality, integrity, and availability

16.3 Threat detection is not a nightmare if detection occurs after attack.

a. True
b. False

16.4 The wavelet transform approach while decomposing the signal from the INSD performs the filtering (denoising) process at the same time in its application to information security.

a. True
b. False

16.5 The idea of cryptography is not to hide the data or the information from possible intruders.

a. True
b. False

16.6 The information encryption, a video steganography method using Haar or Daubechies, or Symlet information wavelet and least significant bits (LSB) are the schemes that can be used to hide the data.

a. True
b. False

16.7 The digital cover image is segmented by using the Non-Uniform Block Adaptive Segmentation on Information (NUBASI).

a. True
b. False

16.8 The Randomized Secret Sharing (RSS) block is not part of the third layer of the three layer securities for hiding secret information into the cover image.

a. True
b. False

16.9 The algorithm used in encrypting the message is known as a *cipher*.

a. True
b. False

16.10 Polyalphabetic substitution ciphers use multiple alphabets to conceal the single letter frequency distribution of the plaintext letters in the ciphertext.

a. True
b. False

Answers: 16.1a, 16.2d, 16.3b, 16.4a, 16.5b, 16.6a, 16.7a, 16.8b, 16.9a, 16.10a

Problems
16.1 Why is the idea of information security important?
16.2 What does confidentiality mean in the information security field?
16.3 What is a possible example of confidentiality?
16.4 Can the confidentiality schemes like biometric verification and encryption be implemented using wavelets?
16.5 (a) What does integrity mean in the security of information? (b) How is it achieved? (c) How is integrity implemented?
16.6 How can availability be best achieved?
16.7 What are some of the best-recommended cybersecurity practices worth adopting for any organization in security information data that wavelets can be applied in the implementation?
16.8 What are some of the detection methods worth adopting for information security?
16.9 What are the five detection schemes that can be used in information security?
16.10 What are the five segments that make up the frameworks for the wavelet analysis in information security?
16.11 Describe data collection and information network security database (INSD).

16.12 Describe the wavelet transform application including the denoised segment.

16.13 What are the necessary steps for the wavelet-based denoising algorithm that can be used in the information security analysis?

16.14 Briefly describe the MATLAB implementation of the wavelet-based analysis algorithms that could be applied to information security.

16.15 Describe the MATLAB algorithm for getting the indices—INDEX in Problem 16.14.

16.16 Describe the MATLAB algorithm for the fast querying from the INSD—SEARCH needed in Problem 16.14.

16.17 What is AES?

16.18 Most AES calculations are done in a particular finite field. Show an example of this finite field.

16.19 Describe the Non-Uniform Block Adaptive Segmentation on Information (NUBASI).

16.20 Describe briefly in four lines the NUBASI implementation algorithm.

16.21 Describe the Randomized Secret Sharing (RSS) used as part of the three layer securities for hiding secret information into the cover image.

16.22 Describe the RSS implementation algorithm.

16.23 Describe the embed implementation algorithm.

16.24 Using the Caesar substitution permutation cipher, encrypt the following information:

INFORMATION SECURITY IS REAL

16.25 Using the information reversal technique, decrypt the following:

EREH SI SAMTSIRHC

16.26 (a) What is cryptography? (b) Describe symmetric cryptography.

16.27 (a) Describe asymmetric cryptography. (b) Does asymmetric cryptography have any advantage over symmetric cryptography?

References

Jian-ping, W., Ping, R., Wu, L., Zhibiao, S., & Donghong, S. (2014, February 28). Analysis of network security data using wavelet transforms. *Journal of Algorithms & Computational Technology, 8*(1), 59–70.

Srinivasan, B., Arunkumar, S., & Rajesh, K. (2015, April). A novel approach for color image, steganography using NUBASI and randomized, secret sharing algorithm. *Indian Journal of Science and Technology, 8*(7), 228–235.

Chapter 17
Application of Wavelets to Biometrics

What lies behind us and what lies before us are tiny matters compared to what lies within us.

– Ralph Waldo Emerson

17.1 Introduction

Biometrics is a term that originated from the Greek words bio (life) and metrikos (measure). Humans as far as we know have always used some of the key body characteristics such as face gait, voice, or signature to recognize each other. It was Alphonse Bertillon, the chief of the criminal identification division of the police department in Paris, who first conceived and actualized the idea of using body measurements for solving crimes more than a century ago. In the past many years, there have been significant developments and studies that have entirely changed the concept of biometrics (Maltoni et al., 2009; Tico & Kuosmanen, 2000). In today's world, a wide variety of applications requires reliable and accurate verification schemes to confirm the identity of individuals based on their physiological and/or behavioral characteristics. Minutiae verification in fingerprint images is one of such biometric schemes. In this chapter, we will discuss the basic concepts of biometrics, the minutiae verification in fingerprint images, and how wavelets and wavelet transforms can be applied to enhance the minutiae verification. Wavelet analysis is applied to biometrics for security and personal identification, and with the advent of Internet of Things (IoT) and big data, it may become the future of privacy and data protection. This chapter covers the application of wavelets and wavelet transforms to biometrics and specifically to minutiae verification in fingerprint images.

© Springer Nature Switzerland AG 2022
C. M. Akujuobi, *Wavelets and Wavelet Transform Systems and Their Applications*,
https://doi.org/10.1007/978-3-030-87528-2_17

17.2   Biometric Characteristics

The key ingredient in biometrics is identifying which biological measurement can be used qualitatively as a biometric characteristic (Phillips et al., 2000). Any human physiological and/or behavioral characteristic can be used as a biometric characteristic if it satisfies the following requirements:

- Collectability: the characteristic can be measured quantitatively.
- Distinctiveness: any two persons should be sufficiently different vis-à-vis the characteristic.
- Permanence: the characteristic should be sufficiently unchangeable (as per the matching criterion) over a period of time.
- Universality: each person should have the characteristic.

However, other issues are considered in practical biometric systems, namely:

- Acceptability, which indicates the degree of acceptance of the usage of a particular biometric characteristic in people's daily lives, for example, the uproar that arose with the recent introduction of biometric capturing schemes at US border crossings for international visitors
- Circumvention, which reflects how the system by fraudulent means can easily be manipulated
- Performance, which implies the attainable recognition accuracy and speed, the resources required to arrive at the desired speed, and accuracy as well as other factors that affect the accuracy and speed

Hence, in the determination of the biometric characteristics, all the issues mentioned in Sect. 17.2 should be given adequate consideration in order to have a robust biometric recognition system. In this chapter, the biometric identifier used is the fingerprint print using wavelets and wavelet transforms as the analysis tools for better enhancements.

17.3   Biometric System

A biometric system is a pattern recognition system that recognizes an individual based on a specific physiological and/or behavioral characteristics possessed by that individual. The recognition is achieved by extracting a feature vector from the physiological and/or behavioral characteristic of an individual, and the featured vector, stored in a database, is used as a reference for comparison purposes.

A biometric system based on the physiological characteristic is more reliable than the one that uses behavioral characteristic, although present-day applications combine for both accuracy and reliability. A biometric recognition system can operate in a verification mode or in an identification mode depending on the intended application (Phillips et al., 2000). The three biometric system basic operational modes are enrollment, verification, and identification.

17.3.1 Enrollment Mode

The enrollment system is as shown in Fig. 17.1. In addition, it shows a verification and an identification system as well. The enrollment process is common to both the verification and the identification systems. This module is responsible for the registration of persons into the system database. It starts with the biometric characteristic of an individual scanned to produce a digital representation of the characteristic, and a quality check is performed to determine if the acquired sample can be

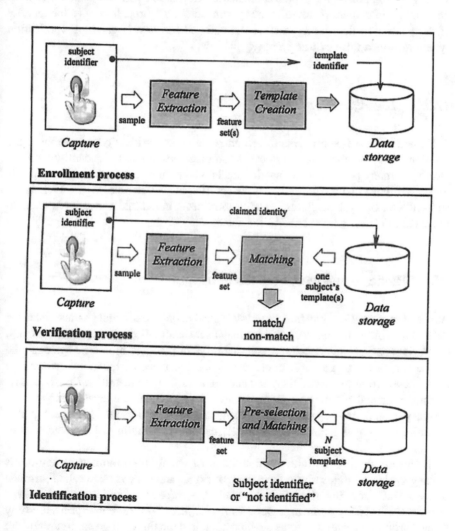

Fig. 17.1 Block diagram of enrollment, verification, and identification schemes. (**a**) Ridge Ending. (**b**) Ride Bifurcation
Source: Maltoni D., Maio D., Jain A.K, Prabhakar S., Handbook of Fingerprint Recognition, Springer, 2008. Reproduced with Permission from Springer Nature

processed by subsequent stages. A feature set is then extracted to generate a template, which can be stored in a central database of the biometric system.

17.3.2   Verification Mode

A biometric verification system as shown in Fig. 17.1 authenticates a person's identity by comparing the captured biometric information with the person's biometric template information stored in the system. In other words, it verifies the identity of the person to determine whether the claimed identity is valid. A verification system returns a false or true identity.

17.3.3   Identification Mode

A biometric identification system recognizes a person's identity by searching the entire template database for a match. It compares an individual's identity with the identity of many people, and a match is made where found applicable. If no match is found, it implies that the individual is not enrolled in the database. Figure 17.1 shows an identification system. Future identification of an individual's identity is made in reference to the template stored in the database.

17.4   Basics of Fingerprint Recognition

As stated in Sect. 17.2, any human physiological and/or behavioral characteristic can be used as a biometric characteristic as long as it satisfies the conditions universality, distinctiveness, permanence, and collectability. In practical biometric systems, the choice of a biometric identifier or characteristics depends on achievable performance using the identifier, acceptability, and circumvention of the identifier. Consequently, a robust practical biometric system should meet an acceptable level of recognition accuracy, speed, and resource requirements and be safe to users, clearly acceptable to the intended user group, and adequately capable of handling attacks and fraudulent intrusion.

Therefore, there is no question that the most widely used biometric identifier is the fingerprint. A fingerprint is the pattern of ridges and valleys (also called furrows) on the surface of a fingertip of a human being. *A ridge is defined as a single curved segment and a valley is the area between two adjacent ridges.* Fingerprints are fully formed around 7 months of fetus development, and the finger ridge configuration are invariant throughout the life of that individual except due to circumstances such as bruises and cuts on the fingertips which can in some cases erode part of the ridge structure. In addition, fingerprints are unique to each individual, and the uniqueness

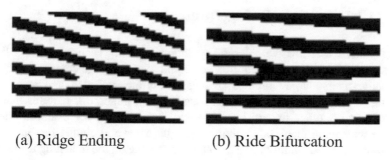

(a) Ridge Ending   (b) Ride Bifurcation

Fig. 17.2 A ridge ending and ridge bifurcation examples

of a fingerprint is exclusively determined by the local ridge characteristics and their relationships.

There are about 150 different local ridge features (islands, short ridges, enclosure). The two most prominent local ridge characteristics, *referred to as minutiae*, are:

(i) Ridge ending
(ii) Ridge bifurcation

The minutiae are the local discontinuities in the ridge flow pattern, and they provide the features that are used in certain types of fingerprint recognition systems. As shown in Fig. 17.2, a ridge ending is the point where the ridge ends abruptly, and a ridge bifurcation is the point where the ridge splits into branch ridges. The white pixels in Fig. 17.2 correspond to the valleys, and the black pixels correspond to the ridges. A good-quality fingerprint image contains about 40–100 minutiae. The matching technique deployed in this Chap. 17 involves the construction of the minutiae images extracted from the fingerprint images. Consequently, there is much emphasis on the successful and reliable extraction of minutiae as discussed in more details later in the chapter.

Fingerprinting has various uses in the area of identification and security. Forensic scientists and the police use fingerprints to identify bodies or criminals, etc. Sometimes they can also be extracted from crime scenes to be checked against the large criminal databases that are used by different governmental organizations. Because of the surge in terrorism, fingerprint identification and verification has seen remarkably used in airports and border posts by immigration officials and other law enforcement officers to identify some of the known terrorists. Advances in technology have resulted in the deployment of fingerprint recognition in homes, with many computers and laptops now coming with optional fingerprint lock that verifies the identity of the authorized user.

17.5   Classification of Fingerprints

In applications that involve a large population segment, such as driver's license registration, forensics, access control, homeland security, large computer networks, and many others, large volumes of fingerprints are collected and stored daily resulting in large databases. The identification of a fingerprint requires that it be matched against other fingerprints stored in the database, for example, the FBI database with more than 200 million fingerprint cards and a daily rate increase of over 30,000–50,000 (Maltoni et al., 2009). Depending on the matching technique, a great deal of time and computational resources is required to achieve reliable and accurate fingerprint matching. In order to reduce the search time and computational complexity, it may be desirable to classify the fingerprint images in an accurate and standardized manner so that the matching of the fingerprint is done with a group of fingerprints in the database that is similar to the fingerprint under test.

Consequently, fingerprint classification can be defined as a technique used to assign a fingerprint to one of several pre-specified types. The traditional approach to classification of fingerprints is based on the information in the global pattern of ridges. Nevertheless, the most successful approaches involve leveraging and incorporating other techniques, such as devising alternative schemes when the fingerprint landmark information cannot be extracted and the use of reliable structural/syntactic pattern recognition methods in addition to statistical methods. Generally, pattern types can classify fingerprints, the size of those patterns, and the position of the patterns on the fingerprints. The ridge structure is normally not changed by injuries since the same pattern comes back when the skin grows.

Fingerprint classification can be viewed as a coarse-level matching of the input fingerprint to one of the pre-specified types and the subsequent comparison to a subset of the database corresponding to the fingerprint type. The first fingerprint classification system was developed in India by Azizul Haque for Sir Edward Henry, the inspector general of police in Bengal, India. Sir Henry got the credit for developing the system, and the classification system quickly spread across India. In 1901, Sir Henry became the first director of the Metropolitan Police's fingerprint department and in 1905 was promoted to the commissioner of police in Scotland Yard in 1905. A recap of the history was made in order to highlight the importance of this classification system, now known as Henry's system of fingerprint classification, which has become the standard for fingerprint classification. The original Henry system relied on the classification of each individual fingerprint into one of three classes: loop, arch, and whorl. Over the years, the FBI and others augmented Henry's system to deal with larger and exhaustive repositories of fingerprint cards. Some of the different types of fingerprint patterns used in classifications are shown in Fig. 17.3.

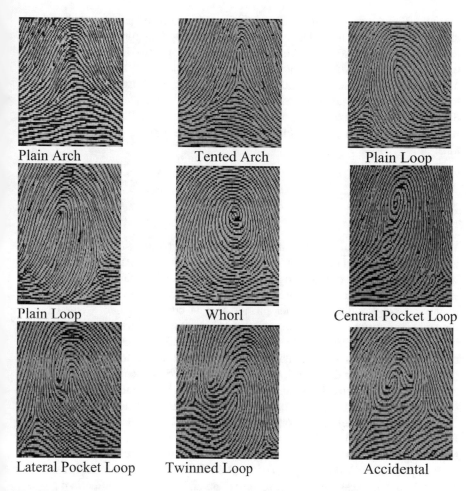

Plain Arch   Tented Arch   Plain Loop

Plain Loop   Whorl   Central Pocket Loop

Lateral Pocket Loop   Twinned Loop   Accidental

Fig. 17.3 Examples of the different types of fingerprint patterns
Source: http://www.policensw.com/info/fingerprints, retrieved 2008

17.5.1   Loops

Loops occur in about 60–70% of the patterns encountered. In a loop pattern, one
or more of the ridges enters on either side of the impression; re-curves, touches, or
crosses the line of the glass running from the delta to the core; and terminates or
tends to terminate on or in the direction of the side where the ridge or ridges entered
(Lennard & Patterson, 2008). A loop pattern is identified with the presence of a delta
and core. The ridge count, in addition to the delta and core, is also used to classify
loop patterns. The ridge count is defined as the number of ridges that intersect or

touch the line drawn from the easily recognized tri-radius (where three ridges meet) to the center of the pattern. There are two types of loop fingerprint patterns, namely, radial and ulnar. These types of loops are named after the two bones of the forearm: the radial and ulnar. The radius joins the hand on the same side as the thumb, and the ulnar on the same side as the little finger. The distinction between ulnar and radial loops depends on the direction of flow of the pattern. If the pattern flow runs in the direction of the radius (toward the thumb), then the loop is a radial loop. If the pattern runs in the direction of the ulnar (toward the little finger), then the loop is an ulnar loop.

17.5.2   Whorls

Whorls occur in about 25–35% of the fingerprints encountered. In a whorl, some of the ridges make a turn through at least one circuit (Lennard & Patterson, 2008). Plain whorls consist of one or more ridges, which make or tend to make a complete circuit with two deltas, between which an imaginary line is drawn and at least one re-curving ridge within the inner pattern area is cut or touched. Central pocket loop whorls consist of at least one re-curving ridge or an obstruction at right angles to the line of flow, with two deltas, between which when an imaginary line is drawn, no re-curving ridge within the pattern area is cut or touched using a red line as an imaginary line. Central pocket loop whorl ridges make one complete circuit which may be spiral, oval, circular, or any variant of a circle. Double loop whorls consist of two separate and distinct loop formations with two separate and distinct shoulders for each core, two deltas, and one or more ridges which make a complete circuit. Between the two, at least one re-curving ridge within the inner pattern area is cut or touched when an imaginary line is drawn. Accidental whorls consist of two different types of patterns with the exception of the plain arch and have two or more deltas or a pattern, which possess some of the requirements for two or more different types, or a pattern, which conforms to none of the definitions of the different types of pattern, elaborated in this chapter.

17.5.3   Arches

Arches occur in only about 5% of the fingerprint patterns encountered. In this type of pattern, the ridges run from one side to the other of the pattern, making no backward turn. Ideally, no delta is associated with this type of pattern. However, if a delta appears, no re-curving ridge must intervene between the core and delta points. There are four types of arches: plain, radial, ulnar, and tented arches. Looking at the plain arch, one will notice that a plain arch has an even flow of ridges from one side of the pattern to the other with a rise in the center. Tented arches have an angle, an up-thrust, or two of the three features of the loop. The tented arch pattern flow makes a significant change or up-thrust in the ridges near the middle that arrange themselves on both sides of a spine or axis toward which the adjoining ridges converge giving the appearance of a tent in outline. It also does not have the same "easy" flow as the plain arch. The arch is an approximation of the loop type, where there is a delta or the appearance of a delta and the ridges slop toward the thumb.

The radial arch might have delta and no re-curving ridge or a delta which is part of a re-curving ridge. Nevertheless, when both of these characteristics are present in a pattern, there must be no ridge count between the core and delta points. The ulnar arch has the same features as the radial arch with the exception that the ridges slop toward the little finger. In the general scheme of fingerprint classification, it is assumed that if a pattern contains no delta, then it is an arch; if it contains one (and only one) delta, it will be a loop. If it contains two or more, it will always be a whorl, and if a pattern contains more than two deltas, it will always be an accidental whorl.

17.6 Fingerprint Matching Techniques

The hearth of a biometric system lies in its ability to correctly match an input identifier or a biometric characteristic with the identifier stored in a database. Fingerprint matching, over the years, have proved difficult mainly due to the variations in different impressions of the same finger (Maltoni et al., 2009). These variations are caused by displacement, rotation, partial overlap, variable pressure, nonlinear distortion, changing skin or damaged skin condition, noise, and feature extraction errors. Consequently, fingerprints from the same finger might look different, while fingerprints from different fingers might look the same. To avoid some of these problems, human fingerprint examiners established some guidelines listed as follows (Maltoni et al., 2009):

- Global pattern configuration, which implies that two fingerprints must be of the same pattern.
- Qualitative concordance, which implies that the corresponding minutiae details must be identical.
- Quantitative factor, which set a guideline for a certain number of corresponding minutiae, must be found (a minimum of 12 according to the forensic guidelines in the United States).
- Corresponding minutiae details must be identically inter-rated.

Advances in technology over the last many decades have necessitated the use of automatic fingerprint matching systems with different design flavors in matching fingerprints. These systems do not have to follow the same guidelines for the manual process, and a lot of them have been mainly designed to be implemented on a computer. Many algorithms have been developed to achieve automatic fingerprint matching. These algorithms can be classified into broad groups such as ridge feature-based matching, correlation-based matching, and minutiae-based matching.

17.6.1   Ridge Feature-Based Matching

The ridge feature-based matching algorithm uses characteristics of the ridge pattern, for example, local orientation and frequency, ridge structure, and texture information for the matching of fingerprints (Avinash, 2006). This approach is very viable when the image quality is very poor. Poor image produces poor minutiae points, and algorithms that use minutiae points, as the basis for fingerprint matching, do not do very well. Ridge-based matching has generated a deal of interest in that it requires less pre-processing efforts. The drawback of this approach is that its distinctiveness is generally low.

17.6.2   Correlation-Based Matching

In the correlation-based matching approach, the template and query fingerprint images are superimposed on each other, and the spatial correlation between them is estimated to establish the degree of similarity. If the rotation and displacement of the query vis-à-vis the template are not known, then the correlation must be computed over all possible rotations and displacements, a task that is computationally expensive (Jain & Nandakumar, 2004). In addition, the presence of noise and nonlinear distortion significantly reduces the global correlation value between two impressions of the same finger. To eliminate these problems, correlation is normally done only in certain regions, for example, regions of high curvature, minutiae information, and many others, of the fingerprint (Jain & Nandakumar, 2004).

17.6.3   Minutiae-Based Matching

The minutiae-based matching methodis the most widely used approach in most applications. The matching algorithm will be discussed in detail, and it centers on the extraction of minutiae points in the fingerprint and subsequent processing. More emphasis will be discussed in the later sections. As discussed in Sect. 17.4, the two major minutiae types are the ridge ending and ridge bifurcation. The core of the minutiae-based system is to effectively extract the minutiae from the fingerprint images for later processing. Consequently, the minutiae-based system can be divided into three different processes: pre-processing, minutiae extraction, and post-processing. These various stages are discussed in Sect. 17.7.

17.7   The Fingerprinting Minutiae Matching System Algorithm Using Wavelet Transform

In most fingerprint images with low quality, additional steps are needed to enhance the quality of the images before the features are extracted. Such additional steps are implemented in the pre-processing stage. This involves obtaining a binary-segmented fingerprint ridge image from an input grayscale image, where the ridges have a value of "1" and the remainder of the image has value "0" (Ratha et al., 1996). After the image has been binarized, the image is then thinned before the minutiae points are extracted. The approach used in this Chap. 17 seeks to construct an image of the feature sets extracted from the fingerprint images, irrespective of the image quality, and store the constructed images in a database for verification purposes. This approach results in less computational time and gives the ability to match fingerprint images of lesser quality.

Figure 17.4 shows a block diagram of the matching system algorithm using the wavelet transform. A skeletal description of the diagram shows a minutiae feature

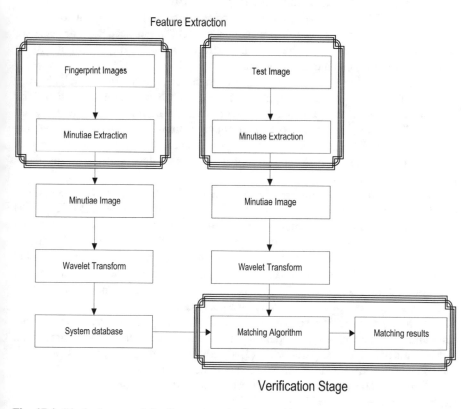

Fig. 17.4 Block diagram of the fingerprint minutiae matching system algorithm using wavelet transform

extraction module, a minutiae image construction module, wavelet transform module, and a system database of the statistical parameters of the wavelet-decomposed minutiae images. The descriptions of the subsystems of Fig. 17.4 are discussed in Sects. 17.7.1, 17.7.2, 17.7.3, and 17.7.4.

17.7.1 Partial Image Enhancement

In an ideal fingerprint, the ridges and furrows alternate and flow in a local constant direction. In this instance, the ridges can easily be detected, and the minutiae points can easily be located in the fingerprint image (Premnath, 2005). In a practical fingerprint image, image quality is degraded due to skin conditions, for example, wet or dry, cut and bruises, noise, and low quality of fingerprints of certain population groups like elderly people, manual workers, and many others. Many minutiae extraction algorithms require that an image enhancement be performed on the images. Some are based on normalization, local frequency estimation, region mask estimation, and Gabor filtering (Hong et al., 1998).

In Sect. 17.7.1, we will discuss a partial image enhancement based on binarization and thinning. A full image enhancement is not necessary because we are constructing an image of the minutiae extracted from the fingerprint and using wavelets to decompose the minutiae image.

The presence of noise and other variations in the fingerprint image results in poor image quality, and the application of wavelets in the matching scheme for the constructed minutiae image is used to filter out these variations. Consequently, full image enhancement is not necessary to obtain optimum results. In addition, the proposed algorithm operates only on binary images, where only two levels of interests are required: the black pixels that represent ridges and the white pixels that represent valleys.

The binarization and thinning of the fingerprint image are achieved using commands inherent in MATLAB. Binarization is the process that converts a gray-level image into a binary image which leads to improved level of contrasts between the ridges and furrows in a fingerprint image. In binarization, the gray-level value of each pixel in the image is examined, and if the value is greater than the global threshold, the pixel is set to binary value of one, and if it is less than the global threshold, then it is set to zero. The result is a binary image containing two levels of information, the foreground ridges and the background valleys.

A thinning process is also performed on the image prior to feature extraction. Thinning is a morphological operation that successively erodes the foreground pixels until they are one pixel wide (Thai, 2003). The thinning process preserves the connectivity of the ridge structures while developing a skeletal version of the binary image. The minutiae are then extracted from the skeleton image, and the procedures are enumerated in Sect. 17.7.2.

17.7.2 Minutiae Extraction

There are many techniques for extracting minutiae in fingerprint images. The most popular technique is the crossing number (CN) concept (Arceli & Baja, 1984). The technique involves the use of the skeleton image where the ridge flow pattern is eight-connected (Thai, 2003). The minutiae are extracted by scanning the neighborhood of each pixel, "p," in the image using a 3×3 window in an anti-clockwise direction as shown in Fig. 17.5. The CN value shown in Eq. 17.1 which is defined as half of the sum of the differences between pairs of adjacent pixels in the eight-neighborhood of "p" is then calculated.

$$CN = \frac{1}{2} \sum_{i=1}^{8} |P_i - P_{i+1}|, \quad P_9 = P_1 \qquad (17.1)$$

where P_i is the pixel value in the neighborhood of P.

The properties of Table 17.1 can be used to determine the ridge pixel, which can be classified as a ridge ending, bifurcation, or non-minutiae point. Crossing numbers 1 and 3 correspond to a ridge ending and a ridge bifurcation, respectively. Table 17.1 explains the other crossing numbers. Figures 17.6 and 17.7 show examples of a ridge ending and ridge bifurcation.

Fig. 17.5 Anti-clockwise scanning of pixel "p" in a 3×3 window

P_4	P_3	P_2
P_5	P	P_1
P_6	P_7	P_8

Table 17.1 Properties of the crossing number

CN	Property
0	Isolated point
1	Ridge ending point
2	Continuing ridge point
3	Bifurcation point
4	Crossing point

Fig. 17.6 CN=1

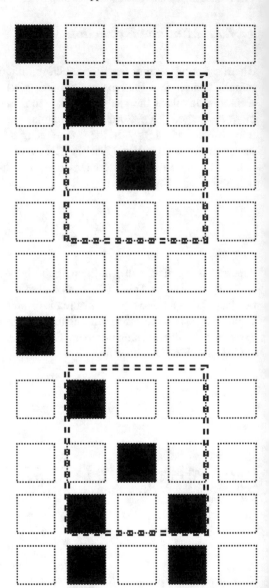

Fig. 17.7 CN=3

17.7.3   Fingerprint Image Post-Processing

With the extraction of the minutiae come spurious minutiae or false minutiae. These false minutiae arise because of noise and image remnants created by the thinning process. These false minutiae must be eliminated using a post-processing stage, which essentially validates the minutiae. We use the validation algorithm proposed by Tico and Kuosmanen (2000). Figure 17.8 shows some examples of the false minutiae. The spike structure generates a false ridge end and false bifurcation and

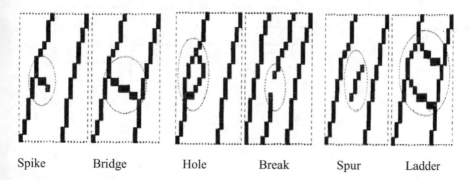

Spike	Bridge	Hole	Break	Spur	Ladder

Fig. 17.8 Examples of false minutiae
Source: Tico, M. Kuosmanen, P., *"An Algorithm For Fingerprint Image Postprocessing" Conference Record Of The Thirty-Fourth Asilomar Conference On Signals*, Systems And Computers. Vol.
2, Page(S): 1735–1739 October 2000

may occur when thinning a non-smooth ridge. The bridge and ladder structures
normally occur between close structures and can generate false bifurcation. The hole
structures can be generated by very wide ridges and can result in false bifurcation.
The break structure can manifest itself in the thinned ridge map image due to the
presence of scars in the fingerprint image and can result in false ridge ending. The
spur can be generated by very wide valley and can result in a false ridge ending.
Many false minutiae are detected close to the boundary of the region of interest
(boundary effect). The boundary effects are nullified by canceling all minutiae below
a certain distance from the boundary of the fingerprint pattern.

In the validation of the minutiae, the skeleton image is scanned and the local
neighborhood around each minutiae point examined. The first step in the validation
process is to create an image M of size $W \times W$, where M corresponds to the $W \times W$
neighborhood centered on the candidate minutiae in the skeleton image (Aliaa et al.,
2007). The central pixel of M corresponds to the minutiae in the skeleton image, and
the pixel is labeled with a value of -1 as shown in Figs. 17.9a and 17.10a. The rest
of the pixels in M are initialized to values of zero. The next steps in the validation
algorithm depend on whether the candidate minutiae point is a ridge ending or a
ridge bifurcation.

17.7.4   Procedure for Validating a Candidate Ridge
Ending Point

In this step, all the pixels in M, which are eight-connected with the ridge ending
point, are labeled with 1 as shown in Fig.17.9b. The next step involves counting the
number of 0 to 1 transitions (T_{01}) along the border of image M. If $T_{01} = 1$, then the
candidate minutiae point is validated as a true ridge ending.

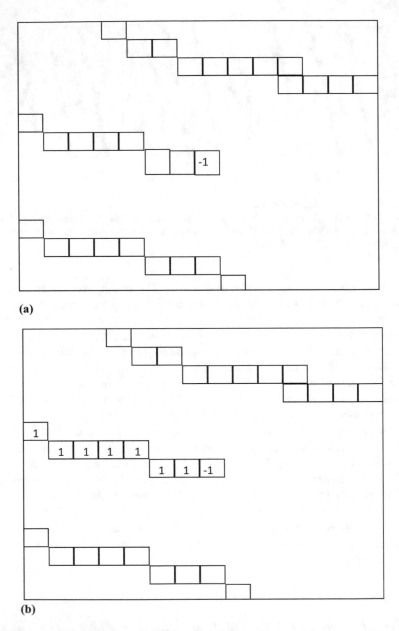

Fig. 17.9 (a) and (b) Examples of validating ridge ending point $T_{01}=1$

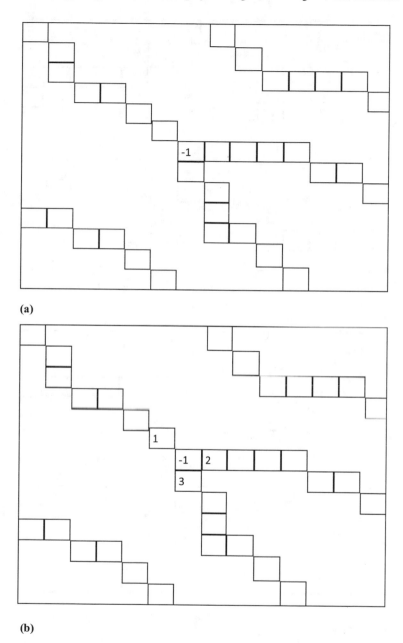

(a)

(b)

Fig. 17.10 (a), (b), (c), (d), (e) Examples of validating a ridge bifurcation point $T_{01}=1$, $T_{02}=1$, and $T_{03}=1$

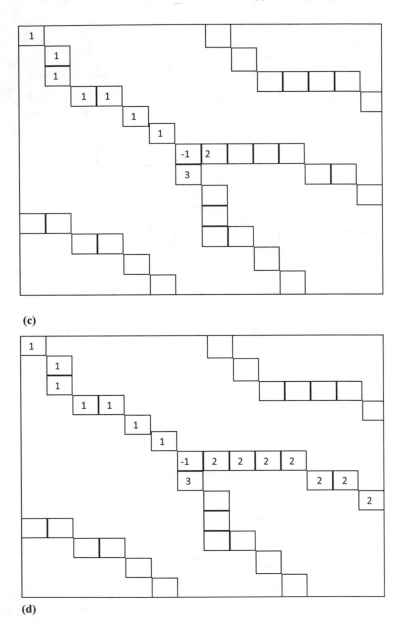

(c)

(d)

Fig. 17.10 (continued)

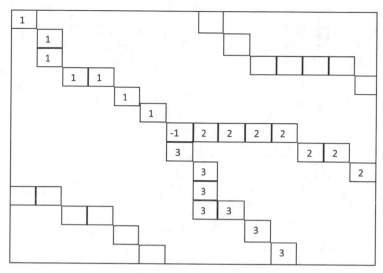

Fig. 17.10 (continued)

17.7.5   Procedure for Validating a Candidate Bifurcation Point

The first step is to examine the eight neighboring pixels surrounding the bifurcation point in a clockwise direction. The three pixels that are connected with the bifurcation point are labeled 1, 2, and 3, respectively, as shown in Fig. 17.10b. The next step is to label the rest of the ridge pixels that are connected to the three pixels. Label each ridge branch connected to the three pixels according to the label on each of the three pixels, respectively, as shown in Fig. 17.10(c), (d), and (e). The last step is to count in a clockwise direction the number of transitions from 0 to 1 (T_{01}), 0 to 2 (T_{02}), and 0 to 3 (T_{03}) along the border of image M.

If the $T_{01} = 1$, $T_{02} = 1$, and $T_{03} = 1$, the candidate minutiae point is validated as a true ridge bifurcation point (Aliaa et al., 2007). Figure 17.11 shows the validation algorithm is able to cancel out two types of false minutiae. Each bifurcation point in the hole structure can be eliminated due to the number of 0 to 3 (T_{03}) transitions along the border of the image window not amounting to 1 as shown in Fig. 17.11(a) (Thai, 2003). Similarly, Fig. 17.11(b) shows that the spur contains two ridge pixels in the center of the image window; hence, the number of 0 to 1 (T_{01}) transitions along the border of the window is zero.

The last stage in the fingerprint post-processing is the construction of the minutiae image after the false minutiae points have been eliminated. The minutiae image is a combination of the minutiae points superimposed on the thinned image. By constructing the minutiae image, we are able to keep track with variations in

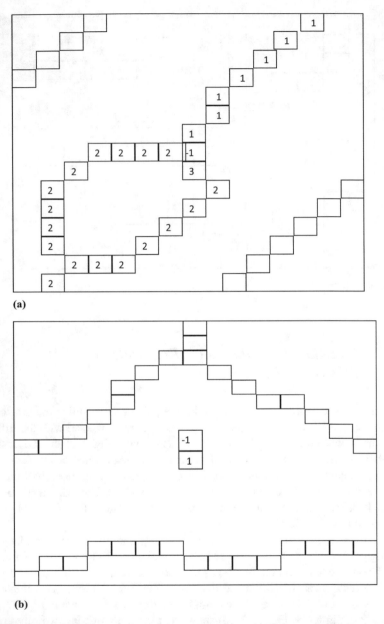

(a)

(b)

Fig. 17.11 (a) and (b) Cancellation of false minutiae using Tico et al.'s algorithm. (a): Hole structure. (b) Spur structure

Source: Tico, M. Kuosmanen, P., *"An Algorithm For Fingerprint Image Postprocessing"* *Conference Record Of The Thirty-Fourth Asilomar Conference On Signals*, Systems And Computers. Vol. 2, Page(S): 1735–1739 October 2000

position, scale, and orientation angle which happens to be a major disadvantage with image-based fingerprint matching techniques. Most minutiae matching algorithms require that the coordinates and the angle of orientation of the minutiae of a test fingerprint be matched with corresponding parameters stored in a template. This process requires that a point of reference needs to be defined so that the matching of the test fingerprint minutiae and its template minutiae points are compared using the reference. This process is not important in the approach used in this chapter since the matching algorithm is based on the matching of the constructed minutiae images and the parameters extracted from the co-occurrence matrix of the constructed images do not change with the position or rotation of the same fingerprint images.

The next stage in the fingerprint image post-processing is the wavelet decomposition of the constructed minutiae images. We implement a level 4 wavelet decomposition with six different wavelets. Figures 17.12 and 17.13, respectively, show the one-level decomposition and reconstruction algorithmic diagrams as samples of the discrete wavelet transforms (DWT) used in the implementation. The four-level decomposition and reconstruction wavelet transform algorithmic diagrams are as shown in Chap. 4, Figs. 4.6 and 4.7, respectively. The types of wavelets used for the minutiae implementations are Haar, Daubechies-4 and Daubechies-8, biorthogonal 5.5, Symlet-8, and Coiflet-4 wavelets. The statistical parameters of the coefficients of the decomposed images are calculated. The statistical parameters and co-occurrence matrix features give an insight into some of the properties of the images. Section 17.8 gives an overview of the parameters used. The fingerprint verification process entails determining the wavelet statistical parameters and co-occurrence matrix features of the given minutiae image and comparing them with those stored in the database using a distance vector formula. The image that has the minimum difference is the verified image.

17.8   Methodology

The procedure involved in image enhancement, minutiae extraction, and post-processing involves a series of steps, which have already been presented in Sects. 17.7.1, 17.7.2, and 17.7.3. In this Sect. 17.8, we will highlight how the developed algorithm as discussed in Sect. 17.7 reduces the amount of computational resources required. The steps are as shown in Fig. 17.14. It shows a typical scheme deployed in most algorithms for fingerprint enhancement. The processing of the extracted minutiae is implemented using different techniques. Some techniques like the one used in this chapter use wavelets (Jain & Hong, 1996; Seok & Nam, 1999; Torii & Okamoto, 2003) like we have and specifically in Sect. 17.7; others use alignment-based matching algorithm.

The fingerprint image enhancement in Fig. 17.14 consists of seven steps. The use of wavelets and wavelet transforms in the matching of the extracted minutiae allows for the filtering of noise and other nonlinear distortion from the images. Consequently, we can safely eliminate five steps from the architecture shown in Fig. 17.14

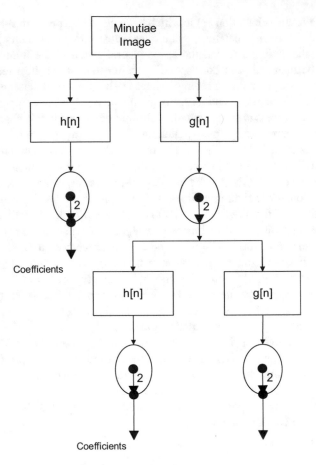

Fig. 17.12 DWT 2D minutiae image decomposition procedure with only one level shown where:

Where: implies down sampling by 2 implies downsampling by 2

g[n] is the low pass filter

h[n] is the high pass filter

g[n] is the low-pass filter
h[n] is the high-pass filter

to the revised approach in Fig. 17.15. The benefit of this approach enables the savings of a considerable amount of computational resources. We therefore employed one of the benefits of using wavelets, which is the reduction of computational complexity (Colestock, 1993).

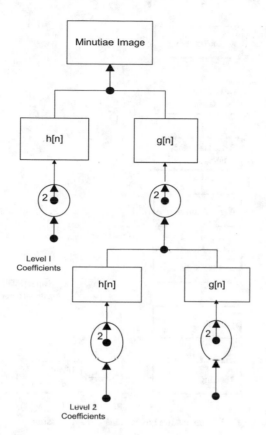

Fig. 17.13 DWT reconstruction procedure with only one level shown where:

Where: implies up sampling

g[n] is the low pass filter

h[n] is the high pass filter

implies upsampling

g[n] is the low-pass filter

h[n] is the high-pass filter

17.8.1   *Performance Metrics*

We use statistical parameters for performance analysis in the collection of data to establish the relationship between individual data points in the data set. The

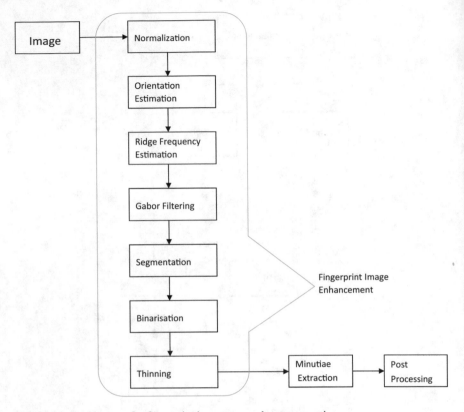

Fig. 17.14 Architecture for fingerprint image pre- and post-processing

statistical parameters of the images are used as a basis for comparison between the test image and the images in the database for verification purposes. An overview of some of the statistical parameters used is as follows:

- *Mean*: For a data set, the mean is the sum of the observations divided by the number of observations. The mean describes the central location of the data. In relation to the images used, the mean is the average of the gray values of the wavelet-decomposed image. The mean is mathematically expressed as shown in Eq. (17.2).

$$\text{Mean}(m) = \frac{1}{N^2} \sum_{i,j=1}^{N} p(i,j) \tag{17.2}$$

- *Standard Deviation*: Often quoted with the mean is the standard deviation. The standard deviation of a data set describes the spread of the data set, and it is defined as the square root of the variance. The variance is the average of the

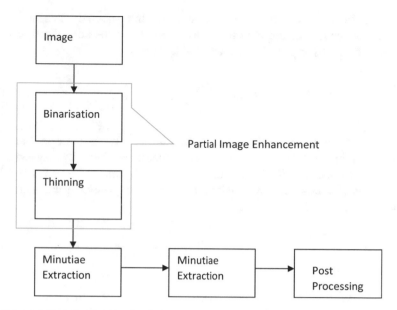

Fig. 17.15 Methodology for the developed algorithm

squared differences between data points and the mean. The standard deviation can be expressed mathematically as shown in Eq. (17.3):

$$\text{Standard Deviation(sd)} = \sqrt{\frac{1}{N^2} \sum_{i,j=1}^{N} [p(i,j) - m]^2} \tag{17.3}$$

where $p(i,j)$ is the transformed value in (i,j) for any sub-band of size N × N for both the mean and standard deviation. The mean and standard deviation are measured over the entire image. They are used in gross adjustment of intensity and contrast (Premnath, 2005).

The co-occurrence matrix is also used to ascertain certain characteristics of the wavelet-decomposed images, and it is created by estimating how often a pixel with intensity value i occurs in a spatial relationship to a pixel of value j in a given matrix. Co-occurrence matrix is often used in measuring the texture of the images. The spatial relationship between pixels is often described as the relationship between the pixel of interest and the pixel that lies horizontally adjacent to it, although it can also be extended to the pixels that are vertical or diagonally adjacent to the pixel of interest (Premnath, 2005). The major advantage of using co-occurrence matrix of an image is evident in the fact that it does not change with the rotation of the image. Some of the features that can be extracted from the co-occurrence matrix are energy, contrast, and entropy.

- *Energy*: Energy is defined as the sum of the squared elements of the co-occurrence matrix and measures the complexity of the normalized matrix. It is mathematically expressed as shown in Eq. (17.4).

$$\text{Energy} = \sum_{i,j=1}^{N} C^2(i,j) \tag{17.4}$$

- *Contrast*: Contrast is the difference in brightness between the bright and dark areas of an image. It can also be defined as a measure of intensity between a pixel and its neighboring pixel over the whole image, and it measures the local variations in a co-occurrence matrix. Contrast is expressed mathematically as shown in Eq. (17.5).

$$\text{Contrast} = \sum_{i,j=1}^{N} (i-j)^2 C(i,j) \tag{17.5}$$

- *Entropy*: Entropy is a statistical measure of randomness that can be used to characterize the texture of an input image. It measures the information contained in the image. A high level of entropy implies that the gray-level changes between pixels are evenly distributed and the image has a high degree of visual texture (Premnath, 2005). Entropy can be mathematically expressed as shown in Eq. (17.6).

$$\text{Entropy} = -\sum_{i,j=1}^{N} C(i,j) \log_2 C(i,j) \tag{17.6}$$

where $C(i,j)$ is the co-occurrence matrix in Eqs. (17.4), (17.5), and (17.6).

17.9   MATLAB Simulation Examples

In Sect. 17.9, we demonstrate using various examples of algorithms MATLAB codes and results of the simulations of the fingerprint minutiae processing and wavelet-based application ideas discussed in this chapter. Samples of the MATLAB codes and helpful operational hints on how to use them can be found in Appendices K and L, respectively.

Example 17.9.1: Creation of Parameters Database
Figure 17.16 shows the steps used in the simulation of the algorithm in MATLAB. Eighty-four (84) grayscale fingerprint images were downloaded from the Biometric System Lab, University of Bologna, Italy (FVC, 2000). Eighty images are enrolled

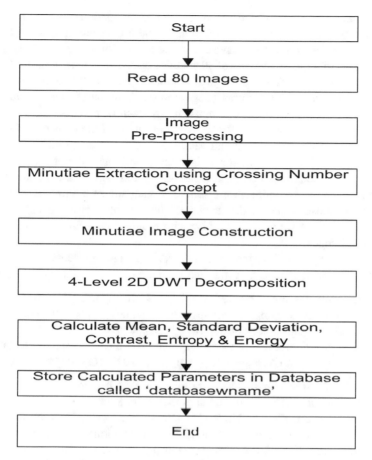

Fig. 17.16 Procedure for the creation of parameters database
Note: "wname" is a string containing the wavelet name. Parameters are mean, standard deviation, contrast, entropy, and energy

in the database, while the other four are not enrolled to test the rejection rate for images not enrolled in the database. A set of samples of the fingerprint images are displayed in Appendix J. The images are subjected to a pre-processing stage which essentially binarized and thinned the images. The next step is the extraction of the minutiae from the fingerprint images using the crossing number concept. After the minutiae are extracted, the false or spurious minutiae are eliminated using Tico and Kuosmanen (2000) algorithm in order to establish the validity of the extracted minutiae. Unlike other algorithms deployed in the past in the literature, where the coordinates and orientation of the minutiae are stored and used as a basis for the verification of the fingerprint by comparison with those of a test image, the approach presented seeks to establish a technique that involves the construction of minutiae images from the extracted minutiae. There are four advantages from using this approach:

a. The comparison of the minutiae coordinates and orientation of a test image has to be done for all the extracted minutiae of all the fingerprint images stored in the database. For example, if we have a database of 10 million fingerprint images and each fingerprint has an average of 50 minutiae points, then to match each minutiae of a test image, the minutiae has to be compared against 500 million possible set of minutiae points. This requires a huge amount of computational resources. With our approach, only 10 million minutiae images will be processed.
b. Since we are using wavelets for the decomposition of the constructed minutiae images, the presence of noise and other image-degrading factors are eliminated. Consequently, it does not matter if the fingerprint images are of low quality.
c. In the comparison of the coordinates and orientation angle of the minutiae, a fixed window needs to be defined for effective comparison of the minutiae of the test image and those stored in the database. This approach is not necessary in our approach because we are not working with minutiae coordinates but rather with minutiae images.
d. Scar, an image-degrading feature in fingerprint images, presents itself as a break structure in the extracted minutiae and is eliminated by validating the minutiae.

The minutiae images are subjected to a level 4, 2D wavelet decomposition. Six wavelets are used in our implementation. The "wname" string in the code specifies the name of the wavelet used. The statistical parameters of mean, standard deviation, and co-occurrence matrix features of energy, entropy, and contrast are calculated and stored in a database. In Example 17.9.2, we show and discuss the verification steps.

Example 17.9.2: Steps for the Verification of a Test Image
Figure 17.17 shows the steps involved in the verification of a test image selected from the database. As explained in the previous sections and in Examples 17.9.1 and 17.9.2, the verification process seeks to establish the validity of a known fingerprint image; hence, we select a known image as our test image. The selected test image is pre-processed using binarization and thinning techniques explained in the preceding sections. The minutiae are extracted from the image using the crossing number concept and a minutiae image constructed. The minutiae image is subjected to a four-level 2D discrete wavelet decomposition. The mean, standard deviation, contrast, entropy, and energy are subsequently calculated.

The "databasewname" of the statistical parameters of all the images is loaded in order to calculate the difference in parameter values between those of the test image and the images in the database using the minimum distance formula shown in Eq. (17.7).

$$D(i) = \sum_{j=i}^{n} abs\left[f_j(x) - f_j(i) \right] \qquad (17.7)$$

where:

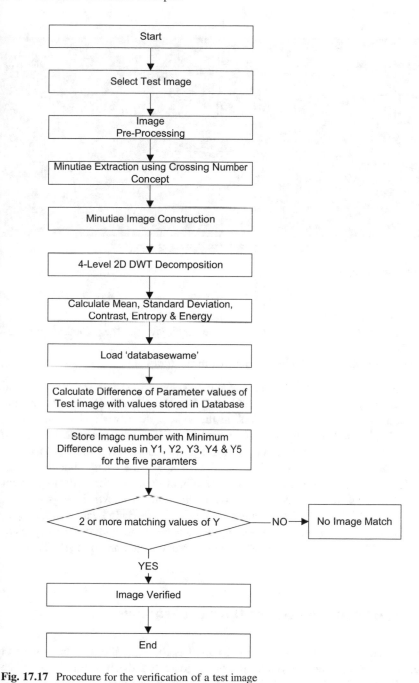

Fig. 17.17 Procedure for the verification of a test image
Note: "wname" is a string containing the wavelet name. Parameter values are mean, standard deviation, contrast, entropy, and energy

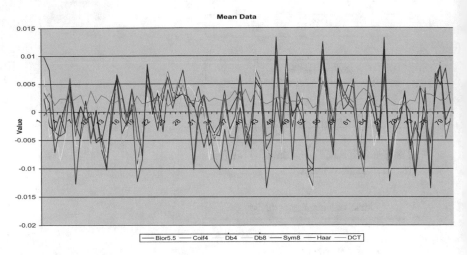

Fig. 17.18 Plot of the mean data

$f_j(x)$ represents the features of the test minutiae image

$f_j(i)$ represents the features of the ith minutiae image in the "databasewname" template

n represents the number of minutiae images in the template

The image numbers with the minimum distance values in mean, standard deviation, contrast, entropy, and energy are stored in Y1, Y2, Y3, Y4, and Y5. To establish the verification of the test image, we look at the values. Any two or more matching values of Y obtained from the MATLAB command line, after the execution of the program codes, show that the selected test image was accurately verified; otherwise, the verification process failed. The results of the parameters obtained using Eqs. (17.2, 17.3, 17.4, 17.5, and 17.6) are stored in the database for all the fingerprint images. The subsequent plots of these parameters are as shown in Figs. (17.18, 17.19, 17.20, 17.21, and 17.22) for mean data, standard deviation data, contrast data, energy data, and entropy data, respectively, for a comparison with different types of wavelets, fast Fourier transform (FFT), and discrete cosine transform (DCT).

17.10   Matching Indices for Different Wavelets

As part of the objective of this chapter, which is to verify the minutiae of a test image, using 6 different wavelets against a generated database of 80 minutiae images, Tables 17.2, 17.3, 17.4, 17.5, and 17.6 show the matching indices of the various parameters of the test images 42, 59, 72, 79, and 81–84 using different wavelets. The images were randomly selected to mimic real-world situation. The readers are expected to choose any number of their own interest randomly as we did.

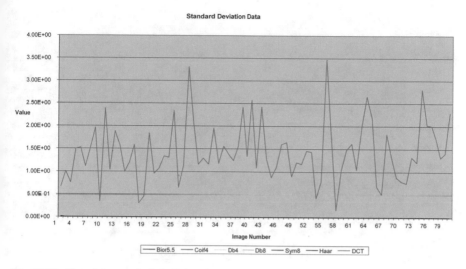

Fig. 17.19 Plot of the standard deviation data
Note: Values of the standard deviation are too small compared to those of the DCT that they do not appear on the plot

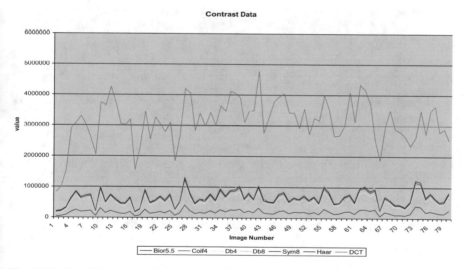

Fig. 17.20 Plot of the contrast data

The value of Y corresponds to the number of the test image that produces a match. The number is displayed in the MATLAB command line when the program codes are executed. Any four matching values of Y imply a correct verification. From the results, we can see that all the wavelets produced a couple of matches. The Haar wavelets were not particularly suitable for the matching algorithm because they produced one non-matching values of Y for image numbers 59 and 79, respectively.

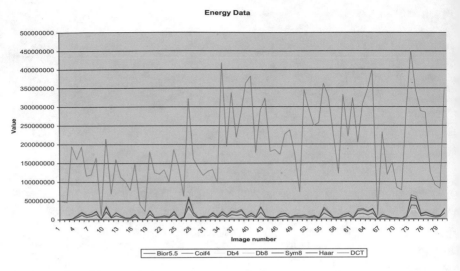

Fig. 17.21 Plot of the energy data

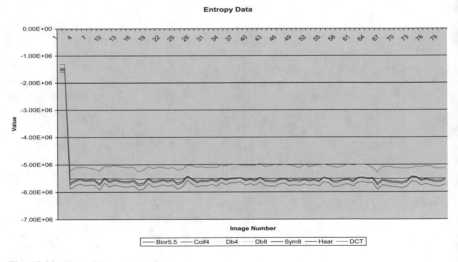

Fig. 17.22 Plot of the entropy data

The other four wavelets produced excellent matching results—100% match. This implies that the randomly selected fingerprint images were correctly verified with all the wavelets except the Haar wavelet. Four images (81–84) as shown in Table 17.6 not enrolled in the database were tested, and the rejection rate was 100%.

The fast Fourier transform (FFT) of an image produces coefficients that are real and complex. To extract the energy, entropy, and contrast from the co-occurrence matrix of an image, the input to the image must be real; hence, FFT cannot be used for analysis. Nevertheless, with FFT, we are able to calculate the mean and standard

Table 17.2 Matching indices for different wavelets of image 42

	Haar	Dau4	Dau8	Bior5.5	Symlet8	Coiflet4	FFT	DCT
Mean (Y1)	*	*	*	*	*	*	+	*
SD (Y2)	*	*	*	*	*	*	+	*
Contrast (Y3)	*	*	*	*	*	*	N/A	*
Energy (Y4)	*	*	*	*	*	*	N/A	*
Entropy (Y5)	*	*	*	*	*	*	N/A	*

Note: * implies a match; + implies no match
N/A implies that this parameter cannot be determined

Table 17.3 Matching indices for different wavelets for image 59

	Haar	Dau4	Dau8	Bior5.5	Symlet8	Coiflet4	FFT	DCT
Mean (Y1)	*	*	*	*	*	*	+	*
SD (Y2)	*	*	*	*	*	*	+	*
Contrast (Y3)	*	*	*	*	*	*	N/A	*
Energy (Y4)	+	*	*	*	*	*	N/A	*
Entropy (Y5)	*	*	*	*	*	*	N/A	*

Note: * implies a match; + implies no match
N/A implies that this parameter cannot be determined

Table 17.4 Matching indices for different wavelets for image 72

	Haar	Dau4	Dau8	Bior5.5	Symlet8	Coiflet4	FFT	DCT
Mean (Y1)	*	*	*	*	*	*	+	*
SD (Y2)	*	*	*	*	*	*	+	*
Contrast (Y3)	*	*	*	*	*	*	N/A	*
Energy (Y4)	*	*	*	*	*	*	N/A	*
Entropy (Y5)	*	*	*	*	*	*	N/A	*

Note: * implies a match; + implies no match
N/A implies that this parameter cannot be determined

Table 17.5 Matching indices for different wavelets for image 79

	Haar	Dau4	Dau8	Bior5.5	Symlet8	Coiflet4	FFT	DCT
Mean (Y1)	*	*	*	*	*	*	+	*
SD (Y2)	*	*	*	*	*	*	+	*
Contrast (Y3)	*	*	*	*	*	*	N/A	*
Energy (Y4)	+	*	*	*	*	*	N/A	*
Entropy (Y5)	*	*	*	*	*	*	N/A	*

Note: * implies a match; + implies no match
N/A implies that this parameter cannot be determined

deviation of the image, and hence our performance parameter is limited to mean and standard deviation. From the results in Tables 17.2, 17.3, 17.4, and 17.5, we see that the dau4, dau8, sysmlet8, coiflet8, and discrete cosine transform (DCT) all produced 100% matching accuracy. The FFT proved very unsuitable for this analysis

Table 17.6 Matching indices for different wavelets for images 81–84

	Haar	Dau4	Dau8	Bior5.5	Symlet8	Coiflet4	FFT	DCT
Mean (Y1)	+	+	+	+	+	+	+	+
SD (Y2)	+	+	+	+	+	+	+	+
Contrast (Y3)	+	+	+	+	+	+	N/A	+
Energy (Y4)	+	+	+	+	+	+	N/A	+
Entropy (Y5)	+	+	+	+	+	+	N/A	+

Note: * implies a match; + implies no match
N/A implies that this parameter cannot be determined

producing 0% match using the parameters of mean and standard deviation as our performance metrics.

From the results of the parameters in Tables 17.3, 17.4, 17.5, and 17.6 and consequently the plots in Figs. 17.18, 17.19, 17.20, 17.21, and 17.22, we note the following observations:

a. The analysis with wavelets produced similar patterns that were consistent and meaningful.
b. The DCT analysis produced the highest values for standard deviation implying a remarkable variance in the DCT coefficients.
c. The contrast of the images was highest with the usage of Db8 wavelet and lowest with those of the Haar wavelets.
d. The deployment of the Db8 wavelet produced the highest energy values and the Db4 produced the lowest energy value.
e. The entropy values were lowest in images decomposed with the coiflet4 wavelets and highest with the Db4 wavelets.
f. The images that were not enrolled in the database produced 100% rejection rate.
g. The most dominant co-occurrence matrix feature was energy.

From the plots and observation of the results, the Daubechies-4 wavelet is the best wavelet for analysis of the minutiae images. The DCT analysis produced results that differ from the patterns observed with wavelets. For example, the standard deviation was extremely high; the energy and contrast data sets were very inconsistent (Fig. 17.23).

Summary
1. Biometrics is a term that originated from the Greek words bio (life) and metrikos (measure).
2. The key ingredient in biometrics is identifying which biological measurement can be used qualitatively as a biometric characteristic.
3. A biometric system based on the physiological characteristic is more reliable than the one that uses behavioral characteristic, although present-day applications combine for both accuracy and reliability. A biometric recognition system can operate in a verification mode or in an identification mode depending on the intended application.

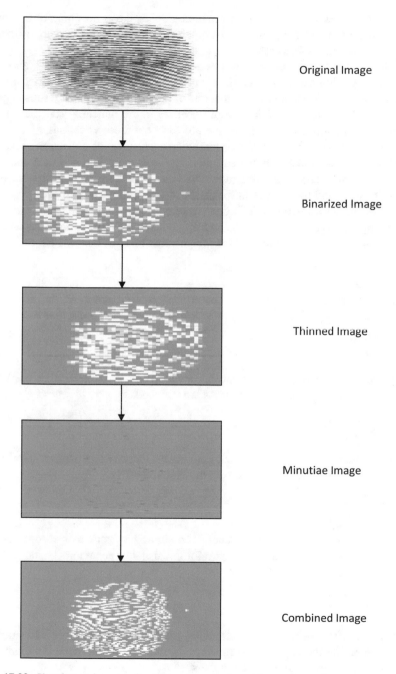

Original Image

Binarized Image

Thinned Image

Minutiae Image

Combined Image

Fig. 17.23 Showing an image as it undergoes pre- and post-image processing

4. Fingerprint classification can be viewed as a coarse-level matching of the input fingerprint to one of the pre-specified types and the subsequent comparison to a subset of the database corresponding to the fingerprint type.
5. The hearth of a biometric system lies in its ability to correctly match an input identifier or a biometric characteristic with the identifier stored in a database.
6. The presence of noise and other variations in the fingerprint image results in poor image quality, and the application of wavelets in the matching scheme for the constructed minutiae image can be used to filter out these variations.
7. We use statistical parameters for performance analysis in the collection of data to establish the relationship between individual data points in the data set. The statistical parameters of the images are used as a basis for comparison between the test image and the images in the database for verification purposes.
8. The fast Fourier transform (FFT) of an image produces coefficients that are real and complex. To extract the energy, entropy, and contrast from the co-occurrence matrix of an image, the input to the image must be real; hence, FFT cannot be used for analysis.

Review Questions

17.1 Humans as far as we know have always used some of the key body characteristics such as face gait, voice, or signature to recognize each other.

    a. True
    b. False

17.2 Can any human physiological and/or behavioral characteristic be used as a biometric if certain characteristics are satisfied.

    a. True
    b. False

17.3 Distinctiveness in the biometric field means that any two persons should not be sufficiently different vis-à-vis the characteristics.

    a. True
    b. False

17.4 A biometric system is a pattern recognition system that recognizes an individual based on a specific physiological and/or behavioral characteristics possessed by that individual.

    a. True
    b. False

17.5 A biometric verification system does not authenticate a person's identity by comparing the captured biometric information with the person's biometric template information stored in the system.

    a. True
    b. False

17.6 The ridge feature-based matching algorithm uses characteristics of the ridge pattern, for example, local orientation and frequency, ridge structure, and texture information for the matching of fingerprints. This approach is very viable when the image quality is very poor.

   a. True
   b. False

17.7 The minutiae-based matching method is not the most widely used approach in most applications.

   a. True
   b. False

17.8 The fingerprint verification process entails determining the wavelet statistical parameters and co-occurrence matrix features of the given minutiae image and comparing them with those stored in the database using a distance vector formula. The image that has the minimum difference is the verified image.

   a. True
   b. False

17.9 The major advantage of using co-occurrence matrix of an image is evident in the fact that it does not change with the rotation of the image.

   a. True
   b. False

17.10 The presence of noise and other variations in the fingerprint image results in poor image quality, and the application of wavelets in the matching scheme for the constructed minutiae image cannot be used to filter out these variations.

   a. True
   b. False

Answers: 17.1a, 17.2a, 17.3b, 17.4a, 17.5b, 17.6a, 17.7b, 17.8a, 17.9a, 17.10b

Problems
   17.1 What are the biometric characteristics?
   17.2 In reference to Problem 17.1, what are the other issues that can be considered in practical biometric systems?
   17.3 (a) Define a biometric system. (b) What are the three key biometric system operational modes?
   17.4 Describe the enrollment mode of the biometric system operations.
   17.5 Describe the verification mode of the biometric system operations.
   17.6 Describe the identification mode of the biometric system operations.
   17.7 What are the basics of fingerprint recognition?

17.8 (a) How many possible local ridge features are possible in a fingerprint? What are the two most prominent local ridges?

17.9 (a) Define minutiae. (b) What is ridge ending? (c) What is ridge bifurcation?

17.10 How do you determine a good fingerprint image?

17.11 (a) How can fingerprint classification be defined? (b) Describe briefly what is involved in fingerprint classification.

17.12 Name some of the different types of fingerprint patterns used in classifications.

17.13 (a) What is a ridge count in relation to loop fingerprint classification? (b) Describe the idea of loop classification of fingerprint images.

17.14 Describe the idea of whorl classification of fingerprint images.

17.15 Describe the idea of arch classification of fingerprint images.

17.16 Why is fingerprint matching important?

17.17 Fingerprints from the same finger might look different, while fingerprints from different fingers might look the same. How can these problems be resolved?

17.18 Describe ridge feature-based matching.

17.19 Describe correlation-based matching.

17.20 Describe minutiae-based matching.

17.21 What is the process required for fingerprint partial image enhancement?

17.22 How can minutiae extraction be performed?

17.23 How is fingerprint image post-processing implemented?

17.24 What types of performance metrics can be used in the statistical parameters of the images that are used as a basis for comparison between the test image and the images in the database for verification purposes?

17.25 What is the usefulness of the co-occurrence matrix in the wavelet-based biometric application?

17.26 (a) What are some of the features that can be extracted from the co-occurrence matrix? Describe each of them.

17.27 What are the advantages of using a technique that involves the construction of minutiae images from the extracted minutiae?

17.28 What are the steps that can be used in the verification of a test image in the wavelet-based biometric application?

References

Aliaa, A. A., Youssif M., Howdhury, S. R., & Nafaa, H.Y. (2007, July 11–13). Fingerprint recognition System Using Hybrid Matching Techniques, techniques. In 6th IEEE/Acis International conference on computer and information science, pp. 234–240.

Arceli, & Baja. (1984). A width independent fast thinning algorithm. *IEEE Transaction on Pattern Analysis and Machine Intelligence*.

Avinash H. R. R. (2006, May). *Fingerprint recognition using wavelets and principal component analysis*. Master's degree thesis, Texas A&M University-Kingsville.

Christopher, J. L., & Patterson, T. (2008). http://www.Policensw.Com/Info/Fingerprints. New South Wales Police Service.

Colestock, M.A. (1993, October 25–28). *Wavelets- A new tool for signal processing analysts*. In Digital avionics systems conference, 12th Dasc. Aiaa/IEEE, pp. 54–59.

FVC. (2000). *Fingerprint verification competition*. http://bias.csr.unibo.it/fvc2000

Hong, L., Wan, Y., & Jain, A. (1998, August). Fingerprint image enhancement: Algorithm and performance evaluation. *IEEE Transactions on Pattern Analysis and Machine Intelligence, 20*(8).

Jain, A., & Hong, L. (1996, August 22–29). *Online fingerprint verification*. In 13th International conference on pattern recognition, vol 3, pp. 596–600.

Jain, A. K., & Nandakumar, K. (2004). *Local correlation-based fingerprint matching*. In Indian conference on computer vision, graphics and image processing (Icvgip).

Lennard, C. J., & Patterson, T. (2008). http://www.Policensw.Com/Info/Fingerprints. New 758 South Wales Police Service.

Maltoni, D., Maio, D., Jain, A. K., & Prabhakar, S. (2009). *Handbook of fingerprint recognition* (2nd ed.). Springer.

Phillips, P. J., Martin, A., Wilson, C. L., & Przybocki, M. (2000, February). An introduction evaluating biometric systems. *Computer, 33*(2), 56–63.

Premnath, A. (2005, May). *Identification of fingerprint using wavelet transform*. Master's degree thesis, Texas A&M University-Kingsville.

Ratha, N. K., Karu, K., Cen, S., & Jain, A. K. (1996, August). A real-time matching system for large fingerprint database. *IEEE Transactions on Pattern Analysis and Machine Intelligence, 18*(8).

Seok, W. L., & Nam, B. (1999). *Fingerprint recognition using wavelet transform and probabilistic neural network*. In International joint conference on neural networks, vol. 5, pp. 3276–3279, 10-16, 1999.

Thai, R. (2003). *Fingerprint image enhancement and minutiae extraction*, Research Submission To School Of Computer Science And Software Engineering, The University of Western Australia.

Tico, M., & Kuosmanen, P. (2000, October). *An algorithm for fingerprint image post processing*. In Conference record of the thirty-fourth Asilomar conference on signals, systems and computers, vol. 2, pp. 1735–1739.

Torii, K., Okamoto, N. (2003, May 4–7). *An effective algorithm for detecting valley of fingerprint based on wavelet transform*. In Canadian conference on electrical and computer engineering, IEEE Ccece 2003, vol. 2, pp. 1223–1226.

Chapter 18
Wavelet Application to Blockchain Technology Systems

Never leave that till tomorrow which you can do today.
                                        – Benjamin Franklin

18.1 Introduction

In this chapter, part of the discussion is on the blockchain technology systems. In addition, the other part of the discussion is on the application of wavelets and wavelet transform to blockchain technology systems. Blockchain is a decentralized or peer-to-peer network that consists of nodes or parties (Bora, 2020; Eze et al., 2019). All nodes in the system maintain a copy of the blockchain (i.e., have the same software, keep a history of the transactions, and receive the same transactions). The blockchain uses cryptographic techniques and consensus algorithms to make transactions secure and records immutable transactions in a distributed fashion. Blockchain technology is a highly disruptive technology and also known as Distributed Ledger Technology (DLT). It is important to emphasize that blockchain is still in its early stages of development and therefore faces some issues. The blockchain concept started initially with the cryptocurrency system popularly known as bitcoin. Today, blockchain technology systems are used in various applications across different domains to serve multiple purposes (Braghin et al., 2019; Dai et al., 2019). In the industries, blockchain integrates into the various processes: manufacturing, supply chain to provide provenance and traceability in Internet of Things, cost saving, process improvements, and many others. Other systems that leverage blockchain technology systems are security for the Internet of Things (IoT) devices, identity systems, digital voting, distributed cloud storage, smart contracts, etc. (Dai et al., 2019; Braghin et al., 2019). Blockchain technology systems' potential to be successful across domains and methods relies on their distributed, immutable nature and their ability to be reprogrammed.

Blockchain as a technology with numerous application potentials may not have a specific definition that is all encompassing. We present some definitions found within the scope of this chapter—blockchain as a peer-to-peer network that is characterized by distinct and irreversible data transfer capabilities. Blockchain can

427
C. M. Akujuobi, *Wavelets and Wavelet Transform Systems and Their Applications*,
https://doi.org/10.1007/978-3-030-87528-2_18

also be defined as a data structure that uses a peer-to-peer system to store information in a computer file in the same way an enterprise network allows every employee access and can store data. Blockchain is an immutable ledger comprising blocks forming the fundamental building block for bitcoin and sustains transactions on the network. The blockchain is a fault-tolerant system and a means of achieving a tamper-proof ledger that is verifiable. It can also be presented as a potential hosting platform for IoT data. In many areas, blockchain is perceived as just a data structure having a distributed multi-version concurrency control. Blockchain therefore can be said to be a decentralized network of databases in the form of blocks capable of holding and transferring any digital asset or data in a tamper-proof manner.

However, in addition to the discussion on blockchain technology systems in this chapter, we explore how wavelets can be applied to blockchain. We discuss the possible advantages and disadvantages of the application of wavelets and wavelet transforms to blockchain technology systems. In Sect. 18.2, we discuss the capabilities and limitations of blockchain technology systems.

18.2   Capabilities and Limitations of Blockchain

In this section, we discuss the capabilities and limitations of blockchain technology systems starting with the capabilities and followed with the limitations.

18.2.1   Capabilities of Blockchain Technology

The blockchain is distributed, immutable, and programmable and self-heals. These properties make blockchain powerful. The core concept in blockchain technology is the autonomous nature, consensus algorithms such as proof of work, smart contracts, and cryptographic hashes. Some of the key capabilities are as follows:

- *Security:* The blockchain technology is built on recursive encryption to protect transaction data. The blockchain protocol encrypts data blocks in a recursive or cascaded manner, i.e., the encryption result of the previous block is used by the protocol in encrypting the current block. The blockchain uses public key cryptography, where every peer generates its public-private key pairs for its encryption process.
- *Autonomous:* The blockchain supports smart contracts and is programmed according to any business logic. Therefore, blockchain technology is a peer-to-peer network that replaced the need for a trusted third party with smart contracts. Smart contracts are contractual rules encoded and executed as a computer program within the blockchain, and they open the blockchain to many application areas.

- *Decentralized*: The blockchain network possesses a distributed architecture such that every node on the network retains a copy of the blockchain. This design eliminates the chances of the central point of failure vulnerability that hackers can exploit to compromise the system. The blockchain is therefore fault-tolerant.
- *Immutable*: A collaborative effort of common interest among the existing participants is called consensus, an algorithm that authenticates any participating entity in the blockchain network. A legitimate participant is hashed in a transaction and committed to the system as a member node, whereas an invalid object is discarded to be malicious.

18.2.2 Limitations of Blockchain Technology

The key limitations of the blockchain technology systems are as follows:

- *The Issue with Scalability:* Scalability is one of the shortcomings of the current blockchain implementation. The problem of scalability in the context of blockchain integration with Internet of Things is caused by (i) a large number of IoT devices to be connected, (ii) the various limitations of blockchain (block size, block generation time, and consensus), and (iii) the limitations inherent in a typical Internet of Things device. With the help of AI, federated learning, a new decentralized learning system is used along with other data shading techniques to make the blockchain system more efficient.
- *The Issue with Interoperability:* Interoperability is the ability to transact and share data across blockchain and non-blockchain systems. Interoperability is a big issue facing the blockchain technology and its application in many industries such as healthcare, finance, and smart or intelligence operations. Blockchain is initially designed to operate with computers with high computational powers on the Internet. Most application areas of blockchain require low power or energy consumption such as the Internet of Things (IoT), and therefore blockchain may not be efficient when considered within the IoT applications.
- *The Issue with Computational Power:* Blockchain is slow in computing codes and smart contracts. The process of mining in blockchain requires a lot of computational power. Therefore, miners or specialized hardware is necessary to carry out blockchain mining in a typical blockchain solution. This issue could lead to inefficiencies and extra investment, which constitute a considerable limitation in blockchain technology.
- *The Issue with Security*: Blockchain is said to be immutable and hack resistant; however, in a scenario where some applications or platforms have extra layers of applications, security could be a significant concern. This limitation also stands in the way of blockchain integration with other technologies like the Internet of Things, AI, and the cloud. Another source of security concern in the blockchain is a buggy smart contract.

- *The Issue with Efficiency:* Inefficiency in the blockchain is caused by the redundant behavior of nodes, each performing the same tasks as every other node on its copy of the data in an attempt to be the first to solve a mathematical puzzle. This inefficiency is a particular concern with blockchain using the proof of work (PoW) consensuses like bitcoin and Ethereum.

18.3   Blockchain-Based Strategies

Many industries are using blockchain to solve most of the IoT issues, such as security and identity management. Combining these two technologies gives extended advantages to these industries. The following are the use cases of combining blockchain and Internet of Things.

- *Automobile Industry:* Automobiles are becoming highly equipped with sensors and Internet capabilities making them part of the IoT ecosystem. Connecting smart cars to the blockchain network will enable the trusted exchange of information, improved connectivity and security, as well as accurate vehicle records such as trip information, service, and fault information.
- *Access Control and Identity Management:* Smart contracts significantly extend the features of a blockchain to adapt to a variety of use cases. A smart contract for access control will enforce access control of the various resources on the blockchain network. In addition, smart contract for identity management will enhance the identity management capability of the blockchain for multiple IoT devices on the system.
- *Secure Update of Edge Devices in IoT:* Blockchain combined with the Internet of Things can be used to provide firmware or software update in scenarios like the smart cities and smart homes. Smart contracts are used to define the update conditions and the secure nature of blockchain that makes them resistant to cyberattacks.
- *Logistics and Supply Chain Management:* In IoT-enabled supply chain, where vehicles and cargoes are equipped with sensors, combining blockchain and Internet of Things enables near-real-time access to status information regarding shipment, increasing visibility, and reliability within the supply chain.
- *Sharing Economy:* As the sharing economy is rapidly growing in adoption, blockchain can enable a decentralized application on a shared economy, making the exchange of value, goods, and services seamless at a reduced cost.
- *Agriculture:* Sensor data from farms stored on the blockchain can provide useful information regarding the provenance of products, improved transparency in the agricultural supply chain, and informed decision-making for farmers and customers.
- *Micropayments:* Micropayments in IoT will involve either machine-to-machine or person-to-machine transactions using cryptocurrencies without involving centralized third parties like the banks. Examples are a smart connected electric

vehicle making payment to a charging station and a person making payment for a product from an Internet-connected vending machine. Micropayment enables faster and cheaper payment among the parties involved.

- *Data Integrity in IoT:* Combining IoT and blockchain will ensure IoT data integrity automatically using the digital signature and hashing techniques that are inbuilt by design on the blockchain. This is especially very useful in scenarios where multiple parties are involved, as the smart energy grid to eliminate fraud and rip-offs by the participants in energy trading.
- *Healthcare Industry*: Blockchain can drastically reduce the problem of fake medicines (one such application is a MediLedger project) by enabling transparency and traceability in the pharmaceutical supply chain, Also, patients using monitoring healthcare devices connected to the blockchain can choose who to share their data and be guaranteed that only healthcare professionals responsible for their care can have access to such data.

18.4   How Blockchain Powers Applications Such as Bitcoin and Other Token-Based Initiatives

Blockchain technology systems are the building principles of all cryptocurrencies. Some examples of cryptocurrency are Bitcoin, Ethereum, Corda, Hyperledger, Komodo, and many others. All blockchain-based cryptocurrency platforms are used to exchange value digitally, and they work on the same backbones of blockchains. *A node or wallet is a single entity in a blockchain network, which can be any device capable of meeting the computational requirements for running the blockchain software or client, creating accounts, and storing the blockchain transaction data.* A node has a unique IP address and port number, a network ID, and other configurations. The IP address and port numbers are used to peer with another node in the network. The network ID is a unique identifier of the system used to secure the network from replay attacks.

A node in a blockchain contains as many accounts as it can create, identified by a unique 42-digit alphanumeric address when prefixed with "0x" or 40-digit alphanumeric address when non-prefixed. Note that the formatting and length of account addresses are different among blockchain. All actions to be performed in a blockchain or information to be retrieved from a blockchain such as sending a transaction or receiving a transaction and deploying smart contracts are carried out through accounts. When an account is used to implement the smart contract, it will contain another type of account within the account called an intelligent contract account.

Blockchain logs all transactions and records in a block (Bora, 2020). An empty block is a block that contains no transaction. When a transaction is sent from an account to a remote account in the network, the mining process writes the transaction in a block and broadcasts to every account active on the network. The mining

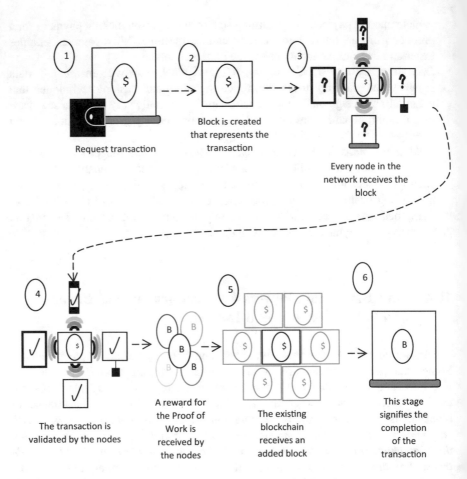

Fig. 18.1 A basic example of the architectural workings of a blockchain

process is a computationally intensive process. A block that contains a transaction is known as a transaction block. Figure 18.1 shows the basic example of the architectural workings of a blockchain. It first starts with the request for a transaction. The block that represents the transaction is created once the request is made. The block is then sent to every node in the network. The nodes proceed to validate the transaction. A reward is given to the nodes for the proof of work. The block is then added to the existing blockchain and that completes the transaction process.

18.5   Different Types of Blockchain Models

The blockchains are categorized as private or permissioned blockchains, public blockchains, or unpermissioned blockchain and consortium blockchain. A public blockchain is free to the public to join without restrictions, but the private blockchain is restricted and can only be connected to with valid credentials. An example of permission blockchain is Hyperledger Fabric and Ethereum.

18.5.1   Hyperledger

Hyperledger that started in December 2015 can be defined as a global enterprise blockchain that offers the necessary framework, standards, guidelines, and tools, to build open-source blockchains and related applications for use across various industries. Hyperledger does not need proof of work/proof of stake. It does not need computers solving problems day and night and does not need people to have cryptocurrency to reach consensus. Many industries currently use Hyperledger technology.

18.5.2   Fabric

Fabric idea, which began in 2015, is a modular blockchain framework that acts as a foundation for developing blockchain-based products, solutions, and applications using plug-and-play components that are aimed for use within private enterprises. Many industries use the fabric blockchain technology.

18.5.3   Ethereum

Ethereum, which was launched in 2015, is a distributed public blockchain network that focuses on running programming code of any decentralized application. It is not just a platform but also a programming language running on a blockchain, helping developers to build and publish distributed applications.

18.6   Wavelet Transform Analysis

As already discussed in Chap. 2, wavelets are functions with the translation (transient, s) and scaling (dilation, u) parameter as their very essential properties. Wavelets overcome the shortcomings of the Fourier transform. The Fourier transform is more suited for stationary signals. The wavelet transform as discussed in Chaps. 2 and 4 is suited for non-stationary signals. It overcomes the shortcomings of the Fourier transforms with their translation and scaling properties. Wavelets, therefore, offer solutions to the problems of poor resolutions, global nature, and fixed window size and can handle both stationary and transient signals. The wavelet (ψ) is therefore formed by the linear combination of scaled and shifted versions of the scaling function. The wavelet $\psi_{u,\,s}(t)$ representation is as shown in Eq. (18.1).

$$\psi_{u,s}(t) = \frac{\psi\left(\frac{t-u}{s}\right)}{\sqrt{s}} \tag{18.1}$$

The wavelet transform $W_{u,\,s}$ representation, which has been discussed in detail in Chap. 4, is as shown in Eq. (18.2). This equation is similar to Eq. (4.8) in Sect. 4.3 in Chap. 4. The complex conjugate of $\psi(.)$ is ψ^* as shown in Eq. (18.2).

$$(W_{u,s}) = K \int_{-\infty}^{+\infty} \psi^*\left(\frac{x-s}{u}\right) f(x)\,dx \tag{18.2}$$

There are two main classes of wavelets, namely, the continuous wavelet and discrete wavelets. The continuous wavelet is limited by impracticality and redundancy issues and is therefore discretized for various applications. Wavelet transform is used to decompose the blockchain information into multiple resolutions using its translation and dilation properties to represent the information in both time and frequency domains. We have two classes of wavelet transforms: the continuous wavelet transform (CWT) and the discrete wavelet transform (DWT). In continuous wavelet transform (CWT), the translation and dilation parameter can be set to an arbitrary value. However, in the discrete wavelet transform (DWT), the translation and dilation parameters have predetermined values. The idea of the wavelet transform as shown in Fig. 18.2 is to analyze a given signal such as the blockchain-based signals. DWT uses the low-pass (*h₀*) and the high-pass (*h₁*) coefficients. The coefficients relate to the scaling and wavelet functions, respectively, to decompose the signal into its different frequency bands (coarse approximations and detailed information). The signal is decimated and interpolated by 2 in the decomposition (analysis) and the reconstruction (synthesis) processes. More discussion of the wavelet transform processes can be found in Sect. 4.3 in Chap. 4.

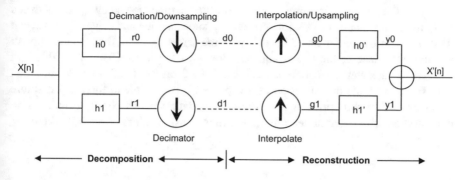

Fig. 18.2 Wavelet analysis and synthesis

18.7 Wavelet Transform Analysis of Blockchain Systems

Wavelet technique is a technology that has numerous signal processing applications that range from multimedia data compression to signal denoising and can be applied to blockchain technology systems information. On the other hand, blockchain is a new technology in various stages of adoption. Blockchain technology systems have the potential of transferring any digital content. Blockchain has all the great possibilities to be applied to high-quality video rights management systems. An important area is how to treat content (video) securely in a proper management system and the issue of coupling video with license information taking into account the video file sizes. Values or signals stored or exchanged within the blockchain can be analyzed using wavelet techniques. An example is using wavelet coherence analysis to examine how certain online factors influence bitcoin price (Kristoufek, 2005; Phillips & Gorse, 2018).

In addition, recent research efforts have proved wavelet-based watermarking very useful (Anumol & Karthigaikumar, 2011). Watermarking is the process of embedding data in a digital media file. The discrete wavelet transform is an ideal choice for efficient watermarking. The watermarking scheme is the best way to provide accountability in blockchain applications in the area of media file distribution. Another use of wavelets in the blockchain system is applied to the multimedia blockchain, which makes use of compressive sensing and self-embedding watermarking.

18.8 Wavelets and Bitcoin

A significant characteristic of the bitcoin market is price dynamics that can be analyzed with wavelet coherence application to unveil some drivers of bitcoin price, which is presumably in line with standard economic theory governing money value such as demand and supply (Kristoufek, 2005). This type of analysis is a typical case of examining the relationship in time-frequency space between two-time series that has a possible connection such as price and driver. The wavelet transform can be implemented on the two-time series to reveal their relative power and current phase in the time-frequency domain using their CWT. In addition, a measure of the wavelet coherence can be defined on the CWT to find important consistency suitable in a situation of low power. It is possible to forecast the future behaviors of bitcoin price in the short term with wavelet-based techniques in combination with machine learning algorithms to assist future decision-making.

18.9   Main Drivers of the Bitcoin Price as Evidenced from Wavelet Coherence Analysis

Wavelet coherence as applied to blockchain technology systems is used to measure or characterize patterns or relationships between two-time series signal to determine how one time series drives or influences the other time series or how some unobserved factor affects both time series. Wavelet coherence is therefore applied to monitor changing temporal relationships occurring over the short, medium, and long term in bitcoin price and online factors (Kristoufek, 2005; Phillips & Gorse, 2018). It is possible to study what happens to the association during specific regimes such as bubble, whether the relationship strengthens or weakens. This helps to find whether the relationship between bitcoin price fluctuations and factors such as online factors has an essential application in diversity risk estimations. Figure 18.3 shows an example of the wavelet coherence scalogram. Figure 18.4 shows the scalogram for Ethereum and Monero, while Fig. 18.5 shows the scalogram for Litecoin and Bitcoin. The time appears in the horizontal axis, while the period occupies the vertical axis. Various colors show the extent of correlation, for example, from dark blue (0, no coherence) to yellow (1, strong coherence). The arrows indicate the phase of association. Left arrow is an anti-phase or negative correlation, whereas the right arrow is an in-phase or positive correlation. An arrow pointing downward means the first time series is leading the second, whereas an indicator pointing upward means the second time series is leading the first. Using wavelets and the wavelet transform tools, the financial time series data for Bitcoin, Ethereum, Litecoin, and Monero can be examined (Kristoufek, 2005; Phillips & Gorse, 2018). The result presents varying relationships such as short-term, medium-term, and long-term relationships and shows what happen over time as the cryptocurrency goes through various market regimes.

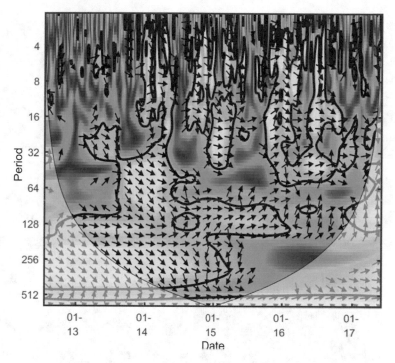

Fig. 18.3 Wavelet coherence scalogram
Source: Phillips and Gorse (2018). Phillips, R.C. and Gorse, D., "Cryptocurrency Price Drivers: Wavelet Coherence Analysis Revisited", PLOS ONE, Vol. 13, No. 4:e0195200, https://doi.org/10.1371/journal.pone.0195200, p. 8, April 18, 2018. *Permission: Creative Commons Attribution License*

18.10 Advantages of Using Wavelets in Blockchain Systems

- *Scalability*: Wavelets could be used to compress the blockchain signals and significantly increase the number of transactions per block of the blockchain. The wavelet solution to the scalability challenge will lead to an increase in application areas within the blockchain.
- *Efficiency*: Wavelet can be used to analyze blockchain data and can be used to improve resource waste in blockchains such as memory and computational power. Also, some prediction could be made on blockchain systems based on wavelet analysis.
- *Enhanced Security*: Wavelet can be used for strengthening the security in blockchain systems through a reduction in complexity and the introduction of an extra layer of protection.

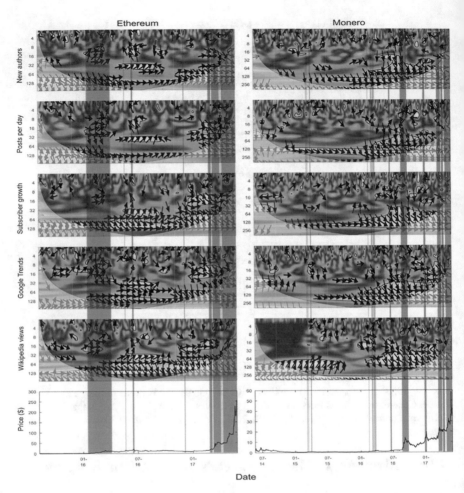

Fig. 18.4 Wavelet coherence scalograms for Ethereum and Monero for different online factors against price. *Source*: Phillips and Gorse (2018). Phillips, R.C. and Gorse, D., "Cryptocurrency Price Drivers: Wavelet Coherence Analysis Revisited", PLOS ONE, Vol. 13, No. 4:e0195200, https://doi.org/10.1371/journal.pone.0195200, p. 11, April 18, 2018. *Permission: Creative Commons Attribution License*

18.11   Disadvantages of Using Wavelets in Blockchain Systems

- Wavelet is a signal processing tool and technology whose integration with the emerging blockchain technology might be very challenging. This challenge could be from the standpoint of integrating the two technologies and getting them to work efficiently together.
- Wavelets and blockchain are two different and sophisticated technologies that would be very costly to maintain at the same time with the right experts.

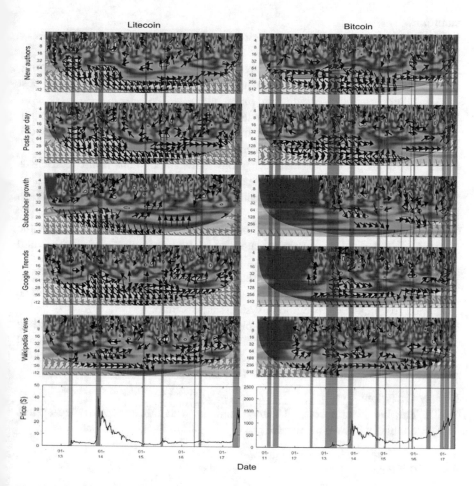

Fig. 18.5 Wavelet coherence scalograms for Litecoin and Bitcoin for different online factor against price. *Source*: Phillips and Gorse (2018). Phillips, R.C. and Gorse, D., "Cryptocurrency Price Drivers: Wavelet Coherence Analysis Revisited", PLOS ONE, Vol. 13, No. 4:e0195200, https://doi.org/10.1371/journal.pone.0195200, p. 12, April 18, 2018. *Permission: Creative Commons Attribution License*

- Wavelet transform seems to be demanding (expensive) in terms of computational power and thus would add to the inefficiency of an already complex and computationally intensive blockchain system very competitive among alternatives in meeting the requirement of IoT applications. Wavelets in combination with blockchain need to be optimized to consume less power and be more efficient for other application.
- Blockchain works with real-time applications such as bitcoin and other cryptocurrencies, and efficient implementation of wavelets should be appropriate within real-time scenarios.

Summary
1. Blockchain is a decentralized or peer-to-peer network that consists of nodes or parties.
2. In the industries, blockchain integrates into the various processes: manufacturing, supply chain to provide provenance and traceability in Internet of Things, cost saving, process improvements, and many others.
3. Blockchain can also be defined as a data structure that uses a peer-to-peer system to store information in a computer file in the same way an enterprise network allows every employee access and can store data.
4. The core concept in blockchain technology is the autonomous nature, consensus algorithms such as proof of work, smart contracts, and cryptographic hashes.
5. Many industries are using blockchain to solve most of the IoT issues, such as security and identity management.
6. A public blockchain is free to the public to join without restrictions, but the private blockchain is restricted and can only be connected to with valid credentials.
7. Wavelet technique is a technology that has numerous signal processing applications that range from multimedia data compression to signal denoising and can be applied to blockchain technology systems information.
8. Wavelet coherence as applied to blockchain technology systems is used to measure or characterize patterns or relationships between two-time series signal to determine how one time series drives or influences the other time series or how some unobserved factor affects both time series.

Review Questions
18.1 The blockchain concept started initially with the cryptocurrency system popularly known as bitcoin.

(a) True
(b) False

18.2 In the industries, blockchain integrates into the various processes: manufacturing, supply chain to provide provenance and traceability in Internet of Things, cost saving, process improvements, and many others.

(a) True
(b) False

18.3 The blockchain technology is not built on recursive encryption to protect transaction data.

(a) True
(b) False

18.4 Scalability is one of the shortcomings of the current blockchain implementation today.

(a) True
(b) False

18.5 Interoperability is the ability to transact and not share data across blockchain and non-blockchain systems.

(a) True
(b) False

18.6 Inefficiency in the blockchain is not caused by the redundant behavior of nodes, each performing the same tasks as every other node on its copy of the data in an attempt to be the first to solve a mathematical puzzle.

(a) True
(b) False

18.7 Blockchain logs all transactions and records in a block.

(a) True
(b) False

18.8 Wavelet transform is used to decompose the blockchain information into multiple resolutions using its translation and dilation properties to represent the information in both time and frequency domains.

(a) True
(b) False

18.9 One use of wavelets in the blockchain system is not to apply it to the multimedia blockchain, which makes use of compressive sensing and self-embedding watermarking.

(a) True
(b) False

18.10 Wavelets can be used for strengthening the security in blockchain systems through a reduction in the complexity and the introduction of an extra layer of protection.

(a) True
(b) False

Answers: 18.1a, 18.2a, 18.3b, 18.4a, 18.5b, 18.6b, 18.7a, 18.8a, 18.9b, 18.10a

Problems
18.1 How can you describe a blockchain technology system?
18.2 How can blockchain be used as a security technology system mechanism?
18.3 18.3 Describe the autonomous property that the blockchain technology system can provide.
18.4 Describe the security, autonomous, decentralized, and immutable capabilities of the blockchain technology system.
18.5 Blockchain has limitation in scalability; what is the limitation?
18.6 How does the issue of interoperability affect the operation of a blockchain technology system?

18.7 How does the issue of computational power affect the operation of a blockchain technology system?

18.8 How does the issue with security affect the operation of a blockchain technology system?

18.9 How does the issue with efficiency affect the operation of a blockchain technology system?

18.10 What are the limitations of blockchain technology?

18.11 What are the basic blockchain-based strategies?

18.12 Describe Hyperledger, Fabric, and Ethereum types of blockchain platforms.

18.13 How can wavelet transform idea be used in blockchain technology systems?

18.14 How can the wavelet coherence analysis be the main drivers of the bitcoin price in blockchain technology systems?

18.15 What are the advantages of using wavelets in blockchain technology systems?

18.16 What are the disadvantages of using wavelets in blockchain technology systems?

References

Anumol, T. J., & Karthigaikumar, P. (2011). *DWT based Invisible image watermarking algorithm for color images.* IJCA special issue on computational science – New dimensions & perspectives, NCCSE, pp.76–79.

Bora, D. (2020, September 15). *Blockchain technology in information security.* https://www.cybrary.it/blog/blockchain-technology-in-information-security/.

Braghin, C., Cimato, S., Cominesi, S. R., Damiani, E., & Mauri, L. (2019). *Towards blockchain-based E-voting systems.* In W. Abramowicz & R. Corchuelo (Eds.), BIS 2019 workshops, LNBIP 373 (pp. 274–286). Cham: Springer. https://doi.org/10.1007/978-3-030-36691-9_24.

Dai, H., Zheng, Z., & Zhang, Y. (2019, October). Blockchain for Internet of Things: A survey. *IEEE Internet of Things Journal, 6*(5).

Eze, K., Akujuobi, C. M., Sadiku, M. N. O., Chouikha, M., & Alam, S. (2019). *Internet of Things and blockchain integration: Use cases and implementation challenges.* In W. Abramowicz & R. Corchuel (Eds.), BIS 2019 workshops, LNBIP 373 (pp. 287–298). Cham: Springer. https://doi.org/10.1007/978-3-030-36691-9_25.

Kristoufek, L. (2005). What are the main drivers of the bitcoin price? Evidence from wavelet coherence analysis. *PLOS ONE, 10*(4), 1–15, e0123923. https://doi.org/10.1371/journal.pone.0123923.

Phillips, R. C., & Gorse, D. (2018, April 18). Cryptocurrency price drivers: Wavelet coherence analysis revisited. *PLOS ONE, 13*(4), 1–21, e0195200. https://doi.org/10.1371/journal.pone.0195200.

Part VII
Wavelet and Wavelet Transform Application to Detection, Discrimination and Estimation

Chapter 19
Wavelet-Based Signal Detection, Identification, Discrimination, and Estimation

No act of kindness, no matter how small, is ever wasted.

Aesop

19.1 Introduction

Wavelets are powerful signal processing tools. They are good decorrelators, capable of separating signals from noise. Several signal detection methods have been used in the past; however, wavelet transform seems to have the capability of producing better results. The identification of the modulation type of a communication signal has many areas of application such as electronic warfare, surveillance, interference detection, signal demodulation, and threat analysis. Several techniques have been used in the past to identify signal modulation types including wavelet. In this chapter, we will discuss how wavelet transform can be used to extract the transient characteristics in a digital modulation signal and how the distinct pattern in the wavelet transform domain can be employed in identifying the signals. We will explore how to use wavelets in the development of wavelet-based detection and measurement algorithms. The goal is to improve typical frequency domain-based and time domain-based methods in detecting and measuring common communication signals/modulations. We discuss how to select specific communications signals/modulations, focusing on specific modulations commonly found that may yield significant improvement in detection and measurement over Fourier methods. We also discuss in this chapter identification, discrimination, and estimation issues using wavelets and wavelet transforms.

© Springer Nature Switzerland AG 2022
C. M. Akujuobi, *Wavelets and Wavelet Transform Systems and Their Applications*,
https://doi.org/10.1007/978-3-030-87528-2_19

19.2   Wavelet Transform

Fast Fourier transform (FFT) and windowed FFT are used for the analysis of signals traditionally. These methods transform signal waveforms into infinite summation (the integral) of sinusoids with the frequency differences between adjacent sinusoids infinitesimally small. The formula for Fourier transform is as shown in Eq. (19.1).

$$S(f) = \int_{-\infty}^{\infty} s(t)\ell^{-i2\pi ft} dt \tag{19.1}$$

where $s(t)$ is the information that is to be transformed.

Fast Fourier transforms (FFTs) and cross-power spectra can be used for frequency resolution, phase measurements, and error calculations. FFT functions are localized in the frequency domain; hence, it possesses the inability to handle transient signals very well. Wavelet, on the other hand, has the capability of localizing signals in both time and frequency domains Mallat (1999). This peculiar feature of wavelet has made it possible for it to be used in various applications such as image processing, biomedical image system, electromagnetic, astronomy, nuclear engineering, etc. By definition, a wavelet ψ is a function of zero average as shown in Eq. (19.2).

$$\int_{-\infty}^{\infty} \psi(t) dt = 0 \tag{19.2}$$

which is dilated with a scale parameter s and translated by u as shown in Eq. (19.3).

$$\psi_{u,s}(t) = \frac{1}{\sqrt{s}} \psi \left(\frac{t-u}{s} \right) \tag{19.3}$$

The wavelet transform of a function or signal f, at the scale s and position u, is computed by correlating f with a wavelet atom as shown in Eq. (19.4).

$$Wf(u,s) = \int_{-\infty}^{\infty} f(t) \frac{1}{\sqrt{s}} \psi^* \left(\frac{t-u}{s} \right) dt \tag{19.4}$$

Wavelets are adjustable and adaptable. Since there are various wavelet bases, they can be designed to fit into different applications. Wavelet transform (WT) can be either continuous or discrete. If the signal is in itself a sequence of numbers, or samples of a function of continuous variables, then the wavelet expansion of the signal can be referred to as discrete time wavelet transform. The algorithm for the discrete wavelet transform is much simpler than the integral function of the continuous wavelet transform. It makes use of low-pass and high-pass filter coefficients

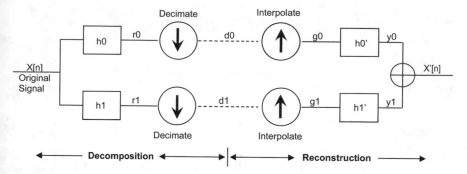

Fig. 19.1 Discrete wavelet transform block diagram

generated from the wavelet functions. Figure 19.1 is the diagram of the algorithm of a discrete wavelet transform, showing the decomposition and reconstruction sections of the signal. It illustrates how a signal is separated into its low-pass coefficients and high-pass coefficients.

The decomposition process involves the convolution of the signal using the respective low-pass (h_0) and high-pass (h_1) of given wavelet coefficients. Thereafter, the signal is taken through a decimation procedure (downsampling by 2) which concludes the decomposition process. The reconstruction process involves the interpolation of the decomposed low-pass and high-pass coefficients. In the interpolation process, you upsample by 2. The result is convoluted with a transformed low-pass and high-pass matrices; the signal is then recombined. There are several types of wavelets among which are Haar, Daubechies, Symlet, Morlet, and Meyer. In this chapter, we use the Haar and Morlet wavelets.

19.2.1   Haar Wavelet Characteristics

The Haar wavelet is the oldest and simplest of all the wavelets. It is defined as shown in Eq. (19.5).

$$\psi(t) = \begin{cases} 1 & 0 \le t \le 0.5 \\ -1 & 0.5 \le t \le 1 \\ 0 & otherwise \end{cases} \tag{19.5}$$

The Haar wavelet is one of the orthogonal types of wavelets. Figure 19.2 shows the plot of the Haar wavelet function.

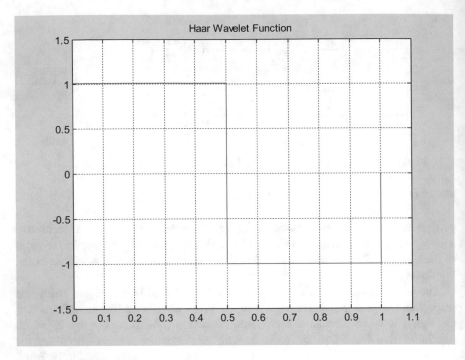

Fig. 19.2 Haar wavelet function

19.2.2   Morlet Wavelet Characteristics

The Morlet wavelet is named after the originator L. Morlet, a physical scientist. This type of wavelet is a combination of Gaussian window with a sinusoidal function. It has good time and frequency localization features (Hernandez et al., 2004). The Morlet wavelet is one of the most widely used wavelets, and it is defined as shown in Eq. (19.6).

$$\psi(t) = \frac{1}{\sqrt{2\pi}} \ell^{-t^2/2} \ell^{-j\omega_0 t} \tag{19.6}$$

The real part of Morlet wavelet is defined as shown in Eq. (19.7).

$$\psi(t) = C\ell^{-t^2/2} \cos(5t) \tag{19.7}$$

The Morlet wavelet function plot is as shown in Fig. 19.3.

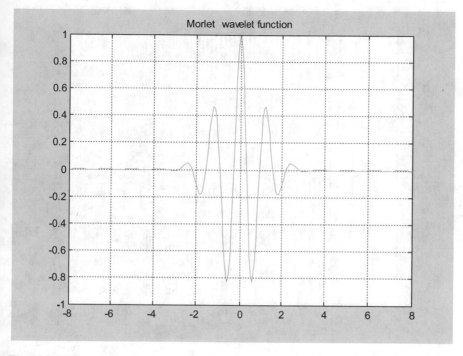

Fig. 19.3 Morlet wavelet function

19.3   Overview of the Wavelet-Based Signal Detection

The overview of the wavelet-based signal detection is discussed in this section. The communication signals considered are the chirp signal, frequency-shift keying (FSK) signals, phase-shift keying (PSK) signals, and the quadrature amplitude modulation (QAM) signals.

19.3.1   Chirp Signal

Chirp signals do occur in lots of signal processing tasks among which are acoustics, communications, radar, and sonar. Several methods have been used in the past to detect the presence of a chirp signal embedded in an additive white noise. The spectrogram of a noisy signal can be interpreted to decide whether a chirp signal is present or not. Also classical methods of detection such as maximum likelihood (ML), Wigner distribution (Yasotharan & Thayaparan, 2006), and line integral transform (LIT) (Ristic & Boashash, 1993; Hong & Ho, 1999) either are computationally intensive or require some prior information. A chirp signal $s(t)$ in complex form can be represented as shown in Eq. (19.8).

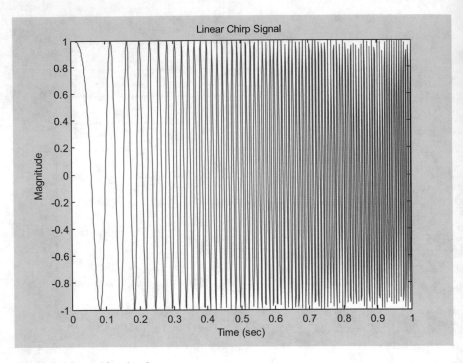

Fig. 19.4 Linear chirp signal

$$s(t) = A\ell^{j\pi kt^2}, \; -T/2 \le t \le T/2 \tag{19.8}$$

The real part of the signal is represented as shown in Eq. (19.9).

$$s(t) = A\cos\left(\pi kt^2\right) \tag{19.9}$$

where A is the amplitude of the chirp, k is the slope rate, and T is the time duration. Figure 19.4 is a representation of a linear chirp signal.

19.3.2   Frequency-Shift Keying (FSK) Signals

FSK signal is a digitally modulated signal where the frequency varies as a function of the digital modulating signal. FSK signal is represented mathematically in complex and real forms as shown in Eqs (19.10 and 19.11), respectively.

$$V_{FSK} = V_c \ell^{\; j(2\pi(\; f_c t + f_n(t - nT_s)) + \theta_c)} \tag{19.10}$$

$$V_{FSK}(t) = V_c \sin [2\pi(\; f_c + v_m(t)\Delta f)t] \tag{19.11}$$

Figure 19.5a and b is a representation of the FSK signal.

19.3.3   Phase-Shift Keying (PSK) Signals

In a PSK signal, the phase angle of the signal varies as a function of the digital modulating signal. A PSK signal is represented in complex and real form, respectively, as shown in Eqs. (19.12 and 19.13).

$$V_{PSK}(t) = V_c \ell^{\; j(2\pi f_c t + \theta_c + \theta_n)} \tag{19.12}$$

$$V_{PSK}(t) = V_c \sin [2\pi f_c t + \theta_n] \tag{19.13}$$

Figure 19.6a and b shows the PSK signal and scatter plot for 8-PSK.

19.3.4   Quadrature Amplitude Modulation (QAM) Signals

In a QAM signal, the signal amplitude, as well as the phase angle of the signal, varies as a function of the digital modulating signal. It is represented as shown in Eqs. (19.14 and 19.15), in complex and real form, respectively.

$$V_{QAM}(t) = V_c \left(a_{In} + j a_{Qn} \right) \ell^{\; j(2\pi f_c t + \theta_c)} \tag{19.14}$$

$$V_{QAM}(t) = V_c \sin (2\pi f_c t + \theta) \tag{19.15}$$

Figure 19.7a and b shows the QAM signal and scatter plot for 8-QAM, respectively.

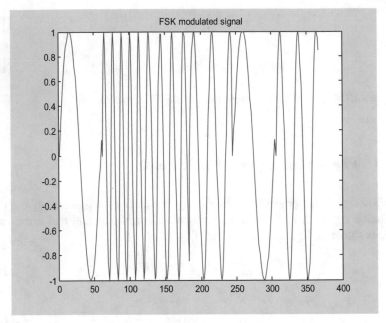

a

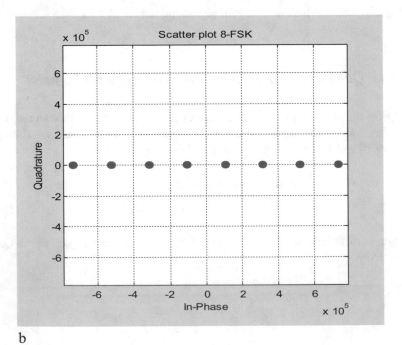

b

Fig. 19.5 (a) FSK signal. (b) 8-FSK signal scatter plot

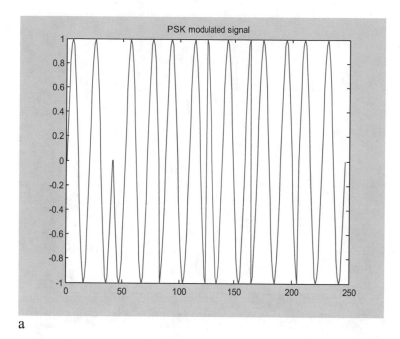

a

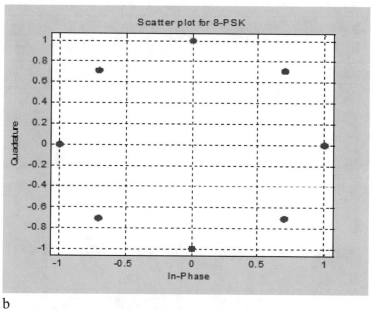

b

Fig. 19.6 (a) PSK signal. (b) Scatter plot for 8-PSK

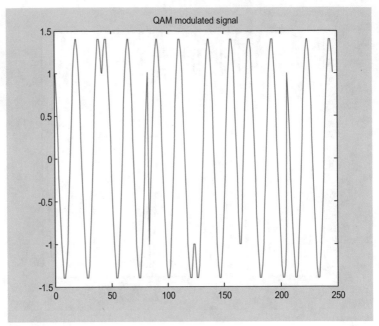

a

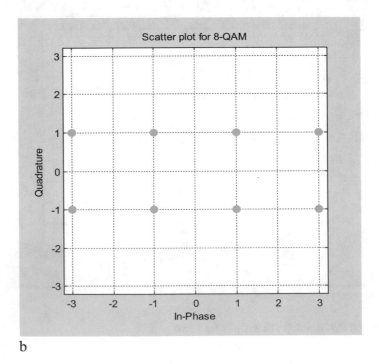

b

Fig. 19.7 (a) QAM signal. (b) Scatter plot for 8-QAM signal

19.4 Overall Detection System Model

Figure 19.8 shows the block diagram of the overall system model for the detection of the signals we considered in this chapter. Whenever an input signal is received, the variance of the signal is computed, V0; at the same time, a continuous wavelet transform of the signal is carried out using Morlet wavelet function, after which the variance is also computed, V1. The two variances are compared to each other; if V1-V0 is greater than zero, then a chirp signal is present; otherwise, a chirp signal is not present. If the presence of a chirp signal is not detected, the process goes to branch *a*, in order to test for the presence of PSK, FSK, or QAM signal. A detail description of the processes in each branch is discussed in the next section.

19.5 Chirp Signal Detection

The ability of the Morlet wavelet function to localize signals effectively in both time and frequency domains allows us to map the signal into the time-frequency domain. It is discovered that at the appropriate scale a, using Morlet wavelet, the variance of the noisy chirp signal is lower than the variance of the signal after continuous

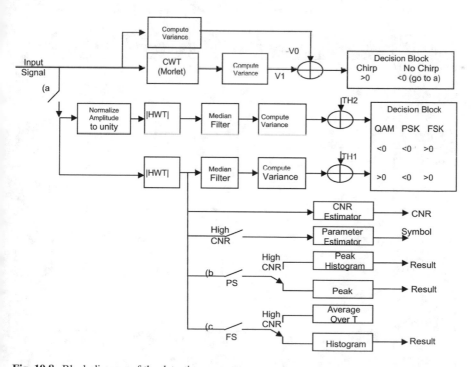

Fig. 19.8 Block diagram of the detection system

wavelet transform. The hypothesis that the variance of a noisy chirp signal after continuous wavelet transform is greater than the variance of the noisy signal before the wavelet transform is tested with four other communication signals, namely, sinusoidal, phase-shift keying (PSK) signal, frequency-shift keying (FSK) signal, and quadrature amplitude modulation (QAM) signal.

Example 19.1

Let a noisy signal suspected to contain chirp signal serve as input into the system shown in Fig. 19.8. The variance (V0) of the incoming noisy chirp signal is estimated. Thereafter, continuous wavelet transform of the noisy signal is carried out on the signal using the Morlet wavelet. The variance (V1) of the continuous wavelet-transformed coefficients is estimated as well.

Table 19.1 Percentage of correct identification for the chirp signal

Signal-to-noise ratio (dB)	Correct identification %
10	100
8	100
5	99.4
4	97.6
3	94.8
2	89.8
1	81.2

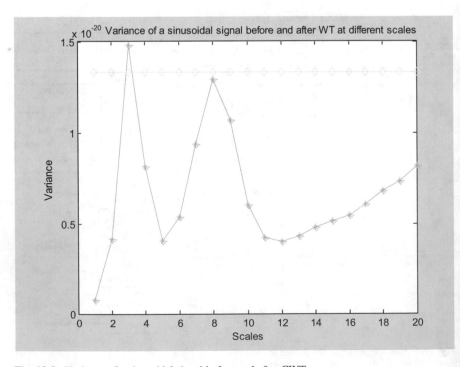

Fig. 19.9 Variance of a sinusoidal signal before and after CWT

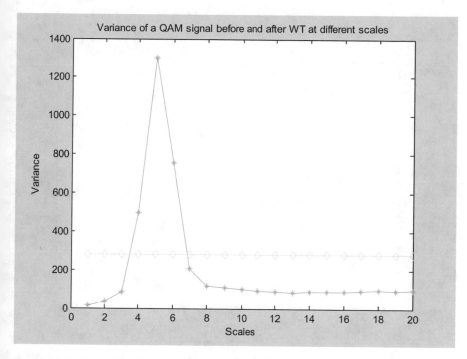

Fig. 19.10 Variance of a QAM signal before and after CWT

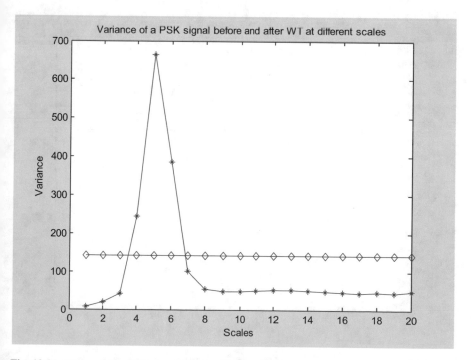

Fig. 19.11 Variance of a PSK signal before and after CWT

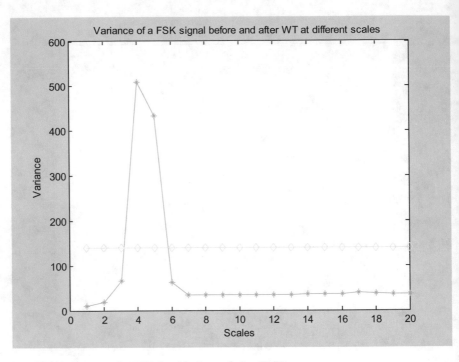

Fig. 19.12 Variance of a FSK signal before and after CWT

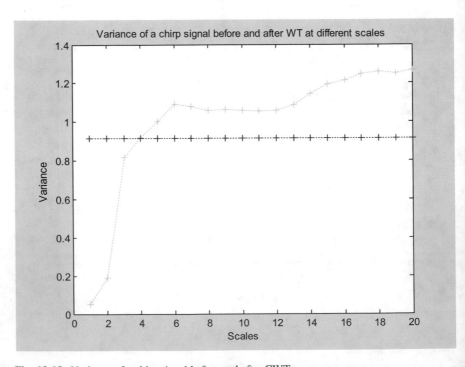

Fig. 19.13 Variance of a chirp signal before and after CWT

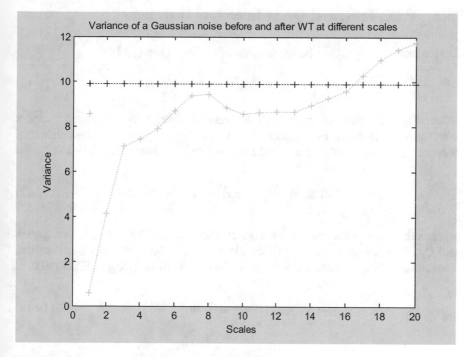

Fig. 19.14 Variance of a Gaussian noise before and after CWT

Solution

The decision rule is as follows:

If V1 > V0, chirp signal is present in the received signal.
However, if V1 < V0, then no chirp signal is present.

Using the Morlet wavelet and a scale of 10, it is possible to run several processes as in the case discussed in this section. The percentage of the correct identification for the noisy chirp signal at different SNR is tabulated in Table 19.1. The results are the average of 1000 independent runs.

Figures 19.9, 19.10, 19.11, 19.12, and 19.13 show variance plots before and after the continuous wavelet transform (CWT) at different scales for each of the signals under consideration. Figure 19.14 shows a similar plot for a white Gaussian noise without any signal. For each of the graphs shown, the straight line represents the variance of the signal before the CWT. It is a constant because it does not depend on the scale of the CWT. It can be seen from the graphs that beyond a scale of 8, the chirp signal is the only one whose variance after the CWT is greater than the variance of the signal before the CWT.

19.6   PSK, FSK, and QAM Interclass Detection

Let the received signal *r(t)* be represented as shown in Eq. (19.16).

$$r(t) = s(t) + n(t) = \overline{s}(t)\ell^{\,j(\omega_c t + \theta_c)} + n(t) \tag{19.16}$$

where *n(t)* is the noise component of the signal and *s(t)* is the modulated complex waveform, ω_c is the carrier frequency, and θ_c is the carrier phase. The continuous wavelet transform (CWT) of a signal *f(t)* is defined as shown in Eq. (19.17).

$$CWT(a, \tau) = \int_{-\infty}^{+\infty} s(t)\frac{1}{\sqrt{a}}\psi * \left(\frac{t-\tau}{a}\right)dt \tag{19.17}$$

where *s* is the scale and *τ* is the translation. The |HWT| of PSK signal, FSK signal, and QAM signal is as shown in Eqs. (19.18, 19.19, and 19.20), when the Haar wavelet is within a symbol time (Ristic & Boashash, 1993; Hong & Ho, 1999).

$$\left|CWT_{PSK(a,\tau)}\right| = \frac{4\sqrt{S}}{\sqrt{a}\omega_c} \sin^2\left(\frac{\omega_c a}{4}\right) \tag{19.18}$$

$$\left|CWT_{FSK(a,\tau)}\right| = \frac{4\sqrt{S}}{\sqrt{a}(\omega_c + \omega_i)} \sin^2\left[\frac{(\omega_c + \omega_i)a}{4}\right] \tag{19.19}$$

$$\left|CWT_{QAM(a,\tau)}\right| = \frac{4\sqrt{S}}{\sqrt{a}\omega_c} \sin^2\left(\frac{\omega_c a}{4}\right) \tag{19.20}$$

Equation (19.20) is for QAM with the amplitude normalized. Without the amplitude normalized, the |HWT| of a QAM signal becomes a multi-step function, with distinct peaks due to phase changes. These properties are used to distinguish between the different types of digital modulation.

The algorithm for the identification of digital modulation technique involves a similar process carried out in two branches, one with the received signal's amplitude normalized and the other without amplitude normalization (Jin et al. 2004). For the branch without amplitude normalization, the continuous wavelet transform (using Haar wavelet) of the signal is carried out, and the magnitude is computed. Thereafter, the result is passed through a median filter in order to remove the peaks. Then the variance of the output of the median filter is computed. The variance is compared with a selected threshold.

A similar process is carried out in the second branch, only that the amplitude is normalized before carrying out the process. The results are passed through the

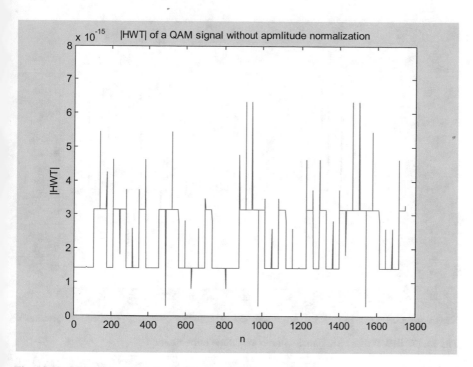

Fig. 19.15 |HWT| of a QAM signalwithout amplitude normalization

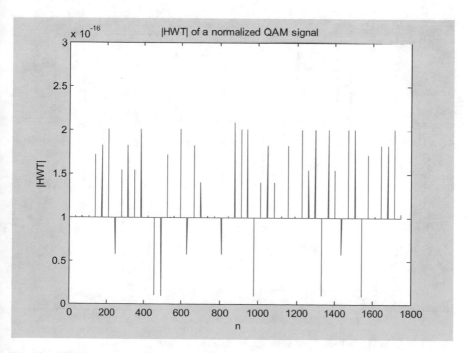

Fig. 19.16 |HWT| of a QAM signal with amplitude normalization

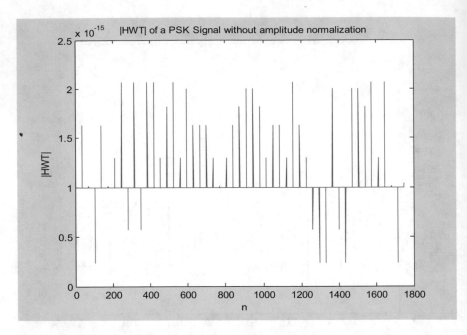

Fig. 19.17 |HWT| of a PSK signal without amplitude normalization

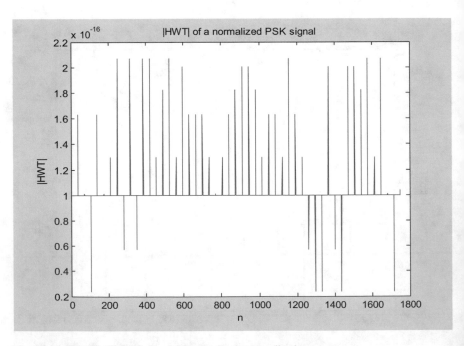

Fig. 19.18 |HWT| of a PSK signal with amplitude normalization

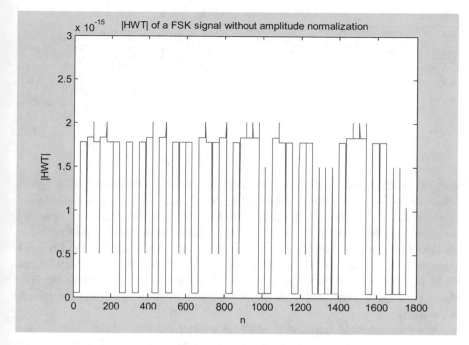

Fig. 19.19 |HWT| of a FSK signal without amplitude normalization

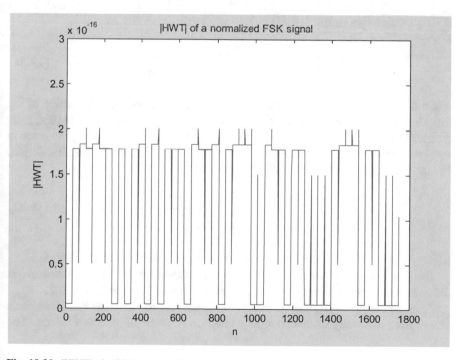

Fig. 19.20 |HWT| of a FSK signal with amplitude normalization

Table 19.2 Percentage of correct identification for QAM, PSK, and FSK signals

Out	CNR = 15dB			CNR = 10dB			CNR = 5dB		
In	QAM	PSK	FSK	QAM	PSK	FSK	QAM	PSK	FSK
QAM8	100	0	0	90.5	9.5	0	88.5	11.5	0
8-PSK	0	100	0	0.7	99.3	0	1.6	98.4	0
8-FSK	0	0	100	0	0	100	0	0	100
QPSK	0	100	0	0	100	0	0	100	0
QFSK	0	0	100	0	0	100	0	0	100

decision block, which is used to determine whether the received signal is QAM, PSK, or FSK. If the variance computed from the branch with amplitude normalization is lower than the threshold while the other variance is higher than the threshold, we classify the input as QAM signal. If both variances are higher than the thresholds, the input is classified as an FSK signal. If both variances are lower than the threshold, the input is classified as a PSK signal. Figures 19.15, 19.16, 19.17, 19.18, 19.19, and 19.20 show the plot of the output of the continuous wavelet transform (using Haar wavelet) of the signal, both with amplitude normalization and without amplitude normalization.

Optimum threshold selection is based on the fact that the wavelet transform is a linear transform; hence, the Haar wavelet transform of a white Gaussian noise is normally distributed with the same mean and variance as the white Gaussian noise itself. Therefore, the probability density function (pdf) of r (a random variable in |HWT|) is as shown in Eq. (19.21).

$$f_R(r) = \frac{r}{\sigma_n^2} \exp\left[-\frac{r^2 + A^2}{2\sigma_n^2}\right] I_0\left(\frac{Ar}{\sigma_n^2}\right) \tag{19.21}$$

where A is the true |HWT| without noise. $I_0(.)$ is the modified Bessel function of the first kind and zero order. When the median filter input is the downsampled PSK wavelet-transformed (WT) magnitude, a Gaussian distribution, whereby the mean is equal to the theoretical median, can approximate the PDF of the median filter output, and the variance shown in Eq. (19.22) equals

$$\sigma_v^2 = \frac{1}{4\left\{L_v - 1 + \left[4f_R^2(\bar{v})\sigma_n^2\right]^{-1}\right\}f_R^2(\bar{v})} \tag{19.22}$$

where $f_R^2(\bar{v})$ is the pdf value evaluated at $r = v$. L_v is the median filter length.

The percentage of correct identification for QAM, PSK, and FSK signals at different carrier-to-noise ratio (CNR) is computed and shown in Table 19.2.

The receiver operating characteristic (ROC) curves are plotted as well. It shows a plot of the probability of detection as a function of probability of false alarm. The

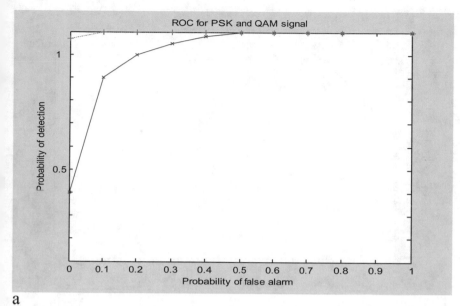

a

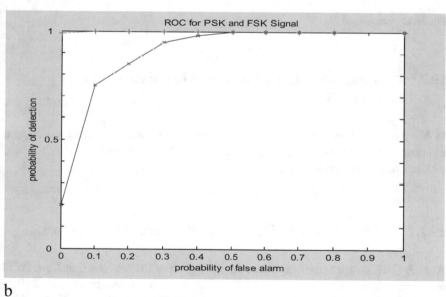

b

Fig. 19.21 (**a**): ROC for PSK and QAM signal. (**b**): ROC for PSK and FSK signal. (**c**) ROC for QAM and FSK signal

plot is generated for carrier-to-noise ratios (CNR) of 5dB and 2dB. Figure 19.21 a, b, and c shows the curves.

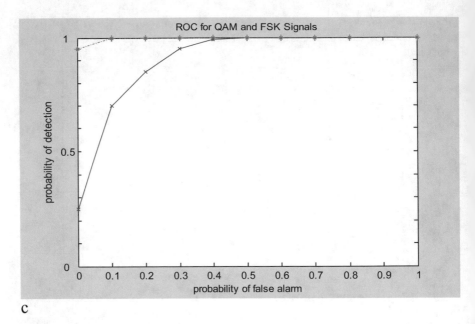

c

Fig. 19.21 (continued)

19.7   PSK and FSK Intraclass Detection, Estimation, and Identification

In this section, we discuss the PSK and FSK intraclass detection, identification, and estimation. This discussion explores the M-ary PSK and M-ary FSK identification, estimation, and detection processes.

19.7.1   M-ary PSK Identification, Estimation, and Detection

At high CNR (carrier-to-noise ratio), the histogram of the wavelet transform (WT) magnitude peaks can easily be used for identification. However, at low CNR, a maximum likelihood approach is needed for accurate identification. With a high CNR, the peaks resulting from phase changes in the WT magnitude can easily be distinguished from noise. The input is declared as M-ary if there exist from M/2 to M-1 Gaussians in the histogram of the peaks. However, for low CNR, it is difficult to differentiate the peaks due to phase changes from within the WT magnitude. The symbol period and synchronization time have to be estimated. Assuming a Rician fading channel model, the received signal can be represented as shown in Eq. (19.23).

$$r(t) = (x + A) \cos \omega_0 t - y \sin \omega_0 t \qquad (19.23)$$

where x and y are Gaussian distributed terms, with $\sigma^2{}_x = \sigma^2{}_y = \sigma^2$.
We can represent

$$x' = x + A \qquad (19.24)$$

Hence, the pdf of x' can be represented as shown in Eq. (19.25).

$$f(x') = \frac{\ell^{-(x'-A)^2/2\sigma^2}}{\sqrt{2\pi\sigma^2}} \qquad (19.25)$$

The received signal envelope is expressed as shown in Eq. (19.26).

$$r^2 = x'^2 + y^2 = (x + A)^2 + y^2 \qquad (19.26)$$

while its phase can be represented by

$$\theta = \tan^{-1}\frac{y}{x'} = \tan^{-1}\frac{y}{(x + A)} \qquad (19.27)$$

Using transformation and the expression for the pdf of a pair of random variables, the pdf of x' and y independent random variables transformed unto r and θ is as shown in Eq. (19.28).

$$
\begin{aligned}
f(r,\theta)dr, d\theta = f(x',y)dx'dy &= \frac{\ell^{-(x'-A)^2/2\sigma^2}}{\sqrt{2\pi\sigma^2}}dx\frac{\ell^{-y^2/2\sigma^2}}{\sqrt{2\pi\sigma^2}}dy \\
&= \frac{\ell^{-\left[(x'-A)^2+y^2\right]/2\sigma^2}}{2\pi\sigma^2}dxdy \qquad (19.28)\\
&= \frac{\ell^{-A^2/2}r\ell^{-\left(r^2-2rA\cos\theta\right)/2\sigma^2}}{2\pi\sigma^2}drd\theta
\end{aligned}
$$

Integrating Eq. (19.28) over all values of θ, we have

$$f(r) = \frac{\ell^{-A^2/2\sigma^2}r\ell^{-r^2/2\sigma^2}}{2\pi\sigma^2}\int\limits_{0}^{2\pi}\ell^{(rA\cos\theta)/\sigma^2}d\theta \qquad (19.29)$$

Using the Bessel function,

$$I_0(z) = \frac{1}{2\pi} \int\limits_0^{2\pi} \ell^{z \cos \theta} d\theta. \tag{19.30}$$

Equation (19.29) can be rearranged with ($z = rA/\sigma^2$), and thus the probability density function (pdf) of r (a random variable in |HWT|) can be given as shown in Eq. (19.31).

$$f_R(r) = \frac{r}{\sigma_\varepsilon^2} \exp\left[-\frac{r^2 + A^2}{2\sigma_\varepsilon^2}\right] I_0\left(\frac{Ar}{\sigma_\varepsilon^2}\right) \tag{19.31}$$

where A is the true |HWT| without noise and σ_ε^2 is the Gaussian noise power. $I_0(.)$ is the modified Bessel function of the first kind and zero order. With $p(i)$ as r, and p_m as A, from the above equation, the wavelet transform (WT) magnitude at phase change has a PDF equal to the sum of M Rician functions. Therefore, the pdf of the WT of a PSK signal can be written as shown in Eq. (19.32).

$$f_p(p(i)|M - aryPSK) = \frac{1}{M} \sum_{m=1}^{M} \frac{p(i)}{\sigma_\varepsilon^2} \exp\left\{-\frac{p^2(i) + \bar{p}_m^2}{2\sigma_\varepsilon^2}\right\} I_0\left(\frac{\bar{p}_m p_i}{\sigma_\varepsilon^2}\right) \tag{19.32}$$

where i represents the symbol period count, $p(i)$ is the random variable, and $\bar{p}_m s$ (the mean or average) can be estimated according to Eq. (19.33).

$$\bar{p}_m = \left|d\,(a) \cos\left(\alpha_m/2\right) - c\,(a) \sin\left(\alpha_m/2\right)\right| \tag{19.33}$$
$$\alpha_m = (m - 1)2\pi/M \quad 1 \le m \le M$$

where

$$d\,(a) = \left\{\frac{1}{N_a - L} \sum_{i=0}^{L-1} \sum_{n=\lceil iT/a \rceil}^{\lceil (i+1)T/a \rceil - 2} \left|WT_p\left(a, iT + a\left(\frac{1}{2} + n\right)\right)\right|^2 - 2\sigma_\varepsilon^2\right\}^{1/2} \tag{19.34}$$

and

$$\hat{c}(a) = \left\{\frac{1}{N_a - L} \sum_{i=0}^{L-1} \sum_{n=\lceil iT/a \rceil}^{\lceil (i+1)T/a \rceil - 2} \left|AVG_p\left(a, iT + a\left(\frac{1}{2} + n\right)\right)\right|^2 - 2\sigma_\varepsilon^2\right\}^{1/2} \tag{19.35}$$

$$AVG_p(a, n) \equiv \frac{1}{\sqrt{a}} \sum_{i=-a/2}^{a/2} S(n + i)$$

where $N_a = \lfloor N/a \rfloor$ and L is the number of symbols. The identification involves computing the likelihood value for all possible M.

$$\Lambda_M = \prod_{i=0}^{L-1} f_p(p(i)|M - ary \quad PSK) \qquad (19.36)$$

The M that yields the largest Λ_M identifies the modulationlevel of M-ary PSK.

19.7.2   M-ary FSK Identification, Estimation, and Detection

Since the WT magnitude of M-ary FSK is an M-step function, identification can be achieved by determining the number of DC levels in the WT magnitude. In a situation whereby the CNR is high, a histogram of the WT magnitude can be used to classify the M-ary FSK. The input is M-ary if there exist from $M/2$ to M peaks in the histogram. When the CNR is low, the noise affects the peaks in the WT magnitude histogram. Since the DC level is constant within a symbol time, averaging the magnitude within a symbol time will improve the result. However, averaging requires an estimation of the symbol period and synchronization time in order to know the duration to average.

Example 19.2
Let us define a random variable $q(i)$ as shown in Eq. (19.37).

$$q(i) = \frac{1}{\lceil (i+1)T/a \rceil - \lceil iT/a \rceil - 1} \sum_{n=\lceil iT/a \rceil}^{\lceil (i+1)T/a \rceil - 2} \left| WT_F \left(a, iT + a \left(\frac{1}{2} + n\right)\right) \right| \quad (19.37)$$

where i is the symbol period count. From central limit theorem and given the mean, $q(i)$ is approximately Gaussian when the symbol period T is large. Therefore, the pdf of $q(i)$ is

$$f_F(q(i)) = \frac{1}{M} \sum_{m=1}^{M} \frac{1}{\sqrt{2\pi\sigma_\varepsilon'^2}} \exp\left\{ -\frac{1}{2\sigma_\varepsilon'^2} (q(i) - \bar{q}_m)^2 \right\} \qquad (19.38)$$

where $\sigma_\varepsilon'^2$ is the corresponding noise power. Decisions are made by observing the number of Gaussians in the histogram of $q(i)$; this is due to the fact that the mean of $\bar{q}_m$ is dependent on an unknown modulation frequency; hence, it is not possible to compute the likelihood value for Eq. (19.38).

Table 19.3 Percentage of correct identification for M-ary PSK

In/out	BPSK	QPSK	8-PSK
BPSK	100	0	0
QPSK	0	90.4	9.6
8-PSK	0	9.4	90.6

Table 19.4 Percentage of correct identification for M-ary FSK

In/out	BFSK	QFSK	8-FSK
BFSK	100	0	0
QFSK	0	98.6	1.4
8-FSK	0	9.1	90.9

Solution

The input is M-ary if there exist $M/2 + 1$ to M peaks in the histogram. The percentages of correct identifications of M-ary PSK and FSK at CNR $= 15$ dB are shown in Tables 19.3 and 19.4.

19.8   Overview of the GUI-Based Simulation Interface

In this section, we discuss the simulation-based GUI that we developed for the demonstration and testing of the algorithm. The MATLAB program written for the detection, identification, discrimination, and estimation of signals is in Appendix M. It should be noted that the MATLAB program codes might require some editing and modifications as you adapt the codes for your own work. The reason is that the MATLAB version used in developing the codes may be different from what is currently in existence now. Figure 19.22 shows the GUI-based interface. It shows a sample simulation test run for the detection of a QAM signal. The operation of the interface is as shown in Example 19.3. The signals included for detection are chirp, FSK, PSK, and QAM.

Example 19.3: Demonstration

(a) Select the type of signal to simulate from the drop-down menu "Select Signal Type"; there are choices of chirp, PSK, FSK, and QAM.
(b) Select M, the level, i.e., whether BPSK, QPSK, or 8-PSK, from the drop-down menu.
(c) Choose the noise—additive white Gaussian noise (AWGN)—level in dB to be added to the signal.
(d) Clicking on the generate button, this generates the signal with noise added. A sample signal plot with constellation plot is shown in the figure section.
(e) Click on the detect button; this is used to detect the signal of interest.

This shows that it is possible to develop a system that is capable of detecting a linear chirp, FSK, PSK, or QAM signal in the presence of additive white Gaussian noise, without any prior information about the signal using wavelets and wavelet

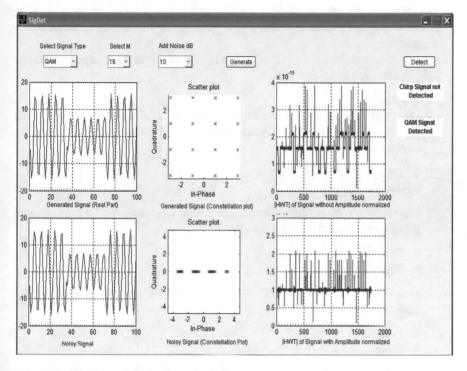

Fig. 19.22 Signal detection simulation GUI interface

transform. It is anticipated that the following can possibly be implemented using similar techniques.

- Implementing optimized solution of the system and procedures using other types of wavelet functions
- Estimation of the various parameters of the signals such as amplitude, frequency, and phase in order to demodulate the signal
- Exploring the possibility of detection in the presence of various types of interferences other than Gaussian noise

Summary
1. The identification of the modulation type of a communication signal has many areas of application such as electronic warfare, surveillance, interference detection, signal demodulation, and threat analysis.
2. Wavelet transform can be used to extract the transient characteristics in a digital modulation signal, and it can be employed using the distinct patterns in the wavelet transform domain in identifying signals.
3. Wavelets are adjustable and adaptable.
4. Chirp signals do occur in lots of signal processing tasks among which are acoustics, communications, radar, and sonar.

5. The ability of the Morlet wavelet function to localize signals effectively in both time and frequency domains allows us to map the signal into the time-frequency domain.
6. In a situation whereby the CNR is high, a histogram of the WT magnitude can be used to classify the M-ary FSK.

Review Questions

19.1 Electronic warfare is not one of the applications of the identification of the modulation type of a communication signal using wavelet transform techniques.

   a. True
   b. False

19.2 Identification, discrimination, and estimation of signals can be implemented using wavelets and wavelet transforms.

   a. True
   b. False

19.3 Fast Fourier transform (FFT) and windowed FFT are usually used for the analysis of signals.

   a. True
   b. False

19.4 Wavelet transform cannot be used to extract the transient characteristics in a digital modulation signal.

   a. True
   b. False

19.5 Wavelet transform can be employed using the distinct patterns in the wavelet transform domain in identifying signals.

   a. True
   b. False

19.6 Wavelets are adjustable and not adaptable.

   a. True
   b. False

19.7 Chirp signals do occur in lots of signal processing tasks among which are acoustics, communications, radar, and sonar.

   a. True
   b. False

19.8 In a PSK signal, the phase angle of the signal does not vary as a function of the digital modulating signal.

a. True
b. False

19.9 In a situation whereby the CNR is high, a histogram of the WT magnitude cannot be used to classify the M-ary FSK.

a. True
b. False

19.10 The ability of the Morlet wavelet function to localize signals effectively in both time and frequency domains allows us to map the signal into the time-frequency domain.

a. True
b. False

Answers: 19.1b, 19.2a, 19.3a, 19.4b, 19.5a, 19.6b, 19.7a, 19.8b, 19.9b, 19.10a

Problems

19.1 What are some of the areas of application of the identification of the modulation types of a communication system signals using wavelets and wavelet transform?

19.2 Describe two of the wavelets that can be used in the detection, identification, discrimination, and estimation of signals.

19.3 Briefly describe, using the overall system model shown as Fig. 19.8 in the textbook, how signal can be detected using the system.

19.4 In the M-ary FSK, how can the identification process be implemented?

19.5 In the M-ary PSK, how can the identification process be implemented?

19.6 Using the developed GUI of Fig. 19.22 in the textbook, state briefly the operation of the interface steps for the detection of chirp, FSK, PSK, and QAM signals.

19.7 Is it possible to detect a linear chirp, FSK, PSK, or QAM signals in the presence of additive white Gaussian noise, without any prior information about the signal?

19.8 Based on the answer to Problem 19.7, what other similar areas can be anticipated to be implemented using similar techniques?

References

Mallat, S. (1999). *A wavelet tour of signal processing*. Academic press.
Hernandez, G., Mendoza, M., Reusch, B., Salinas, L., et al. (2004). Shiftability and filter bank design using Morlet wavelet. *24th International Conference of the Chilean Computer Science Society*, 141–148.

Yasotharan, A., & Thayaparan, T. (2006). Optimum time frequency distribution for detecting a discrete-time chirp signal in noise. *Vision, Image and Signal Processing, IEEE Proceedings, 153*(2), 132–140.

Ristic, B., and Boashash, B, (1993). "Kernel design for time-frequency signal analysis, using the radon transform", IEEE Transmission Signal Process, Vol. 41, No. 5, pp. 1996-2008, 1993.

Jin, J. D., Kwak, Y. J., Lee, K. W., Lee, K. H., Ko, S. J., et al. (2004). Modulation type classification method using wavelet transform for adaptive demodulator. In *Proceedings of 2004 international symposium on intelligent signal processing and communication systems*.

Hong, L., & Ho, K. C. (1999). Identification of digital modulation types using the wavelet transform. *Military Communications Conference Proceedings, 1*, 427–431.

Chapter 20
Wavelet-Based Identification, Discrimination, Detection, and Parameter Estimation of Radar Signals

> *Always be a first-rate version of yourself, instead of a second-rate version of somebody else.*
>
> Judy Garland

20.1 Introduction

The growing density of communication signals that are used by the military and commercial applications, as well as the numerous terrorist attacks recorded globally, has necessitated a high demand in detecting, discriminating, and parameter estimating these communication signals in the presence of noise or some interfering waveform (Prokopiw et al., 2000). In this chapter, we discuss the development of wavelet-based algorithms for the detection, discrimination, and parameter estimation of radar signals. We use a testbed called the Advanced Microwave Receiver Technology Development System (AMRTDS) graphical user interface (GUI) developed by Northup-Grumman in conjunction with Los Alamos National Laboratory (LANL) that we modified with a wavelet-based algorithm. The AMRTDS started as a Miniaturized Satellite Test Reporting System (MSTRS) before the name changed to AMRTDS. The wavelet-based technique added to the testbed enhanced it from being an FFT-based to a wavelet-based testbed (Akujuobi et al., 2002; Akujuobi & Lian, 2001).

Fast Fourier transform is a type of time-frequency analysis, which analyzes the spectral content of a signal as time elapses (Polchlopek & Noonan, 1997). However, most Fourier transform techniques are incapable of handling transient and wideband signals (Noonan et al., 1993; Ovanesova & Saurez, 1999; Scholl et al., 1999). In addition, we explored the use of Neyman-Pearson-based theory to detect and discriminate radar-based signals as well as Bayes' theorem.

In this chapter, part of the objective is to consider the pulse modulated radar signals (i.e., sine wave) and carrier-based communication signals such as the Space Ground Link Subsystem (SGLS). For the SGLS signal, we measure the following parameters:

© Springer Nature Switzerland AG 2022
C. M. Akujuobi, *Wavelets and Wavelet Transform Systems and Their Applications*,
https://doi.org/10.1007/978-3-030-87528-2_20

 (i) Frequency
 (ii) Channel-to-channel phase difference
(iii) Estimated channel-to-channel phase measurement error
(iv) Reference channel amplitude
 (v) Baseline channel amplitude

The parameters for a pulse-modulated signal are the same as above but also included the following parameters that are specific to a pulse signal:

 (i) Pulse repetition interval
 (ii) Pulse width
(iii) FM modulation of signal

Along with the estimation of signal parameters, we wanted to test another method of signal detection using Bayes' theorem. Bayes' detection method must show results of a receiver operating characteristic (ROC) curve. The ROC curve shows the detection error of Bayes' algorithm over a range of detection thresholds.

20.2   AMRTDS

The Advanced Microwave Receiver Technology Development System (AMRTDS) is designed for a small satellite program to serve as a reporting system in the case of radar signal uplinks being jammed with additive white Gaussian noise (AWGN) or some interfering waveform. The Los Alamos National Laboratories developed the graphical user interface, which is the AMRTDS testbed interface shown in Fig. 20.1, allowing for real-time simulation of their primary information signals. This section discusses an overview of the AMRTDS interface and the various types of signals that the graphical user interface (GUI) generates. The AMRTDS GUI is used to create the radar-based signals that is analyzed using wavelets (Akujuobi et al., 2002; Akujuobi & Lian, 2001).

We use as the primary waveform a SGLS radar signal. This Space Ground Link Subsystem (SGLS) operates at a frequency range between 1750 and 1850 MHz and has a 1, 2, or 10 Kb per second transmission rate. The SGLS signal's physical representation is similar to that of a cosine waveform. The uplink signal carrier is phase modulated by a frequency-shift key signal. The AMRTDS GUI gives the user a selection of three different programming options. The first option additive white Gaussian noise (AWGN) generates random additive white Gaussian noise. Gaussian white noise is random in nature and is added to the information signal waveform. AWGN is chosen because noise is a given parameter in signal communication systems and is one of the easiest to analyze and detect. There are two types of AWGN that can be added to the system, wideband and narrowband noise. Once noise is added, the waveform is referred to as a received signal. Signals transmitted through space are likely to be corrupted by some interfering waveform or signal. For such cases, the AMRTDS interface has the option to add one of the three different

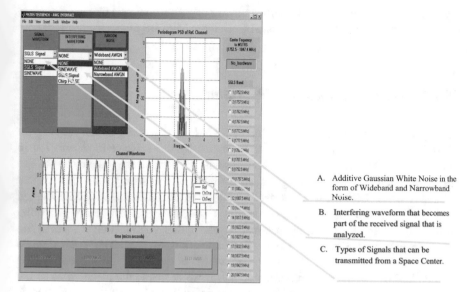

Fig. 20.1 The AMRTD testbed interface with parameters for adding noise or an interfering waveform

types of interfering signals and allows the user to set the frequency for each. The three types of interfering waveforms are sine wave, SGLS, or a chirp. When analyzing data with an interfering chirp signal, existing algorithms did not have the capabilities to overcome the challenges of detection and geo-location of the source of the interfering signal. Another signal that is generated by the AMRTDS interface is a sine wave.

The AMRTDS hardware and software are set up to detect received signals from a space satellite developed by Northrop-Grumman, which contains an earth-based transmission signal, AWGN, and an interfering waveform. The AMRTDS hardware has three receiving antennas; each antenna receives a signal at three different time intervals. These three different time intervals allow engineers to calculate the phase difference between each received transmission, which allows them to geo-locate the origin of the signal transmission. There are currently two prototypes that have been tested since 1997.

20.3   Wavelet-Based Detection of Signals Using Pattern Recognition

Using Daubechie-4 wavelet coefficients discussed in Chap. 2, we create a wavelet transform algorithm that decomposes the received signal X into wavelet coefficients, which are analyzed, to determine if a signal is present. From the graphical representation of these wavelet coefficients, we can discriminate a distinct signature pattern for each signal that is generated from the AMRTDS GUI.

20.3.1   Wavelet Transform Algorithm

The signals that are created in the AMRTDS GUI are first sent through the wavelet transform algorithm. The signal is decomposed by the wavelet transform into a group of different levels. Each level corresponds to a particular frequency band. The energy of each different level gives an estimate of the signal's strength in that frequency band. Level 3 wavelet decomposition is chosen because decomposition levels below 3 are insufficient for detection and all levels above are too time-consuming to analyze. The approximation and detail coefficients at level 3 are shown in Fig. 20.2. From the wavelet decomposition, we can observe the signal's wavelet coefficients. Note that Gaussian white noise has a zero mean and is independent of the signal information signal. If we calculate the mean value of the approximation and detail coefficients, we can determine which of these coefficients contain the location of the information signal as shown in Table 20.1. This coefficient estimate is used to determine the hypothesis testing threshold λ. Different thresholds are used to find a range of detection parameters. If a different level contains mostly noise, then we have a relatively small estimated mean value.

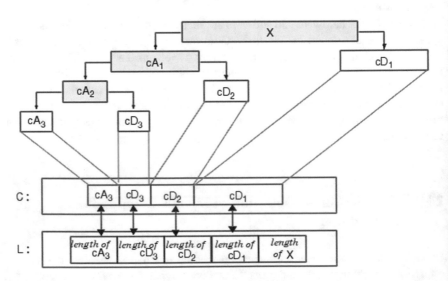

Fig. 20.2 Decomposition level 3 of signal **X** into wavelet approximation and detail coefficients

Table 20.1 Absolute value of the mean of all approximation and detail coefficients for each level

Approx. or Det. coefficients	Absolute value of mean
ca0	1.5386×10^{-6}
ca1	1.9568×10^{-6}
ca2	1.4140×10^{-5}
ca3	4.9150×10^{-5}
cd1	4.3548×10^{-7}
cd2	0.0012
cd3	0.3901

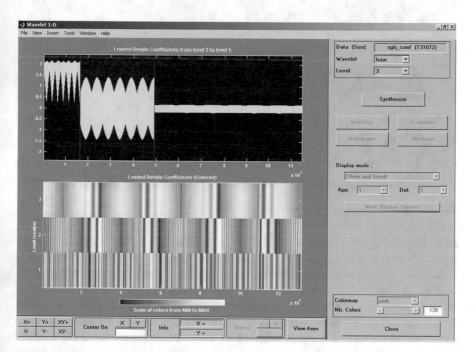

Fig. 20.3 SGLS signal signature pattern

20.3.2   Discrimination of Signals Using Signal Pattern Recognition

We can discriminate different signals from the different types of signals that are generated from the AMRTDS GUI. The received signal is processed with the wavelet algorithm using the Daubechie-4, level 3 decomposition. Then we analyze the wavelet coefficients. From the graphical representation of these coefficients, we can observe that there are distinct patterns for each signal that can be created from the AMRTDS GUI. These distinct patterns are used to identify the type of signals that are transmitted (i.e., SGLS, sine wave, and chirp signals), as shown in Figs. 20.3, 20.4 and 20.5. This type of signal identification technique is only valid if we have a

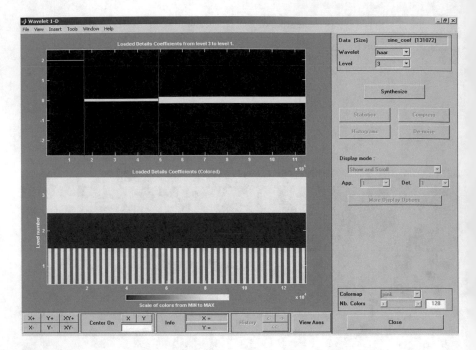

Fig. 20.4 Sine wave signal signature pattern

priori knowledge of the transmitted signal. For unknown signal transmissions, we use a different process to identify the presence of the information signal that could possibly be interfered with the white Gaussian noise or some interfering waveform.

The wavelet coefficients are decomposed into a group of different levels with different frequency bands. We then analyze the coefficients in each different level by estimating the mean value of each level. We know that white Gaussian noise has a mean value of zero and is independent of the information signal. Levels that have a relatively small mean value contain a majority of Gaussian noise. A different level that has the largest mean value yields a large contribution to the information signal. From this information, we can develop a hypothesis testing threshold. A high threshold corresponds to a lower probability of false alarm, and a low threshold represents a high probability of false alarm. This process allows the creation of a detection algorithm based on a range of threshold values. We explore the use of a Bayes' detection algorithm for this purpose, and it is discussed in the next section.

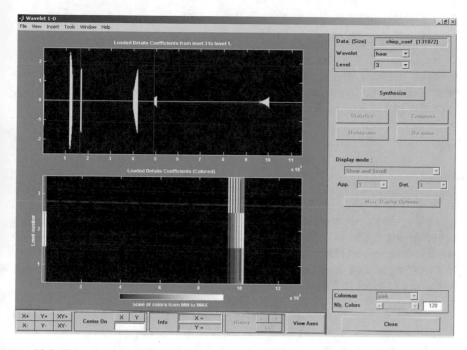

Fig. 20.5 Chirp signal signature pattern

20.4   Wavelet-Based Detection of Signals Using Bayes' Theorem

There are real-world applications that have motivated researchers and this author to explore Bayes' detection method for statistical analysis of geographical coverage, wavelet-based signal analysis, or ecological analysis (Frisch & Messer, 1992; Scholl et al., 1999; Daubechies, 1992). Pilots can test an entire area of interest with long-range radar at the expense of either low probability of detection or many false alarms. Alternatively, a small sub-region can be examined with higher probability of detection and lower false alarm rate (Scholl et al., 1999). Bayesian analysis is useful in some circumstances, which is typical on conservation biology and applied ecology. Ecologists use Bayes' as an indicator for the presence of biological species within a localized area (Daubechies, 1992). Oil companies use various geophysical measurements to try to determine the probability that a potential reservoir has oil using Bayes' theorem (Sinno et al., 1999; Anderson, 1998; Cochran et al., 1999). In this section, we show how Bayes' theorem can be used to determine the probabilities of detecting that a signal is present and is transmitted from a radar ground-based system.

20.4.1   Signal Detection Using Bayes' Theorem and Wavelet Transform

In Bayes' theorem, we use a priori knowledge of the received signal to determine the probabilities. The hypotheses considered are as shown in Eqs. (20.1) and (20.2), respectively.

$$H_0 : \mathbf{X} = \mathbf{N} \tag{20.1}$$

$$H_1 : \mathbf{X} = S + \mathbf{N} \tag{20.2}$$

where S is the known transmitted signal and N is a zero mean white Gaussian noise signal having a known variance σ^2 (Lee et al., 1996). H_0 is the hypothesis that no signal is detected, and H_1 is the hypothesis that there is an information signal present. In Bayes' algorithm, the received signal $\mathbf{X}$ is broken into ten sampled segments to be analyzed as shown in Eq. (20.3).

$$\mathbf{X} = \{x_1 + x_2 + x_3 + \ldots + x_{10}\} \tag{20.3}$$

Take each segment through the wavelet transform for analysis using Daubechies-4 at level 3 decomposition. The wavelet coefficients extracted can be represented as shown in Eq. (20.4).

$$\mathbf{C} = \{c_1 + c_2 + c_3 \ldots + c_{10}\} \tag{20.4}$$

$\mathbf{C}$ is the sum vector of the wavelet coefficients of each coefficient segment. The mean value of levels of approximation and detail coefficients are observed for the presence of a signal. As mentioned in Sect. 20.3, white Gaussian noise has a zero mean value; therefore, for any mean value greater than zero or a determined threshold, we can conclude that there is a signal present. If we graph the wavelet coefficient $\mathbf{C}$, we observe that the majority of the transmitted information signal is contained within the level 3 detail coefficients highlighted in Figs. 20.6 and 20.7, respectively. The level 3 detail coefficients of each wavelet coefficient segment are extracted and observed as shown in Eq. (20.5).

$$\mathbf{C} = \{c_1 + c_2 + c_3 \ldots + c_{10}\}$$
$$\downarrow \tag{20.5}$$
$$\mathbf{CD3} = \{cd3_1 + cd3_2 + cd3_3 \ldots + cd3_{10}\}$$

Since we know that the mean value of Gaussian noise is zero, this information is used to determine the threshold λ value that is used to decide if a transmitted signal is present. A range of threshold values is tested to determine the most efficient

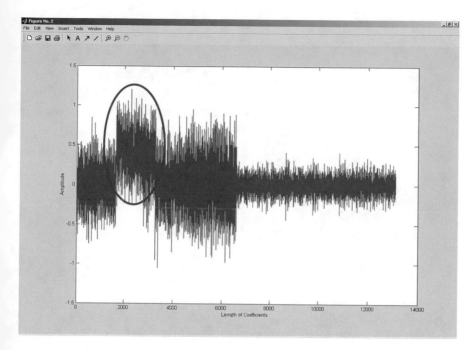

Fig. 20.6 Graph of level 3 detail coefficients of c_1

probability error. Each segment containing level 3 detail coefficients is then analyzed using a threshold range (0.000–0.030), and the results are shown in Table 20.2.

From Table 20.2, we then calculate our probabilities to find the optimal threshold to use for detection. For each threshold value, we create a decision table and estimate the total probabilities at each threshold value (Anderson, 1998). Here are some examples shown in Tables 20.3 and 20.4, respectively.

20.5 Calculation Examples

We use the table results from threshold = (0.018), found in Table 20.4, as an example to calculate the probabilities. The probability of the presence of the NULL signal and of a signal present can be calculated as shown in Eqs. (20.6) and (20.7), respectively. The a priori probability of NULL signal present is as shown in Eq. (20.6).

$$p(H_0) = 20/40 = 0.5 \qquad (20.6)$$

The a priori probability of signal present is as shown in Equation (20.7).

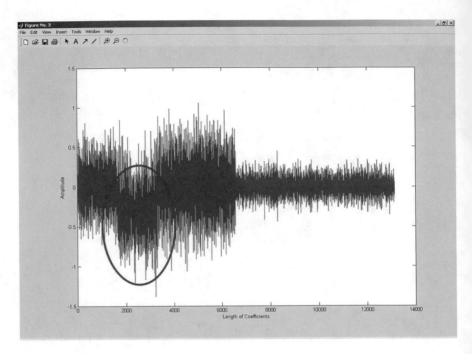

Fig. 20.7 Graph of level 3 detail coefficients of c_2

$$p(H_1) = 20/40 = 0.5 \tag{20.7}$$

Equations (20.8) and (20.9) show the calculations of the probability of detection and probability of false alarm.

$$p(D_1|H_1) = P_D = 12/20 = 0.6 \tag{20.8}$$

$$p(D_1|H_0) = P_{FA} = 1/20 = 0.05 \tag{20.9}$$

From these probabilities, we then compute the missed detection probability and the NULL signal probability as shown in Eqs. (20.10) and (20.11), respectively.

$$P_M = 1 - P_D = 1 - 0.6 = 0.4 \tag{20.10}$$

$$p(D_0|H_0) = 1 - P_{FA} = 1 - 0.05 = .95 \tag{20.11}$$

We can then determine the probability of error, as shown in Eq. (20.12), for the information presented for each threshold value's detection table. The results for the probability of errors for each threshold value are as shown in Table 20.5.

Table 20.2 Results of threshold values (0.009 and 0.018)

NB(1–5MHz)	H0 (no signal present)										
Threshold	0.000	0.003	0.006	0.009	0.012	0.015	0.018	0.021	0.024	0.027	0.030
Decide (H0)	0	0	1	3	6	9	9	9	10	10	10
Decide (H1)	10	10	9	7	4	1	1	1	0	0	0
Total	10	10	10	10	10	10	10	10	10	10	10

WB (0dB)	H0 (no signal present)										
Threshold	0.000	0.003	0.006	0.009	0.012	0.015	0.018	0.021	0.024	0.027	0.030
Decide (H0)	0	7	8	9	9	9	10	10	10	10	10
Decide (H1)	10	3	2	1	1	1	0	0	0	0	0
Total	10	10	10	10	10	10	10	10	10	10	10

SGLS + NB	H1 (signal present)										
Threshold	0.000	0.003	0.006	0.009	0.012	0.015	0.018	0.021	0.024	0.027	0.030
Decide (H0)	0	0	0	1	2	3	4	4	4	4	4
Decide (H1)	10	10	10	9	8	7	6	6	6	6	6
Total	10	10	10	10	10	10	10	10	10	10	10

SGLS + WB	H1 (signal present)										
Threshold	0.000	0.003	0.006	0.009	0.012	0.015	0.018	0.021	0.024	0.027	0.030
Decide (H0)	0	0	1	1	3	3	4	4	4	4	4
Decide (H1)	10	10	9	9	7	7	6	6	6	6	6
Total	10	10	10	10	10	10	10	10	10	10	10

Table 20.3 Calculated probabilities of threshold values

Threshold = 0.009	H_0 No signal	H_1 signal present	Marginal total
Decide H_0 NULL Signal	12	2	14
Decide H_1 Signal Present	8	18	26
Marginal Total	20	20	40

Table 20.4 Calculated probabilities of threshold values

Threshold = 0.018	H_0 No signal	H_1 signal present	Marginal total
Decide H_0 NULL Signal	19	8	27
Decide H_1 Signal Present	1	12	13
Marginal Total	20	20	40

Table 20.5 Results for P_{ERR} for each threshold value

Thresh	P_{FA}	P_D	P_M	P_{NULL}	P_{Err}
0.000	1	1	0	0	0.5
0.003	0.65	1	0	0.35	0.33
0.006	0.55	0.95	0.05	0.45	0.31
0.009	0.4	0.9	0.1	0.6	0.26
0.012	0.25	0.75	0.25	0.75	0.28
0.015	0.1	0.7	0.3	0.9	0.23
0.018	0.05	0.6	0.4	0.95	0.27
0.021	0.05	0.6	0.4	0.95	0.27
0.024	0	0.6	0.4	1	0.24
0.027	0	0.6	0.4	1	0.24
0.030	0	0.6	0.4	1	0.24

$$P_{ERR} = P_M * p(H_1) + P_{FA} * p(H_0)$$
$$= 0.4 * 0.5 + 0.05 * 0.5 \qquad (20.12)$$
$$= 0.225 \text{ or } 22.5\% error$$

20.6   Receiver Operating Characteristic (ROC) Curve

The receiver operating characteristic (ROC) curve is a graphical representation of relationship between maximizing the probability of detection (P_D) and minimizing the probability of false alarm (P_{FA}) based on the testing of the hypotheses H_0 and H_1. The threshold, denoted by λ, specifies the conditions on **X** for which Bayes' detection algorithm declares a signal to be present. This region is also called the critical region (Hero, 1997). The critical region completely specifies the operation of Bayes' detection. The plot of the probability of detection against the probability of false alarm over the range of thresholds - $\infty < \lambda < \infty$ creates the ROC curve, which

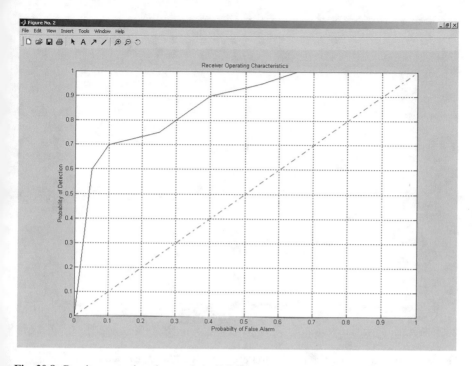

Fig. 20.8 Receiver operating characteristic (ROC) curve

describes the error rate of Bayes' detection as a function of λ shown in Fig. 20.8. Good detection algorithms have ROC curves, which have a high slope of P_D at a low P_{FA}.

20.7   Estimation Parameter and Theory

As can be seen from Sect. 20.4, we presented how Bayes' probability theorem is used to detect an information signal in the presence of noise or an interfering waveform. After detecting the information signal, we must estimate the parameters of the received signal to determine if the information signal received is the signal that was originally transmitted. In this section, we present the techniques used to estimate the parameters of the detected information signal.

20.7.1   Frequency and Power Estimation

One important estimation parameter is the power and frequency spectrum of the detected information signal. The power and frequency spectrum analysis are used to

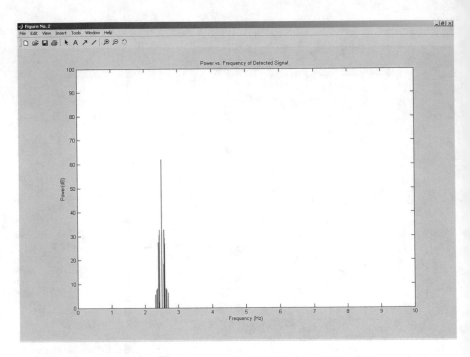

Fig. 20.9 Graph of the frequency and power of a SGLS signal at 2.5Hz, 80dB

determine if the information signal received is the original signal transmitted. Let x represent the received signal with some period T, sampled at intervals $T_s = T/N$. N represents the length of the signal. X is the discrete Fourier transform of received signal x (Brian & Breiner, 1999). The average of x is $X(h)/N$, which is referred to as the DC component of the signal. $X(h)$ is sometimes called the **h-th bin** of X. The power of signal x is represented by Eq. (20.13).

$$Power = \frac{\mathbf{X} * conjugate(\mathbf{X})}{\mathbf{N}} \qquad (20.13)$$

This array of elements is called the power spectrum of x. The square of the signal or a multiple of it represents its power. The sample frequency is defined as N/T Hz or $1/T_s$. The frequency of signal x is calculated by Eq. (20.14).

$$Freq. = \frac{(\#\mathbf{bins} - 1)}{\mathbf{N}} f_s \qquad (20.14)$$

The sampled frequency is represented by f_s. The plot of the frequency of signal x versus the power of the signal x is known as the power spectrum, shown in Figs. 20.9 and 20.10. The sample MATLAB-based codes for the frequency and power estimation algorithm are as listed in Appendix N, Number A.

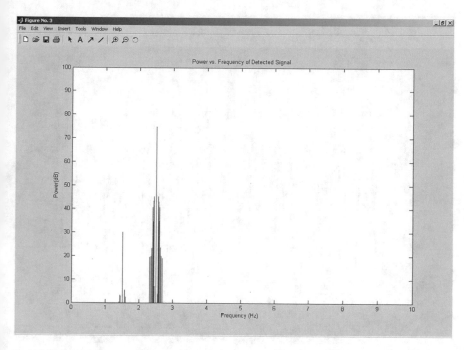

Fig. 20.10 Graph of the frequency and power of a SGLS signal at 2.5Hz, 75dB, and an interfering SGLS waveform at 1.5Hz, 30db

20.8   Frequency Modulation

We can obtain the frequency modulation (FM) of the detected signal using the carrier frequency found in Sect. 20.7 using the fast Fourier transform algorithm. In Fig. 20.11, we compare the modulated signal to the received transmitted signal to display the difference in amplitude. In Fig. 20.12, we plot the original transmitted signal with the modulated signal to compare the signal's frequencies. The sample MATLAB codes are as listed in Appendix N, Number B. From the results, frequency modulating has filtered most the Gaussian white noise, but there is a peak-to-peak offset between the transmitted signal and the modulated signal.

20.9   Channel-to-Channel Phase Estimation

Each signal created by the AMRTDS GUI contains three channels, a reference channel, channel 1, and channel 2. Three channels represent each antenna that transmits the information signal in order to geo-locate the origin of the signal. In order to geo-locate the signal, we must find the phase difference between the reference channel and the other two channels. Appendix N, Number C, gives the

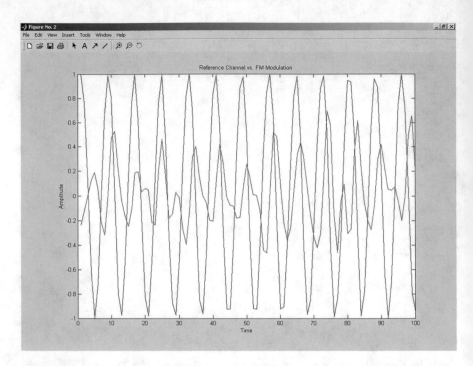

Fig. 20.11 Received signal with noise (red) vs. a modulation of the received signal (blue)

MATLAB program codes that can be adapted for use. We found that it is very inaccurate to calculate the channel-to-channel phase estimation using peak-to-peak phase calculations. In order to calculate the phase difference, we must use center-to-center phase estimation as shown in Fig. 20.13. The center-to-center phase calculations for each channel are as shown in Eqs. (20.15, 20.16, 20.17, 20.18, 20.19, 20.20, 20.21, and 20.22). The channel 1 phase estimation is as shown in Eqs. 20.15 and 20.16, respectively. The phase 1 error estimation calculations are as shown in Eqs. 20.17 and 20.18, respectively. The channel 2 phase estimation and the phase 2 error estimation are as shown in Equations (20.19, 20.20, 20.21, and 20.22), respectively.

Channel 1 Phase Estimation

$$\Phi_{I=[(t_r - t_1)/(\text{Period})]} * 360^{\cdot\cdot} \tag{20.15}$$

$$\Phi_{I=}(8.8 - 8.6)/(12.6 - 5.0)] * 360^{\cdot\cdot} = 9.47^{\cdot\cdot} \tag{20.16}$$

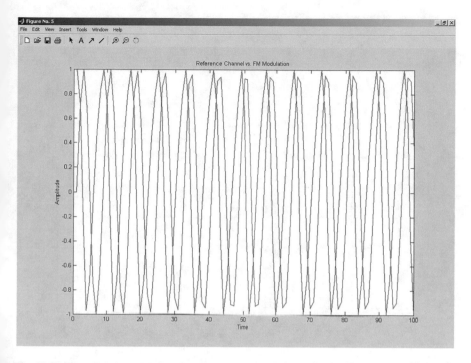

Fig. 20.12 A modulation of the received signal with noise (red) vs. SGLS signal with no noise or interfering waveform (blue)

Phase 1 Error

$$\Phi_1 \text{err} = [(\text{actual phase} - \Phi_1)/\text{actual phase}] * 100 \qquad (20.17)$$

$$\Phi_{1\text{err}} = [(10 - 9.47)/10] * 100 = 5.3\% \text{err} \qquad (20.18)$$

Channel 2 Phase Estimation

$$\Phi_2 = [(t_r - t_2)/\text{Period}] * 360^{''} \qquad (20.19)$$

$$\Phi_2 = [(8.8 - 8.4)/(12.6 - 5.0)] * 360^{''} = 18.95^{''} \qquad (20.20)$$

Phase 2 Error

$$\Phi_1 \text{err} = [(\text{actual phase} - \Phi_1)/\text{actual phase}] * 100 \qquad (20.21)$$

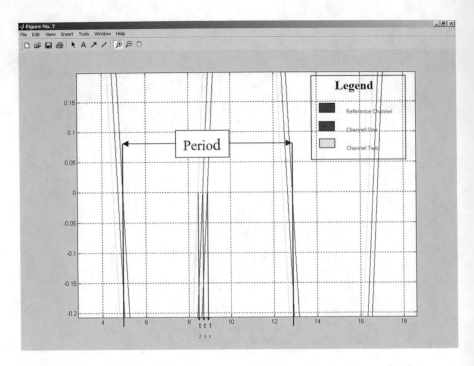

Fig. 20.13 Channel-to-channel phase measurement using zero-crossing phase estimation

$$\Phi_1 \text{err} = [(20 - 18.95)/20] * 100 = 5.25\%\text{err} \qquad (20.22)$$

20.10   Important Issues to Note

Each information signal that is detected operates at a carrier frequency of 2.5 MHz and about 78dB. The frequency of any interfering waveform may operate very close to the carrier frequency and uses lower power than the information signal. Once the carrier frequency is found using the FFT algorithm, you can perform a frequency modulation of the received signal. The channel-to-channel phase calculation of the received signal could not be estimated by the typical peak-to-peak method due to the inaccurate plotting of MATLAB. Instead, you can use a center-to-center method of measurement to calculate the phase difference. The phase difference between channel 1 and the reference channel is about a 5.3% error, 9.47 degrees, and the difference between channel 2 and the reference channel is a 5.25% error, at about 18.95 degrees.

Summary

1. The Advanced Microwave Receiver Technology Development System (AMRTDS) graphical user interface (GUI) developed by Northup-Grumman in conjunction with Los Alamos National Laboratory (LANL). We modified with a wavelet-based algorithm.
2. The AMRTDS started as a Miniaturized Satellite Test Reporting System (MSTRS) before the name changed to AMRTDS.
3. The wavelet-based technique added to the testbed enhanced it from being an FFT-based to a wavelet-based testbed.
4. The Space Ground Link Subsystem (SGLS) operates at a frequency range between 1750 and 1850 MHz and has a 1, 2, or 10 Kb per second transmission rate.
5. The AMRTDS hardware and software are set up to detect received signals from a space satellite developed by Northrop-Grumman, which contains an earth-based transmission signal, AWGN, and an interfering waveform.
6. The wavelet coefficients are decomposed into a group of different levels with different frequency bands. We then analyze the coefficients in each different level by estimating the mean value of each level.
7. Levels that have a relatively small mean value contain a majority of Gaussian noise. A different level that has the largest mean value yields a large contribution to the information signal.
8. The receiver operating characteristic (ROC) curve is a graphical representation of relationship between maximizing the probability of detection (PD) and minimizing the probability of false alarm (PFA) based on the testing of the hypotheses H0 and H1.
9. Each information signal that is detected operates at a carrier frequency of 2.5 MHz and about 78dB.
10. The frequency of any interfering waveform may operate very close to the carrier frequency and uses lower power than the information signal.

Review Questions

20.1 The Space Ground Link Subsystem (SGLS) can be regarded as a carrier-based communication signal.

    a. True
    b. False

20.2 Which of these are part of the parameters we measure for the SGLS signal?

    a. Frequency
    b. Channel-to-channel phase difference
    c. Both (a) and (b)
    d. None

20.3 Which parameters are specific to a pulse signal?

    a. Pulse repetition interval
    b. Pulse width

c. FM modulation of signal

d. All of the above

20.4 The SGLS signal's physical representation is not similar to that of a cosine waveform.

a. True

b. False

20.5 The signal is decomposed by the wavelet transform into a group of different levels.

a. True

b. False

20.6 Each level of the decomposition process corresponds to a particular frequency band.

a. True

b. False

20.7 White Gaussian noise does not have a mean value of zero and is not independent of the information signal.

a. True

b. False

20.8 A high threshold corresponds to a lower probability of false alarm, and a low threshold represents a high probability of false alarm.

a. True

b. False

20.9 In Bayes' theorem, we use a priori knowledge of the received signal to determine the probabilities.

a. True

b. False

20.10 The receiver operating characteristic (ROC) curve is not a graphical representation of relationship between maximizing the probability of detection (PD) and minimizing the probability of false alarm (PFA) based on the testing of the hypotheses H0 and H1.

a. True

b. False

Answers: 20.1a, 20.2c, 20.3d, 20.4b, 20.5a, 20.6a, 20.7b, 20.8a, 20.9a, 20.10b

Problems
20.1 What parameters are measured in the SGLS?

20.2 What are the parameters that can be measure for the pulse signal?

20.3 In Problem 2, which of those parameters are specific to pulse signals?

20.4 What is the desired function of the wavelet transform algorithm?

20.5 Describe how you can discriminate between different types of signals using the idea of pattern recognition.

20.6 How can you identify or discriminate unknown signal transmissions?

20.7 How can Bayes' theorem and wavelet transform be used in the detection of signals?

20.8 If the probability of missed detection (P_M) is 0.5, the probability of signal present ($p(H_1)$) is 0.4, the probability of false alarm is (P_{FA}) is 0.03, and the probability of signal not present ($p(H_0)$) is 0.7, calculate the probability of errors for each threshold.

20.9 What is the receiver operating characteristic (ROC) curve?

20.10 Why are the frequency and power estimations important in a communication system?

20.11 How can the frequency modulation of a signal of interest be detected?

20.12 Why is channel-to-channel phase estimation important in any communication system?

References

Akujuobi, C. M., Lian, J., Davis, J., & Rogers, B. et al. (2002). Wavelet-based algorithm development research into detection, discrimination and parameter estimation of signals for AMRTD system at Prairie View A&M University, *CECSTR Report, 2002*.

Akujuobi, C. M., & Lian, J. (2001). Wavelet-based algorithm development research into the detection, discrimination and parameter estimation of signals for MSTRS system at Prairie View A&M University: A literature review, *CECSTR Paper, 2001*.

Anderson, J. L. (1998). Embracing uncertainty: The interface of Bayesian statistics and cognitive psychology. *Conservation Ecology, 2*(1), 1–29.

Brian, A., & Breiner, M. (1999). *MATLAB 5 for engineers* (2nd ed., pp. 582–598). Addison-Wesley Pub Co.

Cochran, D., Sinno, D., & Cluasen, A. (1999). Source detection and localization using a multi-mode detector: A Bayesian approach. In *Proceedings of IEEE international conference on acoustics, speech, and signal processing*.

Daubechies, I. (1992). *Ten lectures on wavelets*. SIAM.

Frisch, M., & Messer, H. (1992). The use of the wavelet transform in the detection of an unknown transient signal. *IEEE Transactions on Information Theory, 38*(2).

Hero, A. (1997). Signal detection and classification. In *Chapter in CRC the digital signal processing handbook* (Vol. 13, pp. 1–7). CRC Press/IEEE Press.

Lee, N., Huynh, Q., & Schwartz, S. (1996). New methods of linear time-frequency analysis for signal detection. Proceedings of IEEE-SP international symposium, pp. 13–16.

Noonan, J. P., Polchlopek, H. M., & Varteresian, M. (1993). A hypothesis testing technique for the wavelet transform in the presence of noise. *Digital Signal Processing, 3*, 89–96.

Ovanesova, A. V., & Saurez, L. E. (1999). *Wavelet application to structural dynamics*. CRC Press.

Polchlopek, H. M., & Noonan, J. P. (1997). Wavelets, detection, estimation, and sparsity. *Digital Signal Processing, 7,* 28–36.

Prokopiw, W., Ho, K. C., & Chan, Y. T. (2000). Modulation identification of digital signals by the wavelet transform. *IEE Proceedings-Radar, Sonar and Navigation, 147*(4), 169–176.

Scholl, J. F., Agre, J. R., Clare, L. P., Gill, M. C. et al. (1999). A low power impulse signal classifier using the Haar Wavelet Transform. *Proceedings of. SPIE 3577, sensors, C3I, information, and training technologies for law enforcement.* https://doi.org/10.1117/12.336958.

Sinno, D., Cochran, D., & Morrell, D. (1999). A Bayesian risk Approach to multi-mode detection. *Defence Applications of Signal Processing Proceedings,* 187–192.

Chapter 21
Application of Wavelets to Vibration Detection in an Aeroelastic System

What would life be if we had no courage to attempt anything?
— Vincent Van Gogh

21.1   Introduction

In this chapter, we discuss the application of wavelets to vibration detection in aeroelastic systems. We know that in aeroelastic systems, fatigue and breakdown of aircraft structure are common. Thus, we made efforts in discussing how to investigate and improve the monitoring and diagnostics of aircraft systems constantly. These improvements have led to interesting and creative approaches to fault monitoring of aeroelastic systems such as aircrafts. The detection of vibration signals in the wings and fuselage enables the identification of potentially catastrophic faults in airplanes due to aerodynamic forces. This underlies the keen interest in fault detection and identification through vibration analysis.

Since vibration signals contain important information about dynamic mechanical parameters, this non-stationary waveform is useful for fault analysis. For this reason, processing this waveform is crucial. Classically, Fourier analysis is used for signal processing of this kind of information (Lepik, 2001; Luo et al., 2002). However, wavelet analysis seems better suited for detecting vibration signals (Rioul & Vetterli, 1991; Klempnow et al., 2000). Fourier analysis provides information about a signal that is independent of time—the signal's frequency components. Generally, with Fourier analysis, the signal under consideration has to be continuous and stationary. However, vibration is non-stationary. Some examples of these non-stationary signals are the sound pressure from a speaker during playback of voice or music, the occurrences of transient impulses during vibration monitoring, and the vibration generated at engine start-up (Zou et al., 2001). By contrast, wavelet analysis provides information about the scale of the signal under investigation, which is independent of frequency.

Fourier analysis involves representing a signal as the summation of its constituent sine waves at various frequencies. If a signal contains a transient of a finite time interval, its Fourier transform includes the contribution from the transient pulse, but

C. M. Akujuobi, *Wavelets and Wavelet Transform Systems and Their Applications*, https://doi.org/10.1007/978-3-030-87528-2_21

the information about the transient is lost on the time axis. However, only wavelet analysis can adequately handle signals with transients that have both high- and low-frequency components (Zou et al., 2001). Wavelet analysis accomplishes this by dividing a signal into shifted and scaled versions of a fixed function, the original (or mother) wavelet. Thus, wavelet analysis eliminates the restrictions of Fourier techniques by considering a signal locally—i.e., in time and frequency—by "windowing in" only on a small portion of the signal.

Vibration detection is important in aeroelastic systems (Liu et al., 1997; Ansari & Baig, 1998; Butler & Newland, 2000; Lee & White, 2000; Goumas et al., 2001; Giacomin et al., 2002). Because of the non-stationary nature of vibration signals, these waveforms can be used for fault diagnosis. For example, they contain pertinent information about the random, degrading processes occurring in machinery (Ansari & Baig, 1998). Usually, an increase in vibration amplitude is a precursor of mechanical failure; and, once a fault is detected, analysis of the vibration frequencies is used to identify the type of breakdown (Ansari & Baig, 1998). Considering the environment in which they are generated, vibration signals are very noisy. We remove the noise to evaluate the signals correctly. In mechanical processes, analysis of vibration lends itself well to the prediction of aircraft robustness in mechanical processes. We perform feature extraction of vibration data from noise successfully using wavelet analysis, as shown in this chapter. We show in this chapter how we can use wavelet techniques to filter out noise from vibration in order to provide clear indicators of the status of effects of the vibration in an aeroelastic system and for fuselage stability (Brenner and Lind, 1998; Brenner, 1997). Clearly, this is an important role in aeronautics for aircraft stability prediction (Brenner et al., 1997a, 1997b; Brenner & Lind, 1997). Thus, the objective of this chapter is to explore three different wavelets for vibration detection in aeroelastic systems such as in an aircraft.

21.2   Overview of Wavelet Analysis for the Aeroelastic Systems

In this section, we discuss the type of wavelets that may be suitable in the analysis and detection of vibration and the types of wavelet transforms that are used.

21.2.1   The Wavelet Transform

A wavelet is a waveform with a small, finite time interval and an average zero value (Daubechies, 1992). There are several different wavelets. They include Coiflet, Haar, Daubechies, Morlet, Shannon, etc. In wavelet analysis, we divide the signal into shifted and scaled versions of the particular wavelet (the *mother* wavelet) chosen for analysis. The mother wavelet determines the shape of the constituent components of the signal generated during analysis. Scaling and translation produce the generated wavelets of the mother wavelet. For instance, Eq. (21.1) shows the scaling function corresponding to the Daubechies-4 (or db4) wavelet.

$$\phi(t) = c_0\phi(2t) + c_1\phi(2t - 1) + c_2\phi(2t - 2) + c_3\phi(2t - 3), \tag{21.1}$$

where:

$$c_0 = \frac{1 + \sqrt{3}}{4\sqrt{2}}, \tag{21.2}$$

$$c_1 = \frac{3 + \sqrt{3}}{4\sqrt{2}}, \tag{21.3}$$

$$c_2 = \frac{3 - \sqrt{3}}{4\sqrt{2}}, \tag{21.4}$$

$$c_3 = \frac{1 - \sqrt{3}}{4\sqrt{2}}. \tag{21.5}$$

The db4 mother wavelet, $\psi(t)$, can be calculated from Eqs. (21.1, 21.2, 21.3, 21.4 and 21.5) by Eq. (21.6).

$$\psi(t) = d_0\phi(2t) + d_1\phi(2t - 1) + d_2\phi(2t - 2) + d_3\phi(2t - 3), \tag{21.6}$$

where:

$$d_0 = -c_3 = -\frac{1 - \sqrt{3}}{4\sqrt{2}},$$

$$d_1 = c_2 = \frac{3 - \sqrt{3}}{4\sqrt{2}},$$

$$d_2 = -c_1 = -\frac{3 + \sqrt{3}}{4\sqrt{2}},$$

$$d_3 = c_0 = \frac{1 + \sqrt{3}}{4\sqrt{2}}.$$

The generated group of wavelets, $\psi_{j,\,k}$, from this example is specified by dilations (scaling) and translations of itself.

Equations (21.1, 21.2, 21.3, 21.4, 21.5 and 21.6) represent a specific example of a pair of functions, $\varphi(t)$ and $\psi(t)$, called the scaling function and wavelet function, respectively. They are also referred to as the *father* and *mother* wavelets in wavelet theory. One of the main reasons that wavelets have so many successful applications is because of their high order of vanishing moments. It is well known that the db4 mother wavelet $\psi(t)$ in Eq. (21.6) has four vanishing moments. In particular, when $M = 2$, the mother wavelet $\psi(t)$ in Eq. (21.8) yields the classical Haar wavelets. Our example in Eq. (21.6) corresponds to $M = 4$ or the db4 mother wavelet.

Also, in the filter design process, $\{c_0, c_1, c_2, c_3\}$ is the low-pass filter, and $\{d_0, d_1, d_2, d_3\}$ is the high-pass filter. The low-pass filter captures the low-frequency components of the signal, while the high-pass filter extracts the high-frequency components. A more general form of the scaling function, for M coefficients c_k, $k = 0, \ldots, M-1$, where M is even, is determined from Eq. (21.7).

$$\phi(t) = \sum_{k=0}^{M-1} c_k \phi(2t - k) \tag{21.7}$$

Correspondingly, the mother wavelet is given as shown in Eq. (21.8).

$$\psi(t) = \sum_{k=0}^{M-1} d_k \phi(2t + k - M + 1), \tag{21.8}$$

where:

$$d_k = (-1)^{k+1} c_{M-1+k}, \quad k = 0, 1, 2, \ldots, M - 1.$$

For all positive even integers M, the scaling functions $\varphi(t)$ in Eq. (21.7) and their corresponding wavelets $\psi(t)$ in Eq. (21.8) are indeed the Daubechies scaling functions and wavelets, respectively. The continuous wavelet transform (CWT), the discrete wavelet transform (DWT), and types of wavelets are discussed in Chap. 2.

In practice, the DWT is achieved by using the fast wavelet transform (FWT). Compared to the fast Fourier transform (FFT), the computations required for N points using the FWT algorithm are reduced from 2NlogN to 2 N, where log is the base-2 logarithm. Mallat developed this code-saving algorithm in 1988. Succinctly, this algorithm is based on the use of quadrature mirror filters (QMF). Decomposition is performed by repeatedly passing the desired signal through low- and high-pass filters and dyadically downsampling the filters' results. Then, reconstruction (the inverse discrete wavelet transform (IDWT)) is achieved by passing the previously iteratively filtered and downsampled signals through several stages of dyadic upsampling and low- and high-pass filters. Finally, the output from the last set of filters is summed to retrieve the original signal. Figure 21.1 shows a multi-level decomposition and reconstruction process.

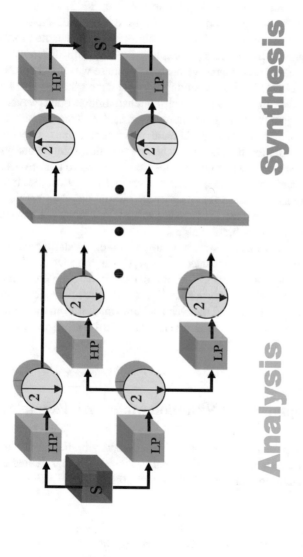

Fig. 21.1 Analysis (decomposition) and synthesis (reconstruction) based on the FWT

21.2.2   The Vibration Signal Analysis Techniques

One of the advantages of wavelet-based analysis over Fourier-based analysis is the use of different waveforms—other than sines and cosines—that "break up" the signal under consideration. Due to the effectiveness of wavelets, it is used in signal analysis, image enhancement/compression, target detection, vibration analysis, and many other areas (Akujuobi, 1998; Akujuobi & Hu, 2002) as exemplified in this textbook. The common approach to wavelet-based vibration analysis is to use a wavelet family as a supplemental filtering technique. In this chapter, multiresolution analysis of three wavelet families is explored along with different thresholding schemes. Additionally, a comparative analysis of these wavelets is presented.

Adaptive techniques for implementing wavelet analysis have also been used to develop a technique that obtains a well-matched orthonormal wavelet basis for representing a signal within a set number of scales. This can be extended to representing a specific signal with M-band wavelet functions. However, both of these methods do not provide a flexible scheme for multiresolution analysis. Instead, these techniques are restricted to decomposition of no more than two scales. However, resolution (and scale) is adapted to the wavelet family without imposing such limits. The wavelets utilized in this chapter are chosen based on ease of implementation, flexibility, and the need for wavelets to match or "look" like the signal under consideration, which is a vibration signal.

The initial step is to establish a working vibration model. Once this model is formulated, the wavelet families used for analysis are introduced. However, before analysis is performed, the level of decomposition is determined to maximize the efficiency and accuracy of the algorithm. Then, the threshold is determined in order to effectively suppress noise and recover vital information from the vibration signal. Finally, the result of this process is a viable wavelet-based algorithm for vibration detection.

21.3   Development of a Vibration Model: An Example

In order to develop an algorithmic model that detects vibration in airplanes, we first develop an accurate signal model. Classically, vibration is represented as a sinusoid and normal distribution white noise (Gaussian noise) as shown in Eq. (21.9).

$$v = s(t) + \sigma n \; , \tag{21.9}$$

where v is the vibration signal that consists of $s(t)$ the signal of interest and σ is the scaling factor of the additive white noise n, in which the signal is lost. We chose a "multiple source" or multi-frequency signal to better approximate the actual vibration environment of an aircraft. The multi-frequency model is as shown in Eq. (21.10).

$$v = \sin(2\pi f_1 t) + \sin(2\pi f_2 t) + \text{noise}, \tag{21.10}$$

where f_1 and f_2 are the frequencies that provide information about the mechanical condition of the plane and *noise* is unscaled, additive white Gaussian noise (AWGN). For this model, the vibration signal is entirely corrupted by noise. This waveform is simulated using the "**mat**rix **lab**oratory" (or MATLAB) programming language as shown in Eq. (21.11).

$$vibsg = \sin(2 * pi * f1 * t) + \sin(2 * pi * f2 * t) + randn, \tag{21.11}$$

where *f1* and *f2* are frequencies of interest and *randn* is AWGN in which the signal is lost.

21.4 Wavelet Families Used for Vibration Detection

There are several wavelets: Symlet, Coiflet, biorthogonal spline, Shannon, complex Gaussian, frequency B-spline, Mexican hat, etc. In this chapter, we discuss the use of the Haar, Daubechies, and Morlet wavelets for vibration analysis.
A wavelet is a waveform of finite duration and zero mean value. It is defined by two functions: a scaling function (φ) and a wavelet function (ψ). First, we present the Haar wavelet. It is the simplest wavelet. Also, it is discontinuous, and it resembles a step function. For the Haar wavelet, the scaling and wavelet basis functions are given in Eq. (21.12). The diagrammatic representations of the Haar, Daubechies, and Morlet wavelets are illustrated in Chap. 2.

$$\begin{aligned}
\psi(x) &= 1 & 0 \leq x \leq 0.5 \\
\psi(x) &= -1 & 0.5 \leq x \leq 1 \\
\psi(x) &= 0 & x = 0.5 \\
\varphi(x) &= 1 & 0 \leq x \leq 1 \\
\varphi(x) &= 0 & x = 0.5
\end{aligned} \tag{21.12}$$

The mathematical representation of the Daubechies wavelet is as shown in Eq. (21.13).

$$\psi(t) = \sqrt{2} \sum_{k=0}^{L-1} g_k \varnothing (2x - k). \tag{21.13}$$

where $\psi(t)$ is the wavelet, g_k is the wavelet coefficient, and $\varnothing$ is the scaling function.
Most Daubechies wavelets (dbN) are not symmetrical. These wavelets are orthogonal and compactly supported. Also, the support length for $\psi(x)$ and $\varphi(x)$ is

2 N − 1. The Morlet wavelet is continuous, and it has no scaling function. The Morlet wavelet function is represented as shown in Eq. (21.14).

$$\psi(x) = \sqrt{\pi f_b} e^{2i\pi f_c x} e^{-\frac{x^2}{f_b}}$$ (21.14)

Wavelets are very useful tools for vibration signal analysis because they can perform local analysis; i.e., they can show both the frequency and time information of the vibration signal.

21.5 Development of the Vibration Detection Algorithm

In this section, we discuss the various stages in the development of the vibration detection algorithm. The initial stage starts at the decomposition level and the thresholding

21.5.1 Initial Consideration: Decomposition Level

Since wavelets can represent signals locally in time and frequency, there are certain parameters to consider in order to ensure proper signal processing. Thus, before analysis is performed using the Haar, Daubechies, and Morlet wavelets, the decomposition level must be determined. The level of decomposition is directly related to the outcome of the analysis. The maximum decomposition level for each wavelet is calculated using a MATLAB function named *wmaxlev*. The input to the function is the matrix size of the desired signal and the wavelet used for analysis. The output of the function is the largest possible number of levels for which, at the last level, at least one coefficient is correct. The maximum level for decomposition is calculated for all the wavelet families mentioned in Sect. 21.4.

In order to verify the results initially, decomposition is performed on a signal at level 10 using the Daubechies-4 (db4) wavelet. The detail coefficient information from this analysis is as shown in Fig. 21.2. Upon inspection, you will find that Fig. 21.2 reveals that the energy at levels 6–10 quickly diminishes. In fact, the energy begins to decline starting at level 5. Therefore, a comparison of the energy of the approximation and detail coefficients at levels 4 and 5 is made to determine at which level decomposition should end. For db4, there is a 65% loss of energy at level 5 compared to level 4. This loss in energy represents a significant decrease in information retained from the analyzed signal as shown in Table 21.1.

Likewise, a decrease of this magnitude in data recovered from the vibration signal would translate into poor results for the algorithm. The process of maximum level decomposition and energy comparison is repeated for all the wavelet families. Table 21.1 shows the energy loss from these comparisons. Consequently, the results

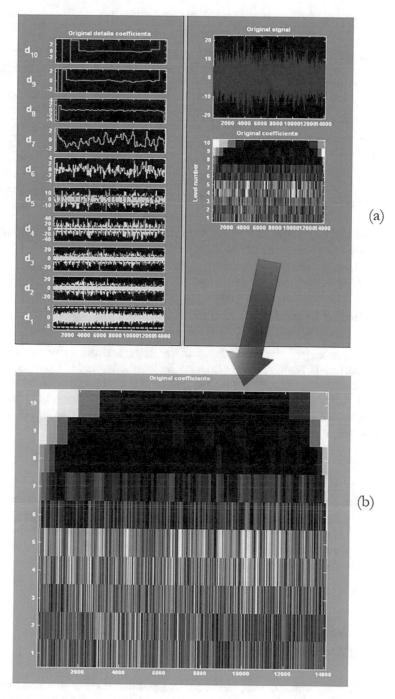

Fig. 21.2 (a) Decomposition up to level 10 for Daubechies-4 wavelet (top). (b) Zoom of the scalogram, which shows the energy of the detail coefficients by level (bottom)

Table 21.1 Maximum decomposition level and associated energy loss

Wavelet	Maximum level	Energy (%)		Energy loss (%)
		Level 4	Level 5	
Haar	13	9.7009173	2.0457016	78.9122870
db4	10	6.0991095	2.1253305	65.1534291
db6	10	4.2305878	1.5464805	63.4452574
db8	9	2.8492018	0.6465958	77.3060734
db10	9	2.5644885	0.3927499	84.6850588
db12	9	2.4456750	0.3477328	85.7817265
db14	9	1.6641354	0.3206463	80.7319571
db16	8	1.2600817	0.6753683	46.4028162
db18	8	1.2552096	0.5008998	60.0943302

indicate that decomposition should be performed up to level 4 for both computational efficiency and accuracy of the algorithm.

21.5.2   Initial Consideration: Threshold

Another requirement of the algorithm that we must address before performing vibration analysis is the threshold (Donoho & Johnstone, 1994; Donoho, 1995). This is a key attribute of the algorithm. The threshold is vital because it enables noise suppression. Noise suppression is accomplished by applying a threshold to the detail coefficients. A "fixed form" threshold is used to denoise the vibration signal, and the threshold is calculated using Eq. (21.15).

$$thr = \sqrt{2 * \log(n)} * \sigma, \qquad (21.15)$$

where n is the length of the signal (based on matrix size); σ is the standard deviation of the noise in the vibration signal; and $\sigma = 1$. Also, the threshold is applied globally, that is, at all levels. The objective of this preparatory process is to evaluate and ascertain key parameters for this application of signal analysis that leads to a detection scheme that provides accurate results.

21.5.3   Vibration Detection Algorithm: An Example
of the Procedure

Finally, after establishing the appropriate decomposition level as discussed in Sect. 21.5.1 and determining the threshold as discussed in Sect. 21.5.2, the complete algorithm for vibration detection process is described as follows:

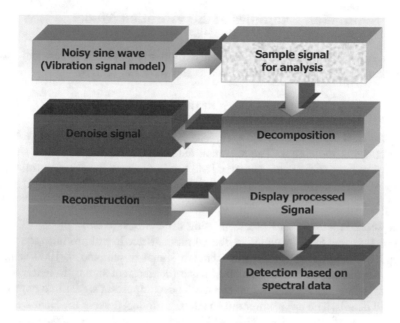

Fig. 21.3 Flowchart of the wavelet-based vibration detection algorithm

- Loading the vibration model. In this case, the signal is sampled above the Nyquist frequency to avoid aliasing.
- A wavelet transform (i.e., the Haar, Daubechies, or Morlet) is applied to the sampled signal at a predetermined level.
- The noise in vibration signal is suppressed by applying the calculated threshold.
- The processed vibration signal is reconstructed from its constituent approximation and details.
- The frequency spectrum of the recovered signal is evaluated to verify detection; specifically, the spectrum of the processed signal is compared to a particular reference signal (the vibration model) to confirm detection of pertinent characteristic frequencies.

As discussed in Sect. 21.6, an increase in vibration amplitude is an indicator of potential mechanical malfunction. Also, the frequency information of the vibration signal provides a method of categorizing the type of breakdown after a fault occurs. Figure 21.3 illustrates the flowchart of the wavelet-based vibration detection algorithm. This algorithm and the results in the example can be extended in a broader sense to real-time health monitoring/diagnostic regime for precise prediction of aircraft robustness.

21.6   Simulation Examples of the Vibration Model

In this section, three wavelet families are explored: Haar, Daubechies, and Morlet. The vibration signal model is expressed as shown in Eq. (21.16).

$$vibsg = \sin\left(2 * pi * f1 * t\right) + \sin\left(2 * pi * f2 * t\right) + randn, , \qquad (21.16)$$

where $t = 0, 0.001, 0.002,\ldots, 2$ s; $f1 = 35$ Hz; $f2 = 75$ Hz; and *randn* is AWGN, which corrupts the signal. The vibration detection algorithm flowchart shown in Fig. 21.3 utilizes these three wavelets, and the procedures used in determining the results of the wavelet analysis are described in detail in Examples 21.1, 21.2 and 21.3.

Example 21.1: Vibration Detection Using the Haar Wavelet

We use first the Haar wavelet in the vibration detection algorithm (see O_1 in Appendix O). In the program, the vibration signal is sampled at 1000 samples/s for 2 s. Simulation is based on a short pulse (or transient signal) to better approximate real vibration signals. Wavelet analysis decomposes a signal into approximations and details. The approximation and details produced using the Haar wavelet at level 4 are as shown in Fig. 21.4, which does not show details D2 and D1. The approximations and details reveal the vibration signal and the noise. In wavelet analysis, the approximations are the output of the low-pass filters, and the details are the output of the high-pass filters. Since noise is a signal of mostly high frequencies, much of the undesired information is contained in the details. The approximations normally contain much of the desired signal. The approximations and details of the Haar wavelet transform effectively separate the signal from the noise. The Haar wavelet is implemented using the discrete wavelet transform (DWT). The DWT provides accurate results without excessive calculations. Computation of the approximations and details is a recursive process.

The reconstructed signal as shown in Fig. 21.5 is accomplished by adding the highest-level approximation and the details. Moreover, the reconstruction step is completed with minute error (i.e., 2.220446×10^{-15}). By applying a threshold to the detail coefficients, the noise is removed from the signal. This global threshold (GT) is calculated using Eq. (21.15). For this vibration model, the threshold is equal to 3.824453003. The denoising signal shown in Fig. 21.6 is accomplished by adding the highest-level approximation and the modified details (i.e., all of the details whose coefficients are thresholded). As expected, the frequency spectra of the noisy vibration signal and the reconstructed vibration signal (before denoising) are identical. The spectra of these signals reveal frequencies at 35 Hz and 75 Hz, respectively, as shown in Fig. 21.7. In Fig. 21.8, the frequency spectrum of the denoised signal reveals part of the content of the vibration signal. Also, it contains extraneous frequencies due to the noise. Noise artifacts appear near the 35 Hz frequency. The processed signal spectrum shows that the 75 Hz frequency is significantly attenuated, which is attributed to the threshold in the denoising step.

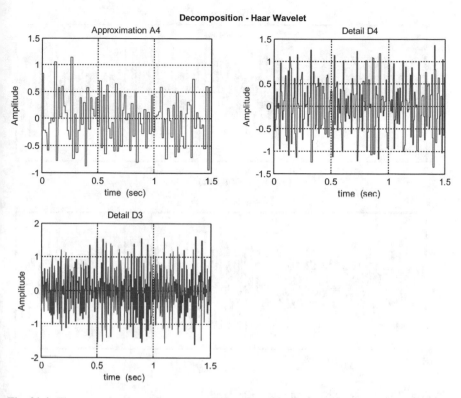

Fig. 21.4 The approximation and details of the signal using the Haar wavelet at level 4

That is, it is probable that the threshold setting allowed an increase in the noise floor. A different way of looking at the detected frequencies is using the scalogram.

A scalogram is a "map" of the frequency and time information of a signal. It is an image—not a graph. Actually, a scalogram is a pictorial representation of the energy at various frequencies versus time. The magnitude of the energy is represented as blue to yellow to red, which is from minimum to maximum. In Fig. 21.9a, the scalogram on top shows the frequencies of interest, that is, the frequencies at 35 Hz and 75 Hz in bold red lines. The scalogram of the denoised signal shows 35 Hz in red along with miscellaneous noise, but 75 Hz appears "weaker" or fainter in the image (see the blue arrows in Fig. 21.9b). This is similar to the information that the frequency spectrum provided above in Fig. 21.8. The spectrum of the processed signal confirms that only part of the desired information was recovered from the noisy vibration signal. Moreover, analysis produced considerable reduction of energy for the 75 Hz frequency. Thus, detection was not successful.

Example 21.2: Vibration Detection Using the Daubechies Wavelet
In this Example 21.2, the vibration detection algorithm is employed using a family of Daubechies wavelets. In the program O-2 of Appendix O, the vibration signal is sampled at 1000 samples/s for a period of 2 s. We use the Daubechies wavelet (db4,

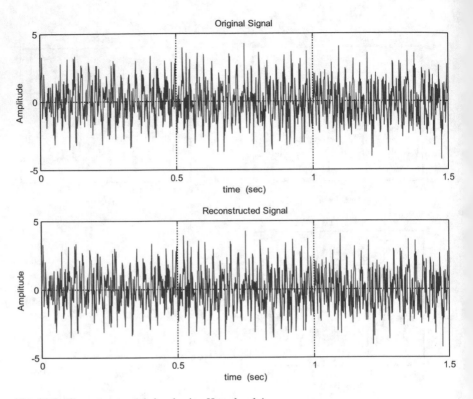

Fig. 21.5 The reconstructed signal using Haar, level 4

initially) at level 4 to analyze the vibration signal. As in the Haar program, the approximations and details contain the pertinent information about the vibration signal. Figure 21.10 illustrates the decomposition step, showing the approximation and details at level 4 for the db4 wavelet. The last details of the decomposition using the Daubechies-4 (db4) wavelet are as shown in Fig. 21.11. The reconstruction as shown in Fig. 21.12 is performed through an iterative process of downsampling, upsampling, and summing the results of the approximation and detail coefficients. Using db4, at level 4, produces complete reconstruction of the vibration signal with a negligible error of only $1.1667556 \times 10^{-11}$.

Applying the fixed form threshold to the detail coefficients and then summing the highest-level approximation and the modified details yield a noiseless vibration signal as shown in Fig. 21.13. In this case, the frequency spectra of the noisy vibration signal and the reconstructed vibration signal (before denoising) are identical as shown in Fig. 21.14. Their frequency spectra reveal the 35 Hz and 75 Hz frequencies. Figure 21.15 shows the frequency spectrum of the processed signal using db4 at level 4. The processed signal retains the 35 Hz frequency, but most of the 75 Hz frequency was lost. Similar to the results of the analysis using Haar, the Daubechies (db4) wavelet causes a large reduction of the 75 Hz frequency. The

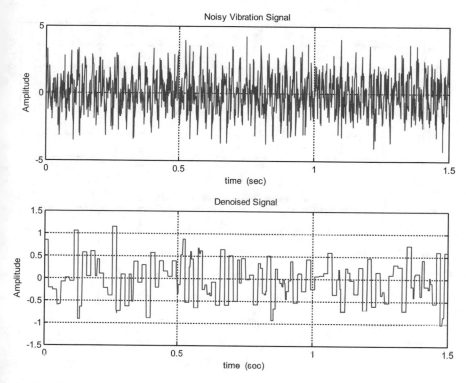

Fig. 21.6 The processed vibration signal using Haar, level 4

threshold is considered to be the cause of this unwanted spectral limitation. Also, noise artifacts appear near 35 Hz as before.

In Fig. 21.16a, the top scalogram shows the frequencies of interest (35 Hz and 75 Hz). The scalogram of the processed signal (Fig. 21.16b (bottom)) shows the dominant 35 Hz frequency in the spectrum. The 75 Hz frequency is shown faintly in red (see the blue arrows). We surmise that the reduction of the higher frequency is most likely caused by thresholding. Therefore, the requirement for detection is not met as a result of the analysis. It turns out that using Daubechies 6–18 at level 4 and the other dbN produced similar results. We found considerable attenuation of the 75 Hz frequency due to the denoising process. Therefore, we know there is a loss of power at this frequency. However, the statistical error, in general, associated with reconstruction decreases as the value of N increases for the Daubechies family.

Example 21.3: Vibration Detection Using the Morlet Wavelet

In Example 21.3, the vibration detection algorithm uses the Morlet wavelet. In the program O_3 of Appendix O, the vibration signal is sampled at 1000 samples/s for 2 s. A very short period is chosen in order to provide a good approximation of a real vibration signal. In Fig. 21.17, two sinusoids are shown with noise, i.e., s1 (blue), s2 (purple), and *noise* (red). Signal s1 has a frequency of 35 Hz, and s2 has a frequency of 75 Hz. The noise is random, white Gaussian noise. The combined signals of s1,

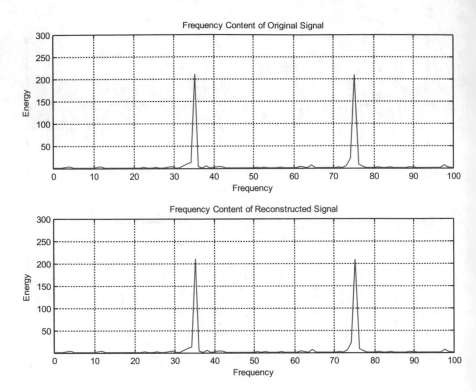

Fig. 21.7 The frequency spectrum of the noisy vibration signal and the reconstructed signal

s2, and noise from the vibration signal, "s," is shown in blue in Fig. 21.18. Remember that the Morlet wavelet is continuous. This means that it does not translate well for computational analysis. To compensate for this property of the Morlet basis function, the programming code based on a modified version of the Mallat algorithm is used to perform the Morlet wavelet for vibration analysis. Unlike signal processing using the Haar and Daubechies wavelets, removing the noise from the vibration signal with the Morlet wavelet requires thresholding the wavelet coefficients $C(a,b)$ shown in Eq. (21.17).

$$C(a,b) = \int\limits_{R} s(t)\frac{1}{\sqrt{a}}\psi\left(\frac{t-b}{a}\right)dt, \quad a \in R^{+} - \{0\}, \quad b \in R \qquad (21.17)$$

Analysis of the vibration signal using the Morlet wavelet is a redundant process of continuous scaling and translation of its wavelet function over the entire period of the vibration signal. The coefficients are generated in this manner. Then, the fixed form threshold is applied to these coefficients. After denoising, a small interval of the processed signal (in purple) is compared to the original signal (in blue) as shown in Fig. 21.19. In the scalogram shown in Fig. 21.20, the 35 Hz and 75 Hz frequencies

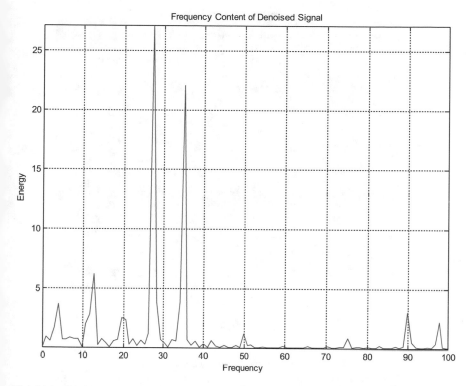

Fig. 21.8 The frequency spectrum of the vibration signal analyzed with the Haar wavelet

are shown with no decrease of either frequency. The fixed form threshold equals 3.824453003. The error associated with the processed signal is 1.6130903, which is the smallest post-processing error of the wavelet families using the fixed form threshold. In addition, detection is confirmed by the recovery of the 35 Hz and 75 Hz frequencies that appear in the scalogram of Fig. 21.20 of the denoised vibration signal.

Table 21.2 summarizes the performance of the wavelets. While detection is unsuccessful for the Haar and Daubechies wavelets, detection is accomplished using the Morlet wavelet. Global thresholding resulted in undesired attenuation of the 75 Hz frequency for the simulated vibration model using the Haar and Daubechies wavelets. The noise artifacts present in the denoised signal spectra are evidence of ineffective suppression of noise and loss of significant signal data. Also, the results from analysis using the vibration detection algorithm produced negative signal-to-noise ratio (SNR) values for the processed signal. At this point, the performance of the algorithm needed improvement. Therefore, we have to explore further noting the importance of thresholding in vibration analysis in order to enhance algorithm performance.

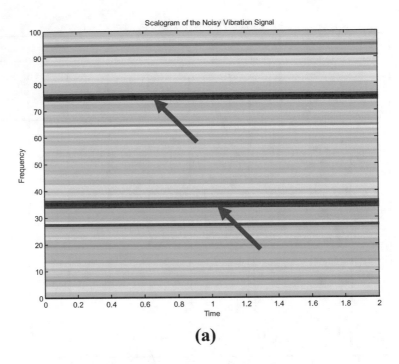

(a)

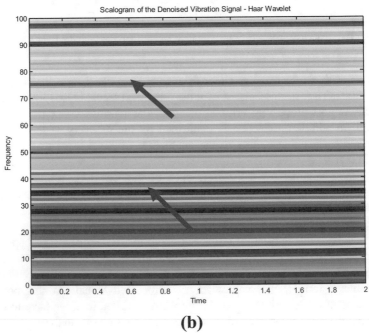

(b)

Fig. 21.9 For the noisy vibration signal, the frequencies (35 and 75 Hz) appear as very bold lines in (**a**) (top); for the processed signal, using Haar at level 4, 35 Hz is bold, but 75 Hz appears slightly faint in (**b**) (bottom)

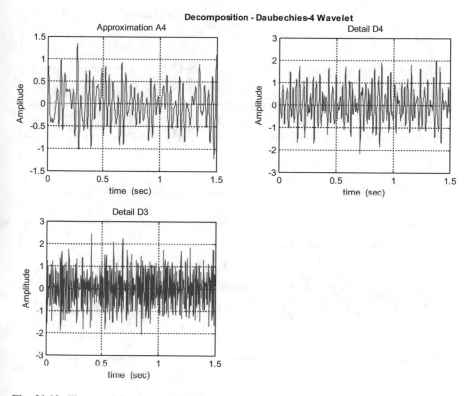

Fig. 21.10 The approximation and details of the signal using the Daubechies (db4) wavelet at level 4

21.7   Threshold Experimentation

The selection of the appropriate threshold value is the key to the proper detection optimization of the vibration information as demonstrated in Donoho and Johnstone (1994) and Donoho (1995). In this section, we explain the trials conducted to obtain the optimal threshold for denoising the vibration signal. Initially in Sect. 21.5, we accomplish the denoising by using a global threshold; that is, the same threshold is applied to the detail coefficients at all levels. This resulted in the partial to complete elimination of the 75 Hz frequency. Since this is an undesirable effect, we conducted several trials to change this outcome. The trials described herein represent the results from experiments using different thresholds in the simulations.

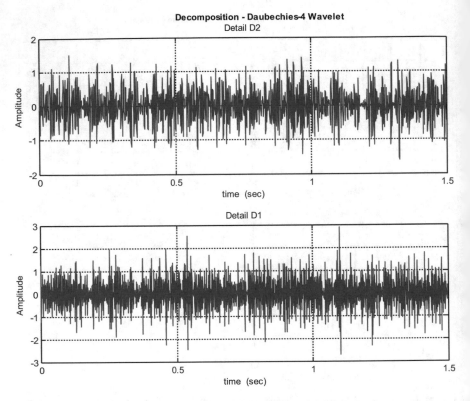

Fig. 21.11 Last details of the decomposition using the Daubechies (db4) wavelet

21.7.1   Global Thresholding

We show the numerous simulations performed using the Daubechies-4 wavelet at level 4. From the results, Daubechies-4 reveal fewer noise artifacts in the processed signal's frequency spectra, which demonstrates that the wavelet lends itself well to quick parametric changes. Based on the vibration model, the fixed form, global threshold has a calculated value of 3.824453003. Figure 21.21 shows the result of denoising the vibration signal using this threshold value. In the subsequent trials, we used fractions of the fixed form threshold. At 70% of the threshold (i.e., thr = 2.677117102), the algorithm produced the results shown in Fig. 21.22. Then at 50% of the threshold (1.912226502), this trial yielded the processed signal shown in Fig. 21.23. Using 50% of the threshold (thr = 1.912226502) produced a denoised signal with a 75% increase in amplitude compared to the results for the denoised signal at 70% of the threshold (thr = 2.677117102).

Overall, the results using 50% and 70% of the threshold are mixed. At 70% of the threshold, the noise artifacts are virtually eliminated from the frequency spectra, but both frequencies (35 Hz and 75 Hz) are nearly cutoff as shown in Fig. 21.24. In addition, at 50% of the threshold, 35 Hz is recovered, but the 75 Hz frequency is

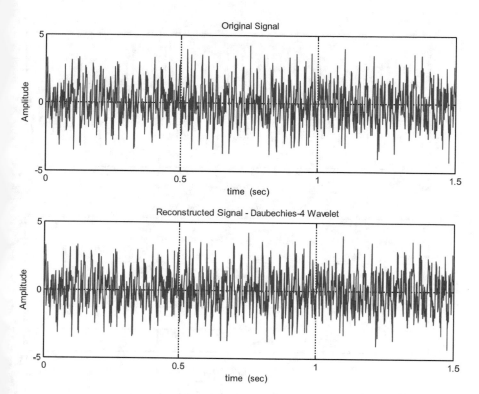

Fig. 21.12 The reconstructed vibration signal using db4 at level 4

suppressed, and more noise artifacts are introduced as shown in Fig. 21.25. Even though the algorithm recovered part of the key information from the noisy vibration signal, global thresholding still did not provide detection. Calculation of the global threshold is automated to determine whether the program would obtain better results than manual calculation provided. The MATLAB function named *ddencmp* computed the threshold according to the signal matrix size, the type of thresholding (i.e., hard or soft), and the contribution of the approximation coefficients. The threshold that the program computed was thr = 3.911167193. Notice that the frequency spectrum of the processed vibration signal as shown in Fig. 21.26 shows the 75 Hz completely clipped and the noise floor is still dominant. Detection is not accomplished using the computer-generated threshold either.

Many different threshold experiments were conducted, but they provided no improvement to the main four experimental trials. In fact, we cannot complete detection as desired, and the performance of the vibration detection algorithm is as regarded to be unacceptable. The summaries of the results from the main global threshold trials are as shown in Tables 21.3, 21.4, 21.5, and 21.6, respectively.

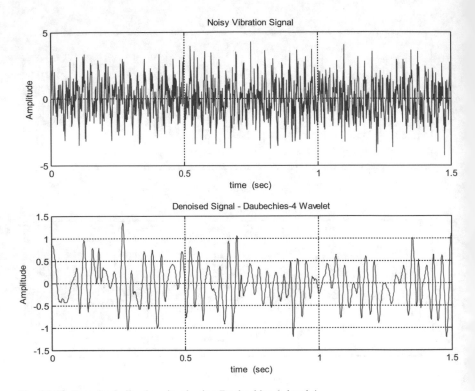

Fig. 21.13 Denoised vibration signal using Daubechies 4, level 4

21.7.2 Per-Level Thresholding

Changing the threshold directly affected the efficacy of the detection algorithm. However, global thresholding did not completely remove noise and cutoff desired frequency information. Therefore, per-level thresholding (PLT) is investigated to correct the shortcomings of the global threshold scenario. A different threshold is applied to level 1, level 2, and so on. This new threshold method is explored using the Haar and Daubechies wavelets. The thresholding scheme is as follows:

Level 1: First, the "sqtwolog" threshold is used on the detail coefficients because this detail contains most of the noise. The threshold value is usually the largest.
Level 2: Again, the "sqtwolog" threshold is applied to the detail coefficients at level 2 in order to remove more noise.
Level 3: "Heursure" threshold provides a small threshold value or large one depending on the noisiness of the signal at this level.
Level 4: "Rigrsure" threshold generates the smallest value for denoising, since it is based on the risk involved with losing information from the details.

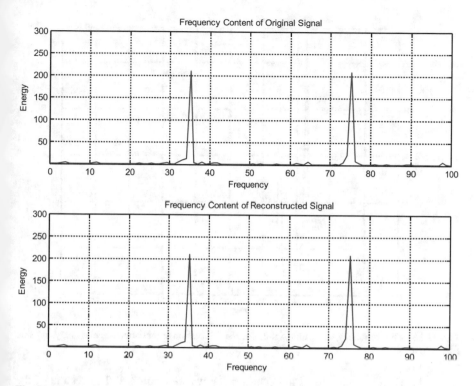

Fig. 21.14 The frequency spectrum for the noisy vibration signal and the reconstructed vibration signal (db4, level 4)

Based on the decomposition process, most of the noise is found in the details. Additionally, in multiresolution wavelet analysis, the details at the lower levels contain more noise than the details at higher levels. Thus, the "fixed form" threshold, *sqtwolog*, is applied to both levels 1 and 2. Threshold *heursure* is applied to level 3 because the threshold value increases or decreases based on the amount of noise at this level. Finally, since level 4 contains the least amount of noise, the *rigrsure* threshold is applied; this threshold has the smallest value. By using a different threshold for each level, the noise that appeared in Figs. 21.25 and 21.26 is eliminated, and both the low and high frequencies are wholly recovered as shown in Fig. 21.27. Additionally, the scalogram of the processed signal revealed that the 35 Hz and 75 Hz frequencies were dominant in the frequency spectrum, as shown by the bold red lines on the scalogram shown in Fig. 21.28. The dark, bold red lines in the figure indicate that the signal energy is concentrated at these key frequencies (see the blue arrows).

We therefore now have a thresholding scheme that effectively suppressed noise and enabled the detection of pertinent frequency information from the noisy vibration signal model (see Appendix O_4 of O). However, since a continuous wavelet does not use levels for analysis, per-level thresholding is not applicable to the Morlet wavelet. Still, the per-level thresholding technique is applied in many other trials

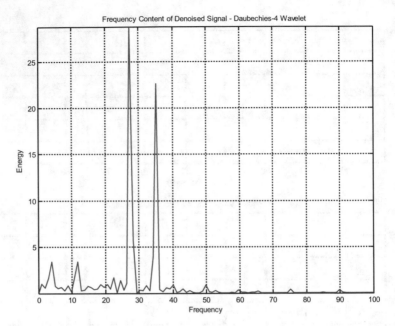

Fig. 21.15 The frequency spectrum of the processed vibration signal after wavelet analysis using db4 at level 4

using the Haar wavelet and Daubechies 6–18 wavelet families. The results of the per-level trials are shown in Table 21.5. Of particular note, the SNR values, though small, are much improved from the global threshold (GT) case. Then the (percent) improvement of the algorithm performance is listed in Table 21.7. Finally, the result of signal analysis using per-level thresholding is consistent detection of the corrupted vibration signal. Table 21.8 shows the results obtained from the algorithm performance improvement of the per-level thresholding (PLT).

21.8 Application of the PLT Algorithm on an Aeroelastic Vibration Data for Verification

In this section, we apply the vibration detection algorithm to real data from flight test conducted on the F-15B/836 Flight Research testbed airplane conducted at NASA Dryden Flight Research Center in Edwards, CA. The MATLAB program for the implementation of the PLT algorithm is as listed in O_5 of Appendix O. The F-15B is equipped with an advanced data system that includes a research air data system; a custom global positioning/navigation system (GPS); a radome with an air data probe; a digital data recorder; and telemetry antennas. The F-15B testbed aircraft is used for a variety of flight research missions, but emphasis is placed on flight flutter tests (e.g., the aerostructures test wing experiment). Triaxial accelerometers

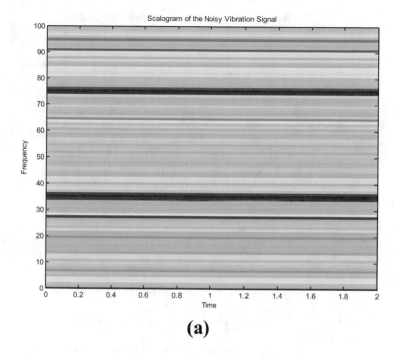

(a)

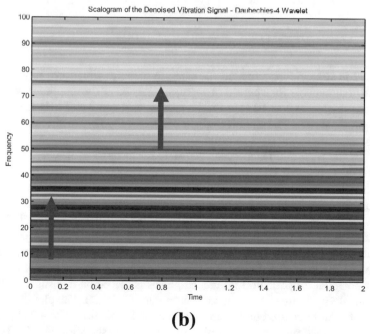

(b)

Fig. 21.16 (**a**) The noisy vibration signal shows 35 Hz and 75 Hz in bold red lines (top); (**b**) the processed vibration signal, using db4 at level 4, shows a strong 35 Hz frequency, but the 75 Hz frequency is virtually cutoff, shown by a faint red line (bottom)

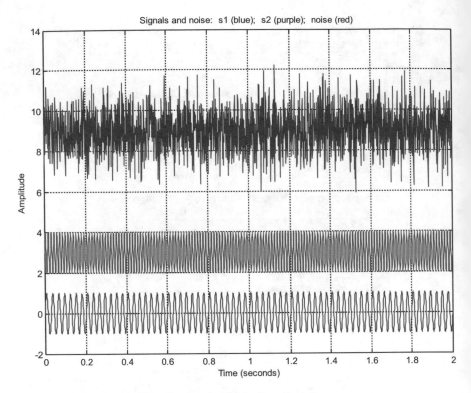

Fig. 21.17 Components of the vibration signal (s1, s2, and noise)

are placed in strategic locations on the plane to measure vibrational input. Data are recorded continuously for a period of approximately 25 s. The amplitude response of the accelerometers is 2 to 10 kHz for the normal axis and 2 to 7 kHz for the lateral and longitudinal axes. The takeoff of flight 256 of the F-15B/836 aircraft is recorded, and data are sampled at 7042 samples/s.

We use the Haar and Daubechies wavelets for analysis of this real test data. In addition, data collected from the sensors are supplied from NASA Dryden for this particular flight. This information served as a reference for verifying the results of the vibration detection algorithm. The results of the detection analysis are shown in Table 21.9. The vibration signal from takeoff of the F-15B/836 for test flight 256 is shown in Fig. 21.29 for each axis. The vibration in the normal axis turned out to be worse than the vibration in the lateral and longitudinal axes. However, since the vibration in the normal axis has the greatest amplitude, which translates into the greatest potential for structural failure, emphasis is placed on investigating the data for this axis. The figure of the data is as shown in Fig. 21.30. The plot in Fig. 21.30 shows that events occurred in this 25-second period. The first spike in the plot is the "event of the brake release." A snapshot of the event is shown in Fig. 21.31.

The brake release event occurred in 2 s. Specifically, the vibrational event occurred from 6 to 8 s. The vibration detection algorithm is used to suppress noise

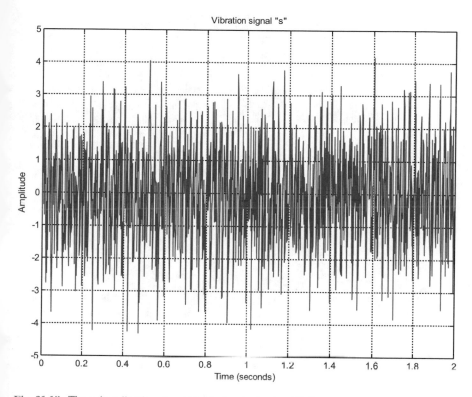

Fig. 21.18 The noisy vibration signal that is denoised using Morlet

and recover the frequency information from this "event." Initially, the event signal is processed using the Dabuechies-4 wavelet (at level 4). The same PLT scheme, as described in Sect. 21.7.2, is used in this signal processing application. The recovered signal (bottom plot) is shown in Fig. 21.32.

Analysis is implemented using db4 at level 4 with PLT. There is a tradeoff of maximum data retrieval for maximum noise suppression. The initial results from the simulations show that it is possible to suppress more noise, but the outcome showed a significant loss of the signal. However, the goal is to recover as much frequency information as possible for detection. Thus, the appearance of the processed signal is similar to the original noisy event. Still, the vibration detection algorithm recovered the frequency information of the noisy brake release event.

There is some attenuation in the upper part of the spectrum of the processed event signal as shown in Fig. 21.33. This slight attenuation is acceptable because the frequency information is successfully recovered from the noisy event and because the dominant frequency is identified in the spectrum (i.e., approximately 300 Hz). Specifically, some attenuation is acceptable because the eventual production of a real-time health monitoring system would rely on identifying maximum vibration levels and their associated frequencies for prediction of structural faults in aeroelastic systems. Figure 21.34 shows the PSD of the noisy event and the processed event

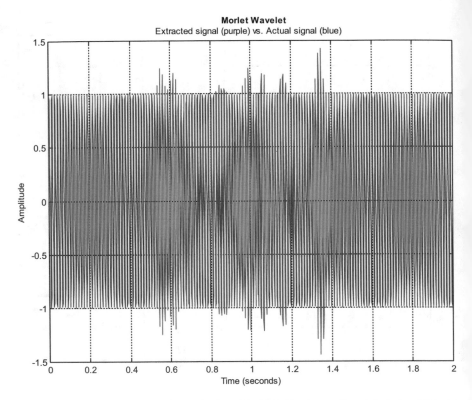

Fig. 21.19 Comparison of short interval of the denoised vibration to the original signal (Morlet wavelet)

signals, while Fig. 21.35 shows the scalogram of the processed event signal. The frequency of interest is detected (around 300 Hz) as a dark red line. The dark red color indicates (see arrow) that most of the vibration signal energy is located at this particular frequency. The results from the vibration detection algorithm agree with the signal reference which verifies that detection is achieved.

21.9   Remarks on the Wavelets Used for the Vibration Signal Detection

As discussed in Sect. 21.1, one of the objectives of this chapter is to explore how vibration signals corrupted by noise can be detected using wavelets. In this section, we remark on the outcomes because of the exploration. These results are based on decomposition and reconstruction processes, denoising, and statistical and spectral data. The confirmation of detection is based on the recovery of frequency spectra of the corrupted vibration signal. This is accomplished by using a simulated vibration

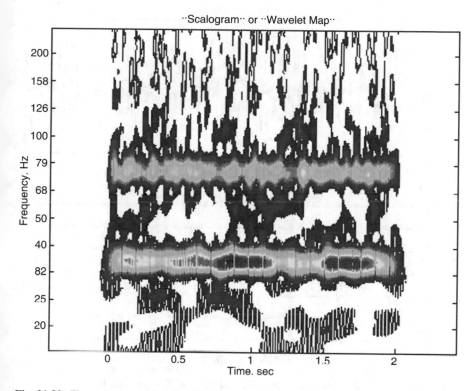

Fig. 21.20 The scalogram of the vibration signal analyzed with the Morlet wavelet

Table 21.2   Summary of vibration detection algorithm performance

Wavelet[a]	Reconstruction error	Processing error	% Processing error	MSE	SNR
Haar	$2.2204456 \times 10^{-15}$	2.164	108.189	2.081×10^{-6}	−6.674
db4	$1.6675560 \times 10^{-11}$	2.461	123.070	2.693×10^{-6}	−7.344
db6	$8.4268148 \times 10^{-12}$	2.342	117.098	2.438×10^{-6}	−6.715
db8	$2.8593128 \times 10^{-11}$	2.162	108.089	2.077×10^{-6}	−6.632
db10	$3.6464165 \times 10^{-11}$	2.499	124.937	2.775×10^{-6}	−8.642
db12	$6.6702199 \times 10^{-13}$	4.289	114.342	2.324×10^{-6}	−8.060
db14	$5.9809935 \times 10^{-12}$	2.354	117.684	2.462×10^{-6}	−7.543
db16	$3.3148819 \times 10^{-11}$	2.405	120.267	2.571×10^{-6}	−9.295
db18	$6.8737016 \times 10^{-11}$	2.310	115.504	2.372×10^{-6}	−9.142
Morlet	4.4321971	1.613	80.655	2.082×10^{-6}	−2.694

[a]*Note*: db is an abbreviation for Daubechies (e.g., db4, db6, and so on)

model and a real data from an aeroelastic system—flight data from in-flight vibration experiments. In order to achieve this goal, a novel thresholding scheme is exploited that provides minimum loss of the desired signal and maximum suppression of the undesired noise.

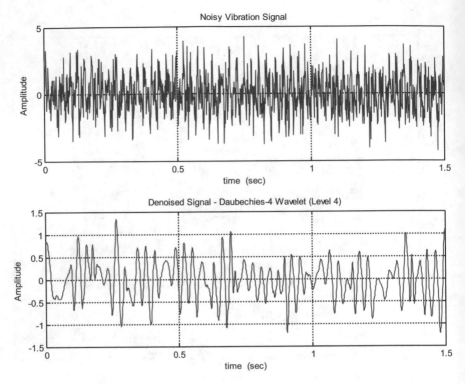

Fig. 21.21 The vibration model denoised using db4 (level 4) with threshold 1

The Haar wavelet yields the smallest error of the three wavelet families. Yet, since the Haar wavelet is not continuous because it has a step function shape, it does not provide "smoothing." Smoothing is a property of a wavelet that is continuously differentiable by d times, where d is an integer and $d \geq 1$. The Daubechies wavelet performs well with decomposition and reconstruction. The error from reconstruction is lowest for db12 at level 4. Daubechies is a very flexible wavelet family with the capability to support analysis beyond db20. Also, Daubechies is continuously differentiable, which makes it a good smoothing function for signal analysis. Generally, as N increases for db$N$ wavelet families, the performance of the algorithm improves. This is the case up to db14, after this value of N; there is no appreciable improvement of the algorithm for this application.

The Morlet wavelet is also a good smoothing function. Notwithstanding, the error associated with reconstruction is high. Like Daubechies, the Morlet wavelet effectively analyzes the signal of interest. The Morlet is symmetrical. It "looks" like or closely matches the signal under evaluation. The high error is attributed to the threshold selection (i.e., global threshold values). The issue of appropriate thresholding is addressed by two schemes: global thresholding (GT) and per-level thresholding (PLT). These threshold methods were applied to the vibration model. The GT scenario is applied first. The best case (Threshold 3 shown in Table 21.5) for

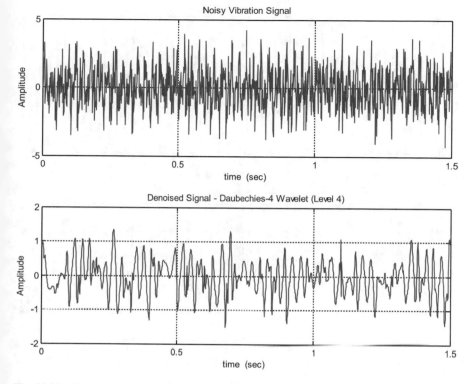

Fig. 21.22 The vibration model denoised using db4 with threshold 2

GT produced a processing percent error of nearly 114%. Without question, the detection of the signal is not possible due to the near or complete attenuation of the 75 Hz frequency of the vibration model. The effect of the GT scheme is unacceptable because of the performance by the vibration detection algorithm.

We continuously refined the algorithms, which lead to the PLT scheme. This threshold method resulted in the complete recovery of the frequency spectrum of the signal vibration model. PLT enabled successful detection. However, the highest processing percent error is associated with PLT with 95% for the simulated vibration model and 23% for F-15B/836 vibration signal. The high percent error for the PLT that is applied to the simulated signal is due to the size of the sample. Since percent error is a statistical parameter, the sample size affects the outcome. For example, the vibration signal has the dimension in MATLAB: simulated vibration model, 1500 × 1 matrix, and F-15B/836 vibration model signal, 14,085 × 1 matrix. In addition, varying the sampling rate plays a role in the statistical computations because the larger the sampling rate, the larger the sample size of the signal under analysis.

The following observations are made after reviewing the performance of the wavelets in the algorithm: thresholding is an important factor in the efficacy of the algorithm. Trial and error revealed that effective thresholding is accomplished using

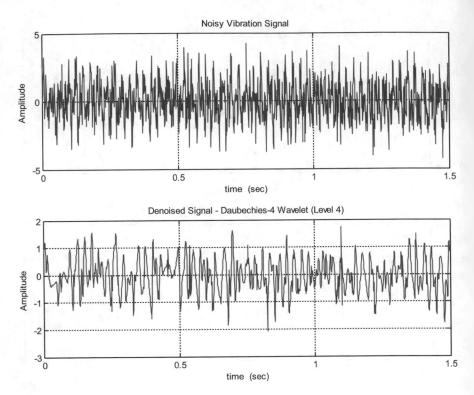

Fig. 21.23 The vibration model denoised using db4 with threshold 3

a per-level thresholding (PLT) scheme. Not only is detection successful using the Haar, Daubechies, and Morlet wavelets with PLT, but also the accuracy of the program is controllable based on the values of the level thresholds. It is important to note that the Morlet wavelet is useful for signal analysis, but its capability is limited because it does not support multiple-level thresholding since it is a continuous wavelet. An approach could be devised to use the Morlet with PLT, but this procedure would not be as straightforward as applying PLT with the (discrete) Haar or Daubechies wavelets. The Daubechies wavelets overwhelmingly outperformed the other wavelets. Moreover, the results showed that db10, db12, and db14 were the most effective wavelets in the algorithm.

Summary
1. Vibration signals contain important information about dynamic mechanical parameters, and this non-stationary waveform is useful for fault analysis.
2. A wavelet is a waveform with a small, finite time interval and an average zero value.
3. Fourier analysis involves representing a signal as the summation of its constituent sine waves at various frequencies.

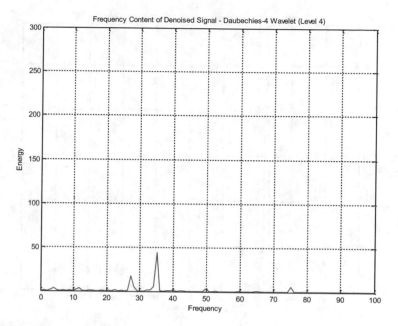

Fig. 21.24 Frequency spectrum of the denoised signal at threshold 2

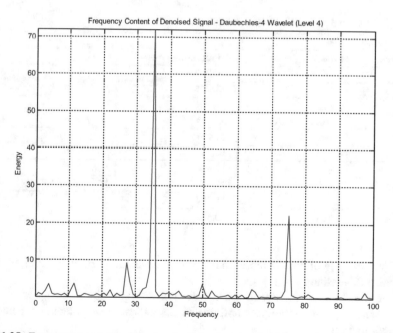

Fig. 21.25 Frequency spectrum of the denoised signal at threshold 3

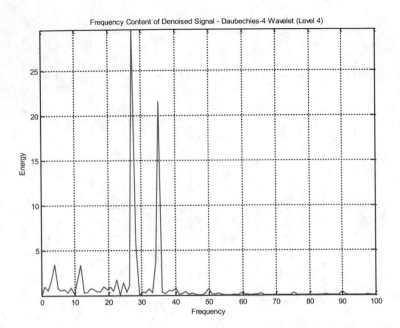

Fig. 21.26 Frequency spectrum of the processed signal at threshold 4

Table 21.3 Global threshold results (Threshold 1)

thr = 3.824453003

Wavelet	Processing error	% Processing error	MSE	SNR
Haar	2.164	108.189	2.081×10^{-6}	−6.674
db4	2.461	123.070	2.693×10^{-6}	−7.344
db6	2.342	117.098	2.438×10^{-6}	−6.715
db8	2.162	108.089	2.077×10^{-6}	−6.632
db10	2.499	124.937	2.775×10^{-6}	−8.642
db12	4.289	114.342	2.324×10^{-6}	−8.060
db14	2.354	117.684	2.462×10^{-6}	−7.543
db16	2.405	120.267	2.571×10^{-6}	−9.295
db18	2.310	115.504	2.372×10^{-6}	−9.142
Morlet	1.613	80.655	2.082×10^{-6}	−2.694

4. Classically, vibration is represented as a sinusoid and normal distribution white noise (Gaussian noise).
5. The threshold is vital because it enables noise suppression.
6. An increase in vibration amplitude is an indicator of potential mechanical malfunction.
7. The frequency information of the vibration signal provides a method of categorizing the type of breakdown after a fault occurs.
8. The Haar wavelet yields the smallest error of the three wavelet families.

Table 21.4 Global threshold results (threshold 2)

thr = 2.677117102

Wavelet	Processing error	% Processing error	MSE	SNR
Haar	2.164	108.189	2.081×10^{-6}	-5.704
db4	2.419	120.971	2.602×10^{-6}	-5.967
db6	2.321	116.068	2.395×10^{-6}	-5.613
db8	2.079	103.840	1.921×10^{-6}	-5.497
db10	2.419	120.958	2.601×10^{-6}	-6.942
db12	2.114	105.701	1.986×10^{-6}	-6.624
db14	2.249	112.462	2.248×10^{-6}	-6.359
db16	2.337	116.865	2.428×10^{-6}	-7.463
db18	2.301	115.070	2.354×10^{-6}	-7.295
Morlet	3.283	164.127	1.073×10^{-5}	-3.154

Table 21.5 Global threshold results (threshold 3)

thr = 1.912226502

Wavelet	Processing error	% Processing error	MSE	SNR
Haar	2.028	101.407	1.828×10^{-6}	4.132
db4	2.275	113.763	2.300×10^{-6}	-4.183
db6	2.058	102.895	1.882×10^{-6}	-3.986
db8	2.080	104.005	1.923×10^{-6}	-3.909
db10	2.193	109.647	2.137×10^{-6}	4.667
db12	2.011	100.546	1.797×10^{-6}	-4.521
db14	2.096	104.785	1.952×10^{-6}	-4.456
db16	2.128	106.378	2.012×10^{-6}	4.980
db18	2.230	111.486	2.210×10^{-6}	-4.861
Morlet	1.536	76.818	2.706×10^{-6}	-3.006

Table 21.6 Global threshold results (threshold 4)

thr = 3.911167193

Wavelet	Processing error	% Processing error	MSE	SNR
Haar	2.164	108.189	2.081×10^{-6}	-6.701
db4	2.470	123.520	2.712×10^{-6}	-7.395
db6	3.306	165.300	4.858×10^{-6}	-6.646
db8	3.096	154.803	4.260×10^{-6}	-6.686
db10	2.765	138.270	3.399×10^{-6}	-8.532
db12	3.211	160.553	4.583×10^{-6}	-7.695
db14	3.122	156.118	4.333×10^{-6}	-7.361
db16	2.786	139.322	3.451×10^{-6}	-8.778
db18	3.088	154.392	4.238×10^{-6}	-8.383
Morlet	0.913	91.330	4.484×10^{-6}	-2.838

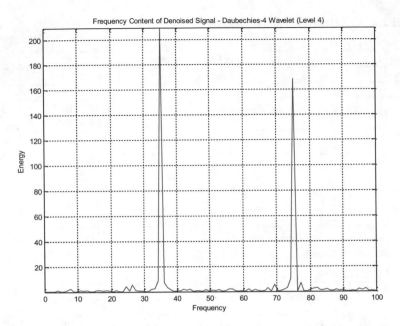

Fig. 21.27 Frequency spectrum of the denoised signal using db4 with per-level thresholding

9. The Daubechies wavelet performs well with decomposition and reconstruction.
10. Not only is detection successful using the Haar, Daubechies, and Morlet wavelets with PLT, but also the accuracy of the program is controllable based on the values of the level thresholds.
11. It is important to note that the Morlet wavelet is useful for signal analysis, but its capability is limited because it does not support multiple-level thresholding since it is a continuous wavelet.

Review Questions

21.1 In aeroelastic systems, fatigue and breakdown of aircraft structure are common.

(a) True
(b) False

21.2 Vibration signals contain important information about dynamic mechanical parameters, and this non-stationary waveform is useful for fault analysis.

(a) True
(b) False

21.3 Feature extraction of vibration data from noise is not done successfully using wavelet analysis.

(a) True
(b) False

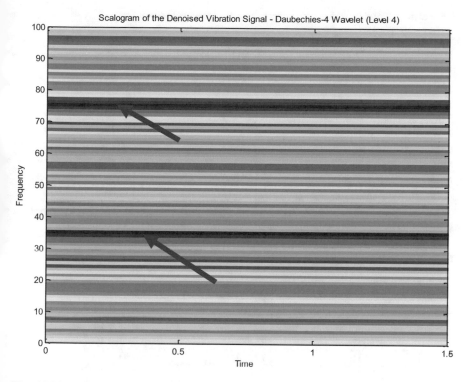

Fig. 21.28 Scalogram of the processed vibration signal using db4 (level 4) with per-level thresholding

Table 21.7 Per-level threshold results—simulated vibration model

Per-level threshold				
Wavelet	Processing error	% Processing error	MSE	SNR
Haar	1.912	95.595	1.625×10^{-6}	-0.434
db4	1.576	78.815	1.104×10^{-6}	0.066
db6	1.602	80.096	1.141×10^{-6}	-0.363
db8	1.599	79.944	1.136×10^{-6}	0.069
db10	1.599	79.967	1.137×10^{-6}	0.454
db12	1.567	78.342	1.091×10^{-6}	0.357
db14	1.334	66.701	7.909×10^{-7}	0.488
db16	1.621	81.052	1.168×10^{-6}	-0.083
db18	1.601	80.032	1.139×10^{-6}	0.158

21.4 Classify the following wavelets, Haar, Daubechies, and Morlet.

  (a) Haar, Daubechies, and Morlet are all continuous wavelets.
  (b) Haar is a continuous wavelet.
  (c) Daubechies is a continuous wavelet.

Table 21.8 Algorithm performance improvement—per-level thresholding

Performance improvement (in percentage)[a]

Wavelet	Process error	% Process error	MSE	SNR
Haar	5.720	5.731	11.105	89.497
db4	30.725	30.720	52.000	98.422
db6	22.157	22.158	39.373	90.893
db8	23.125	23.134	40.926	98.235
db10	27.086	27.069	46.795	90.272
db12	22.079	22.083	39.288	92.104
db14	36.355	36.345	59.483	89.048
db16	23.825	23.808	41.948	98.333
db18	28.206	28.213	48.462	96.750

[a]Note: The comparison is made based on the best global threshold values (i.e., Threshold 3, Table 21.5) versus the per-level threshold values from analysis of the vibration model

Table 21.9 Results—F-15B/836 flight research

Per-level threshold

Wavelet	Processing error	% Processing error	MSE	SNR
Haar	4.319	18.902	9.405×10^{-8}	12.650
db4	5.295	23.170	1.413×10^{-7}	11.413
db6	5.001	21.883	1.260×10^{-7}	13.259
db8	4.441	19.433	9.940×10^{-8}	12.492
db10	4.127	18.059	8.585×10^{-8}	13.728
db12	5.125	22.429	1.324×10^{-7}	12.565
db14	4.238	18.544	9.052×10^{-8}	13.398
db16	3.910	17.112	7.708×10^{-8}	14.422
db18	4.044	17.696	8.242×10^{-8}	14.314

(d) Morlet is a continuous wavelet.

(e) None is a continuous wavelet.

21.5 The mother wavelet determines the shape of the constituent components of the signal, which are generated during analysis.

(a) True

(b) False

21.6 Compared to the fast Fourier transform (FFT), the computations required for N points using the FWT algorithm are reduced from 2NlogN to 2 N, where log is the base-2 logarithm.

(a) True

(b) False

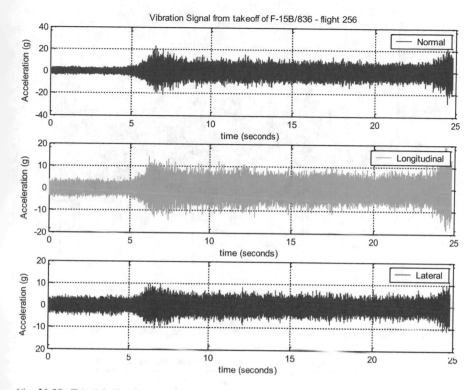

Fig. 21.29 Triaxial vibration signal from takeoff of flight 256 (F-15B/836)

21.7 Decomposition is not performed by repeatedly passing the desired signal through low- and high-pass filters and dyadically downsampling the filters' results.

(a) True
(b) False

21.8 The threshold is vital because it enables noise suppression.

(a) True
(b) False

21.9 Classically, vibration is represented as a sinusoid and normal distribution white noise (Gaussian noise).

(a) True
(b) False

21.10 Not only is detection successful using the Haar, Daubechies, and Morlet wavelets with PLT, but also the accuracy of the program is controllable based on the values of the level thresholds.

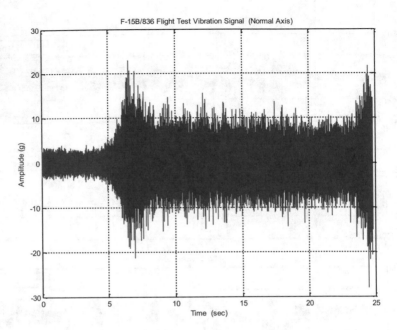

Fig. 21.30 Vibration signal measured on the normal axis during takeoff (F-15B/836)

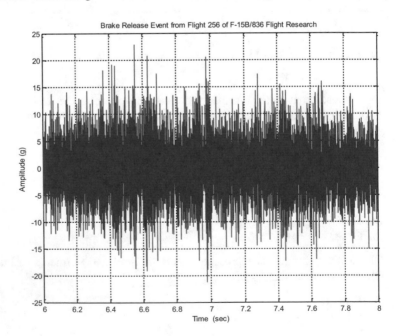

Fig. 21.31 Noisy brake release event (F-15B/836 flight test)

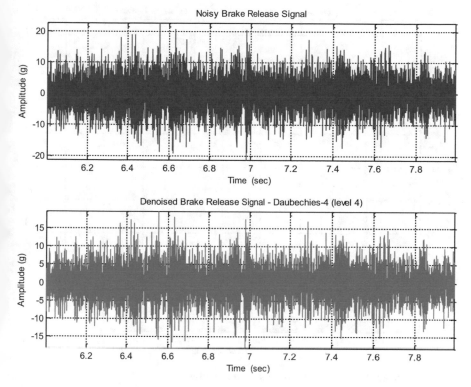

Fig. 21.32 The processed signal using db4 (level 4) with PLT

(a) True

(b) False

Answers: 21.1a, 21.2a, 21.3b, 21.4d, 21.5a, 21.6a, 21.7b, 21.8a, 21.9a, 21.10a.

Problems

21.1 What are some of the key common failure issues found in aeroelastic systems such an aircraft structures that demand the constant improvement of the monitoring and diagnostic systems of the aircrafts?

21.2 Why are vibration signals important in the health monitoring of aeroelastic systems such as aircrafts?

21.3 What are the classical tools that can be used in the signal processing of vibration signals?

21.4 Why are wavelets important in the analysis of vibration signals?

21.5 Why is vibration detection important in aeroelastic systems?

21.6 Name some of the key wavelets that can be used in the analysis and detection of vibration information.

21.7 Why is the idea of thresholding important in the detection of vibration systems information?

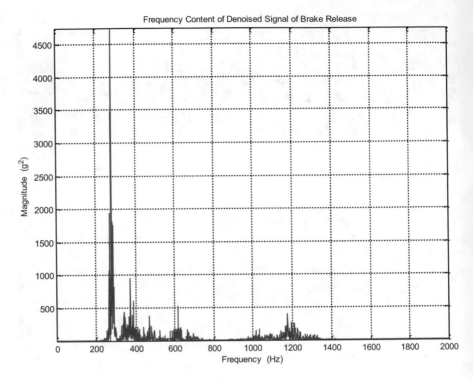

Fig. 21.33 Frequency spectrum of the processed event signal

21.8 What is the process for the complete algorithm for vibration detection process?

21.9 How can a vibration model be modeled?

21.10 What is the difference between global thresholding and per-level thresholding?

21.11 While there are different wavelets discussed in this chapter, which wavelet outperformed the others?

21.12 Why is the Haar type of wavelet not necessarily capable of providing "smoothing" in the analysis even though it yielded the smallest error?

21.13 Is the Morlet type of wavelet a good smoothing function?

21.14 How can the issue of thresholding be addressed?

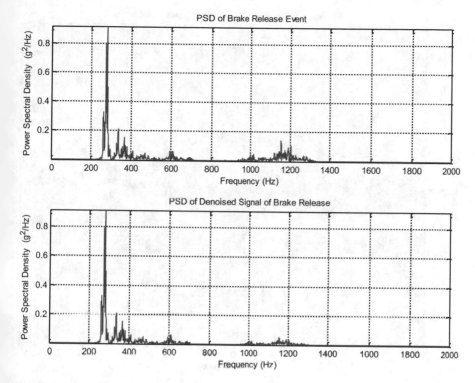

Fig. 21.34 The PSD of the noisy event and the processed event signals

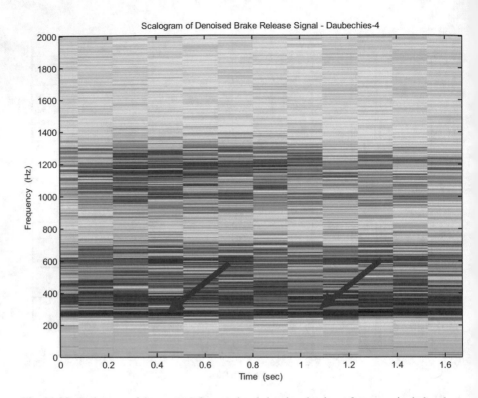

Fig. 21.35 Scalogram of the processed event signal showing dominant frequency in dark red

References

Akujuobi, C. M. (1998, September 13–16). *Implementation of wavelet-based solutions to signal processing applications*. In Proceedings of DSP World and ICSPAT, Toronto, Canada.

Akujuobi, C. M., & Hu, L. (2002, November 18–20), *A novel parametric test method for communication systems mixed-signal circuits using discrete wavelet transform*. In Proceedings of the IASTED international conference, communications, internet, and information technology, St. Thomas, US Virgin Islands, pp. 132–l35.

Ansari, & Baig. (1998). A PC based vibration analyzer for condition monitoring of process machinery. *IEEE Transactions on Instrumentation and Measurement, 47*(2), 378–383.

Brenner, M. (1997, April). *Wavelet analyses of F/A-18 aeroelastic and aeroservoelastic flight test data*. NASA, technical report. TM-4793 (H-2164), pp. 1–28.

Brenner, M., & Lind, R. (1997, September). *On-line robust modal stability prediction using wavelet processing*. NASA TM-1998-206550, pp. 1–14.

Brenner, M., & Lind, R. (1998, March). *Wavelet filtering to reduce conservatism in aeroservoelastic robust stability margins*. NASA, technical report. TM-1998-206545 (H-2222), pp. 1–12.

Brenner, M. J., Lind, R. C., & Voracek, D. F. (1997a, April). *Overview of recent flight flutter testing research at NASA Dryden*. NASA, technical report. TM-4792 (H-2165), pp. 1–27.

Brenner, M., Haley, S. M., & Lind, R. (1997b, November). *Estimation of modal parameters using a wavelet-based approach*. NASA TM-97-206300, pp. 1–9.

Butler, G. D., & Newland, D. E. (2000). Application of time-frequency analysis to transient data from centrifuge earthquake testing. *Shock and Vibration, 7*, 195–202.

Daubechies, I. (1992). *Ten lectures on wavelets*. Society for Industrial and Applied Mathematics.

Donoho, D. L. (1995, May). De-noising by soft-thresholding. *IEEE Transactions on Information Theory, 41*(3), 613–627.

Donoho, D. L., & Johnstone, I. M. (1994). *Threshold selection for wavelet shrinkage of noisy data*. In Proceedings of the 16th annual IEEE conference, Vol. 1, pp. A24–A25.

Giacomin, J. A., Staszewski, W. J., & Steinwolf, A. (2002). *On the need for bump event correction in vibration test profiles representing road excitations in automobiles*. In Proceedings of the institution of mechanical engineers, Vol. 216, part D, pp. 279–295.

Goumas, S., Zervakis, M., Pouliezos, A., & Stavrakakis, G. S. (2001). Intelligent on-line quality control of washing machines using discrete wavelet analysis features and likelihood classification. *Engineering Applications of Artificial Intelligence, 14*, 655–666.

Klempnow, A., Lescano, V., & Piñeyro, J. (2000). Effectiveness of new spectral tools in the anomaly detection of rolling element bearings. *Journal of Alloys and Compounds, 310*, 276–279.

Lee, S.K., & White, P.R. (2000). *Application of wavelet analysis to the impact harshness of a vehicle*. In Proceedings of the Institution of mechanical engineers, Vol. 214, part C, pp. 1331–1338.

Lepik, U. (2001). *Application of wavelet transform techniques to vibration studies*. In Proceedings of the Estonian Academy of Science, Physics, and Mathematics, Vol. 50, No. 3, pp. 155–168.

Liu, B., Ling, S., & Men, Q. (1997). Machinery diagnosis based on wavelet packets. *Journal of Vibration and Control, 3*, 5–17.

Luo, G. Y., Osypiw, D., & Irle, M. (2002). Vibration modeling with fast Gaussian wavelet algorithm. *Advances in Engineering Software, 33*, 191–197.

Rioul, O., & Vetterli, M. (1991, October). Wavelets and signal processing. IEEE. *Signal Processing, 8*(4), 14–38.

Zou, J., Chen, J., & Geng, Z. M. (2001). *Application of wavelet packets algorithm to diesel engines' vibroacoustic signature extraction*. In Proceedings of the Institution of mechanical engineers, Vol. 215, part D, pp. 987–993.

Appendices

Appendix A: Wavelet Coefficients

In this appendix, we show the different wavelet filter coefficients of different types of wavelets.

```
%------------------------------------------------------------
% Biorthogonal Coefficients -- these represent the SCALING functions
%------------------------------------------------------------

% Biorthogonal set 1.  From Vetterli and Herley.  Both length 18
% Analysis scaling coeffs.
bio1A = [
0    0.00122430
1   -0.00069860
2   -0.01183749
3    0.01168591
4    0.07130977
5   -0.03099791
6   -0.22632564
7    0.06927336
8    0.73184426
9    0.73184426
10   0.06927336
11  -0.22632564
12  -0.03099791
13   0.07130977
14   0.01168591
15  -0.01183749
16  -0.00069860
17   0.00122430
];

% Synthesis scaling coeffs
bio1S = [
```

© Springer Nature Switzerland AG 2022

C. M. Akujuobi, *Wavelets and Wavelet Transform Systems and Their Applications*,

https://doi.org/10.1007/978-3-030-87528-2

```
0   0.00122430
1   0.00069979
2  -0.01134887
3  -0.01141245
4   0.02347331
5   0.00174835
6  -0.04441890
7   0.20436993
8   0.64790805
9   0.64790805
10   0.20436993
11  -0.04441890
12   0.00174835
13   0.02347331
14  -0.01141245
15  -0.01134887
16   0.00069979
17   0.00122430
];

% Biorthogonal set 2.  From Vetterli and Herley.
% Analysis is length 24, Synthesis is length 20
% Analysis scaling coeffs -- 24
bio2A = [
0   0.00133565
1  -0.00201229
2  -0.00577577
3   0.00863853
4   0.01279957
5  -0.02361445
6  -0.01900852
7   0.04320273
8  -0.00931630
9  -0.12180846
10   0.05322182
11   0.41589714
12   0.41589714
13   0.05322182
14  -0.12180846
15  -0.00931630
16   0.04320273
17  -0.01900852
18  -0.02361445
19   0.01279957
20   0.00863853
21  -0.00577577
22  -0.00201229
23   0.00133565
];

% Synthesis scaling coeffs -- 20, make same length as above w/zeros
bio2S = [
0   0
```

```
0   0
0   0.00465997
1   0.00702071
2   -0.01559987
3   -0.02327921
4   0.05635238
5   0.10021543
6   -0.06596151
7   -0.13387993
8   0.38067810
9   1.10398118
10   1.10398118
11   0.38067810
12   -0.13387993
13   -0.06596151
14   0.10021543
15   0.05635238
16   -0.02327921
17   -0.01559987
18   0.00702071
19   0.00465997
0   0
0   0
];
```

```
% Biorthogonal set 2. From Cheong, et..al.
% Analysis is length 5, Synthesis is length 3.
% Note that the indices do not start at 0.
% These are the scaling function coefficients.

% Analysis scaling coefficients -- 5
bio3A = [
-2   -0.125
-1   0.25
 0   0.75
 1   0.25
 2   -0.125
];
```

```
% Synthesis scaling coefficients -- 3, make same length as above w/zeros
bio3S = [
-2   0
-1   0.25
 0   0.50
 1   0.25
 2   0
];
```

```
%----------------------------------------------------------------
% The Daubechies Coefficients -- these represent the SCALING functions
%----------------------------------------------------------------
coeff4 = [
0   (1+sqrt(3))./(4*sqrt(2))
```

```
1   (3+sqrt(3))./(4*sqrt(2))
2   (3-sqrt(3))./(4*sqrt(2))
3   -(-1+sqrt(3))./(4*sqrt(2))
];
```

```
% This version has more quantization error than above
%coeff4 = [
%0   0.482962913145
%1   0.836516303738
%2   0.224143868042
%3   -0.129409522551
%];
```

```
coeff6 = [
0   0.332670552950
1   0.806891509311
2   0.459877502118
3   -0.135011020010
4   -0.085441273882
5   0.035226291882
];
```

```
coeff8 = [
0   0.230377813309
1   0.714846570553
2   0.630880767930
3   -0.027983769417
4   -0.187034811719
5   0.030841381836
6   0.032883011667
7   -0.010597401785
];
```

```
coeff10 = [
0   0.160102397974
1   0.603829269797
2   0.724308528438
3   0.138428145901
4   -0.242294887066
5   -0.032244869585
6   0.077571493840
7   -0.006241490213
8   -0.012580751999
9   0.003335725285
];
```

```
coeff12 = [
0   0.111540743350
1   0.494623890398
2   0.751133908021
3   0.315250351709
4   -0.226264693965
```

```
5   -0.129766867567
6    0.097501605587
7    0.027522865530
8   -0.031582039318
9    0.000553842201
10   0.004777257511
11  -0.001077301085
];


coeff14 = [
0    0.077852054085
1    0.396539319482
2    0.729132090846
3    0.469782287405
4   -0.143906003929
5   -0.224036184994
6    0.071309219267
7    0.080612609151
8   -0.038029936935
9   -0.016574541631
10   0.012550998556
11   0.000429577973
12  -0.001801640704
13   0.000353713800
];


coeff16 = [
0    0.054415842243
1    0.312871590914
2    0.675630736297
3    0.585354683654
4   -0.015829105256
5   -0.284015542962
6    0.000472484574
7    0.128747426620
8   -0.017369301002
9   -0.044088253931
10   0.013981027917
11   0.008746094047
12  -0.004870352993
13  -0.000391740373
14   0.000675449406
15  -0.000117476784
];


coeff18 = [
0    0.038077947364
1    0.243834674613
2    0.604823123690
3    0.657288078051
4    0.133197385825
5   -0.293273583279
6   -0.096840783223
```

```
7   0.148540749338
8   0.030725681479
9  -0.067632829061
10  0.000250947115
11  0.022361662124
12 -0.004723204758
13 -0.004281503682
14  0.001847646883
15  0.000230385764
16 -0.000251963189
17  0.000039347320
];

coeff20 = [
0   0.026670057901
1   0.188176800078
2   0.527201188932
3   0.688459039454
4   0.281172343661
5  -0.249846424327
6  -0.195946274377
7   0.127369340336
8   0.093057364604
9  -0.071394147166
10 -0.029457536822
11  0.033212674059
12  0.003606553567
13 -0.010733175483
14  0.001395351747
15  0.001992405295
16 -0.000685856695
17 -0.000116466855
18  0.000093588670
19 -0.000013264203
];

% Haar Wavelet coefficients
 g = [1 1]/sqrt (2);
 h = [-1 1]/sqrt (2);

% Lemarie-Battle SCALING coefficients
coeff = [
0   0.7661300537597422
1   0.4339226335893024
2  -0.0502017246714322
3  -0.1100370183880987
4   0.0320808974701767
5   0.0420683514407039
6  -0.0171763154919797
7  -0.0179823209809603
8   0.0086852948130698
9   0.0082014772059938
10 -0.0043538394577629
```

```
11   -0.0038824252655926
12    0.0021867123701413
13    0.0018821335238871
14   -0.0011037398203844
15   -0.0009271987314557
];

name = 'Mallat 23 coefficient symmetric';

coeff = [
-11   -0.002
-10   -0.003
-9    0.006
-8    0.006
-7   -0.013
-6    0.012
-5   -0.030
-4    0.023
-3   -0.078
-2   -0.035
-1    0.307
0     0.542
1     0.307
2    -0.035
3    -0.078
4     0.023
5    -0.030
6     0.012
7    -0.013
8     0.006
9     0.006
10   -0.003
11   -0.002
];

name = 'Vetterli-Herley 22 coefficient';

% An orthonormal wavelet (not biorthogonal) -- 22 scaling coefficients
coeff = [
0     0.055739
1     0.288322
2     0.614682
3     0.608634
4     0.113646
5    -0.290892
6    -0.131805
7     0.162510
8     0.085330
9    -0.099666
10   -0.042965
11    0.060044
12    0.015233
13   -0.032323
```

```
14   -0.001634
15    0.014199
16   -0.002305
17   -0.004433
18    0.001808
19    0.000646
20   -0.000577
21    0.000111
];
```

Appendix B: Riesz Basis

In Chap. 4, we discussed the wavelet transform and the multiscale wavelet analysis and representation technique and the role the Riesz bases play. In this appendix, we define a Riesz basis. Several ways can be used to define a Riesz basis. However, two useful characterizations are as follows:

i. $(u_n)_{n \in N}$ *is a Riesz basis in a Hilbert space H iff:*

* * The closure of the finite linear span of the u_n is H*
     and
* * $\exists A > 0, B < \infty$ so that*

$$A \, \Sigma_n |c_n|^2 \leq \| \Sigma_n c_n u_n \|^2 \leq B \, \Sigma_n |c_n|^2 \qquad \text{(B1)}$$

$$\forall c = (c_n)_{n \in N} \in l^2(N).$$

ii. $(u_n)_{n \in N}$ *is a Riesz basis iff:*

* * The u_n are independent, i.e., no u_{n0} lies within the closure of the finite span of the other u_n,*
     and
* * $\exists A > 0, B < \infty$ so that*

$$A \|f\|^2 \leq \Sigma_n |<f, u_n>|^2 \leq B \|f\|^2 \qquad \text{(B2)}$$

$$\forall f \in H.$$

Appendix C: The QR Factorization (QRF)

In Chap. 6, we discussed the test point selection using wavelet transforms for digital-to-analog converters and the multiscale wavelet analysis and representation technique and the role the QR factorization (QRF) can play. In this appendix, we discuss the QRF. We will mention some of the other ways in which the QRF can be computed.

An $m \times n$ matrix A in the QRF can be written as follows:

$$A = QR, \tag{C1}$$

where:

Q is an $m > n$ is a unitary matrix
R is the upper triangle in $m \times n$

There are at least four ways in which the QRF can be computed. They are as follows:

1. Householder transformations—which can be used to zero out all the elements of a vector except for one component. That is, for the vector $x = [x_1 \, x_2 \, x_3 \ldots \ldots x_n]^T$, there is a vector v in the Householder transformation H_v such that:

$$H_v x = [a \, 0 \, 0 \, 0 \ldots \ldots 0]^T \tag{C2}$$

for some scalar a. Since H_v is unitary, $\|x\|_2 = \|H_v x\|_2$; hence, $a \pm \|x\|_2$.

2. Givens rotations—in this case, quite unlike the Householder transformation which zeros out the entire columns at a stroke, the Givens technique selectively picks the zeros one element at a time, using a rotation.
3. The Gram-Schmidt algorithm—this technique provides an orthonormal basis spanning the column space of A. The Gram-Schmidt algorithm can also be used to determine the dimension of the space spanned by a set of vectors, since a vector linearly independent on other vectors examined a priori in the procedure yields a zero vector.
4. The modified Gram-Schmidt algorithm—is a modified version of the Gram-Schmidt algorithm.

The most important application of the QRF is to the full rank least squares problems.

Appendix D: Signal Power: Parseval's Relation to Fast Fourier Transform

In the context of signal power, Parseval has relationship to fast Fourier transform (FFT). Supposing we consider a signal $S(t)$ and we assume that it is causal, the output in relation to the input signal is as shown in Eq. (D1).

$$S_0(t) = \int_0^t s(\tau)h(t-\tau)d\tau. \tag{D1}$$

At the instant of time t_0, we have the following as shown in Eq. (D2).

$$S_0(t_0) = \int_0^{t0} s(\tau)h(t0-\tau)d\tau. \tag{D2}$$

Supposing that if we assume in addition that $s(t)$ is supported only over $[0, t_0)$ (so that we are using the entire signal s to make our decision), then we can write

$$S_0(t_0) = \int_{-\infty}^{\infty} s(\tau)h(t0-\tau)d\tau. \tag{D3}$$

Let $\omega(t) = h(t0-\tau)$. Then,

$$S_0(t_0) = \int_{-\infty}^{\infty} s(\tau)\widetilde{\omega}(\tau)dt = \int_{-\infty}^{\infty} S(f)W(f)dt. \tag{D4}$$

where $S(f)$ and $W(f)$ are the Fourier transforms, respectively, of $s_0(t_0)$ and $\omega(\tau)$ and where the equality follows by Parseval's theorem. Using the definition of ω, we have

$$S_0(t_0) = \int_{-\infty}^{\infty} S(f)H(f)e^{j2\pi ft}{}_0 dt. \tag{D5}$$

The signal power S at some time instant t_0 is $|s_0(t_0)|^2$, or

$$\left| \int_0^{\infty} S = S(f)H(f)e^{j2\pi ft}{}_0 df \right|^2. \tag{D6}$$

Appendix E: Automation Process Testing GUIs Operation Manual for Mixed Signal Systems Using DWT

Operation Manual

In this document, an overview of ADC and DAC testing GUIs is illustrated. From the start menu, select National Instrument LabView 8.6 which is the version used. If you update to the latest version, use that latest version and update accordingly. Once LabView program is opened, select "Testing.vi" to launch the main GUIs as seen in Fig. E1.

First, select ADC testing tab (1) or DAC testing tab (2) as shown in Fig. E1, in order to configure testing parameters for the DUT. As illustrated in Fig. E2, for DAC DUT, select pattern generator as in (3), and set the number of channels as in (4). In (5), select the proper waveform pattern, amplitude, and number of samples to be generated. To set up the digitizer (oscilloscope), select Dev 4 as in (6) and the over sample rate if needed as in (7). Select channel 1 or 0 for the digitizer input as in (8). Also, for the digitizer setup, select the desired stimulus waveform vertical range and trigger level as in (9 and 10). Select the DUT number of bit (11) and clocking frequency as in (12). In (14), select type of DWT, and set the device voltage range PK-PK as in (13). Also, per the devise specification, set the device range for DNL if desired. Finally, in (16), select run test.

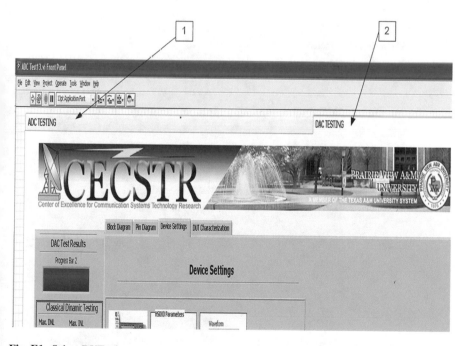

Fig. E1 Select DUT tab

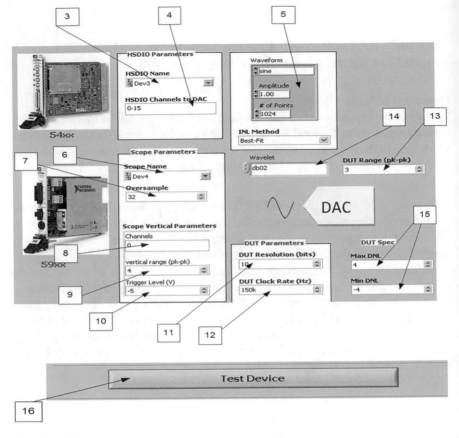

Fig. E2 DUT testing parameters setup

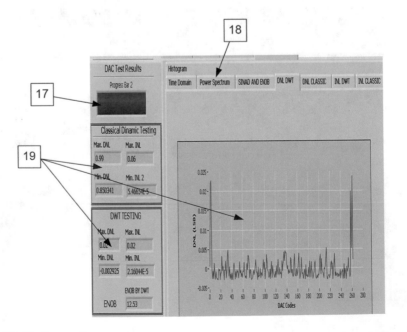

Fig. E3 Testing results

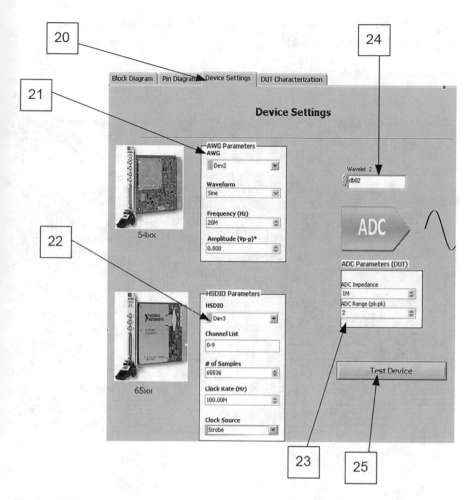

Fig. E4 ADC testing setup

Once the test is in progress, as in Fig. E3, the status bar in (17) will indicate the testing progress. When testing is completed, select the desired testing parameters measured as in (18). Also, testing results can be viewed as in (19).

Meanwhile, in testing ADC, select ADC tab as in (1); then from Fig. E4, select testing parameters setup as in (20). Arbitrary analog waveform generator can be set up in (24) by selecting the device, frequency, and amplitude. Then in (22), set the logic analyzer parameters in terms of total collected samples and clocking source. In (23), select the DUT impedance. In (24), select type of desired DWT, and then select (25) to run the test. Output results can be viewed as in Fig. E3.

Appendix F: 12-Bit and 14-Bit ADCs DWT Testing Results (Figs. F1, F2, F3, and F4)

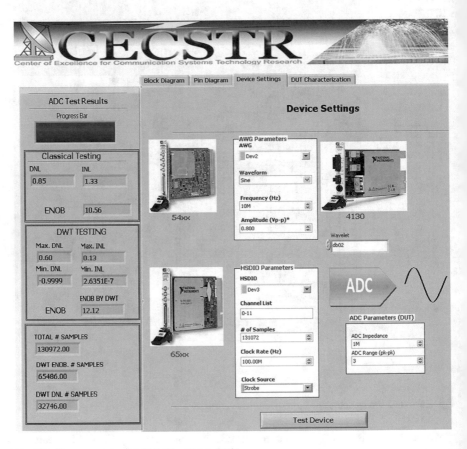

Fig. F1 Test parameters at 10 MHz, 12-bit ADC

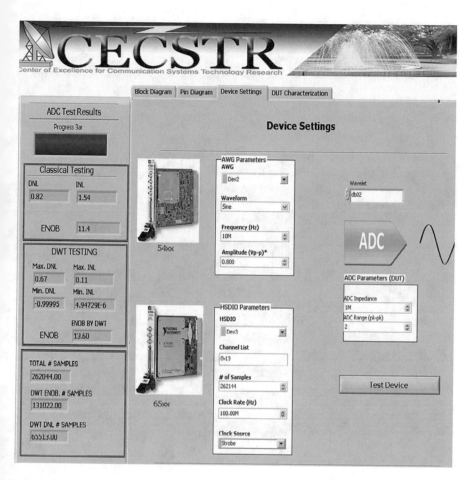

Fig. F2 Test parameters at 10 MHz, 14-bit ADC

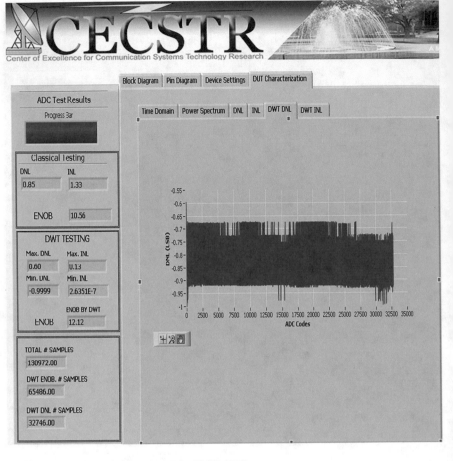

Fig. F3 Instantaneous DNL at 10 MHz, 12-bit ADC

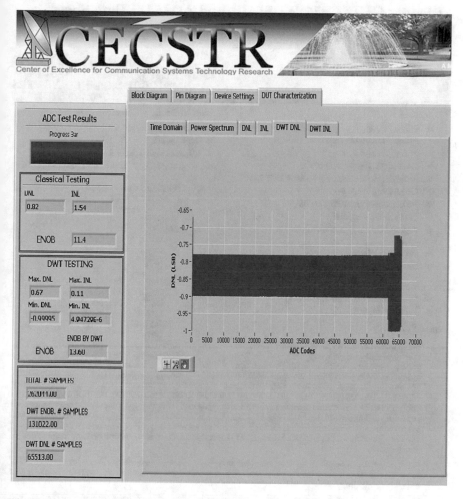

Fig. F4 Instantaneous DNL at 10 MHz, 14-bit ADC

Appendix G: TLC876, ADS5410, and ADS5423 Data Sheet Manual

TLC876M, TLC876I, TLC876C
10-BIT 20 MSPS PARALLEL OUTPUT CMOS
ANALOG-TO-DIGITAL CONVERTERS
SLAS140E – JULY 1997 – REVISED OCTOBER 2000

features

- 10-Bit Resolution 20 MSPS Sampling Analog-to-Digital Converter (ADC)
- Power Dissipation . . . 107 mW Typ
- 5-V Single Supply Operation
- Differential Nonlinearity . . . ±0.5 LSB Typ
- No Missing Codes
- Power Down (Standby) Mode
- Three State Outputs
- Digital I/Os Compatible With 5-V or 3.3-V Logic
- Adjustable Reference Input
- Small Outline Package (SOIC), Super Small Outline Package (SSOP), or Thin Small Outline Package (TSOP)
- Pin Compatible With the Analog Devices AD876

applications

- Communications
- Multimedia
- Digital Video Systems
- High-Speed DSP Front-End . . . TMS320C6x

DB, DW, OR PW PACKAGE
(TOP VIEW)

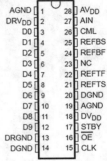

AGND	1	28	AV$_{DD}$
DRV$_{DD}$	2	27	AIN
D0	3	26	CML
D1	4	25	REFBS
D2	5	24	REFBF
D3	6	23	NC
D4	7	22	REFTF
D5	8	21	REFTS
D6	9	20	DGND
D7	10	19	AGND
D8	11	18	DV$_{DD}$
D9	12	17	STBY
DRGND	13	16	OE
DGND	14	15	CLK

NC – No internal connection

description

The TLC876 is a CMOS, low-power, 10-bit, 20 MSPS analog-to-digital converter (ADC). The speed, resolution, and single-supply operation are suited for applications in video, multimedia, imaging, high-speed acquisition, and communications. The low-power and single-supply operation satisfy requirements for high-speed portable applications. The speed and resolution ideally suit charge-coupled device (CCD) input systems such as color scanners, digital copiers, electronic still cameras, and camcorders. A multistage pipelined architecture with output error correction logic provides for no missing codes over the full operating temperature range. Force and sense connections to the reference inputs provide a more accurate internal reference voltage to the reference resistor string.

A standby mode of operation reduces the power to typically 15 mW. The digital I/O interfaces to either 5-V or 3.3-V logic and the digital output terminals can be placed in a high-impedance state. The format of the output data is straight binary coding.

A pipelined multistaged architecture achieves a high sample rate with low power consumption. The TLC876 distributes the conversion over several smaller ADC sub-blocks, refining the conversion with progressively higher accuracy as the device passes the results from stage to stage. This distributed conversion requires a small fraction of the 1023 comparators used in a traditional flash ADC. A sample-and-hold amplifier (SHA) within each of the stages permits the first stage to operate on a new input sample while the second through the fifth stages operate on the four preceding samples.

The TLC876C is characterized for operation from 0°C to 70°C, the TLC876I is characterized for operation from –40°C to 85°C, and the TLC876M is characterized for operation over the full military temperature range of –55°C to 125°C.

Please be aware that an important notice concerning availability, standard warranty, and use in critical applications of Texas Instruments semiconductor products and disclaimers thereto appears at the end of this data sheet.

TEXAS
INSTRUMENTS

TLC876M, TLC876I, TLC876C
10-BIT 20 MSPS PARALLEL OUTPUT CMOS
ANALOG-TO-DIGITAL CONVERTERS
SLAS140E – JULY 1997 – REVISED OCTOBER 2000

operating characteristics at $AV_{DD} = DV_{DD} = 5$ V, $DRV_{DD} = 3.3$ V, $V_{I(REFT)} = 3.6$ V, $V_{I(REFB)} = 1.6$ V, $f_{CLK} = 20$ MSPS (unless otherwise noted)

dc accuracy

PARAMETER	TEST CONDITIONS	MIN	TYP	MAX	UNIT
Integral nonlinearity (INL)			±1.5		LSB
Differential nonlinearity (DNL) (see Note 1)			±0.5	<±1	
Offset error			−0.4		%FSR
Gain error			0.2		%FSR

NOTE 1:   A differential nonlinearity error of less than ±1 LSB ensures no missing codes.

analog input

PARAMETER		TEST CONDITIONS	MIN	TYP	MAX	UNIT
C_i	Input capacitance			5		pF

reference input

PARAMETER		TEST CONDITIONS	MIN	TYP	MAX	UNIT
R_{ref}	Reference input resistance		350	500		Ω
I_{ref}	Reference input current			4		mA
	Reference top offset voltage			35		mV
	Reference bottom offset voltage			35		mV

dynamic performance†

PARAMETER		TEST CONDITIONS		MIN	TYP	MAX	UNIT
Effective number of bits (ENOB)	All suffixes	$f_I = 1$ MHz			8.5		Bits
	All suffixes	$f_I = 3.58$ MHz, $T_A = 25°C$		8	8.5		
	C and I suffixes	$f_I = 3.58$ MHz, T_A = Full Range		8	8.5		
	M suffix				7.5		
	All suffixes	$f_I = 10$ MHz			8.1		
Signal-to-total harmonic distortion+noise (S/(THD+N))	All suffixes	$f_I = 1$ MHz			53		dB
	All suffixes	$f_I = 3.58$ MHz, $T_A = 25°C$		50	53		
	C and I suffixes	$f_I = 3.58$ MHz, T_A = Full Range		50	53		
	M suffix				47		
	All suffixes	$f_I = 10$ MHz			51		
Total harmonic distortion (THD)		$f_I = 1$ MHz			−63		dB
		$f_I = 3.58$ MHz			−62	−56	
		$f_I = 10$ MHz			−61		
Spurious free dynamic range		$f_I = 3.58$ MHz			−64		dB
BW	Analog input full-power bandwidth				200		MHz
	Differential phase				0.5		degrees
	Differential gain				1%		

† The voltage difference between AV_{DD} and DV_{DD} cannot exceed 0.5 V to maintain performance specifications. At input clock rise times less than 20 ns, the offset full-scale error increases approximately by a factor of $(20/t_r)0.5$ where t_r equals the actual rise time in nanoseconds.

TEXAS
INSTRUMENTS

TEXAS INSTRUMENTS
www.ti.com

ADS5410

SLAS346 – JUNE 2002

12-BIT, 80 MSPS CommsADC™
ANALOG-TO-DIGITAL CONVERTER

FEATURES

- 80-MSPS Maximum Sample Rate
- 12-Bit Resolution
- No Missing Codes
- 360-mW Power Dissipation
- CMOS Technology
- On-Chip S/H
- 75 dB Spurious Free Dynamic Range at 100 MHz IF
- 1-GHz Bandwidth Differential Analog Input
- On-Chip References
- 2s Complement Digital Output
- 3.3-V Analog, 1.8-V Digital Supply
- 1.8 V–3.3 V I/O

APPLICATIONS

- Cellular Base Transceiver Station Receive Channel
  - IF Sampling Applications
  - TDMA: GSM, IS-136, EDGE/UWC-136
  - CDMA: IS-95, UMTS, CDMA2000
  - Wireless Local Loop
  - LMDS, MMDS
  - Wideband Baseband Receivers
- Medical Imaging:
  - Ultrasound
  - Magnetic Resonant Imaging
- Portable Instrumentation

DESCRIPTION

The ADS5410 is a high-speed, high-performance pipelined analog-to-digital converter with exceptionally low-noise and high spurious-free dynamic range. The ADS5410 high input bandwidth makes it ideal for IF subsampling solutions where digital I/Q demodulators are used. Its high dynamic range makes it well suited for GSM, IS-95, UMTS, and IS-136 digital receivers. Its linearity and low DNL make it ideal for medical imaging applications. Low power consumption makes the ADS5410 ideal for applications in compact pico- and micro-base stations and in portable designs.

FUNCTIONAL BLOCK DIAGRAM

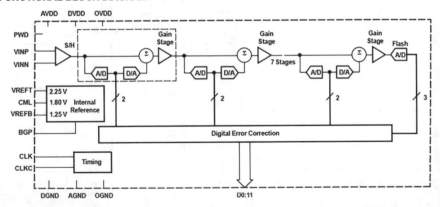

⚠ Please be aware that an important notice concerning availability, standard warranty, and use in critical applications of Texas Instruments semiconductor products and disclaimers thereto appears at the end of this data sheet.

CommsADC is a trademark of Texas Instruments.

Texas
Instruments
www.ti.com

ADS5410

SLAS346 – JUNE 2002

DC ELECTRICAL CHARACTERISTICS

over operating free-air temperature range, clock frequency = 80 MSPS, 50% clock duty cycle (AVDD = 3.3 V, DVDD = 1.8 V, OVDD = 1.8 V) (unless otherwise noted)

	PARAMETER	TEST CONDITIONS	MIN	TYP	MAX	UNIT
DC Accuracy[1]						
	No missing codes	Fs = 88 MSPS[2]		Assured		
DNL	Differential nonlinearity		-0.9	±0.5	1	LSB
INL	Integral nonlinearity		-2	±1.5	2	LSB
E_O	Offset error			3		mV
E_G	Gain error			0.5		%FS
Power Supply						
$I_{(AVDD)}$	Analog supply current			105		
$I_{(DVDD)}$	Digital supply current	Fs = 80 MSPS, A_i = FS, f_i = 2 MHz		1		mA
$I_{(OVDD)}$	Digital output driver supply current			3.5		
	Power dissipation			360	450	mW
	Power down dissipation	PWDN = high		30	45	mW
PSRR	Power supply rejection ratio			±0.3		mV/V
References						
$V_{ref(VREFB)}$	Reference bottom		1.1	1.25	1.4	V
$V_{ref(VREFT)}$	Reference top		2.1	2.25	2.4	V
	$V_{REFT} - V_{REFB}$			1.06		V
	$V_{REFT} - V_{REFB}$ variation (6σ)			0.06		V
$V_{OC(CML)}$	Common mode output voltage			1.8		V
Digital Inputs (PWD)						
I_{IH}	High-level input current	V_i = 1.6 V	-10		10	µA
I_{IL}	Low-level input current	V_i = 0.3 V	-10		10	µA
V_{IH}	High-level input voltage		1.8			V
V_{IL}	Low-level input voltage				0.8	V
Digital Outputs						
V_{OH}	High-level output voltage	I_{OH} = -50 µA	1.4			V
V_{OL}	Low-level output voltage	I_{OL} = 50 µA			0.4	V

(1) Fs = 80 MSPS, sinewave input, f_i = 2 MHz
(2) Speed margin test

 **TEXAS INSTRUMENTS**
www.ti.com

ADS5423

SLWS160A – FEBRUARY 2005 – REVIISED JANUARY 2010

14 Bit, 80 MSPS
Analog-to-Digital Converter

FEATURES

- 14 Bit Resolution
- 80 MSPS Maximum Sample Rate
- SNR = 74 dBc at 80 MSPS and 50 MHz IF
- SFDR = 94 dBc at 80 MSPS and 50 MHz IF
- 2.2 V_{pp} Differential Input Range
- 5 V Supply Operation
- 3.3 V CMOS Compatible Outputs
- 1.85 W Total Power Dissipation
- 2s Complement Output Format
- On-Chip Input Analog Buffer, Track and Hold, and Reference Circuit
- 52 Pin HTQFP Package With Exposed Heatsink
- Pin Compatible to the AD6644/45
- Industrial Temperature Range = −40°C to 85°C

APPLICATIONS

- Single and Multichannel Digital Receivers
- Base Station Infrastructure
- Instrumentation
- Video and Imaging

RELATED DEVICES

- Clocking: CDC7005
- Amplifiers: OPA695, THS4509

DESCRIPTION

The ADS5423 is a 14 bit 80 MSPS analog-to-digital converter (ADC) that operates from a 5 V supply, while providing 3.3 V CMOS compatible digital outputs. The ADS5423 input buffer isolates the internal switching of the on-chip Track and Hold (T&H) from disturbing the signal source. An internal reference generator is also provided to further simplify the system design. The ADS5423 has outstanding low noise and linearity, over input frequency. With only a 2.2 V_{PP} input range, simplifies the design of multicarrier applications, where the carriers are selected on the digital domain.

The ADS5423 is available in a 52 pin HTQFP with heatsink package and is pin compatible to the AD6645. The ADS5423 is built on state of the art Texas Instruments complementary bipolar process (BiCom3) and is specified over full industrial temperature range (−40°C to 85°C).

FUNCTIONAL BLOCK DIAGRAM

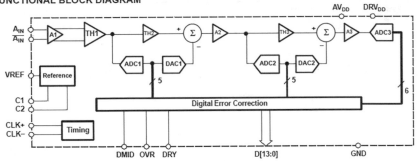

TEXAS INSTRUMENTS
www.ti.com

SLWS160A – FEBRUARY 2005 – REVIISED JANUARY 2010

ELECTRICAL CHARACTERISTICS

Over full temperature range (T_{MIN} = –40°C to T_{MAX} = 85°C), sampling rate = 80 MSPS, 50% clock duty cycle, AV_{DD} = 5 V, DRV_{DD} = 3.3 V, –1 dBFS differential input, and 3 V_{PP} differential sinusoidal clock, unless otherwise noted

PARAMETER	TEST CONDITIONS	MIN	TYP	MAX	UNIT
Resolution			14		Bits
Analog Inputs					
Differential input range			2.2		V_{PP}
Differential input resistance	See Figure 30		1		$k\Omega$
Differential input capacitance	See Figure 30		1.5		pF
Analog input bandwidth			570		MHz
Internal Reference Voltages					
Reference voltage, V_{REF}			2.4		V
Dynamic Accuracy					
No missing codes			Tested		
Differential linearity error, DNL	f_{IN} = 5 MHz	–0.95	±0.5	1.5	LSB
Integral linearity error, INL	f_{IN} = 5 MHz		±1.5		LSB
Offset error		–5	0	5	mV
Offset temperature coefficient			1.7		ppm/°C
Gain error		–5	0.9	5	%FS
PSRR			1		mV/V
Gain temperature coefficient			77		ppm/°C
Power Supply					
Analog supply current, I_{AVDD}	V_{IN} = full scale, f_{IN} = 70 MHz		355	410	mA
Output buffer supply current, I_{DRVDD}	V_{IN} = full scale, f_{IN} = 70 MHz		35	42	mA
Power dissipation	Total power with 10-pF load on each digital output to ground, f_{IN} = 70 MHz		1.85	2.2	W
Power-up time			20	100	ms
Dynamic AC Characteristics					
Signal-to-noise ratio, SNR	f_{IN} = 10 MHz		74.6		dBc
	f_{IN} = 30 MHz	73	74.3		
	f_{IN} = 50 MHz		74.2		
	f_{IN} = 70 MHz	73	74.1		
	f_{IN} = 100 MHz		73.6		
	f_{IN} = 170 MHz		72		
	f_{IN} = 230 MHz		71.5		
Spurious-free dynamic range, SFDR	f_{IN} = 10 MHz		94		dBc
	f_{IN} = 30 MHz	85	93		
	f_{IN} = 50 MHz		94		
	f_{IN} = 70 MHz		90		
	f_{IN} = 100 MHz		86		
	f_{IN} = 170 MHz		73		
	f_{IN} = 230 MHz		64		

Appendix H: 12-Bit and 14-Bit DWT DACs Testing (Figs. H1, H2, H3, H4, and H5)

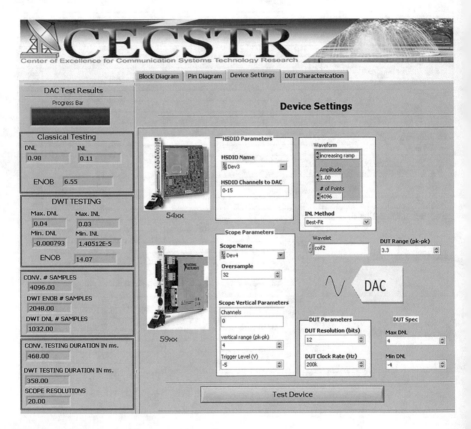

Fig. H1 Test parameters setup, 12-bit DAC

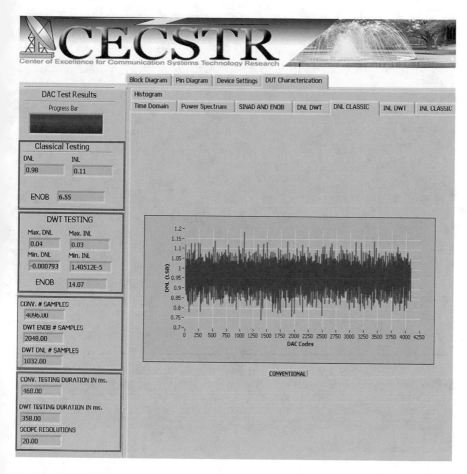

Fig. H2 Classical DNL, 12-bit DAC

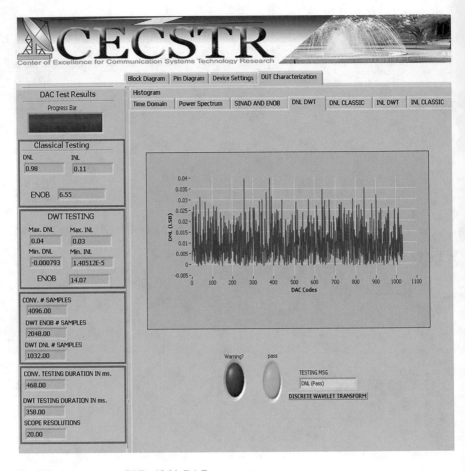

Fig. H3 Instantaneous DNL, 12-bit DAC

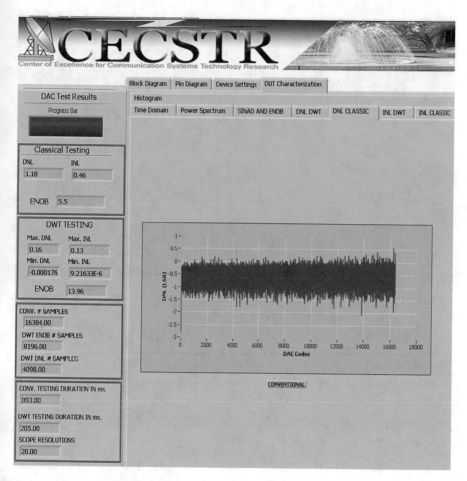

Fig. II4 Classical DNL, 14-bit DAC

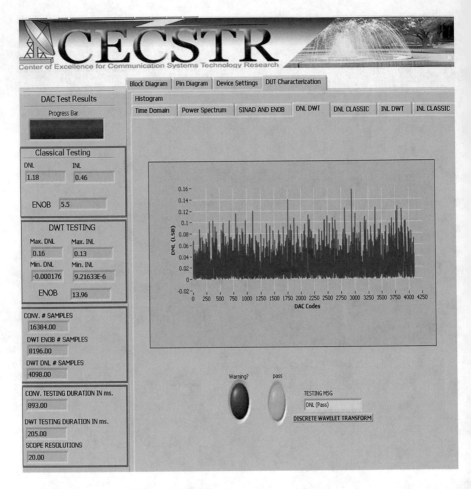

Fig. H5 Instantaneous DNL, 14-bit DAC

Appendix I: DAC2900, DAC2902, and DAC2904 Data Sheet Manual

 **TEXAS INSTRUMENTS**

DAC2900

www.ti.com

SBAS166C – JUNE 2001 – REVISED SEPTEMBER 2008

Dual, 10-Bit, 125MSPS
DIGITAL-TO-ANALOG CONVERTER

FEATURES

- 125MSPS UPDATE RATE
- SINGLE SUPPLY: +3.3V or +5V
- HIGH SFDR: 68dB at f_{OUT} = 20MHz
- LOW GLITCH: 2pV-s
- LOW POWER: 310mW at +5V
- INTERNAL REFERENCE
- POWER-DOWN MODE: 23mW

APPLICATIONS

- COMMUNICATIONS:
  Base Stations, WLL, WLAN
  Baseband I/Q Modulation
- MEDICAL/TEST INSTRUMENTATION
- ARBITRARY WAVEFORM GENERATORS (ARB)
- DIRECT DIGITAL SYNTHESIS (DDS)

DESCRIPTION

The DAC2900 is a monolithic, 10-bit, dual-channel, high-speed Digital-to-Analog Converter (DAC), and is optimized to provide high dynamic performance while dissipating only 310mW on a +5V single supply.

Operating with high update rates of up to 125MSPS, the DAC2900 offers exceptional dynamic performance, and enables the generation of very high output frequencies suitable for *Direct IF* applications. The DAC2900 has been optimized for communications applications in which separate I and Q data are processed while maintaining tight gain and offset matching.

Each DAC has a high-impedance differential-current output, suitable for single-ended or differential analog output configurations.

The DAC2900 combines high dynamic performance with a high throughput rate to create a cost-effective solution for a wide variety of waveform-synthesis applications:

- Pin compatibility between family members provides 10-bit (DAC2900), 12-bit (DAC2902), and 14-bit (DAC2904) resolution.
- Pin compatible to the AD9763 dual DAC.
- Gain matching is typically 0.5% of full-scale, and offset matching is specified at 0.02% max.
- The DAC2900 utilizes an advanced CMOS process; the segmented architecture minimizes output glitch energy, and maximizes dynamic performance.
- All digital inputs are +3.3V and +5V logic compatible. The DAC2900 has an internal reference circuit, and allows use of an external reference.
- The DAC2900 is available in a TQFP-48 package, and is specified over the extended industrial temperature range of –40°C to +85°C.

 Please be aware that an important notice concerning availability, standard warranty, and use in critical applications of Texas Instruments semiconductor products and disclaimers thereto appears at the end of this data sheet.

All trademarks are the property of their respective owners.

 TEXAS INSTRUMENTS
www.ti.com

Copyright © 2001-2008, Texas Instruments Incorporated

ABSOLUTE MAXIMUM RATINGS

+V_A to AGND	–0.3V to +6V
+V_D to DGND	–0.3V to +6V
AGND to DGND	–0.3V to +0.3V
+V_A to +V_D	–6V to +6V
CLK, PD to DGND	–0.3V to V_D + 0.3V
D0-D9 to DGND	–0.3V to V_D + 0.3V
I_{OUT}, $\overline{I_{OUT}}$ to AGND	–1V to V_A + 0.3V
BW, BYP to AGND	–0.3V to V_A + 0.3V
REF_{IN}, FSA to AGND	–0.3V to V_A + 0.3V
$\overline{INT}/EXT$ to AGND	–0.3V to V_A + 0.3V
Junction Temperature	+150°C
Case Temperature	+100°C
Storage Temperature	+125°C

ELECTROSTATIC DISCHARGE SENSITIVITY

This integrated circuit can be damaged by ESD. Texas Instruments recommends that all integrated circuits be handled with appropriate precautions. Failure to observe proper handling and installation procedures can cause damage.

ESD damage can range from subtle performance degradation to complete device failure. Precision integrated circuits may be more susceptible to damage because very small parametric changes could cause the device not to meet its published specifications.

PACKAGE/ORDERING INFORMATION[1]

PRODUCT	PACKAGE	PACKAGE DESIGNATOR	SPECIFIED TEMPERATURE RANGE	PACKAGE MARKING	ORDERING NUMBER	TRANSPORT MEDIA, QUANTITY
DAC2900Y	TQFP-48	PFB	–40°C to +85°C	DAC2900Y	DAC2900Y/250	Tape and Reel, 250
"	"	"	"	"	DAC2900Y/1K	Tape and Reel, 1000

NOTE: (1) For the most current package and ordering information, see the Package Option Addendum at the end of this document, or see the TI website at www.ti.com.

PRODUCT	EVM ORDERING NUMBER	COMMENT
DAC2900	DAC2900-EVM	Fully populated evaluation board. See user manual for details.

ELECTRICAL CHARACTERISTICS

At T_{MIN} to T_{MAX}, +V_A = +5V, +V_D = +3.3V, differential transformer coupled output, and 50Ω doubly-terminated, unless otherwise noted. Independent Gain mode.

PARAMETER	TEST CONDITIONS	DAC2900Y MIN	DAC2900Y TYP	DAC2900Y MAX	UNIT
RESOLUTION			10		Bits
Output Update Rate (f_{CLOCK})			125		MSPS
STATIC ACCURACY[1]					
Differential Nonlinearity (DNL)	T_A = +25°C		±0.25		LSB
	T_{MIN} to T_{MAX}	–1.0		+1.0	LSB
Integral Nonlinearity (INL)	T_A = +25°C		±0.25		LSB
	T_{MIN} to T_{MAX}	–1.0		+1.0	LSB
DYNAMIC PERFORMANCE					
Spurious-Free Dynamic Range (SFDR)	To Nyquist				
f_{OUT} = 1MHz, f_{CLOCK} = 50MSPS	0dBFS Output	70	80		dBc
	–6dBFS Output		75		dBc
	–12dBFS Output		70		dBc
f_{OUT} = 1MHz, f_{CLOCK} = 26MSPS			80		dBc
f_{OUT} = 2.18MHz, f_{CLOCK} = 52MSPS			80		dBc
f_{OUT} = 5.24MHz, f_{CLOCK} = 52MSPS			80		dBc
f_{OUT} = 10.4MHz, f_{CLOCK} = 78MSPS			75		dBc
f_{OUT} = 15.7MHz, f_{CLOCK} = 78MSPS			71		dBc
f_{OUT} = 5.04MHz, f_{CLOCK} = 100MSPS			80		dBc
f_{OUT} = 20.2MHz, f_{CLOCK} = 100MSPS			68		dBc
f_{OUT} = 20.1MHz, f_{CLOCK} = 125MSPS			61		dBc
f_{OUT} = 40.2MHz, f_{CLOCK} = 125MSPS			56		dBc
Spurious-Free Dynamic Range within a Window					
f_{OUT} = 1.0MHz, f_{CLOCK} = 50MSPS	2MHz Span		86		dBc
f_{OUT} = 5.02MHz, f_{CLOCK} = 50MSPS	10MHz Span		80		dBc
f_{OUT} = 5.03MHz, f_{CLOCK} = 78MSPS	10MHz Span		80		dBc
f_{OUT} = 5.04MHz, f_{CLOCK} = 125MSPS	10MHz Span		80		dBc
Total Harmonic Distortion (THD)					
f_{OUT} = 1MHz, f_{CLOCK} = 50MSPS			–77	–68	dBc
f_{OUT} = 5.02MHz, f_{CLOCK} = 50MSPS			–74		dBc
f_{OUT} = 5.03MHz, f_{CLOCK} = 78MSPS			–73		dBc
f_{OUT} = 5.04MHz, f_{CLOCK} = 125MSPS			–70		dBc
Multitone Power Ratio	8 Tone with 110kHz Spacing				
f_{OUT} = 2.0MHz to 2.99MHz, f_{CLOCK} = 65MSPS	0dBFS Output		80		dBc

2

TEXAS INSTRUMENTS
www.ti.com

DAC2900
SBAS166C

DAC2902

www.ti.com

SBAS167C– JUNE 2001 – REVISED SEPTEMBER 2008

Dual, 12-Bit, 125MSPS
DIGITAL-TO-ANALOG CONVERTER

FEATURES

● **125MSPS UPDATE RATE**
● **SINGLE SUPPLY: +3.3V or +5V**
● **HIGH SFDR: 70dB at f_{OUT} = 20MHz**
● **LOW GLITCH: 2pV-s**
● **LOW POWER: 310mW**
● **INTERNAL REFERENCE**
● **POWER-DOWN MODE: 23mW**

APPLICATIONS

● **COMMUNICATIONS:**
  Base Stations, WLL, WLAN
  Baseband I/Q Modulation
● **MEDICAL/TEST INSTRUMENTATION**
● **ARBITRARY WAVEFORM GENERATORS (ARB)**
● **DIRECT DIGITAL SYNTHESIS (DDS)**

DESCRIPTION

The DAC2902 is a monolithic, 12-bit, dual-channel, high-speed Digital-to-Analog Converter (DAC), and is optimized to provide high dynamic performance while dissipating only 310mW.

Operating with high update rates of up to 125MSPS, the DAC2902 offers exceptional dynamic performance, and enables the generation of very high output frequencies suitable for *Direct IF* applications. The DAC2902 has been optimized for communications applications in which separate I and Q data are processed while maintaining tight gain and offset matching.

Each DAC has a high-impedance differential-current output, suitable for single-ended or differential analog output configurations.

The DAC2902 combines high dynamic performance with a high throughput rate to create a cost-effective solution for a wide variety of waveform-synthesis applications:

• Pin compatibility between family members provides 10-bit (DAC2000), 12-bit (DAC2002), and 14-bit (DAC2004) resolution.

• Pin compatible to the AD9765 dual DAC.

• Gain matching is typically 0.5% of full-scale, and offset matching is specified at 0.02% max.

• The DAC2902 utilizes an advanced CMOS process; the segmented architecture minimizes output glitch energy, and maximizes the dynamic performance.

• All digital inputs are +3.3V and +5V logic compatible. The DAC2902 has an internal reference circuit, and allows use of an external reference.

• The DAC2902 is available in a TQFP-48 package, and is specified over the extended industrial temperature range of –40°C to +85°C.

 Please be aware that an important notice concerning availability, standard warranty, and use in critical applications of Texas Instruments semiconductor products and disclaimers thereto appears at the end of this data sheet.

All trademarks are the property of their respective owners.

 TEXAS
INSTRUMENTS
www.ti.com

ABSOLUTE MAXIMUM RATINGS

$+V_A$ to AGND	–0.3V to +6V
$+V_D$ to DGND	–0.3V to +6V
AGND to DGND	–0.3V to +0.3V
$+V_A$ to $+V_D$	–6V to +6V
CLK, PD, WRT to DGND	–0.3V to V_D + 0.3V
D0-D11 to DGND	–0.3V to V_D + 0.3V
I_{OUT}, $\overline{I_{OUT}}$ to AGND	–1V to V_A + 0.3V
GSET to AGND	–0.3V to V_A + 0.3V
REF_{IN}, FSA to AGND	–0.3V to V_A + 0.3V
Junction Temperature	+150°C
Case Temperature	+100°C
Storage Temperature	+125°C

⚠ ELECTROSTATIC DISCHARGE SENSITIVITY

This integrated circuit can be damaged by ESD. Texas Instruments recommends that all integrated circuits be handled with appropriate precautions. Failure to observe proper handling and installation procedures can cause damage.

ESD damage can range from subtle performance degradation to complete device failure. Precision integrated circuits may be more susceptible to damage because very small parametric changes could cause the device not to meet its published specifications.

PACKAGE/ORDERING INFORMATION[1]

PRODUCT	PACKAGE	PACKAGE DESIGNATOR	SPECIFIED TEMPERATURE RANGE	PACKAGE MARKING	ORDERING NUMBER	TRANSPORT MEDIA, QUANTITY
DAC2902Y	TQFP-48	PFB	–40°C to +85°C	DAC2902Y	DAC2902Y/250	Tape and Reel, 250
"	"	"	"	"	DAC2902Y/1K	Tape and Reel, 1000

NOTE: (1) For the most current package and ordering information, see the Package Option Addendum at the end of this document, or see the TI web site at www.ti.com.

PRODUCT	EVM ORDERING NUMBER	COMMENT
DAC2902	DAC2902-EVM	Fully populated evaluation board. See user manual for details.

ELECTRICAL CHARACTERISTICS

At T_{MIN} to T_{MAX}, $+V_A$ = +5V, $+V_D$ = +3.3V, differential transformer coupled output, and 50Ω doubly-terminated, unless otherwise noted. Independent Gain mode.

		DAC2902Y			
PARAMETER	CONDITIONS	MIN	TYP	MAX	UNIT
RESOLUTION			12		Bits
Output Update Rate (f_{CLOCK})			125		MSPS
STATIC ACCURACY[1]					
Differential Nonlinearity (DNL)	T_A = +25°C	–2.0	±1	+2.0	LSB
	T_{MIN} to T_{MAX}	–2.5		+2.5	LSB
Integral Nonlinearity (INL)	T_A = +25°C	–2.0	±1	+2.0	LSB
	T_{MIN} to T_{MAX}	–3.0		+3.0	LSB
DYNAMIC PERFORMANCE					
Spurious-Free Dynamic Range (SFDR)	To Nyquist				
f_{OUT} = 1MHz, f_{CLOCK} = 50MSPS	0dBFS Output	72	82		dBc
	–6dBFS Output		77		dBc
	–12dBFS Output		72		dBc
f_{OUT} = 1MHz, f_{CLOCK} = 26MSPS			81		dBc
f_{OUT} = 2.18MHz, f_{CLOCK} = 52MSPS			81		dBc
f_{OUT} = 5.24MHz, f_{CLOCK} = 52MSPS			81		dBc
f_{OUT} = 10.4MHz, f_{CLOCK} = 78MSPS			77		dBc
f_{OUT} = 15.7MHz, f_{CLOCK} = 78MSPS			71		dBc
f_{OUT} = 5.04MHz, f_{CLOCK} = 100MSPS			80		dBc
f_{OUT} = 20.2MHz, f_{CLOCK} = 100MSPS			70		dBc
f_{OUT} = 20.1MHz, f_{CLOCK} = 125MSPS			72		dBc
f_{OUT} = 40.2MHz, f_{CLOCK} = 125MSPS			64		dBc
Spurious-Free Dynamic Range within a Window					
f_{OUT} = 1.0MHz, f_{CLOCK} = 50MSPS	2MHz Span	80	90		dBc
f_{OUT} = 5.02MHz, f_{CLOCK} = 50MSPS	10MHz Span		88		dBc
f_{OUT} = 5.03MHz, f_{CLOCK} = 78MSPS	10MHz Span		88		dBc
f_{OUT} = 5.04MHz, f_{CLOCK} = 125MSPS	10MHz Span		88		dBc
Total Harmonic Distortion (THD)					
f_{OUT} = 1MHz, f_{CLOCK} = 50MSPS			–79	–70	dBc
f_{OUT} = 5.02MHz, f_{CLOCK} = 50MSPS			–77		dBc
f_{OUT} = 5.03MHz, f_{CLOCK} = 78MSPS			–76		dBc
f_{OUT} = 5.04MHz, f_{CLOCK} = 125MSPS			–75		dBc
Multitone Power Ratio	8 Tone with 110kHz Spacing				
f_{OUT} = 2.0MHz to 2.99MHz, f_{CLOCK} = 65MSPS	0dBFS Output		80		dBc

Texas Instruments
www.ti.com

DAC2902
SBAS167C

TEXAS INSTRUMENTS

DAC2904

www.ti.com   SBAS198C – AUGUST 2001 – REVISED OCTOBER 2009

Dual, 14-Bit, 125MSPS
DIGITAL-TO-ANALOG CONVERTER

Check for Samples: DAC2904

FEATURES

- **125MSPS UPDATE RATE**
- **SINGLE SUPPLY: +3.3V or +5V**
- **HIGH SFDR: 78dB at f_{OUT} = 10MHz**
- **LOW GLITCH: 2pV-s**
- **LOW POWER: 310mW**
- **INTERNAL REFERENCE**
- **POWER-DOWN MODE: 23mW**

DESCRIPTION

The DAC2904 is a monolithic, 14-bit, dual-channel, high-speed Digital-to-Analog Converter (DAC), and is optimized to provide high dynamic performance while dissipating only 310mW.

Operating with high update rates of up to 125MSPS, the DAC2904 offers exceptional dynamic performance, and enables the generation of very-high output frequencies suitable for "Direct IF" applications. The DAC2904 has been optimized for communications applications in which separate I and Q data are processed while maintaining tight-gain and offset matching.

Each DAC has a high-impedance differential-current output, suitable for single-ended or differential analog-output configurations.

APPLICATIONS

- **COMMUNICATIONS:**
  - **Base Stations, WLL, WLAN**
  - **Baseband I/Q Modulation**
- **MEDICAL/TEST INSTRUMENTATION**
- **ARBITRARY WAVEFORM GENERATORS (ARB)**
- **DIRECT DIGITAL SYNTHESIS (DDS)**

The DAC2904 combines high dynamic performance with a high update rate to create a cost-effective solution for a wide variety of waveform-synthesis applications:

- Pin compatibility between family members provides 10-bit (DAC2900), 12-bit (DAC2902), and 14-bit (DAC2904) resolution.
- Pin compatible to the AD9767 dual DAC.
- Gain matching is typically 0.5% of full-scale, and offset matching is specified at 0.02% max.
- The DAC2904 utilizes an advanced CMOS process, the segmented architecture minimizes output-glitch energy, and maximizes the dynamic performance.
- All digital inputs are +3.3V and +5V logic compatible. The DAC2904 has an internal reference circuit, and allows use in a multiplying configuration.

The DAC2904 is available in a TQFP-48 package, and is specified over the extended industrial temperature range of –40°C to +85°C.

TEXAS
INSTRUMENTS

DAC2904

www.ti.com   SBAS198C –AUGUST 2001–REVISED OCTOBER 2009

ELECTRICAL CHARACTERISTICS

T_{MIN} to T_{MAX}, $+V_A$ = +5V, $+V_D$ = +3.3V, differential transformer coupled output, and 50Ω doubly-terminated, unless otherwise noted. Independent Gain Mode.

PARAMETER	TEST CONDITIONS	DAC2904			UNIT
		MIN	TYP	MAX	
RESOLUTION					
Resolution			14		Bits
Output Update Rate (f_{CLOCK})			125		MSPS
STATIC ACCURACY[1]					
Differential Nonlinearity (DNL)	T_A = +25°C		±4.0		LSB
Integral Nonlinearity (INL)	T_A = +25°C		±5.0		LSB
DYNAMIC PERFORMANCE					
Spurious-Free Dynamic Range (SFDR)	To Nyquist				
	0dBFS Output	71	82		dBc
f_{OUT} = 1MHz, f_{CLOCK} = 50MSPS	–6dBFS Output		77		dBc
	–12dBFS Output		72		dBc
f_{OUT} = 1MHz, f_{CLOCK} = 26MSPS			82		dBc
f_{OUT} = 2.18MHz, f_{CLOCK} = 52MSPS			81		dBc
f_{OUT} = 5.24MHz, f_{CLOCK} = 52MSPS			81		dBc
f_{OUT} = 10.4MHz, f_{CLOCK} = 78MSPS			78		dBc
f_{OUT} = 15.7MHz, f_{CLOCK} = 78MSPS			72		dBc
f_{OUT} = 5.04MHz, f_{CLOCK} = 100MSPS			80		dBc
f_{OUT} = 20.2MHz, f_{CLOCK} = 100MSPS			69		dBc
f_{OUT} = 20.1MHz, f_{CLOCK} = 125MSPS			69		dBc
f_{OUT} = 40.2MHz, f_{CLOCK} = 125MSPS			64		dBc
Spurious-Free Dynamic Range within a Window					
f_{OUT} = 1MHz, f_{CLOCK} = 50MSPS	2MHz span	80	90		dBc
f_{OUT} = 5.24MHz, f_{CLOCK} = 52MSPS	10MHz span		88		dBc
f_{OUT} = 5.26MHz, f_{CLOCK} = 78MSPS	10MHz span		88		dBc
f_{OUT} = 5.04MHz, f_{CLOCK} = 125MSPS	10MHz span		88		dBc
Total Harmonic Distortion (THD)					dBc
f_{OUT} = 1MHz, f_{CLOCK} = 50MSPS			–79	–70	dBc
f_{OUT} = 5.24MHz, f_{CLOCK} = 52MSPS			–77		dBc
f_{OUT} = 5.26MHz, f_{CLOCK} = 78MSPS			–76		dBc
f_{OUT} = 5.04MHz, f_{CLOCK} = 125MSPS			–75		dBc
Multitone Power Ratio	Eight tone with 110kHz spacing				
f_{OUT} = 2.0MHz to 2.99MHz, f_{CLOCK} = 65MSPS	0dBFS output		80		dBc
Signal-to-Noise Ratio (SNR)					
f_{OUT} = 5.02MHz, f_{CLOCK} = 50MHz	0dBFS output		68		dBc
Signal-to-Noise and Distortion (SINAD)					
f_{OUT} = 5.02MHz, f_{CLOCK} = 50MHz	0dBFS output		67		dBc
Channel Isolation					
f_{OUT} = 1MHz, f_{CLOCK} = 52MSPS			85		dBc
f_{OUT} = 20MHz, f_{CLOCK} = 125MSPS			77		dBc

(1) At output I_{OUT}, while driving a virtual ground.

Appendix J: Samples of Fingerprint Images

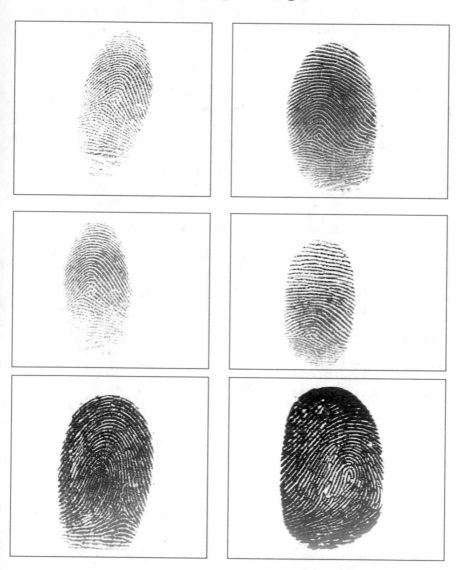

Source: FVC (2000). FVC 2000 "Fingerprint Verification Competition," http://bias.csr.unibo.it/
fvc2000

Appendix K: MATLAB Program Listings for Fingerprint Minutiae Processing Using Six Different Types of Wavelets

Six different wavelets are used in the analysis. The programs are the same for all wavelets except the wavelet name used in each program. The programs for the Bior5.5 wavelet are listed in Appendices K1 and K2.

Appendix K1: Program to Create Database of Statistical Parameters for 80 Fingerprint Images

```
%PROGRAM FOR FINGERPRINT VERIFICATION

%We create a database of 80 images extracted from the Fingerprint
%verification competition (FVC2000) website: http://bias.csr.unibo.
it/fvc2000/

%The Images are storage in C:\DATA. USE ADDPATH TO ADD C:\DATA TO MATLAB
SEARCH
%PATH for our experiment. NOTE: When using this program, you have to
create %your own MATLAB PATH and make the necessary changes in the
program.
%The approach is to extract the Minutiae of all the fingerprints,
construct a %minutiae image and use wavelets for fingerprint verification

clear all;
clc;
addpath c:\data
addpath c:\data\Minu
addpath c:\data\Fingerwave

N=80;
for a=1:80
b=int2str(a);
c=strcat(b,'.tif');
I=imread(c);
%Thin image to 1 pixel
IW = edge(I, 'canny'); % Returns a binary ridge map image
% where the ridge and valley pixels have a value 1 and 0 respectively.

% We proceed to count the minutiae points
Mcount = 1; % We start from the first row of pixel
M(Mcount, :) = [0,0,0,0]; % Reset the counter to zero
% We parse through the thinned image, ignoring at least 10 pixels for
border %because to detect the minutiae points in a pixel, we need to define
a 3 x 3 %window around the pixel and use the crossing number method. If we
examine the
```

```
% whole image, the software will run into problems examining the pixels at
% the borders due to boundary effect.

[m,n]=size(I);
for y=10:n - 10
  for x=10:m - 10
    if (IW(x, y) == 1)
      CN = 0;
      for i = 1:8
        CN = CN + abs (D(IW, x, y, i) - D(IW, x, y, i + 1));
      end
      CN = CN / 2;

      if (CN == 1) | (CN == 3)
        theta = EhisFindTheta(IW, x, y, CN);
        M(Mcount, :) = [x, y, CN, theta];
        Mcount = Mcount + 1;
      end
    end
  end
end

% We construct minutiae image
M_img = uint8(zeros(m,n,3));

for i=1:Mcount - 1
  x1 = M(i, 1);
  y1 = M(i, 2);
  if (M(i, 3) == 1)
    M_img(x1, y1,:) = [255, 0, 0]; % Use red for ridge endings
  else
    M_img(x1, y1,:) = [0, 0, 255]; % Use blue for ridge bifurcation
  end

end
Z=M_img;
  filename=(['c:\data\Minu\Z',num2str(a),'.tif']);
  imwrite(Z, filename );

x=imread(filename);
cc=rgb2gray(x);
d=double(cc);
[e,f]=size(d);

%We perform a level 4 2D wavelet decomposition on the images

[F,k]=wavedec2(d,4,'bior5.5'); %Bio-orthogonal wavelet is used for
the %decoposition
%The co-efficients of Approximation and details components are
concatenated %into the one vector F; k keeps track of the sizes of each
component
```

```
%We extract the multilevel approximation coefficients

fA1=appcoef2(F,k,'bior5.5',1); % Level 1
fA2=appcoef2(F,k,'bior5.5',2); % Level 2
fA3=appcoef2(F,k,'bior5.5',3); % Level 3
fA4=appcoef2(F,k,'bior5.5',4); % Level 4

% We extract the multilevel detail coefficients as follows
fH1=detcoef2('h',F,k,1); % Level 1 Horizontal detail coeffients
fH2=detcoef2('h',F,k,2); % Level 2 Horizontal detail coeffients
fH3=detcoef2('h',F,k,3); % Level 3 Horizontal detail coeffients
fH4=detcoef2('h',F,k,4); % Level 4 Horizontal detail coeffients

fV1=detcoef2('v',F,k,1); % Level 1 Vertical detail coeffients
fV2=detcoef2('v',F,k,2); % Level 2 Vertical detail coeffients
fV3=detcoef2('v',F,k,3); % Level 3 Vertical detail coeffients
fV4=detcoef2('v',F,k,4); % Level 4 Vertical detail coeffients

fD1=detcoef2('d',F,k,1); % Level 1 Diagonal detail coeffients
fD2=detcoef2('d',F,k,2); % Level 2 Diagonal detail coeffients
fD3=detcoef2('d',F,k,3); % Level 3 Diagonal detail coeffients
fD4=detcoef2('d',F,k,4); % Level 4 Diagonal detail coeffients

%We now store all the coefficents into one unique matrix

%first, we concatenate all the details coefficients at each level i.e,

f1=[fH1,fV1,fD1]; %Coeficients at level 1
f2=[fH2,fV2,fD2]; %Coeficients at level 2
f3=[fH3,fV3,fD3]; %Coeficients at level 3
f4=[fH4,fV4,fD4]; %Coeficients at level 4

%Then, we proceed to concatenate the coeffcients at all levels
[p,q]=size(f1);
f1=reshape(f1,1,p*q); %Reshape into 1-D matrix

[p,q]=size(f2);
f2=reshape(f2,1,p*q); %Reshape into 1-D matrix

[p,q]=size(f3);
f3=reshape(f3,1,p*q); %Reshape into 1-D matrix

[p,q]=size(f4);
f4=reshape(f4,1,p*q); %Reshape into 1-D matrix

G=horzcat(f1,f2,f3,f4); %Unique matrix of all co-efficients

%We proceed to calculate the mean and the standard deviation
[p,q]=size(G);

I=sum(G);
```

```
mean=I/(p*q);  % Mean

for t=1:q
    u(1,q)=G(1,q)-mean;
    u(1,q)=u(1,q)^2;
end

for t=1:q
   tt=sum(u(1,q))/(p*q);
end

sd=sqrt(tt);  %Standatd Deviation

mean_data(a)=mean;
sd_data(a)=sd;

%To improve the verification rate, we calculate the co-occurrence matrix
%features for the coeffcients of the decomposed images at level 1
F1=[fA1,fH1,fV1,fD1];  %Approximation and detail co-efficients at
level 1
GLCMS=graycomatrix(F1,'offset',[2 0]);

%We find the Contrast of the fingerprint image.
Contrast=0;
for m=1:8
  for n=1:8;
    C=(m-n)^2;
    Cc(m,n)=C*GLCMS(m,n);
    Contrast=Contrast+(Cc(m,n));
  end
end
Contrast;

% We find the energy of the fingerprint image as follows

energy=0;
for m=1:8
 for n=1:8
  Ee(m,n)=GLCMS(m,n)^2;
  Energy=energy+Ee(m,n);
end
end
%We find the entropy of the fingerprint image as follows
Entropy=0;
threshold=1;
for m=1:8
  for n=1:8
if GLCMS(m,n)<=threshold
  GLCMS(m,n)=threshold;
end
```

```
Ent(m,n)=-log2(GLCMS(m,n));
Entr(m,n)=GLCMS(m,n)*Ent(m,n);
Entropy=Entropy+Entr(m,n);
   end
end
Contrast_data(a)=Contrast;
Energy_data(a)=Energy;
Entropy_data(a)=Entropy;
end

%We now save all the performance metrics in a Database for the
Bio-orthogonal %wavelet
save databasebior mean_data sd_data Contrast_data Energy_data
Entropy_data

E=mean_data;
E=reshape(E,N,1);
S=sd_data;
S=reshape(S,N,1);
H=Contrast_data;
H=reshape(H,N,1);
L=Energy_data;
L=reshape(L,N,1);
Q=Entropy_data;
Q=reshape(Q,N,1);

%We now proceed to the fingerprint verification process
%The process entails determining the wavelet statistical parameters and
co-%occurence matrix features.
%of the given %fingerprint image and comparing them with those stored in
the %database using a distance vector formula
%The image that has the minimum difference is the verified image
```

Appendix K2: Program to Verify a Test Image

```
% We proceed to select a fingerprint for verification. We randomly take
image
% number a
clear all;
%clc;
a=9;
b=int2str(a);
c=strcat(b,'.tif');
I=imread(c);

%Thin image to 1 pixel
IW = edge(I, 'canny'); % Returns a binary ridge map image
% where the ridge and valley pixels have a value 1 and 0 respectively.
```

```
% We proceed to count the minutiae points
Mcount = 1; % We start from the first row of pixel
M(Mcount, :) = [0,0,0,0]; % Reset the counter to zero
% We parse through the thinned image, ignoring at least 10 pixels for
border %because to detect the minutiae points in a pixel, we need to define
a 3X3 %window around the pixel and use the crossing number method. If we
examine the
% whole image, the software will runs into problems examining the pixels
at
% the borders due to boundary effect

[m,n]=size(I);
for y=10:n - 10
  for x=10:m - 10
    if (IW(x, y) == 1)
       CN = 0;
       for i = 1:8
         CN = CN + abs (D(IW, x, y, i) - D(IW, x, y, i + 1));
       end
       CN = CN / 2;

       if (CN -- 1) | (CN == 3)
         theta = EhisFindTheta(IW, x, y, CN);
         M(Mcount, :) = [x, y, CN, theta];
         Mcount = Mcount + 1;
       end
    end
  end
end

% We construct minutiae image
M_img = uint8(zeros(m,n,3));

for i=1:Mcount - 1
  x1 = M(i, 1);
  y1 = M(i, 2);
  if (M(i, 3) == 1)
    M_img(x1, y1,:) = [255, 0, 0]; % Use red for ridge endings
  else
    M_img(x1, y1,:) = [0, 0, 255]; % Use blue for ridge bifurcation
  end

end
Z=M_img;
  filename=(['c:\data\Minu\Z',num2str(a),'.tif']);
  imwrite(Z, filename);

x=imread(filename);
cc=rgb2gray(x);
d=double(cc);
[e,f]=size(d);
%We perform a level 4 2D wavelet decomposition on the images
```

```
[F,k]=wavedec2(d,4,'bior5.5'); %The Bio-orthogonal wavelet is used
for the decoposition
%The co-efficients of Approximation and details components are
concatenated into the
%one vector F; k keeps track of the sizes of each component

%We extract the multilevel approximation coefficients

fA1=appcoef2(F,k,'bior5.5',1); % Level 1
fA2=appcoef2(F,k,'bior5.5',2); % Level 2
fA3=appcoef2(F,k,'bior5.5',3); % Level 3
fA4=appcoef2(F,k,'bior5.5',4); % Level 4

% We extract the multilevel detail coefficients as follows
 fH1=detcoef2('h',F,k,1); % Level 1 Horizontal detail coefficients
 fH2=detcoef2('h',F,k,2); % Level 2 Horizontal detail coefficients
 fH3=detcoef2('h',F,k,3); % Level 3 Horizontal detail coefficients
 fH4=detcoef2('h',F,k,4); % Level 4 Horizontal detail coefficients

 fV1=detcoef2('v',F,k,1); % Level 1 Vertical detail coefficients
 fV2=detcoef2('v',F,k,2); % Level 2 Vertical detail coefficients
 fV3=detcoef2('v',F,k,3); % Level 3 Vertical detail coefficients
 fV4=detcoef2('v',F,k,4); % Level 4 Vertical detail coefficients

 fD1=detcoef2('d',F,k,1); % Level 1 Diagonal detail coefficients
 fD2=detcoef2('d',F,k,2); % Level 2 Diagonal detail coefficients
 fD3=detcoef2('d',F,k,3); % Level 3 Diagonal detail coefficients
 fD4=detcoef2('d',F,k,4); % Level 4 Diagonal detail coefficients

%We now store  all the coefficients into one unique matrix

%first, we concatenate all the details coefficients at each level i.e,

f1=[fH1,fV1,fD1]; %Coefficients at level 1
f2=[fH2,fV2,fD2]; %Coefficients at level 2
f3=[fH3,fV3,fD3]; %Coefficients at level 3
f4=[fH4,fV4,fD4]; %Coefficients at level 4

%Then, we proceed to concatenate the coefficients at all levels
[p,q]=size(f1);
f1=reshape(f1,1,p*q); %Reshape into 1-D matrix

[p,q]=size(f2);
f2=reshape(f2,1,p*q); %Reshape into 1-D matrix

[p,q]=size(f3);
f3=reshape(f3,1,p*q); %Reshape into 1-D matrix

[p,q]=size(f4);
f4=reshape(f4,1,p*q); %Reshape into 1-D matrix
```

```
G=horzcat(f1,f2,f3,f4); %Unique matrix of all coefficients

%We proceed to calculate the mean and the standard deviation
[p,q]=size(G);

 I=sum(G);

mean=I/(p*q); % Mean

for t=1:q
    u(1,q)=G(1,q)-mean;
    u(1,q)=u(1,q)^2;
end

for t=1:q
  tt=sum(u(1,q))/(p*q);
end

sd=sqrt(tt); %Standard Deviation

%To improve the verification rate, we calculate the co-occurrence matrix
features for
%the coefficients of the decomposed images at level 1
F1=[fA1,fH1,fV1,fD1]; %Approximation and detail coefficients at level
1
GLCMS=graycomatrix(F1,'offset',[2 0]);

%We find the Contrast of the fingerprint image.
Contrast=0;
for m=1:8
  for n=1:8;
    C=(m-n)^2;
    Cc(m,n)=C*GLCMS(m,n);
    Contrast=Contrast+(Cc(m,n));
  end
end
Contrast;

%We find the energy of the fingerprint image as follows

energy=0;
for m=1:8
 for n=1:8
  Ee(m,n)=GLCMS(m,n)^2;
  Energy=energy+Ee(m,n);
end
end
%We find the entropy of the fingerprint image as follows
Entropy=0;
threshold=1;
for m=1:8
```

```
  for n=1:8
if GLCMS(m,n)<=threshold
  GLCMS(m,n)=threshold;
end

Ent(m,n)=-log2(GLCMS(m,n));
Entr(m,n)=GLCMS(m,n)*Ent(m,n);
Entropy=Entropy+Entr(m,n);
  end
end

%We can determine the differences in the statistical parameters of the
%fingerprint image with the values in the database
load databasebior

for i=1:80
  Mdiff(i)=(mean_data(i)-mean)^2;
  SDdiff(i)=(sd_data(i)-sd)^2;
  Cdiff(i)=(Contrast_data(i)-Contrast)^2;
  ENdiff(i)=(Energy_data(i)-Energy)^2;
  ETdiff(i)=(Entropy_data(i)-Entropy)^2;
end

%We proceed to determine the image that has the minimum difference and
%assign the image number to a value for the different statistical
%parameters

MM=Mdiff(1);
Y1=1;
for i=2:80;
  if Mdiff(i)<MM;
    MM=Mdiff(i);
    Y1=i;
  else
    continue;
  end
end
SS=SDdiff(1);
Y2=1;
for i=2:80;
  if SDdiff(i)<SS;
    SS-SDdiff(i);
    Y2=i;
  else
    continue;
  end
end
CC=Cdiff(1);
Y3=1;
for i=2:80;
  if Cdiff(i)<CC
    CC=Cdiff(i);
```

```
      Y3=i;
    else
      continue;
    end
end
EE=ENdiff(1);
Y4=1;
for i=2:80
    if ENdiff(i)<EE
      EE=ENdiff(i);
      Y4=i;
    else
      continue;
    end
end
ET=ETdiff(1);
Y5=1;
for i=2:80;
    if ETdiff(i)<ET
      ET=ETdiff(i);
      Y5=i;
    else
      continue;
    end
end
Y1
Y2
Y3
Y4
Y5
%Note if there is match for 2 or more values of K, then the image is
%correctly identified as the matching number.
```

%Subroutine
%NOTE: Store Codes in a different file named D

```
% Detect Minutiae points using the Crossing Number Method

function k = D (IW, x, y, i)
% get pixel value based on chart:
% 4 | 3 | 2
% 5 |   | 1
% 6 | 7 | 8
switch (i)
  case {1, 9}
    k = IW(x+1, y);
  case 2
    k = IW(x + 1, y-1);
  case 3
    k = IW(x, y - 1);
  case 4
    k = IW(x - 1, y - 1);
  case 5
    k = IW(x - 1, y);
```

```
  case 6
    k = IW(x - 1, y + 1);
  case 7
    k = IW(x, y + 1);
  case 8
    k = IW(x + 1, y + 1);
end
```

%Subroutine
%NOTE: Store Codes in a different file named EhisFindtheta

```
%We eliminate the false Minutiae points using the method developed by
Tico
%and Kuosmanen

function theta = EhisFindTheta (I, x, y, CN)
theta = -1; % Label the central pixel of image IW as the pixel
corresponding
%to the minutia point in the thinned image

if (CN == 1)
  % Label with 1 all pixels connected to corresponding minutiae point
  for i=1:8
    if (D(I, x, y, i) == 1)
        switch (i)
        case {1, 9}
           theta = 0;
        case 2
           theta = 45;
        case 3
           theta = 90;
        case 4
           theta = 135;
        case 5
           theta = 180;
        case 6
           theta = 225;
        case 7
           theta = 270;
        case 8
           theta = 315;
        end
    end
  end
else
  % ridge bifucation
  if (D(I, x, y, 1) & ~(D(I, x, y, 2) & D(I, x, y, 8)))
     theta = 0;
  end
  if (D(I, x, y, 2) & ~(D(I, x, y, 1) & D(I, x, y, 8) & D(I, x, y, 3) & D(I, x,
y, 4)))
     theta = 45;
  end
```

```
 if (D(I, x, y, 3) & ~(D(I, x, y, 2) & D(I, x, y, 4)))
   theta = 90;
 end
 if (D(I, x, y, 4) & ~ (D(I, x, y, 3) & D(I, x, y, 2) & D(I, x, y, 5) & D(I, x,
y, 6)))
   theta = 135;
 end
 if (D(I, x, y, 5) & ~ (D(I, x, y, 4) & D(I, x, y, 6)))
   theta = 180;
 end
 if (D(I, x, y, 6) & ~ (D(I, x, y, 4) & D(I, x, y, 5) & D(I, x, y, 7) & D(I, x,
y, 8)))
   theta = 225;
 end
 if (D(I, x, y, 7) & ~ (D(I, x, y, 6) & D(I, x, y, 8)))
   theta = 270;
 end
 if (D(I, x, y, 8) & ~ (D(I, x, y, 2) & D(I, x, y, 1) & D(I, x, y, 7) & D(I, x,
y, 6)))
   theta = 315;
 end
end
```

Appendix L: Instruction for MATLAB Programs Execution for Fingerprint Minutiae Processing of Appendix K

The following instructions show the steps for the use of the developed codes in Appendix K.

(1) Install MATLAB 7.01 or higher if MATLAB is not already installed in your computer.
(2) Create directory Data in the C: drive of your hard drive.

   (2a) Create a sub-directory in Data called Minu. Store the minutiae images in this directory.
   (2b) Create a sub-directory in Data called Fingerwave. Store the source files in this directory.

(3) Copy the 80 fingerprint images downloaded from "(FVC2000) website: http://bias.csr.unibo.it/fvc2000/" server into c:\data. *Note*: You may have different numbers of fingerprint images. The key is to store them in this directory and make the necessary changes accordingly.
(4) Copy source codes into c:\data\fingerwave.
(5) Create a database of the parameter values by running the fingerprint "wname" in the c:\data\fingerwave directory. The "wname" is a string containing the wavelet name. If the MATLAB editor requests for a file location or path, click, under to run this file, "add directory to the top of the MATLAB path."

Note: These codes take a lot of processing time, and your computer speed will determine how long it takes. *Please be patient.*

(6) To verify an image, open the corresponding verification "wname" file in c:\data \fingerwave, and select an image number in the file.

(7) The image is verified if there are two or more matching values of Y.

Appendix M: MATLAB Programs for Chap. 19

Wavelet-Based Signal Detection, Identification, Discrimination, and Estimation Demo

This program is written for the detection, identification, discrimination, and estimation of Signals. The Signals included for detection are CHIRP, FSK, PSK, and QAM.

```
function varargout = SigDet(varargin)
% SIGDET M-file for SigDet.fig
% This program is written for the detection of Signals
% Signals included for detection are CHIRP, FSK, PSK and QAM
%

% Edit as necessary to fit your needs, especially modifying the response
to help SigDet.

% Last Modified by GUIDE v2.5 08-Dec-2006 05:30:26

% Begin initialization code - DO NOT EDIT
gui_Singleton = 1;
gui_State = struct('gui_Name',    mfilename, ...
            'gui_Singleton', gui_Singleton, ...
            'gui_OpeningFcn', @SigDet_OpeningFcn, ...
            'gui_OutputFcn', @SigDet_OutputFcn, ...
            'gui_LayoutFcn', [] , ...
            'gui_Callback',  []);
if nargin && ischar(varargin{1})
  gui_State.gui_Callback = str2func(varargin{1});
end

if nargout
  [varargout{1:nargout}] = gui_mainfcn(gui_State, varargin{:});
else
  gui_mainfcn(gui_State, varargin{:});
end
% End initialization code - DO NOT EDIT

% --- Executes just before SigDet is made visible.
function SigDet_OpeningFcn(hObject, eventdata, handles, varargin)
% This function has no output args, see OutputFcn.
```

```
% hObject    handle to figure
% eventdata  reserved - to be defined in a future version of MATLAB
% handles    structure with handles and user data (see GUIDATA)
% varargin   command line arguments to SigDet (see VARARGIN)

% Choose default command line output for SigDet
handles.output = hObject;

%The position for the scatter plot is determined by this
strName = 'Scatter Plot';
posn = get(handles.axes2,'position');
szn = min(posn(3), posn(4));
set(handles.axes2,'Position',[1.1*posn(1) posn(2) 2*szn szn], ...
          'Tag',strName, ...
          'visible','off');

pos = get(handles.scatterplot,'position');
sz = min(pos(3), pos(4));
set(handles.scatterplot,'Position',[1.1*pos(1) pos(2) 2*sz sz], ...
        'Tag',strName, ...
        'visible','off');

% Update handles structure
guidata(hObject, handles);

% UIWAIT makes SigDet wait for user response (see UIRESUME)
% uiwait(handles.figure1);

%  - - - Outputs from this function are returned to the command line.
function varargout = SigDet_OutputFcn(hObject, eventdata, handles)
% varargout  cell array for returning output args (see VARARGOUT);
% hObject    handle to figure
% eventdata  reserved - to be defined in a future version of MATLAB
% handles    structure with handles and user data (see GUIDATA)

% Get default command line output from handles structure
varargout{1} = handles.output;

% --- Executes on selection change in listsig.
function varargout = listsig_Callback(hObject, eventdata, handles)
% hObject    handle to listsig (see GCBO)
% eventdata  reserved - to be defined in a future version of MATLAB
% handles    structure with handles and user data (see GUIDATA)

% Hints: contents = get(hObject,'String') returns listsig contents as
cell array
%      contents{get(hObject,'Value')} returns selected item from listsig

varargout{1} = handles.output;
% --- Executes during object creation, after setting all properties.
function listsig_CreateFcn(hObject, eventdata, handles)
```

```
% hObject    handle to listsig (see GCBO)
% eventdata  reserved - to be defined in a future version of MATLAB
% handles    empty - handles not created until after all CreateFcns called


% Hint: popupmenu controls usually have a white background on Windows.
%      See ISPC and COMPUTER.
if ispc && isequal(get(hObject,'BackgroundColor'), get
(0,'defaultUicontrolBackgroundColor'))
    set(hObject,'BackgroundColor','white');
end


% --- Executes on selection change in SNVal.
function varargout = SNVal_Callback(hObject, eventdata, handles)
% hObject    handle to SNVal (see GCBO)
% eventdata  reserved - to be defined in a future version of MATLAB
% handles    structure with handles and user data (see GUIDATA)


% Hints: contents = get(hObject,'String') returns SNVal contents as
cell array
%        contents{get(hObject,'Value')} returns selected item from SNVal


varargout{1} = handles.output;
% --- Executes during object creation, after setting all properties.
function SNVal_CreateFcn(hObject, eventdata, handles)
% hObject    handle to SNVal (see GCBO)
% eventdata  reserved - to be defined in a future version of MATLAB
% handles    empty - handles not created until after all CreateFcns called


% Hint: popupmenu controls usually have a white background on Windows.
%      See ISPC and COMPUTER.
if ispc && isequal(get(hObject,'BackgroundColor'), get
(0,'defaultUicontrolBackgroundColor'))
    set(hObject,'BackgroundColor','white');
end


% --- Executes on button press in Detect.
% -- This function is used to detect which signal is generated
function Detect_Callback(hObject, eventdata, handles)
% hObject    handle to Detect (see GCBO)
% eventdata  reserved - to be defined in a future version of MATLAB
% handles    structure with handles and user data (see GUIDATA)
%


contents = get(handles.listsig,'String');
sigsel = contents{get(handles.listsig,'Value')};


yg = handles.yg;          %Retrieves the generated signal
varr1 = var(yg);          %Estimate the variance of the signal
wtyg = cwt(yg,12,'morl'); %Performs Continuous wavelet transform
varr2 = var(wtyg);        %Estimate the variance of the signal after cwt
diff = varr2 - varr1;     %Estimate the difference between the two
variances
```

```
if diff > 0          % If the difference is greater than zero, Chirp signal is
present
    set (handles.textdet, 'String', ...
    ['Chirp Signal Detected']);
    set (handles.digdet, 'String', ...
     ['Chirp Signal Detected']);
                  % If diff is less than zero, Chirp is not
                  % present, then determine which one of QAM, PSK
                  % or FSK is present
  else
    set (handles.textdet, 'String', ...
     ['Chirp Signal not Detected']);
    ygn = yg./abs(yg);      %normalise the signal
    CSS = cwt (yq,1,'db1'); % Perform cwt of signal using Haar wavelet
function
    ACSS = abs (CSS);
    CSSn1 = cwt (ygn,1,'db1'); % Perform cwt of normalised signal
    ACSSn1 = abs (CSSn1);

    %downsample
    dss1 = ACSS (1:15:end);
    dssn1 = ACSSn1 (1:15:end);
    Na = size (dss1,2) - 1;

    %Median filtering
    Lv = 5;
    ss1 = medfilt1 (ACSS,Lv);
    ssn1 = medfilt1 (ACSSn1,Lv);

    %Calculate the variance
    v1 = var (ss1);
    vn1 = var (ssn1);

    sv1 = var (dss1);
    svn1 = var (dssn1);

    %Determine the threshold using Chi-Square function
    tt1 = ((Na)*sv1)/(chi2inv(0.99,Na));
    ttn1 = ((Na)*svn1)/(chi2inv(0.99,Na));

    % Compare the variance to the threshold
    qq = v1 - tt1;
    qqn = vn1 - ttn1;

      if (qq>0) & (qqn<0)

        set (handles.digdet, 'String', ...
          ['QAM Signal Detected']);

      elseif (qq>0) & (qqn>0)
```

```
          set(handles.digdet,'String',...
            ['FSK Signal Detected']);
        elseif (qq<0) & (qqn<0)

          set(handles.digdet,'String',...
            ['PSK Signal Detected']);

        else
          disp('UND');
        end

        axes(handles.axes3) % Select the proper axes
        plot(ACSS)
        set(handles.axes3,'XMinorTick','on')
        grid on

        axes(handles.axes4) % Select the proper axes
        plot(ACSSn1)
        set(handles.axes4,'XMinorTick','on')
        grid on
end


% --- Executes on selection change in Mval.
function varargout = Mval_Callback(hObject, eventdata, handles)
% hObject   handle to Mval (see GCBO)
% eventdata reserved - to be defined in a future version of MATLAB
% handles   structure with handles and user data (see GUIDATA)


% Hints: contents = get(hObject,'String') returns Mval contents as cell
array
%      contents{get(hObject,'Value')} returns selected item from Mval
varargout{1} = handles.output;


% --- Executes during object creation, after setting all properties.
function Mval_CreateFcn(hObject, eventdata, handles)
% hObject   handle to Mval (see GCBO)
% eventdata reserved - to be defined in a future version of MATLAB
% handles   empty - handles not created until after all CreateFcns called


% Hint: popupmenu controls usually have a white background on Windows.
%     See ISPC and COMPUTER.
if ispc && isequal(get(hObject,'BackgroundColor'), get
(0,'defaultUicontrolBackgroundColor'))
  set(hObject,'BackgroundColor','white');
end


% --- Executes on button press in Generate.
% -- This function is used to generate the selected signal
function Generate_Callback(hObject, eventdata, handles)
% hObject   handle to Generate (see GCBO)
% eventdata reserved - to be defined in a future version of MATLAB
```

```
% handles    structure with handles and user data (see GUIDATA)
%sigsel = get (handles.listsig, 'Value') ;

%Retrieves the selected signal type
contents = get (handles.listsig, 'String') ;
sigsel = contents{get (handles.listsig, 'Value')} ;
% Retrieves the M value
contenta = get (handles.Mval, 'String') ;
M = str2double (contenta{get (handles.Mval, 'Value')}) ;
% Retrieves the SNR value
contenta = get (handles.SNVal, 'String') ;
SN = str2double (contenta{get (handles.SNVal, 'Value')}) ;
Mag = 5;
wc = 20000;
tt = 0:0.00007:0.0017;


x = randint (50,1,M) ;   %message signal
switch (sigsel)
      % Generate the Chirp Signal
     case 'Chirp'
      t= 0:0.001:5;            % 10 seconds @ 1kHz sample rate
      fo=10;f1=400;            % Start at 10Hz, go up to 400Hz
      y=chirp(t,fo,10,f1,'logarithmic') ;
      yg = awgn(y,SN) ;
      handles.yg = yg;
      guidata(hObject,handles) ; % store the changes

      axes(handles.axes5) % Select the proper axes
      plot(real(y))
      set(handles.axes5,'XMinorTick','on')
      grid on
      axes(handles.axes6) % Select the proper axes
      plot(real(yg))
      set(handles.axes6,'XMinorTick','on')
      grid on

      % Generate PSK Signal
     case 'PSK'
      y3 = pskmod(x,M) ;    %PSK modulation
      yn = awgn(y3,SN) ;
        for u = 1:50
           S2(:,u) = Mag*y3(u)*exp(i*wc*tt) ;
        end
      [a2,b2] = size(S2) ;
      uu2 = a2*b2;
      y = reshape(S2,1,uu2) ;
      yg = awgn(y,SN) ;
      ygn = awgn(y,SN,'measured') ;
      handles.yg = yg;
      guidata(hObject,handles) ; % store the changes
```

```
axes(handles.axes5) % Select the proper axes
plot(real(y(1:100)))
set(handles.axes5,'XMinorTick','on')
grid on
axes(handles.axes6) % Select the proper axes
plot(real(ygn(1:100)))
set(handles.axes6,'XMinorTick','on')
grid on
%%%%%%%%%%%%%SCATTER PLOT%%%%%%%%%%%%%%%%%%%%%%%%%%%%%%%%%%%%%%%
%
n=1;
offset = 0;
plotstring = 'rx';
axes(handles.scatterplot) % Select the proper axes
[r, c] = size(y3);
if r * c == 0
   error('Input variable X is empty.')
end;
if r == 1
   y3 = y3(:);
end;

% don't allow N to be noninteger or less than or equal zero
if ((fix(n) ~= n) | (n <= 0))
   error('N must be a positive integer.')
end

% don't allow offset to be noninteger or less than or equal zero
if ((fix(offset) ~= offset) | (offset < 0))
   error('OFFSET must be a nonnegative integer.')
end

% increment offset to create index into x
offset = offset + 1;

if ~isreal(y3) > 0
   y3 = [real(y3), imag(y3)];
end;

yy = y3(offset : n : size(y3, 1), :);
maxAll = max(max(abs(yy)));

strName = 'Scatter Plot';
   axis([-eps eps -eps eps]);
% figure and axes creation setup constants
pos = get(handles.scatterplot,'position');
sz = min(pos(3), pos(4));
set(handles.scatterplot,'Position', [pos(1) pos(2) 2*sz sz], ...
    'Tag',strName, ...
    'visible','on');
axo = axis;
```

```
% plot the scatter plot
[len_yy, wid_yy]=size(yy);
if wid_yy == 1
  % real data only
  plot(yy, zeros(1,len_yy), plotstring);
elseif wid_yy == 2
  %complex data only
  plot(yy(:,1), yy(:,2), plotstring);
else
 error('Number of columns in the input data, X, cannot exceed 2.');
end
if(~ishold)
  set(handles.scatterplot, 'nextplot', 'replacechildren');
end


% Adjust the limits
limFact = 1.07;
limits = max(max(abs(axo)),maxAll*limFact);
axis equal;
axis([-limits limits -limits limits]);

% Label the plot
ylabel('Quadrature')
xlabel('In-Phase')
title('Scatter plot')

%%%%%%%%%%%%%%%%%%%%%%%%%%%%%%%%%%%%%%%%%%%%%%%%%%%%%%%%%%%%
axes(handles.axes2) % Select the proper axes
[r, c] = size(yn);
if r * c == 0
  error('Input variable X is empty.')
end;
if r == 1
  yn = yn(:);
end;

% don't allow N to be noninteger or less than or equal zero
if ((fix(n) ~= n) | (n <= 0))
  error('N must be a positive integer.')
end

% don't allow offset to be noninteger or less than or equal zero
if ((fix(offset) ~= offset) | (offset < 0))
  error('OFFSET must be a nonnegative integer.')
end

% increment offset to create index into x
offset = offset + 1;

if ~isreal(y3) > 0
  yn = [real(yn), imag(yn)];
end;
```

```matlab
yyn = yn(offset : n : size(yn, 1), :);
maxAlln = max(max(abs(yyn)));

strName = 'Scatter Plot';
  axis([-eps eps -eps eps]);
% figure and axes creation setup constants
posn = get(handles.axes2, 'position');
szn = min(posn(3), posn(4));
set(handles.axes2, 'Position', [posn(1) posn(2) 2*szn szn], ...
    'Tag', strName, ...
    'visible', 'on');
axo = axis;

% plot the scatter plot
[len_yyn, wid_yyn]=size(yyn);
if wid_yyn == 1
  % real data only
  plot(yyn, zeros(1,len_yyn), plotstring);
elseif wid_yyn == 2
  %complex data only
  plot(yyn(:,1), yyn(:,2), plotstring);
else
 error('Number of columns in the input data, X, cannot exceed 2.');
end
if(~ishold)
  set(handles.axes2, 'nextplot', 'replacechildren');
end

% Adjust the limits
limFact = 1.07;
limits = max(max(abs(axo)),maxAlln*limFact);
axis equal;
axis([-limits limits -limits limits]);

% Label the plot
ylabel('Quadrature')
xlabel('In-Phase')
title('Scatter plot')

% Generate FSK Signal and plot the signal
case 'FSK'
 Fs - 200000;
 df = Fs/(M+4);
 freqsep = Fs/(M+4);

 for u = 1:50,

 switch (M)
   case 2
   switch (x(u))
    case 0
     Smfa = -df;
```

```
      case 1
        Smfa = df;
      end
      case 4
      switch (x(u))
      case 0
        Smfa = -df;
      case 1
        Smfa = -3*df;
      case 2
        Smfa = df;
      case 3
        Smfa = 3*df;
      end

      case 8
      switch (x(u))
      case 0
        Smfa = -df;
      case 1
        Smfa = -3*df;
      case 2
        Smfa = -5*df;
      case 3
        Smfa = -7*df;
      case 4
        Smfa = df;
      case 5
        Smfa = 3*df;
      case 6
        Smfa = 5*df;
      case 7
        Smfa = 7*df;
      end
    case 16
      switch (x(u))
      case 0
        Smfa = -df;
      case 1
        Smfa = -3*df;
      case 2
        Smfa = -5*df;
      case 3
        Smfa = -7*df;
      case 4
        Smfa = -9*df;
      case 5
        Smfa = -11*df;
      case 6
        Smfa = -13*df;
      case 7
        Smfa = -15*df;
      case 8
```

```
      Smfa = df;
    case 9
      Smfa = 3*df;
    case 10
      Smfa = 5*df;
    case 11
      Smfa = 7*df;
    case 12
      Smfa = 9*df;
    case 13
      Smfa = 11*df;
    case 14
      Smfa = 13*df;
    case 15
      Smfa = 15*df;
  end
  end
  y3(u) = Smfa*2*pi;
  end
  %y3 = exp(i*fs3);
  un = size(y3,2);
  for n = 1:un
    S4(:,n) = Mag*exp(i*(wc+y3(n))*tt);
  end

  [a4,b4] = size(S4);
  nn4 = a4*b4;
  y = reshape(S4,1,nn4);
  yn = awgn(y3,SN);
  yg = awgn(y,SN);
    handles.yg = yg;
    guidata(hObject,handles); % store the changes
  %%%%%%%%%%Plot the Signal%%%%%%%%%%%%%%%%%%%%%
  axes(handles.axes5) % Select the proper axes
  plot(real(y(1:100)))
  set(handles.axes5,'XMinorTick','on')
  grid on
  axes(handles.axes6) % Select the proper axes
  plot(real(yg(1:100)))
  set(handles.axes6,'XMinorTick','on')
  grid on

  %%%%%%%%%%%%%%SCATTER PLOT%%%%%%%%%%%%%%%%%%%%%%%%%%%%%%%%%%%%%%%%%%%
%
  n=1;
  offset = 0;
  plotstring = 'rx';
  axes(handles.scatterplot) % Select the proper axes
  [r, c] = size(y3);
  if r * c == 0
    error('Input variable X is empty.')
  end;
  if r == 1
```

```
   y3 = y3(:);
end;

% don't allow N to be noninteger or less than or equal zero
if ((fix(n) ~= n) | (n <= 0))
  error('N must be a positive integer.')
end

% don't allow offset to be noninteger or less than or equal zero
if ((fix(offset) ~= offset) | (offset < 0))
  error('OFFSET must be a nonnegative integer.')
end

% increment offset to create index into x
offset = offset + 1;

if ~isreal(y3) > 0
  y3 = [real(y3), imag(y3)];
end;

yy = y3(offset : n : size(y3, 1), :);
maxAll = max(max(abs(yy)));

strName = 'Scatter Plot';
  axis([-eps eps -eps eps]);
% figure and axes creation setup constants
pos = get(handles.scatterplot,'position');
sz = min(pos(3), pos(4));
set(handles.scatterplot,'Position', [pos(1) pos(2) 2*sz sz], ...
    'Tag',strName, ...
    'visible','on');
axo = axis;

% plot the scatter plot
[len_yy, wid_yy]=size(yy);
if wid_yy == 1
  % real data only
  plot(yy, zeros(1,len_yy), plotstring);
elseif wid_yy == 2
  %complex data only
  plot(yy(:,1), yy(:,2), plotstring);
else
 error('Number of columns in the input data, X, cannot exceed 2.');
end
if (~ishold)
  set(handles.scatterplot,'nextplot','replacechildren');
end

% Adjust the limits
limFact = 1.07;
limits = max(max(abs(axo)),maxAll*limFact);
```

```
axis equal;
axis([-limits limits -limits limits]);

% Label the plot
ylabel('Quadrature')
xlabel('In-Phase')
title('Scatter plot')

%%%%%%%%%%%%%%%%%%%%%%%%%%%%%%%%%%%%%%%%%%%%%%%%%%%%%%%%%%%%%%
axes(handles.axes2) % Select the proper axes
[r, c] = size(yn);
if r * c == 0
  error('Input variable X is empty.')
end;
if r == 1
  yn = yn(:);
end;

% don't allow N to be noninteger or less than or equal zero
if ((fix(n) ~= n) | (n <= 0))
  error('N must be a positive integer.')
end

% don't allow offset to be noninteger or less than or equal zero
if ((fix(offset) ~= offset) | (offset < 0))
  error('OFFSET must be a nonnegative integer.')
end

% increment offset to create index into x
offset = offset + 1;

if ~isreal(y3) > 0
  yn = [real(yn), imag(yn)];
end;

yyn = yn(offset : n : size(yn, 1), :);
maxAlln = max(max(abs(yyn)));

strName = 'Scatter Plot';
  axis([-eps eps -eps eps]);
% figure and axes creation setup constants
posn = get(handles.axes2, 'position');
szn = min(posn(3), posn(4));
set(handles.axes2, 'Position', [posn(1) posn(2) 2*szn szn], ...
    'Tag', strName, ...
    'visible', 'on');
axo = axis;

% plot the scatter plot
[len_yyn, wid_yyn] = size(yyn);
if wid_yyn == 1
  % real data only
```

```
          plot(yyn, zeros(1,len_yyn), plotstring);
       elseif wid_yyn == 2
          %complex data only
          plot(yyn(:,1), yyn(:,2), plotstring);
       else
        error('Number of columns in the input data, X, cannot exceed 2.');
       end
       if(~ishold)
          set(handles.axes2,'nextplot','replacechildren');
       end

       % Adjust the limits
       limFact = 1.07;
       limits = max(max(abs(axo)),maxAlln*limFact);
       axis equal;
       axis([-limits limits -limits limits]);

       % Label the plot
       ylabel('Quadrature')
       xlabel('In-Phase')
       title('Scatter plot')

       % Generate QAM Signal
    case 'QAM'
       y3 = qammod(x,M);
          for u3 = 1:50
             S3(:,u3) = Mag*y3(u3)*exp(i*wc*tt);
          end
        [a3,b3] = size(S3);
        uu3 = a3*b3;
        y = reshape(S3,1,uu3);
        yn = awgn(y3,SN);
        yg = awgn(y,SN);
        ygn = awgn(y3,SN);
          handles.yg = yg;
          guidata(hObject,handles); % store the changes

    %%%%%%%%%PLOT THE SIGNAL%%%%%%%%%%%%%%%%%%%
    axes(handles.axes5) % Select the proper axes
    plot(real(y(1:100)))
    set(handles.axes5,'XMinorTick','on')
    grid on
    axes(handles.axes6) % Select the proper axes
    plot(real(yg(1:100)))
    set(handles.axes6,'XMinorTick','on')
    grid on

    %%%%%%%%%%%%SCATTER PLOT%%%%%%%%%%%%%%%%%%%%%%%%%%%%%%%%%%%%%%%%%%%%
%
    n=1;
    offset = 0;
    plotstring = 'rx';
```

```
axes(handles.scatterplot) % Select the proper axes
[r, c] = size(y3);
if r * c == 0
  error('Input variable X is empty.')
end;
if r == 1
  y3 = y3(:);
end;

% don't allow N to be noninteger or less than or equal zero
if ((fix(n) ~= n) | (n <= 0))
  error('N must be a positive integer.')
end

% don't allow offset to be noninteger or less than or equal zero
if ((fix(offset) ~= offset) | (offset < 0))
  error('OFFSET must be a nonnegative integer.')
end

% increment offset to create index into x
offset = offset + 1;

if ~isreal(y3) > 0
  y3 = [real(y3), imag(y3)];
end;

yy = y3(offset : n : size(y3, 1), :);
maxAll = max(max(abs(yy)));

strName = 'Scatter Plot';
  axis([-eps eps -eps eps]);
% figure and axes creation setup constants
pos = get(handles.scatterplot,'position');
sz = min(pos(3), pos(4));
set(handles.scatterplot,'Position', [pos(1) pos(2) 2*sz sz], ...
    'Tag',strName, ...
    'visible','on');
axo = axis;

% plot the scatter plot
[len_yy, wid_yy]=size(yy);
if wid_yy == 1
  % real data only
  plot(yy, zeros(1,len_yy), plotstring);
elseif wid_yy == 2
  %complex data only
  plot(yy(:,1), yy(:,2), plotstring);
else
 error('Number of columns in the input data, X, cannot exceed 2.');
end
if (~ishold)
```

```
  set(handles.scatterplot,'nextplot','replacechildren');
end

% Adjust the limits
limFact = 1.07;
limits = max(max(abs(axo)),maxAll*limFact);
axis equal;
axis([-limits limits -limits limits]);

% Label the plot
ylabel('Quadrature')
xlabel('In-Phase')
title('Scatter plot')

%%%%%%%%%%%%%%%%%%%%%%%%%%%%%%%%%%%%%%%%%%%%%%%%%%%%%%%%%%%%%%%%%%
axes(handles.axes2) % Select the proper axes
[r, c] = size(ygn);
if r * c == 0
  error('Input variable X is empty.')
end;
if r == 1
  ygn = ygn(:);
end;

% don't allow N to be noninteger or less than or equal zero
if ((fix(n) ~= n) | (n <= 0))
  error('N must be a positive integer.')
end

% don't allow offset to be noninteger or less than or equal zero
if ((fix(offset) ~= offset) | (offset < 0))
  error('OFFSET must be a nonnegative integer.')
end

% increment offset to create index into x
offset = offset + 1;

if ~isreal(yg) > 0
  ygn = [real(ygn), imag(ygn)];
end;

yyn = ygn(offset : n : size(ygn, 1), :);
maxAlln = max(max(abs(yyn)));

strName = 'Scatter Plot';
  axis([-eps eps -eps eps]);
% figure and axes creation setup constants
posn = get(handles.axes2,'position');
szn = min(posn(3), posn(4));
set(handles.axes2,'Position',[posn(1) posn(2) 2*szn szn], ...
    'Tag',strName, ...
```

```
          'visible','on');
     axo = axis;

     % plot the scatter plot
     [len_yyn, wid_yyn]=size(yyn);
     if wid_yyn == 1
       % real data only
       plot(yyn, zeros(1,len_yyn), plotstring);
     elseif wid_yyn == 2
       %complex data only
       plot(yyn(:,1), yyn(:,2), plotstring);
     else
      error('Number of columns in the input data, X, cannot exceed 2.');
     end
     if(~ishold)
       set(handles.axes2,'nextplot','replacechildren');
     end

     % Adjust the limits
     limFact = 1.07;
     limits = max(max(abs(axo)),maxAlln*limFact);
     axis equal;
     axis([-limits limits -limits limits]);

     % Label the plot
     ylabel('Quadrature')
     xlabel('In-Phase')
     title('Scatter plot')
end

% --- If Enable == 'on', executes on mouse press in 5 pixel border.
% --- Otherwise, executes on mouse press in 5 pixel border or over
Generate.

% --- Executes during object creation, after setting all properties.
function textdet_CreateFcn(hObject, eventdata, handles)
% hObject   handle to textdet (see GCBO)
% eventdata reserved - to be defined in a future version of MATLAB
% handles   empty - handles not created until after all CreateFcns called
```

Appendix N: MATLAB Programs for Chap. 20

Wavelet-Based Identification, Discrimination, Detection, and Parameter Estimation of Radar Signals

A. Frequency and Power Estimation Algorithm

This algorithm is developed to use existing data files to calculate the center frequency and the amplitude of the power using fast Fourier transform (FFT). The user is encouraged to adapt to whatever data they are using.

The received signal is loaded from the MATLAB data files, where ref_sig is the reference channel signal (sumr), chan_1 is the channel 1 signal, and chan_2 is the channel 2 signal.

load ref_sig;
load chan_1;
load chan_2;

Each loaded signal is assigned to a variable, where x is sumr, x1 is sum1, and x2 is sum2.

x = sumr;
x1 = sum1;
x2 = sum2;

The length of signal **x** is assigned to the variable **N**.

N = length(x);

The number of discrete bin points is represented by variable **b** that will range from 1 to the length of sumr.

b = 1:N;

The sampling period and sampling frequency are calculated based off the given cutoff frequency of the signal.

Ts = 1/20;
fs = 1/Ts;

Discrete FFT of signal **x**

X = fft(x');

The power of the signal is calculated by multiplying the discrete FFT by the conjugate of the discrete FFT of the signal, which is then divided by the length of the original signal.

pwr = X.*conj(X)/N;

The power is then converted into decibels.

powerdB = 20*log10(pwr);

The frequencies at each discrete bin point are found by the number of bin point minus 1 divided by the length of the signal. This is then multiplied by the sampling frequency.

frs = (b-1)/N*fs;

The results are then plotted, the power of the signal versus the frequency of the signal.

```
figure
plot(frs,powerdB)
axis([0 10 0 100])
title('Power vs. Frequency of Detected Signal')
xlabel('Frequency (Hz)');
ylabel('Power(dB)');
```

B. Frequency Modulation Algorithm

Once we find the power and center frequency using discrete FFT, we can then find the frequency modulation of the signal.

The center frequency is found from the minimum frequency with greatest power amplitude.

```
fc = frs(find(powerdB == max(powerdB)));
fc = min(fc);
```

The frequency modulation of the signal is found using the function "amod," which is supported by MATLAB.

```
y = amod(x,fc,fs,'fm');
```

The result of the frequency modulated signal is plotted against the reference channel signal.

```
figure;
plot(y);hold on;plot(x,'r');
axis([0 100 -1 1])
title('Reference Channel vs. FM Modulation')
xlabel('Time');
ylabel('Amplitude');
```

C. Wavelet Transform Algorithm

This algorithm decomposes the received signal using Daubechies-4 wavelets, Level −3, into its difference level components. The mean value is found for each difference level wavelet coefficients to determine the location of the information signal.

The received signal is loaded from the MATLAB data files, where ref_sig is the reference channel signal (sumr), chan_1 is the channel 1 signal, and chan_2 is the channel 2 signal. As a user, you can create data files of interest to you.

```
load ref_sig;
load chan_1;
load chan_2;
```

Each loaded signal is assigned to a variable, where x is sumr, x1 is sum1, and x2 is sum2.

```
x = sumr;
x1 = sum1;
x2 = sum2;
```

The wavelet transform decomposition of the reference channel using Daubechies-4 wavelet coefficients, at level 3, is presented in this segment. The wavelet coefficients are represented by **c**, and **l** is the Daubechies-4 filter coefficients.

```
[c,l] = wavedec(x,3,'db4');
```

The wavelet coefficients are broken into approximation and details coefficients for each difference level represented by **ca0**, **ca1**, **ca2**, **ca3**, **cd1**, **cd2**, and **cd3**.

```
ca0 = appcoef(c,l,'db4',0);
ca1 = appcoef(c,l,'db4',1);
ca2 = appcoef(c,l,'db4',2);
ca3 = appcoef(c,l,'db4',3);
[cd1,cd2,cd3] = detcoef(c,l,[1 2 3]);
```

The mean value is found for each approximation and detail coefficients, and the output is then displayed.

```
mean(ca0)
mean(ca1)
mean(ca2)
mean(ca3)
mean(cd1)
mean(cd2)
mean(cd3)
```

The wavelet coefficient from the wavelet decomposition is reconstructed.

```
X = idwt(c,'db4', 131072);
```

Each difference level of wavelet coefficients is plotted to show the location of the information signal.

```
figure
plot(c)
title('Graph of Wavelet Coefficients')
xlabel('Length of coef. vector');
ylabel('Amplitude');

figure
plot(ca0)
title('ca0 Coefficients')
xlabel('Length of coef. vector');
ylabel('Amplitude');

figure
plot(ca1)
```

```
title('ca1 Coefficients')
xlabel('Length of coef. vector');
ylabel('Amplitude');

figure
plot(ca2)            \
title('ca2 Coefficients')
xlabel('Length of coef. vector');
ylabel('Amplitude');

figure
plot(ca3)
title('ca3 Coefficients')
xlabel('Length of coef. vector');
ylabel('Amplitude');

figure
plot(cd1)
title('cd1 Coefficients')
xlabel('Length of coef. vector');
ylabel('Amplitude');

figure
plot(cd2)
title('cd2 Coefficients')
xlabel('Length of coef. vector');
ylabel('Amplitude');

figure
plot(cd3)
title('cd3 Coefficients')
xlabel('Length of coef. vector');
ylabel('Amplitude');
```

D. Bayes' Detection Algorithm

From the wavelet algorithm, we are able to use Bayes' probability theorem to determine the presence of an information signal.

The received signal is loaded from the MATLAB data files, where ref_sig is the reference channel signal (sumr).

load ref_sig;

The reference signal is assigned to variable **x**.

x = sumr;

Variable N defines the length of the signal. The length of the signal is divided into ten samples.

```
N = length(sumr);
```
% Length of Reference Signal
```
sample_size = N/10;
```
% Calculate the length of each sample size.
Variables *N1* through *N9* represent the end value for each sample size.

```
N1 = 1 + sample_size;
N2 = N1 + sample_size;
N3 = N2 + sample_size;
N4 = N3 + sample_size;
N5 = N4 + sample_size;
N6 = N5 + sample_size;
N7 = N6 + sample_size;
N8 = N7 + sample_size;
N9 = N8 + sample_size;
```

The received signal *x* is divided into ten equal sample sizes.

```
x1 = x(1:N1);
x2 = x(N1:N2);
x3 = x(N2:N3);
x4 = x(N3:N4);
x5 = x(N4:N5);
x6 = x(N5:N6);
x7 = x(N6:N7);
x8 = x(N7:N8);
x9 = x(N8:N9);
x10 = x(N9:N);
```

The wavelet coefficients are determined by taking the wavelet decomposition of each signal sample and assigned to each variable *c1* through *c10*.

```
[c1,l] = wavedec(x1,3,'db4');
[c2,l] = wavedec(x2,3,'db4');
[c3,l] = wavedec(x3,3,'db4');
[c4,l] = wavedec(x4,3,'db4');
[c5,l] = wavedec(x5,3,'db4');
[c6,l] = wavedec(x6,3,'db4');
[c7,l] = wavedec(x7,3,'db4');
[c8,l] = wavedec(x8,3,'db4');
[c9,l] = wavedec(x9,3,'db4');
[c10,l] = wavedec(x10,3,'db4');
```

The level 3 detail coefficients are extracted from each of the wavelet coefficient samples *c1* through *c10* and assigned to variables *cd3_1* through *cd3_10*.

```
[cd3_1] = detcoef(c1,l,[3]);
[cd3_2] = detcoef(c2,l,[3]);
[cd3_3] = detcoef(c3,l,[3]);
```

[cd3_4] = detcoef(c4,l,[3]);
[cd3_5] = detcoef(c5,l,[3]);
[cd3_6] = detcoef(c6,l,[3]);
[cd3_7] = detcoef(c7,l,[3]);
[cd3_8] = detcoef(c8,l,[3]);
[cd3_9] = detcoef(c9,l,[3]);
[cd3_10] = detcoef(c10,l,[3]);

The mean value is found for each level 3 detail coefficient sample. A mean value of zero means that the sample contains mostly noise.

cd3(1) = abs(mean(cd3_1));
cd3(2) = abs(mean(cd3_2));
cd3(3) = abs(mean(cd3_3));
cd3(4) = abs(mean(cd3_4));
cd3(5) = abs(mean(cd3_5));
cd3(6) = abs(mean(cd3_6));
cd3(7) = abs(mean(cd3_7));
cd3(8) = abs(mean(cd3_8));
cd3(9) = abs(mean(cd3_9));
cd3(10) = abs(mean(cd3_10));

Declare the variables signal_detect and nosig_detect, and set them equal to zero.

signal_detect = 0;
nosig_detect = 0;

Inserting a range of threshold values between 0.000 and 0.030, we can determine whether the information signal is present within each of the sampled segments.

```
for i = 1:10

  if cd3(i) > 0.018
    signal_detect = signal_detect + 1;
  else
    nosig_detect = nosig_detect + 1;
  end

end
```

This is the output of the decision made based on the threshold values of whether or not a signal was found in each of the ten signal segments.

signal_detect
nosig_detect

Appendix O: MATLAB Programs for Chap. 21

MATLAB Programs for Application of Wavelets to Vibration Detection in an Aeroelastic System

Programs Using the Vibration Model and Global Thresholding

Simulations are executed using the "**matrix lab**oratory" (or MATLAB) programming language. The Help Library of MATLAB will be an extensive source of information. Custom programs are written to implement the algorithm. These programs implement the wavelet transforms in combination with global thresholding (GT). The outcome is not desirable, but these initial programs laid the foundation for the improvements to the algorithm that followed.

Threshold 3 yielded the best results for the GT scheme.

Once the vibration model is derived, the MATLAB programs were coded to detect this signal that was corrupted by noise. See the Haar, Daubechies-4, and Morlet programs, which use GT threshold 3, on the next page.

Appendix O_1 of N Using Haar Wavelet

```
% HaarWTt3.m (Haar Wavelet Transform)
%
% This program uses the Haar wavelet at level 4 to denoise the
% vibration signal model that is a combination of a sum of sinusoids
% and additive white Gaussian noise (AWGN). The vibration model consists
% of the following:
%
%    s1 = sinusoid at frequency f1 (low)
%    s2 = sinusoid at frequency f2 (high)
%    noise - normal distributed random noise
%
% This program removes the noise from the vibration signal and extracts
pertinent
% information hidden in the noise. Detection is based on the frequency
spectra.
%
%

% Initialize the work space
clear all; close all;

%Create a path here for the Haar wavelet depending on the directory where
it can be found

% path(path, 'c:\...............\haar2')
```

```
% Get the vibration signal
load sign1X

format('long','g')

% Sample the signal for analysis
s = vibsg(1:1500);
Nsg = noise(1:1500);

% Plot vibration signal
figure(1)
plot(t(1:1500), s), grid on; title('Noisy Vibration Signal')
xlabel('time (sec)'), ylabel('Amplitude')

% Perform decomposition of the signal
[C,L] = wavedec(s,4,'haar');
[cD1,cD2,cD3,cD4] = detcoef(C,L,[1,2,3,4]);
cA4 = appcoef(C,L,'haar',4);

% Determine the approximation and details
A4 = wrcoef('a',C,L,'haar',4);
D1 = wrcoef('d',C,L,'haar',1);
D2 = wrcoef('d',C,L,'haar',2);
D3 = wrcoef('d',C,L,'haar',3);
D4 = wrcoef('d',C,L,'haar',4);

% Plot the approximation and details
figure(2)
subplot(221); plot(t(1:1500),A4), grid; title('Approximation A4')
subplot(222); plot(t(1:1500),D4), grid; title('Detail D4')
subplot(223); plot(t(1:1500),D3), grid; title('Detail D3')
figure(3)
subplot(211); plot(t(1:1500),D2), grid; title('Detail D2')
subplot(212); plot(t(1:1500),D1), grid; title('Detail D1')

% Perform reconstruction of the signal
recsig = waverec(C,L,'haar');

% Check for perfect reconstruction
figure(4)     % visually
subplot(211); plot(t(1:1500),s), grid; title('Original Signal')
xlabel('time (sec)'), ylabel('Amplitude')
subplot(212); plot(t(1:1500),recsig), grid; title('Reconstructed
Signal')
xlabel('time (sec)'), ylabel('Amplitude')
n = length(s); % numerically
err = max(abs(recsig-s));
prterr = max(abs(recsig-s))/max(s)*100;
mse = err^2/n^2;

snr1 = 10*log10(max(s)^2/max(abs(s-recsig))^2);
```

```
% Check denoising process by comparing the highest level approximation
% with the noisy vibration signal
figure(5)
subplot(211); plot(t(1:1500),s), grid; title('Noisy Vibration
Signal')
xlabel('time (sec)')
subplot(212); plot(t(1:1500),A4), grid; title('Level
4 Approximation')
xlabel('time (sec)')

% Set the threshold and denoise the signal
thr = 3.82445300326473*0.5;    % fixed threshold (at 50%)
densig = wdencmp('gbl',C,L,'haar',4,thr,'s',1);


snr2 = 10*log10(max(densig)^2/max(abs(Nsg-densig))^2);

% Plot the noisy vibration signal and the denoised signal
figure(6)
subplot(211); plot(t(1:1500),s), grid; title('Noisy Vibration
Signal')
xlabel('time (sec)'), ylabel('Amplitude')
subplot(212); plot(t(1:1500),densig), grid; title('Denoised Signal')
xlabel('time (sec)'), ylabel('Amplitude')

% Compare the frequency content of the vibration signal and the
% recontructed signal
Ns = 1024;
origspec = fft(s,Ns);
spec1 = origspec.*conj(origspec)/Ns;
freq1 = 1000*(0:512)/Ns;
recsigspec = fft(recsig,Ns);
spec2 = recsigspec.*conj(recsigspec)/Ns;
freq2 = 1000*(0:512)/Ns;
figure(7)
subplot(211); plot(freq1,spec1(1:513)), axis([0 100,-Inf 300]), grid
title('Frequency Content of Original Signal')
xlabel('Frequency'), ylabel('Energy')
subplot(212); plot(freq2,spec2(1:513)), axis([0 100,-Inf 300]), grid
title('Frequency Content of Reconstructed Signal')
xlabel('Frequency'), ylabel('Energy')

% Determine the frequency content of the denoised signal
densigspec = fft(densig,Ns);
spec3 = densigspec.*conj(densigspec)/Ns;
freq2 = 1000*(0:512)/Ns;
figure(8)
plot(freq2,spec3(1:513)), axis([0 100,-Inf Inf]), grid
title('Frequency Content of Denoised Signal')
xlabel('Frequency'), ylabel('Energy')

% Compare the PSDs for the vibration signal, the reconstructed signal,
% and the denoised signal
```

```
figure(9)
[Po,fo] = pmtm(s,3.5,Ns,Fs);
subplot(221); psdplot(Po,fo,'Hz','db','PSD of Original Signal'), axis
([0 100,-Inf -5])
[Pr,fr] = pmtm(recsig,3.5,Ns,Fs);
subplot(222); psdplot(Pr,fr,'Hz','db','PSD of Reconstructed
Signal'), axis([0 100,-Inf -5])
[Pd,fd] = pmtm(densig,2,Ns,Fs);
subplot(223); psdplot(Pd,fd,'Hz','db','PSD of Denoised Signal'), axis
([0 100,-50 -5])
[Pns,fns] = pmtm(Nsg,3.5,Ns,Fs);
Pwo = (Fs/Ns)*sum(Po);
Pwd = (Fs/Ns)*sum(Pd);
Pwns = (Fs/Ns)*sum(Pns);
SNR1 = 10*log10(Pwo/Pwns);  % Compute the SNR for the vibration signal
SNR2 = 10*log10(Pwd/Pwns);  % Compute the SNR for the denoised signal

% Generate the scalograms
figure(10)
specgram(s,1024,Fs,[],256), axis([0 2 0 100])
title('Scalogram of the Noisy Vibration Signal')
figure(11)
specgram(densig,1024,Fs,[],256), axis([0 2 0 100])
title('Scalogram of the Denoised Vibration Signal')

% Compare the denoised and vibration signals
figure(12)        % visually
subplot(211); plot(t(1:1500),densig,'m'), axis('tight'), grid
title('Denoised Signal')
xlabel('time (sec)'), ylabel('Amplitude')
subplot(212); plot(t(1:1500),sumsig(1:1500)), axis('tight'), grid
title('Vibration Signal')
xlabel('time (sec)'), ylabel('Amplitude')
ssig = sumsig(1:1500);
m = length(ssig);    % numerically
err2 = max(abs(densig-ssig));
prterr2 = max(abs(densig-ssig))/max(ssig)*100;
mse2 = err2^2/m^2;

% Stop
disp('   ')
disp('   ')
disp('.... End of program ....')
```

Appendix O_2 of O Using Daubechies-4 Wavelet

```
% WTDB4t3.m (Daubechies Wavelet Transform - db4)
%
% This program uses Daubechies wavelet db4 at level 4 to denoise the
% vibration signal model that is a combination of a sum of sinusoids
```

```
% and additive white Gaussian noise (AWGN). The vibration model consists
% of the following:
%
%    s1 = sinusoid at frequency f1 (low)
%    s2 = sinusoid at frequency f2 (high)
%    noise = normal distributed random noise
%
% Denoising the signal is done using a threshold calculated as:
%
%    thr = sqrt(2 * log(length(s)))
%
% This program removes the noise from the vibration signal and extracts
the
% pertinent information hidden in the noise. Detection is based on
frequency spectra.
%
%


% Initialize the work space
clear all; close all;

% Create the correct directory path
% Example, path(path, 'c:\...........\.........._daub\daubt3')

% Get the vibration signal
load signlX

format('long','g')

% Sample the signal for analysis
s = vibsg(1:1500);
Nsg = noise(1:1500);

% Plot vibration signal
figure(1)
plot(t(1:1500), s), grid on; title('Noisy Vibration Signal')
xlabel('time (sec)'), ylabel('Amplitude')

% Perform decomposition of the signal
[C,L] = wavedec(s,4,'db4');
[cD1,cD2,cD3,cD4] = detcoef(C,L,[1,2,3,4]);
cA4 = appcoef(C,L,'db4',4);

% Determine the approximation and details
A4 = wrcoef('a',C,L,'db4',4);
D1 = wrcoef('d',C,L,'db4',1);
D2 = wrcoef('d',C,L,'db4',2);
D3 = wrcoef('d',C,L,'db4',3);
D4 = wrcoef('d',C,L,'db4',4);

% Plot the approximation and details
figure(2)
```

```
subplot(221); plot(t(1:1500),A4), grid; title('Approximation A4')
subplot(222); plot(t(1:1500),D4), grid; title('Detail D4')
subplot(223); plot(t(1:1500),D3), grid; title('Detail D3')
figure(3)
subplot(211); plot(t(1:1500),D2), grid; title('Detail D2')
subplot(212); plot(t(1:1500),D1), grid; title('Detail D1')

% Perform reconstruction of the signal
recsig = waverec(C,L,'db4');

% Check for perfect reconstruction
figure(4)      % visually
subplot(211); plot(t(1:1500),s), grid; title('Original Signal')
xlabel('time (sec)'), ylabel('Amplitude')
subplot(212); plot(t(1:1500),recsig), grid; title('Reconstructed
Signal')
xlabel('time (sec)'), ylabel('Amplitude')
n = length(s);  % numerically
err = max(abs(recsig-s));
prterr = max(abs(recsig-s))/max(s)*100;
mse = err^2/n^2;

snr1 = 10*log10(max(s)^2/max(abs(s-recsig))^2);

% Check denoising process by comparing the highest level approximation
% with the noisy vibration signal
figure(5)
subplot(211); plot(t(1:1500),s), grid; title('Noisy Vibration
Signal')
xlabel('time (sec)')
subplot(212); plot(t(1:1500),A4), grid; title('Level
4 Approximation')
xlabel('time (sec)')

% Set the threshold and denoise the signal
thr = 3.82445300326473*0.5;    % fixed threshold (at 50%)
densig = wdencmp('gbl',C,L,'db4',4,thr,'s',1);

snr2 = 10*log10(max(densig)^2/max(abs(s-densig))^2);

% Plot the noisy vibration signal and the denoised signal
figure(6)
subplot(211); plot(t(1:1500),s), grid; title('Noisy Vibration
Signal')
xlabel('time (sec)'), ylabel('Amplitude')
subplot(212); plot(t(1:1500),densig), grid; title('Denoised Signal')
xlabel('time (sec)'), ylabel('Amplitude')

% Compare the frequency content of the vibration signal and the
% recontructed signal
Ns = 1024;
origspec = fft(s,Ns);
```

```
spec1 = origspec.*conj(origspec)/Ns;
freq1 = 1000*(0:512)/Ns;
recsigspec = fft(recsig,Ns);
spec2 = recsigspec.*conj(recsigspec)/Ns;
freq2 = 1000*(0:512)/Ns;
figure(7)
subplot(211); plot(freq1,spec1(1:513)), axis([0 100,-Inf 300]), grid
title('Frequency Content of Original Signal')
xlabel('Frequency'), ylabel('Energy')
subplot(212); plot(freq2,spec2(1:513)), axis([0 100,-Inf 300]), grid
title('Frequency Content of Reconstructed Signal')
xlabel('Frequency'), ylabel('Energy')

% Determine the frequency content of the denoised signal
densigspec = fft(densig,Ns);
spec3 = densigspec.*conj(densigspec)/Ns;
freq2 = 1000*(0:512)/Ns;
figure(8)
plot(freq2,spec3(1:513)), axis([0 100,-Inf Inf]), grid
title('Frequency Content of Denoised Signal')
xlabel('Frequency'), ylabel('Energy')

% Compare the PSDs for the vibration signal, the reconstructed signal,
% and the denoised signal
figure(9)
[Po,fo] = pmtm(s,3.5,Ns,Fs);
subplot(221); psdplot(Po,fo,'Hz','db','PSD of Original Signal'), axis
([0 100,-Inf -5])
[Pr,fr] = pmtm(recsig,3.5,Ns,Fs);
subplot(222); psdplot(Pr,fr,'Hz','db','PSD of Reconstructed
Signal'), axis([0 100,-Inf -5])
[Pd,fd] = pmtm(densig,2,Ns,Fs);
subplot(223); psdplot(Pd,fd,'Hz','db','PSD of Denoised Signal'), axis
([0 100,-50 -5])
[Pns,fns] = pmtm(Nsg,3.5,Ns,Fs);
Pwo = (Fs/Ns)*sum(Po);
Pwd = (Fs/Ns)*sum(Pd);
Pwns = (Fs/Ns)*sum(Pns);
SNR1 = 10*log10(Pwo/Pwns);  % Compute the SNR for the vibration signal
SNR2 = 10*log10(Pwd/Pwns);  % Compute the SNR for the denoised signal

% Generate the scalograms
figure(10)
specgram(s,1024,Fs,[],256), axis([0 2, 0 100])
title('Scalogram of the Noisy Vibration Signal')
figure(11)
specgram(densig,1024,Fs,kaiser(1024,2.5),256), axis([0 2, 0 100])
title('Scalogram of the Denoised Vibration Signal')

% Compare the denoised and vibration signals
figure(12)          % visually
subplot(211); plot(t(1:1500),densig,'m'), axis('tight'), grid
title('Denoised Signal')
```

```
xlabel('time (sec)'), ylabel('Amplitude')
subplot(212); plot(t(1:1500),sumsig(1:1500)), axis('tight'), grid
title('Vibration Signal')
xlabel('time (sec)'), ylabel('Amplitude')
ssig = sumsig(1:1500);
m = length(ssig);       % numerically
err2 = max(abs(densig-ssig));
prterr2 = max(abs(densig-ssig))/max(ssig)*100;
mse2 = err2^2/m^2;


% Stop
disp('    ')
disp('    ')
disp('.... End of program ....')
```

Appendix O_3 of O Using Morlet Wavelet

```
%
% mvwt_analysis.m
% This program uses the Morlet wavelet to analyze and detect the signal
under consideration. This % program uses the main subroutine "mvsg.m"
for denoising
% and data segmentation. The vibration signal is modeled as follows:
%
%    s1 = low freq. sine
%    s2 = high freq. sine
%    s = s1 + s2 + noise
%
% Also, this function reconstructs "s1" and "s2" from the scalogram of
"s" with noise. This program allows the user to interactively select the
signal ("s1" or "s2") to consider for analysis.
%
%
%
%
%


    warning ('off', 'MATLAB:HandleGraphics:RenamedProperty:
YTickLabels')

% Create the proper directory paths

    % Example, path(path, 'c:\............\grafics')
    % Example, path(path, 'c:\............\lercmwt')
    % Example, path(path, 'c:\............\perifera')


% Initialize the workspace
%
    clear;
    close all;
```

```
    load sign1X

% Basic explanation of keystrokes and mouse movements during program
%
    disp(' ')
    disp(' ')
    disp('        To CONTINUE after each figure please remember to press
"Enter" ')
    disp(' ')
    disp('        It is also possible to use the MOUSE to separate "figures"
from one another for viewing... ')
    disp(' ')
    disp('        But, DO NOT block this MATLAB Command window so you can see
the instructions... ')
    disp(' ')
    disp('        Press "Enter" to continue ')
    pause
    clc


    r=1;    % Noise scaling factor
    noid=(randn(size(t)))*r;    % White Gaussian noise
    figure(1); plot(time,s1,'b',time,s2+3,'m',time,noid+9,'r'), axis
([0 2 -2 14]), grid
    title('Signals and noise: s1 (blue); s2 (purple); noise (red)')    %
Show all signals
    xlabel('Time (seconds)'), ylabel('Amplitude')
    disp(' ')
    disp(' Press "Enter"')
    pause;
    clc
    s = s1 + s2 + noid;    % Vibration model corrupted by noise
    figure(2); plot(time,s), axis([0 2 -5 5]), grid
    title(' Vibration signal "s" ')   % Show noisy vibration signal
    xlabel('Time (seconds)'), ylabel('Amplitude')
    disp(' ')
    disp(' Figure 2 shows the combined signal "s" (s1 + s2 + noise) ')
    disp(' When you are done viewing the graph, Press "Enter" again
PLEASE...')
    pause
    clc
    disp(' ')    % User friendly command prompts
    disp(' You can extract either "s1" or "s2" out by UNFOLDING "s" onto a
"Scalogram" or a "Wavelet Map" ')
    disp(' ')
    disp(' To use this program, please press "Enter" and follow the
instructions in the MATLAB command window...')
    disp(' ')
    disp(' ')
    disp(' Press "Enter"')
    disp(' ')
    pause
```

```
% Call Unfold_Trim_Reconstruct, which yields "sr" as the reconstructed
signal
% that is obtained by using the Morlet wavelet to recover the signal

   [sr, tsr] = unfoldtr(s,time,2,5,4,'rmorlet');

% Which signal best matches with sr ?
%
   mt = max([min(time) min(tsr)]); mxt = min([max(time) max(tsr)]);
   idx = find((time > mt) & (time < mxt));
   idxr = find((tsr > mt) & (tsr < mxt));      % Check the closest matches in
time for the signals
   lm = min([length(idx) length(idxr)]);
   idx = idx([1:lm]); idxr = idxr([1:lm]);   % Locate the closest matches
between the signals
   id1 = find(abs(s1(idx)-sr(idxr)) < 0.1*max(s1(idx)));
   id2 = find(abs(s2(idx)-sr(idxr)) < 0.1*max(s2(idx)));
   splot = s1;
   if (length(id2) > length(id1))
      splot = s2;
   end

% Display the vibration signal extracted from the noise and the original
signal
%
   figure(5); plot(time,splot,'b',tsr,sr,'m'), grid
   title('Extracted signal (purple) vs. Actual signal (blue) ')  %
Compare the reconstructed signal to the original signal
   xlabel('Time (seconds)'), ylabel('Amplitude')
   ids = find((time < 1) & (time > 0.5));
   idr = find((tsr < 1) & (tsr > 0.5));
   figure(6); plot(time(ids), splot(ids),'b',tsr(idr),sr(idr),'m'),
grid
   title('Comparison of Extracted signal (purple) and Actual signal
(blue) Over Short Interval')   % Zoom in on signals
   xlabel('Time (seconds)'), ylabel('Amplitude')
   clc

% Determine pertinent information about the original and denoised
signals
%
   sscl = s(1:length(sr));
   noidscl = noid(1:length(sr));
   n = length(sscl);
   err = max(abs(sscl-sr));
   prterr = max(abs(sscl-sr))/max(sscl)*100;
   mse = err^2/n^2;
   snr1 = 10*log10(max(s)^2/max(noid)^2);
   snr2 = 10*log10(max(sr)^2/max(noid)^2);

% Compare the frequency content of the vibration signal and the
% denoised signal
```

```
   Ns = 1024;
   origspec = fft(s,Ns);
   spec1 = origspec.*conj(origspec)/Ns;
   freq1 = 1000*(0:512)/Ns;
   srspec = fft(sr,Ns);
   spec2 = srspec.*conj(srspec)/Ns;
   freq2 = 1000*(0:512)/Ns;
   figure(7)
   subplot(211); plot(freq1,spec1(1:513)), axis([0 100,-Inf Inf]),
grid
   title('Frequency Content of Original Signal')
   xlabel('Frequency'), ylabel('Energy')
   subplot(212); plot(freq2,spec2(1:513)), axis([0 100,-Inf Inf]),
grid
   title('Frequency Content of Denoised Signal')
   xlabel('Frequency'), ylabel('Energy')

% Compare the PSD of the vibration signal and the denoised signal
%
   figure(8)
   [Po,fo] = pmtm(s,3.5,Ns,Fs);
   subplot(211); psdplot(Po,fo,'Hz','db','PSD of Original Signal'),
axis([0 100,-Inf -5])
   [Pd,fd] = pmtm(sr,2,Ns,Fs);
   subplot(212); psdplot(Pd,fd,'Hz','db','PSD of Denoised Signal'),
axis([0 100,-Inf -5])
   [Pns,fns] = pmtm(noidscl,3.5,Ns,Fs);
   Pwo = (Fs/Ns)*sum(Po);
   Pwd = (Fs/Ns)*sum(Pd);
   Pwns - (Fs/Ns)*sum(Pns);
   SNR1 = 10*log10(Pwo/Pwns);
   SNR2 = 10*log10(Pwd/Pwns);

% Generate the scalograms
%
   figure(9)
   specgram(s,1024,Fs,[],256), axis([0 Inf,0 100])
   title('Scalogram of the Noisy Vibration Signal')
   xlabel('Time (seconds)'), ylabel('Frequency')
   figure(10)
   specgram(sr,1024,Fs,[],256), axis([0 Inf,0 100])
   title('Scalogram of the Denoised Vibration Signal')
   xlabel('Time (seconds)'), ylabel('Frequency')

% Compare the denoised and vibration signals
%
   figure(11)          % visually
   if (splot == s2)
     subplot(211); plot(tsr,sr,'m'), axis('tight'), grid
     title('Denoised Signal')
     xlabel('time (sec)'), ylabel('Amplitude')
     subplot(212); plot(tsr,s2(1:length(sr))), axis('tight'), grid
     title('Original Signal')
```

```
    xlabel('time (sec)'), ylabel('Amplitude')
  else
    subplot(211); plot(tsr,sr,'m'), axis('tight'), grid
    title('Denoised Signal')
    xlabel('time (sec)'), ylabel('Amplitude')
    subplot(212); plot(tsr,s1(1:length(sr))), axis('tight'), grid
    title('Original Signal')
    xlabel('time (sec)'), ylabel('Amplitude')
  end

  if (splot == s2)
    ssig = s2;        % numerically
    lsr = length(sr);
    ss = ssig(1:lsr);
    m = length(ss);
    err2 = max(abs(ss-sr));
    prterr2 = max(abs(ss-sr))/max(ss)*100;
    mse2 = err2^2/m^2;
  else
    ssig = s1;        % numerically
    lsr = length(sr);
    ss = ssig(1:lsr);
    m = length(ss);
    err2 = max(abs(ss-sr));
    prterr2 = max(abs(ss-sr))/max(ss)*100;
    mse2 = err2^2/m^2;
  end

  disp('    ')
  disp('    ')
  disp('.... End of program ....')

  return
```

Appendix O_4 of O

Programs Using the Vibration Model and PLT

Custom programs are written to implement the algorithm developed for the vibration model. These programs apply the wavelet transforms in combination with per-level thresholding (PLT). With the change in the thresholding method, detection is achieved. Further changes are made to the algorithm to improve its results.

Once signal analysis is performed with the appropriate threshold, the algorithm is performed as desired. Through numerous trials and by retuning the programs, the Daubechies-14 wavelet performed best for analyzing the simulated vibration signal using PLT.

The program coded for the db14 wavelet is listed on the following pages.

Appendix O_4 of O Using Daubechies-14 Wavelet

```
% WTDB14t5.m (Daubechies Wavelet Transform)
%
% This program uses wavelet db14 at level 4 to denoise the
% vibration signal model that is a combination of a sum of sinusoids
% and additive white Gaussian noise (AWGN). The vibration model consists
% of the following:
%
%    s1 = sinusoid at frequency f1 (low)
%    s2 = sinusoid at frequency f2 (high)
%    noise = normal distributed random noise
%
% This program removes the noise from the vibration signal.
% Detection is achieved using the frequency spectra.
%
%


% Initialize the work space
clear all; close all;

% Create the path for the appropriate director
% Example, path(path, 'c:\..........\....._daub\daubn')

% Get the vibration signal
load sign1X

format ('long', 'g')

% Sample the signal for analysis
s = vibsg(1:1500);
Nsg = noise(1:1500);

% Plot vibration signal
figure(1)
plot(t(1:1500), s), grid on; title('Noisy Vibration Signal')
xlabel('time (sec)'), ylabel('Amplitude')

% Perform decomposition of the signal
[C,L] = wavedec(s,4,'db14');
[cD1,cD2,cD3,cD4] = detcoef(C,L, [1,2,3,4]);
cA4 = appcoef(C,L,'db14',4);

% Determine the approximation and details
A4 = wrcoef('a',C,L,'db14',4);
D1 = wrcoef('d',C,L,'db14',1);
D2 = wrcoef('d',C,L,'db14',2);
D3 = wrcoef('d',C,L,'db14',3);
D4 = wrcoef('d',C,L,'db14',4);
```

```
% Plot the approximation and details
figure(2)
subplot(221); plot(t(1:1500),A4), grid; title('Approximation A4')
subplot(222); plot(t(1:1500),D4), grid; title('Detail D4')
subplot(223); plot(t(1:1500),D3), grid; title('Detail D3')
figure(3)
subplot(211); plot(t(1:1500),D2), grid; title('Detail D2')
subplot(212); plot(t(1:1500),D1), grid; title('Detail D1')

% Perform reconstruction of the signal
recsig = waverec(C,L,'db14');

% Check for perfect reconstruction
figure(4)     % visually
subplot(211); plot(t(1:1500),s), grid; title('Original Signal')
xlabel('time (sec)'), ylabel('Amplitude')
subplot(212); plot(t(1:1500),recsig), grid; title('Reconstructed
Signal')
xlabel('time (sec)'), ylabel('Amplitude')
n = length(s); % numerically
err = max(abs(recsig-s));
prterr = max(abs(recsig-s))/max(s)*100;
mse = err^2/n^2;

snr1 = 10*log10(max(s)^2/max(abs(s-recsig))^2);

% Check denoising process by comparing the highest level approximation
% with the noisy vibration signal
figure(5)
subplot(211); plot(t(1:1500),s), grid; title('Noisy Vibration
Signal')
xlabel('time (sec)')
subplot(212); plot(t(1:1500),A4), grid; title('Level
4 Approximation')
xlabel('time (sec)')

% Determine the threshold per level for denoising the signal
td1 = thselect(cD1,'sqtwolog'); td2 = thselect(cD2,'sqtwolog');
td3 = thselect(cD3,'heursure'); td4 = thselect(cD4,'rigrsure');
T = [td1,td2,td3,td4];
NC = wthcoef('t',C,L,[1:4],T,'s');
densig = waverec(NC,L,'db14');  % Denoise using thresholds

snr2 = 10*log10(max(densig)^2/max(abs(s-densig))^2);

% Plot the noisy vibration signal and the denoised signal
figure(6)
subplot(211); plot(t(1:1500),s), grid; title('Noisy Vibration
Signal')
xlabel('time (sec)'), ylabel('Amplitude')
subplot(212); plot(t(1:1500),densig), grid; title('Denoised Signal')
xlabel('time (sec)'), ylabel('Amplitude')
```

```
% Compare the frequency content of the vibration signal and the
% recontructed signal
Ns = 1024;
origspec = fft(s,Ns);
spec1 = origspec.*conj(origspec)/Ns;
freq1 = 1000*(0:512)/Ns;
recsigspec = fft(recsig,Ns);
spec2 = recsigspec.*conj(recsigspec)/Ns;
freq2 = 1000*(0:512)/Ns;
figure(7)
subplot(211); plot(freq1,spec1(1:513)), axis([0 100,-Inf 300]), grid
title('Frequency Content of Original Signal')
xlabel('Frequency'), ylabel('Energy')
subplot(212); plot(freq2,spec2(1:513)), axis([0 100,-Inf 300]), grid
title('Frequency Content of Reconstructed Signal')
xlabel('Frequency'), ylabel('Energy')

% Determine the frequency content of the denoised signal
densigspec = fft(densig,Ns);
spec3 = densigspec.*conj(densigspec)/Ns;
freq2 = 1000*(0:512)/Ns;
figure(8)
plot(freq2,spec3(1:513)), axis([0 100,-Inf Inf]), grid
title('Frequency Content of Denoised Signal')
xlabel('Frequency'), ylabel('Energy')

% Compare the PSDs for the vibration signal, the reconstructed signal,
% and the denoised signal
figure(9)
[Po,fo] = pmtm(s,3.5,Ns,Fs);
subplot(221); psdplot(Po,fo,'Hz','db','PSD of Original Signal'), axis
([0 100,-Inf -5])
[Pr,fr] = pmtm(recsig,3.5,Ns,Fs);
subplot(222); psdplot(Pr,fr,'Hz','db','PSD of Reconstructed
Signal'), axis([0 100,-Inf -5])
[Pd,fd] = pmtm(densig,2,Ns,Fs);
subplot(223); psdplot(Pd,fd,'Hz','db','PSD of Denoised Signal'), axis
([0 100,-50 -5])
[Pns,fns] = pmtm(Nsg,3.5,Ns,Fs);
Pwo = (Fs/Ns)*sum(Po);
Pwd = (Fs/Ns)*sum(Pd);
Pwns = (Fs/Ns)*sum(Pns);
SNR1 = 10*log10(Pwo/Pwns); % Compute the SNR for the vibration signal
SNR2 = 10*log10(Pwd/Pwns); % Compute the SNR for the denoised signal

% Generate the scalograms
figure(10)
specgram(s,1024,Fs,[],256), axis([0 1.5,0 100])
title('Scalogram of the Noisy Vibration Signal')
figure(11)
specgram(densig,1024,Fs,[],256), axis([0 1.5,0 100])
title('Scalogram of the Denoised Vibration Signal')
```

```
% Compare the denoised and vibration signals
figure(12)          % visually
subplot(211); plot(t(1:1500),densig,'m'), axis('tight'), grid
title('Denoised Signal')
xlabel('time (sec)'), ylabel('Amplitude')
subplot(212); plot(t(1:1500),sumsig(1:1500)), axis('tight'), grid
title('Vibration Signal')
xlabel('time (sec)'), ylabel('Amplitude')
ssig = sumsig(1:1500);
m = length(ssig);      % numerically
err2 = max(abs(densig-ssig));
prterr2 = max(abs(densig-ssig))/max(ssig)*100;
mse2 = err2^2/m^2;


% Stop
disp('    ')
disp('    ')
disp('.... End of program ....')
```

Appendix O_5 of O

Programs Using Flight Research Data

Programs were written to implement the algorithm developed for real flight data. Two different sets of vibration measurements were analyzed using the detection programs. These programs execute the wavelet transforms in combination with per-level thresholding (PLT). Detection is successful on all counts.

Based on the performance of the programs with the flight data, the Daubechies wavelet proved to be the most flexible and yielded the best results. For the vibration "event" of the F-15B/836 testbed, the best performance is shown by db10, at level 4. The best results are produced by db10 and db14, respectively.

The Daubechies program (db10) for the F-15B/836 is listed on the next page.

```
% WTDB10t5f256.m (Daubechies Wavelet Transform)
%
% This program uses the Daubechies-10 wavelet at level 4 to analyze the
% vibration data collected during takeoff of F-15B/836 flight 256.
%
% This program removes noise from the vibration signal and extracts
% key frequency information hidden in the noise.
%
% The vibration signal is detected using frequency spectra.
%
%===============================================

% Initialize the work space
%***********************************
clear all; close all;
```

```
path(path, 'c:\nasaresearch\nasadfrc\')

% Get the vibration signal
%*************************
load f256df409 % Replace this data with data of interest and then load

format('long','g')

% Sample the signal for analysis
%*******************************
s = eu(42255:56339,1);
tm = t(42255:56339);
Fs = 7042;

% Plot vibration signal
%**********************
figure
plot(t,eu(:,1)), grid; title(' F-15B/836 Flight Test Vibration Signal
(Normal Axis) ')
xlabel('Time  (sec)'), ylabel('Amplitude (g)')

% Plot event portion of signal
%****************************
figure
plot(tm,s), grid; title(' Brake Release Event from Flight 256 of F-15B/
836 Flight Research ')
xlabel('Time  (sec)'), ylabel('Amplitude (g)')

% Perform decomposition
%*********************
[C,L] = wavedec(s,4,'db10');
[cD1,cD2,cD3,cD4] = detcoef(C,L,[1,2,3,4]);
cA4 = appcoef(C,L,'db10',4);

% Determine the approximation and details
%****************************************
A4 = wrcoef('a',C,L,'db10',4);
D1 = wrcoef('d',C,L,'db10',1);
D2 = wrcoef('d',C,L,'db10',2);
D3 = wrcoef('d',C,L,'db10',3);
D4 = wrcoef('d',C,L,'db10',4);

% Plot the approximation and details
%**********************************
figure
subplot(221); plot(tm,A4), axis('tight'), grid; title('Approximation
A4')
xlabel('Time  (sec)'), ylabel('Amplitude (g)')
subplot(222); plot(tm,D4), axis('tight'), grid; title('Detail D4')
xlabel('Time  (sec)'), ylabel('Amplitude (g)')
subplot(223); plot(tm,D3), axis('tight'), grid; title('Detail D3')
xlabel('Time  (sec)'), ylabel('Amplitude (g)')
```

```
figure
subplot(211); plot(tm,D2), axis('tight'), grid; title('Detail D2')
xlabel('Time (sec)'), ylabel('Amplitude (g)')
subplot(212); plot(tm,D1), axis('tight'), grid; title('Detail D1')
xlabel('Time (sec)'), ylabel('Amplitude (g)')

% Perform reconstruction
%***********************
recsig = waverec(C,L,'db10');

% Check for perfect reconstruction
%********************************
figure      % visually
subplot(211); plot(tm,s), axis('tight'), grid; title('Brake Release
Event from F-15B/836 Flight Test')
xlabel('Time (sec)'), ylabel('Amplitude (g)')
subplot(212); plot(tm,recsig), axis('tight'), grid; title
('Reconstructed Signal of Brake Release Event')
xlabel('Time (sec)'), ylabel('Amplitude (g)')
n = length(s); % numerically
err = max(abs(s-recsig));
prterr = max(abs(s-recsig))/max(s)*100;
mse = err^2/n^2;
snr1 = 10*log10(max(s)^2/max(abs(s-recsig))^2);

% Check denoising process
%***********************
figure
subplot(211); plot(tm,s), axis('tight'), grid; title('Noisy Brake
Release Signal')
xlabel('Time (sec)'), ylabel('Amplitude (g)')
subplot(212); plot(tm,A4), axis('tight'), grid; title('Level
4 Approximation')
xlabel('Time (sec)'), ylabel('Amplitude (g)')

% Determine the threshold per level for denoising the signal
%***********************************************************
td1 = thselect(cD1,'sqtwolog'); td2 = thselect(cD2,'minimaxi');
td3 = thselect(cD3,'heursure'); td4 = thselect(cD4,'rigrsure');
T = [td1,td2,td3,td4];
NC = wthcoef('t',C,L,[1:4],T,'s');

% Denoise using the thresholds
%****************************
densig = waverec(NC,L,'db10');

% Plot the noisy vibration and denoised signals
%*********************************************
figure
subplot(211); plot(tm,s), axis('tight'), grid; title('Noisy Brake
Release Signal')
xlabel('Time (sec)'), ylabel('Amplitude (g)')
```

```
subplot(212); plot(tm,densig,'Color',[0 0.502 0.251]), axis
('tight'), grid; title('Denoised Brake Release Signal')
xlabel('Time (sec)'), ylabel('Amplitude (g)')

% Compare frequency content: vibration and reconstructed signals
%*****************************************************************
N = 7168;
origspec = fft(s,N);
spec1 = origspec.*conj(origspec)/N;
freq1 = 1*(0:N/2);
recsigspec = fft(recsig,N);
spec2 = recsigspec.*conj(recsigspec)/N;
freq2 = 1*(0:N/2);
figure
subplot(211); plot(freq1,spec1(1:3585)), axis([0 2000,-Inf Inf]),
grid
title('Frequency Content of Brake Release Event')
xlabel('Frequency (Hz)'), ylabel('Magnitude (g^2)')
subplot(212); plot(freq2,spec2(1:3585)), axis([0 2000,-Inf Inf]),
grid
title('Frequency Content of Reconstructed Signal of Brake Release')
xlabel('Frequency (Hz)'), ylabel('Magnitude (g^2)')

% Determine the frequency content of the denoised signal
%***********************************************************
densigspec - fft(densig,N);
spec3 = densigspec.*conj(densigspec)/N;
freq3 = 1*(0:N/2);
figure
plot(freq3,spec3(1:3585)), axis([0 2000,-Inf Inf]), grid
title('Frequency Content of Denoised Signal of Brake Release')
xlabel('Frequency (Hz)'), ylabel('Magnitude (g^2)')

% Compare the PSDs: original and denoised signals
%***************************************************
figure
[Po,fo] = pmtm(s,2,N,Fs);
subplot(211); psdplot(Po,fo,'Hz','linear','PSD of Brake Release
Event'), axis([0 2000,-Inf Inf])
ylabel('Power Spectral Density (g^2/Hz)')
[Pd,fd] = pmtm(densig,2,N,Fs);
subplot(212); psdplot(Pd,fd,'Hz','linear','PSD of Denoised Signal of
Brake Release'), axis([0 2000,-Inf Inf])
ylabel('Power Spectral Density (g^2/Hz)')
figure
[Pd,fd] = pmtm(densig,2,N,Fs);
psdplot(Pd,fd,'Hz','linear','PSD of Denoised Signal of Brake
Release'), axis([0 2000,-Inf Inf])
ylabel('Power Spectral Density (g^2/Hz)')

% Generate the scalograms
%***********************
figure
```

```
specgram(s,2048,Fs,[],512), axis([0 Inf,0 2000])
title('Scalogram of Brake Release Event')
xlabel('Time (sec)'), ylabel('Frequency (Hz)')
figure
specgram(densig,2048,Fs,kaiser(2048,5),1024), axis([0 Inf,0 2000])
title('Scalogram of Denoised Brake Release Signal')
xlabel('Time (sec)'), ylabel('Frequency (Hz)')


% Compare the denoised and vibration signals
%*******************************************
figure          % visually
plot(tm,s,tm,densig,'m'), axis('tight'), grid
title('Noisy Brake Release Signal vs Denoised Brake Release Signal')
xlabel('Time (sec)'), ylabel('Amplitude (g)')
legend('Brake Release','Denoised Brake Release')
m = length(s);      % numerically
snr2 = 10*log10(max(densig)^2/max(abs(s-densig))^2);
err2 = max(abs(s-densig));
prterr2 = max(abs(s-densig))/max(s)*100;
mse2 = err2^2/m^2;


disp('    ')
disp('    ')
disp('.... End of program ....')


% *** EOF ***
```

Index

A

Access control, 430

Adaptively scanned wavelet difference
      reduction (ASWDR) algorithm,
      259, 260

Adaptive thresholding, 312, 313

ADC testing setup, 555

ADC wavelet-based static testing, 168

Additive white Gaussian noise (AWGN),
      471, 476, 503, 617, 625

Advanced Encryption Standard
      (AES) algorithm, 375, 380

Advanced Microwave Receiver Technology
      Development System
      (AMRTDS), 475–477

Aeroelastic systems
    F-15B/836 Flight Research, 534
    PLT algorithm application, 520, 524
    threshold experimentation
        global thresholding, 516, 517, 530, 531
        per-level thresholding, 518, 519, 533
    vibration detection
        Daubechies wavelet, 509, 511
        decomposition level, 504
        energy loss, 506
        Haar wavelet, 508, 509
        Morlet wavelet, 511, 513
        process, 506
        threshold, 506
        wavelet families, 503, 504
    vibration model development, 502, 503
    vibration signal analysis techniques, 502
    wavelet transform, 498–500
    wavelets used for the vibration
        signal detection, 524

Affine transformations, 94

Agriculture, 430

AMRTDS GUI, 476

AMRTD testbed interface, 477

Analog-to-digital converters (ADCs), 107
    automated DWT-based algorithm, 203
    DNL measurements
        sinusoid histogram, 153
        wavelet transform, 153
    dynamic testing
        Haar wavelet coefficients, 146
        Hilbert transform implementation,
            142, 143
    ENOB measurements, 149–151
    static test (*see* Static test of ADCs)

Anomaly detection, 363, 367, 369, 382

Anti-clockwise scanning, 399

Arbitrary analog waveform generator, 555

Arches, 394, 395

Automated DWT-based algorithm, 205

Automobile industry, 430

B

Baseline/isopotential line, 291

Bayes' probability theorem, 610

Bayes' theorem, 342
    signals using, wavelet-based
        detection of, 481
        signal detection, 482, 483, 485, 486

Berlage wavelets, 24, 25, 95

Beta wavelets, 40, 41

Bi-level image template matching, 321

Binarization, 398

Biomedical engineering, 52

Printed in the United States
by Baker & Taylor Publisher Services